CONTENTS

INTRODUCTORY LECTURES

Production and study of very heavy atoms

Atomic Physics through Astrophysics and Nuclear Physics

PRODUCTION OF THE HEAVIEST ELEMENTS 107 to 109, LIMITATIONS, AND PROSPECTS TO GO BEYOND

P. Armbruster

Gesellschaft für Schwerionenforschung mbH
P.O. Box 110541
D-6100 Darmstadt, FRG

In recent years isotopes of the elements 107 to 109 were discovered. Together with studies on the isotopes of the elements 104-106 it was established that beyond element 105 the main decay mode for isotopes with (N-Z) = (47-49) is α-decay. The trend of strongly increasing instability against spontaneous fission for the heaviest elements is broken. The isotope $^{260}106$ has a partial halflife against spontaneous fission of about 7 ms, which is to be compared to a halflife of 8 ms for $^{256}104$. The long α-chains detected allow to determine the absolute masses. Together with macroscopic mass values the shell correction energies are obtained. It is shown that the fission barrier for the (N-Z) = 48-isotopes of the heaviest elements stay constant at a value of about 6 MeV, in spite of vanishing macroscopic fission barriers. From an analysis of the mass values and the spontaneous fission halflives it follows, that the heaviest isotopes detected are protected against spontaneous fission by a single humped narrow fission barrier, which is due to shell corrections, e.g. the isotope ^{260} 106 is shell stabilized by 15 orders of magnitude in its halflife against spontaneous fission. Paskevich et al. and Møller et al. independently predict the nuclei to be deformed and to have a strong negative β_4 deformation (sausage-like). The isotopes investigated are shell stabilized isotopes of superheavy elements (SHE), in the sense that SHE are elements unstable within macroscopic models but stabilized by shell effects to halflives long enough to be detected still. The SHE made are sausage-like, and not spherical nuclei, they live ms and not millions of years. They would not be detected in a world without shell corrections. The limitation to produce still heavier SHE comes from the production mechanism. Fusion is finally limited by the increasing Coulomb forces in the formation process of a compound system, as well as in its deexcitation. Moreover, nuclear structure effects in all stages of evaporation residue (EVR) formation are shown to be of importance. The wide field of fusion reaction studies and possible experimental techniques is projected onto the task of element synthesis, and only those aspects that are of relevance here are covered.

INTRODUCTION

The heaviest known atomic nucleus is the isotope with mass number 266 of element 109. In the decay chain of the only one observed atom the three

decay modes which are energetically possible in the heavy element region, namely alpha decay, beta decay and fission, occur.

$$^{266}109 \xrightarrow[T_{1/2} = 3\ ms]{\alpha} {}^{262}107 \xrightarrow[T_{1/2} = 5\ ms]{\alpha} {}^{250}105 \xrightarrow[T_{1/2} = 4\ s]{EC} {}^{250}104 \xrightarrow[T_{1/2} = 13\ ms]{sf} A_1 + A_2$$

Alpha and beta decay combine neighboring nuclei with similar nuclear structure in general and, therefore, their halflives can be predicted with reliability[1-3]. In the region of nuclei Z = 104-116 and N = 150-162 the α halflives cover a range from about 1 s (N-Z = 52) to 1 μs (N-Z = 42 to 48). Beta halflives are longer due to the vicinity of the valley of beta stability at N-Z = 56. The beta halflives decrease slowly to 1 s at N-Z = 39 to 45.

The prediction of the liquid drop model gives a vanishing fission barrier near element 110. Already for elements beyond Md halflives of less than 10^{-6} sec would be expected. But nuclear structure effects can change the height and shape of the fission barrier and thus prevent these heavy nuclei from an immediate fission[4,5]. The biggest influence on the value of the WKB integral determining fission halflives have ground state shell effect and ground state deformations. In addition, saddle point shell effects, shell corrections leading to a double humped barrier, and in case of odd and odd-odd nuclei, specialization energies can drastically change the liquid drop predictions of fission halflives. The unexpected deviation of the measured spontaneous fission halflives beyond element 102 from early systematics proves that these microscopic effects cannot be extrapolated with sufficient accuracy[6]. Experimental data and elaborate theoretical calculations are needed to establish the ground-state properties of the heaviest isotopes.

The severest obstacle to produce still heavier elements is not their ground state stability, but the decreasing production cross sections. The production has been shown to be restricted threefold: 1) To overcome the fusion barrier is restricted already in the very first stage of amalgamation of the two fusing nuclei, 2) fusion leads to considerable excitation of the compound system which decays during deexcitation with high probability by fission, 3) the ground state shell stabilization is lost with increasing temperature of the compound system, and it was found that the excitation energy the compound system can take strongly depends on its nuclear structure. The production cross sections go down from 10^{-28} cm^2 for the making of fermium to 10^{-35} cm^2 for the making of element 109. For a more detailed discussion of the limitations by the production mechanisms the reader is referred to a recent review article[7].

EXPERIMENTAL METHOD USED TO MAKE ELEMENTS 106 to 109

General Guidelines

Since the first operation of SHIP[8] the sensitivity of experiments to search for rare isotopes could be steadily improved. Developments in ion source techniques and accelerator acceptance increased the beam currents of titanium to nickel ions up to 5 x 10^{12}/s. Thin lead or bismuth targets (500 μg/cm^2) mounted on a wheel or 30 cm diameter are moved with a velocity of 2 cm/ms through the beam spot of 8 mm diameter. Radiative cooling prevents the low melting targets from melting.

Using targets of lead and bismuth, compound nuclei of the highly fissionable heavy elements can be produced in fusion reactions with a minimum of excitation energy of 15 to 20 MeV (cold fusion)[9]. Moreover, in reactions with the projectiles ^{50}Ti, ^{54}Cr, ^{58}Fe, and ^{62}Ni compound

nuclei of the same N-Z = 50 and 49, respectively, are produced. After evaporation of 1 or 2 neutrons the residues, if α-emitting, belong to one and the same decay chain and unknown heavier elements can be identified by genetic correlations to their known daughters.

The biggest α branching ratios could be expected for odd-odd isotopes, whereas the even-even isotopes were expected to decay by fission with halflives decreasing below 1 μs (the separation time of SHIP) with increasing element number[10]. Therefore, negative results in search experiments for even elements could have two reasons, either too low cross-sections or too short halflives. As there is no enhanced cross section for even-even isotopes to be expected, the detection of odd elements not suffering from the above ambiguities would help to design later experiments aiming at even elements. To start our program of heavy element synthesis experiments on odd elements were chosen.

Detection System

The velocity filter SHIP is a two-stage filter consisting of spatially separated electrostatic and magnetic dipole fields, and two magnetic quadrupole lenses[8]. The separated ions pass a pair of time-of-flight detectors and are finally implanted into an array of position sensitive silicon detectors. A mass resolution of $(\Delta M/M)_{FWHM}$ = 10% further suppresses a small amount of scattered transfer products with masses around the target nucleus passing the velocity filter.

The detection of single decay chains has become possible by correlation analysis. Time and position of subsequent α- or fission-decays are correlated to the time and position signals obtained from the implantation process[11]. Halflives can be determined from the time difference between two subsequent signals even in case of only one decaying atom. The energy and position resolution of Si-surface barrier detectors cooled to 260° K is ΔE_{FWHM} = 18 keV and ΔY_{FWHM} = 210 μm, respectively. These values increase to about 35 keV and 400 μm after implantation of 10^9 particles. Since the implantation depth of the ions is smaller than the range of the decay alphas, there is a chance of about 45% that an alpha particle is observed with only a part of its full energy.

RESULTS ON ELEMENTS 106-109

The Odd Elements 105, 107, and 109 with N-Z = 48

The isotopes ^{258}105, ^{262}107, and ^{266}109 have been observed in the reactions of ^{50}Ti, ^{54}Cr, and ^{58}Fe projectiles with ^{289}Bi targets[12-14]. The experiments on investigation of the odd elements have been carried out in February, 1981 (Z = 105 and 107), August, 1982 (Z = 105 and 109), and March, 1985 (^{261}107).

The measured total halflives are 4.4 s, 5 ms, and 3 ms, respectively, and the main decay mode is α emission. An observed fission activity is assigned to ^{258}104, the daughter after electron capture (33%) of ^{258}105. The number of observed atoms was 129, 6 (in this number the results from March '85 are not yet included), and 1, respectively, in measuring times of 2, 6, and 13 days resulting in cross-sections of 2.9 nb, 200 pb, and 16 pb at projectile energies between 4.7 and 5.1 MeV/u. A compilation of decay and production data is given in Table 1.

The cross-sections are shown in Figure 1. An exponential function fitted to the data gives a cross-section decrease of a factor of 0.27 per element. An extrapolation to element 111 results in σ ∿ 1 pb, a value

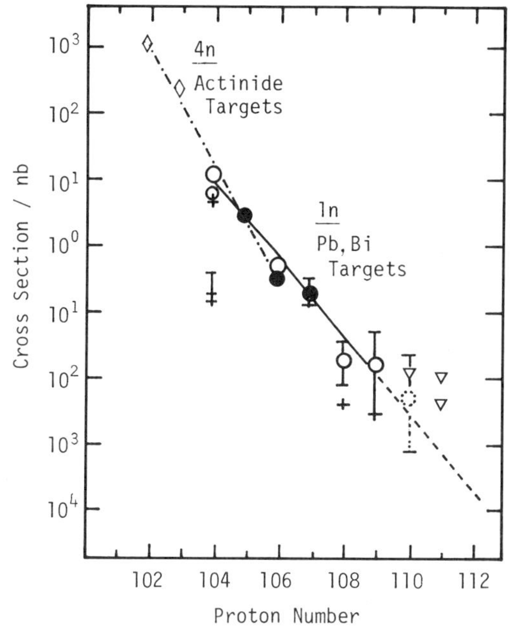

Fig. 1. Comparison of cross sections for production of heavy elements, (◇ Berkeley data, + Dubna data, ○ SHIP data). Only cross section limits are known from experiments to produce element 110 (SHIP) and 111 (Dubna), ▽.

which cannot be reached in a two weeks experiment with our present technology. Cross-section limits for element 111 are published by the Dubna group[15] (see Figure 1). They do not contradict our extrapolations. The cross-section values of the other odd elements given by the Dubna group, determined from the long liver ^{246}Cf α activity or ^{258}104 fission activity[15], agree with our values within the error bars. Why there is a discrepancy for element 109 (and also for element 108, see Sec. 3.2)) is open. Preliminary results of our last experiment to produce ^{261}107 in a 2n evaporation channel gave similar decay properties as for the neighboring isotope ^{262}107. Both are α-emitters. An 80%-spontaneous fission branch in the ms-region discussed by the Dubna group[36] has not been seen in our experiment. Final results will be published after a complete analysis.

The Even Elements 106 and 108

Replacing ^{209}Bi by ^{208}Pb should not significantly change the reaction process and an interpolation of the odd element cross sections is, therefore, a good base to plan experiments to produce isotopes of even elements. These experiments have been carried out in March, 1984 (Z = 106 and 108), and in March, 1985 (Z = 110).

In irradiations of ^{208}Pb with ^{54}Cr 4.80 MeV/u and 4.92 MeV/u we found events which were correlated to the implanted nucleus with an α-energy of 9.56 MeV. In the second generation of decays these events were correlated to the isotope ^{257}104 known to have an α spectrum of several lines. The isotope ^{261}106 is an α emitter, E_α = (9.56±0.03) MeV, with a halflife of $(0.26^{+0.11}_{-0.06})$ s in agreement with the conclusion by Demin et al.[16]. Its

formation cross section in the reaction $^{208}Pb\ (^{54}Cr,n)^{261}106$ is 0.5 nb.

With a beam of ^{56}Fe we found three correlated events at a bombarding energy of 5.02 MeV/u which are assigned to $^{265}108$[17]. Previously known isotopes in the chain are, among others, $^{257}104$ (4.5 s) and ^{253}Es (20 d), the latter being used by the Dubna group to prove the formation of isotopes in the chain[16,18]. The second-generation α energy observed in one event and the measured correlation times agree with the previous finding on the isotope $^{261}106$. All chains are correlated in the third generation to the isotope $^{257}104$ which once more proves that the mother nucleus for these chains is $^{265}108$. The nucleus $^{265}108$ decays by α emission, E_α = (10.36± 0.03) MeV, with a halflife of $(1.8^{+2.2}_{-0.7})$ ms. In the three events no fission was observed in any generation. The obtained production cross section is (19^{+18}_{-11}) pb.

Both cross sections agree with the interpolated values. An extrapolation of the cross sections to element 110 results in values between 1 and 5 pb. The replacement of ^{62}Ni by ^{64}Ni may result in an increase of cross section by a factor of 3 for the production of the (N-Z) = 51 nucleus $^{271}110$. In a 15 day irradiation of ^{268}Pb a dose of 1 × 10[18] ^{64}Ni projectiles (E = 5.06 MeV/) was collected. No decay was observed that could be assigned to an evaporation residue of the compound nucleus $^{272}110$. A cross-section limit for the production of nuclei with $T_{\frac{1}{2}}$ > 20 μs is 12 pb, calculated with 95% confidence. The range of halflives between 2 and 20 μs is not completely analyzed. If an event would be found - but there is only little hope left - the cross-section value could be 5 pb with a 95% confidence interval from 0.8 to 24 pb. Also in this experiment the experimental limit does not contradict the expectations and no further conclusions can be drawn except that the cross sections do not increase.

Coming back to element 106 we want to present data on the neighboring isotopes of $^{261}106$. The identification of element 106 in Reference 9 was based on the detection of $^{259}106$ which was measured to be a 7 ms spontaneous-fission emitter. This finding became questionable after Demin et al.[16] inferred a longer halflife for this isotope. In the $^{287}Pb(^{54}Cr,2n)$ $^{259}106$ reaction we measured a halflife of $(0.48^{+0.28}_{-0.12})$ s for $^{259}106$. It emits an α particle of (9.62±0.03) MeV.

The isotope $^{268}106$ was produced at 4.92 MeV/u in the reaction $^{268}Pb(^{54}Cr,2n)^{268}106$. It decays by emission of α particles to the spontaneous fissioning nucleus $^{256}104$ ($T_{\frac{1}{2}}$ = 7.4 ms). Additionally, a spontaneous fission branch of about 50% has been measured. The halflife of $^{260}106$ is $(3.6^{+0.9}_{-0.6})$ ms, giving a partial spontaneous fission halflife of about 7 ms. This value is larger or at least equal to the spontaneous fission halflife of the daughter isotope reached by α decay. When going from $^{256}104$ to $^{268}106$ the spontaneous fission halflife does not decrease, a finding which contradicts all theoretical predictions, but again is in agreement with the analysis of Demin et al.[16]. The production cross section of the 2n-reaction channel is within a factor 2 equal to the 1n-reaction channel leading to $^{261}106$.

A single α-decay chain of $^{256}104$, we observed in an irradiation of ^{50}Ti at 4.77 MeV/u on ^{208}Pb, allows to connect the even-even isotope $^{260}106$ to the known mass excesses of lighter daughter isotopes.

Table 1. Compilation of Data on Isotopes of Elements 106-109, [ ] revised assignments

Element 106

Isotope	$T_{1/2}$/s	E_α/MeV	b_α	Reaction	E^*_c/MeV	E_1/MeV	Dose/10^{17}	σ/nb	Comment	Ref. -
263	0.9±0.2	9.25	1.0	$^{249}Cf(^{18}O,4n)$	40.2	95	13.4	0.3	decay chain (α)	27
261	$0.26^{+0.11}_{-0.06}$	9.56(0.60) 9.52(0.27) 9.47(0.13)	1.0	$^{208}Pb(^{54}Cr,n)$	22.8	262	1.3	0.5	decay chain (α)	17
	-		>0.5	$^{208}Pb(^{54}Cr,n)$	22.8	300	5	0.3	Detection of ^{253}Es (α)	18
260	(2.5±1.5) ms	-	>0.8	$^{207}Pb(^{54}Cr,n)$	22.4	290	-	0.3	Detection of $^{256}104$	16
				$^{208}Pb(^{54}Cr,2n)$	22.8	290	-	0.4	(sf) and built up	
	$(3.6^{+0.9}_{-0.6})$ ms	9.77	0.5	$^{208}Pb(^{54}Cr,2n)$	22.8	266	1.6	0.3	Decay chain (α,sf)	24,33
	-	-	>0.2	$^{208}Pb(^{54}Cr,2n)$	22.8	300	16	0.4	Detection of $^{256}104$ (sf)	18
[259]	(4 - 10) ms	-	-	$^{207}Pb(^{54}Cr,[2n])$	22.4	280	0.3	1.0	mass and element	34
				$^{208}Pb(^{54}Cr,[3n])$	22.8	280	0.2	1.0	number revised in Ref.16	
259	-	-	~1.0	$^{206}Pb(^{54}Cr,n)$	23.1	290	-	0.4	Detection of $^{255}104$	16
				$^{207}Pb(^{54}Cr,2n)$	22.4	290	-	0.4	(sf)	
	$0.48^{+0.28}_{-0.13}$	9.6	1.0	$^{207}Pb(^{54}Cr,2n)$	22.4	265	1.4	0.3	Decay chain (α)	24,33
	-	-	>0.5	$^{208}Pb(^{54}Cr,3n)$	22.8	300	5	0.02	Detection of $^{255}104$ (sf)	18

Element 107

Isotope	$T_{1/2}$/ms	E_α/MeV	b_α	Reaction	E^*_c/MeV	E_1/MeV	Dose/10^{17}	σ/nb	Comment	Ref.
262	$4.7^{+2.3}_{-1.6}$	10.38	1.0	$^{209}Bi(^{54}Cr,n)$	22.3	262	1.2	0.2	Decay chain (α)	13
	-	-	1.0	$^{209}Bi(^{54}Cr,n)$	22.3	290	5	0.2	Detection of $^{258}104$	15,36
				$^{208}Pb(^{55}Mn,n)$	24.2	310	60	0.1	(sf) and ^{246}Cf (α)	
[261]	1-2	-	[0.8]	$^{209}Bi(^{54}Cr,[2n])$	22.3	290	4.6	0.1	Analysis modified	35
				$^{208}Pb(^{55}Mn,[2n])$	24.2	290	2.2	0.05	later Ref.36	
261	-	-	0.2	$^{207}Pb(^{55}Mn,n)$	24.0	310	2.3	0.03	Detection of $^{257}105$ (sf)	36
260	-	-	1.0	$^{206}Pb(^{55}Mn,n)$	24.1	310	55	∿0.02	Detection of $^{256}104$ (sf)	36

Element 108

Isotope	$T_{1/2}$/ms	E_α/MeV	b_α	Reaction	E^*_c/MeV	E_1/MeV	Dose/10^{17}	σ/nb	Comment	Ref.
265	$1.8^{+2.2}_{-0.7}$	10.36	1.0	$^{208}Pb(^{58}Fe,n)$	20.4	292	6	$0.02^{+0.02}_{-0.01}$	Decay chain (α)	17
	-	-	1.0	$^{208}Pb(^{58}Fe,n)$	20.4	320	30	0.004	Detection of ^{253}Es (α)	18
264	-	-	1.0	$^{207}Pb(^{58}Fe,n)$	20.2	320	22	0.005	Detection of $^{256}104$	18
				$^{208}Pb(^{58}Fe,2n)$	20.4	320	32	0.002	(sf)	
263	-	-	1.0	$^{209}Bi(^{55}Mn,n)$	25.8	300	130	0.002	Detection of $^{255}104$ (sf)	18,36

Element 109

Isotope	$T_{1/2}$/ms	E_α/MeV	b_α	Reaction	E^*_c/MeV	E_1/MeV	Dose/10^{17}	σ/nb	Comment	Ref.
266	$3.5^{+16.6}_{-1.6}$	11.10	1.0	$^{209}Bi(^{58}Fe,n)$	19.8	299	2.8	$0.015^{+0.035}_{-0.012}$	Decay chain (α,sf)	14,37
	-	-	1.0	$^{209}Bi(^{58}Fe,n)$	19.8	320	36	0.003	Detection of $^{258}104$ (sf) and ^{246}Cf (α)	15

DISCUSSION OF GROUND STATE PROPERTIES OF THE HEAVIEST ISOTOPES

Experimental Masses and Shell Corrections

The mass excesses of the heaviest even-even nucleus known until now, $^{260}106$, and its daughter nucleus, $^{256}104$, as determined in our experiments are given in Table 2. They are compared to different mass predictions.

Best agreement is obtained with the mass predictions of Liran and Zeldes[22]. All other approaches fail to reproduce the stability of these nuclei by about 1 to 1.5 MeV. For $^{260}106$, a shell correction energy of -5.1 MeV is obtained using the value of 111.7 MeV for the macroscopic part of the mass excess from Reference 23. Especially in macroscopic-microscopic approaches it seems to be difficult to reproduce the amount of the measured shell stabilization.

From the experimental mass excesses and the liquid drop (spherical) part of the Møller-Nix mass formula[23] we determined shell corrections for the N-Z = 48 isotopes from ^{232}U to $^{260}106$ (Figure 2a). Included are shell effects for the odd-mass isotopes with the same N-Z (open circles). The shell effects show a continuous increase with proton number, from 0.5 MeV for uranium to more than 5 MeV for Z > 104 elements.

Table 2. Experimental Mass Excess/MeV of $^{260}106$ and $^{256}104$ Compared to Predictions of Different Mass Tables

Nucleus	Ref.19	20	21	22	23	this work
$^{260}106$	108.27	107.29	107.7	106.94	108.13	106.58±0.10
$^{256}104$	95.90	94.84	95.6	94.37	95.77	94.24±0.07

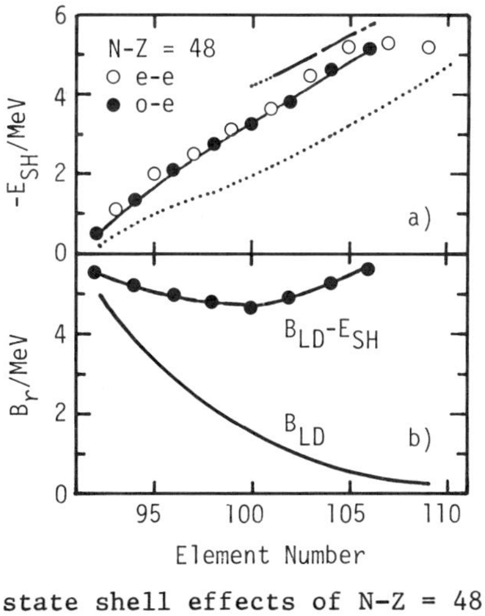

Fig. 2. a) Ground state shell effects of N-Z = 48 nuclei. The reference surface is the macroscopic part of the Møller-Nix mass formula[23]. Dashed line: Microscopic part of the Møller-Nix mass formula[3]. Dotted line: Calculations from Cwiok et al.[25]. b) Liquid drop fission barrier (B_{LD})[30] and experimental fission barriers ($B_{LD}-E_{SH}$)[24].

The shell corrections calculated by Møller-Nix[3,23] are smaller by about 1 MeV than the experimental values. But both curves suggest that the next heavier elements 110 and 111 with N-Z $\sim$ 48 have shell correction energies of at least 5 MeV[24] pointing to an increased stability against fission. Recent calculations by Cwiok et al.[25], giving values even higher than the experimental data, support this assumption.

Experimental data cannot yet answer the question whether the shell effects continue to increase monotonically further to the superheavy region. The N-Z = 48 odd elements rather suggest a flattening. This would confirm calculations[3,25] predicting an island of deformed nuclei of higher stability extended in front of the classical spherical island of superheavy nuclei. The latter gain their stability from spherical shell closures at Z = 114 and N = 184[4]. Shell corrections increase up to a maximum value of nearly 9 MeV for the nuclei Z = 114, N = 178[3]. The island around Z = 109, N = 162 consists of nuclei which are deformed (ε_2 = + 0.20, ε_4 = + 0.09). In Figure 3 we have plotted the results of the recent shell model calculations[3,25]. The figure suggests that the investigated N-Z = 48 and 49 nuclei are crossing the island of increased stability west from the maximum, and ^{266}109 is already on the side of decreasing shell corrections.

A hope is that the increasing groundstate quadrupole deformation may help to stabilize the nuclear system when approaching the center of deformed nuclei around Z = 108, N = 162. Nuclei with maximum shell corrections below element 109 where the cross sections are expected to be higher than for elements 110 and 111, cannot be made with lead or bismuth targets. Experiments to produce element 108 with radium targets[26] by ^{226}Ra(^{48}Ca,3n)^{271}108 failed probably because of too low sensitivity. The heaviest element that could be identified with actinide targets was found by Ghiorso et al.[27] in the reaction ^{249}Cf(^{18}O,4n)^{263}106. The isotope ^{263}106 is an α emitter with a halflife of 0.9 s and the production cross section was 0.3 nb. This cross section is similar to the value measured for element 106 in reactions with lead targets (Figure 1). Recent experiments[28] to produce nuclei in the spherical superheavy element region (October 1982 and March 1983) failed. The cross-section limits are 240 pb

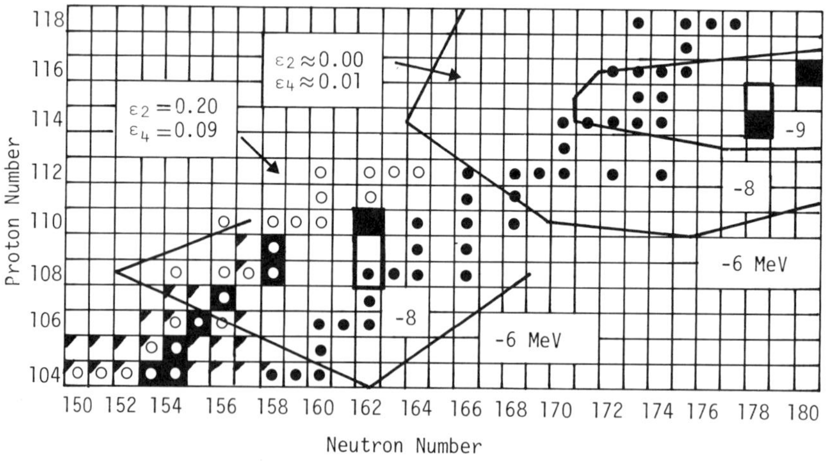

Fig. 3. Ground state shell corrections as calculated in Reference 25 (lower left part) and Reference 3 (upper right part). Open circles: Compound nuclei of reactions (Ti to Zn) + ^{208}Pb $\rightarrow$ Z = 104 to 112. Dots: Compound nuclei of reactions (Mg to Fe) + ^{236}U $\rightarrow$ Z = 104 to 118. Full triangles: Known nuclei. Shaded: Compound systems investigated with SHIP.

for halflives between 10^{-5} and 10^{6} s and 20 pb for halflives around 10^{7} s. The investigated reaction $^{48}Ca+^{248}Cm$ gives compound nuclei close to a region of highest shell stabilization (Figure 3). We do not think the halflives are out of the experimental range but the cross sections are below the experimental limits (see discussion on "Limitation on Heavy Element Production").

Fission Barriers and Halflives

Fission barriers of N-Z = 48 nuclei have been determined as a sum of the liquid drop barrier[30] and the experimental ground state shell effects (Figure 2b). Saddle point shell effects are neglected. Although the liquid drop barrier vanishes for the heavy elements the increasing shell effects keep the fission barriers constant near 5 MeV.

Figure 4a shows the systematic decrease of experimental fission halflives up to nobelium by 2 or 3 orders of magnitude per element at equal N-Z. The overlaying structure with maximum halflives of N = 152 isotones is due to a subshell closure at this neutron number. As the barrier height is constant for the N-Z = 48 nuclei the decrease of the halflives by more than 20 orders of magnitude is due to the decrease of the liquid drop barrier widths. This trend changes at element 104 from where on the constant halflives suggest that beside the barrier heights also the shapes of the barriers remain constant. For N-Z = 50 nuclei the shapes have changed from a double humped barrier to a less broad single humped barrier between element 102 and element 104[6].

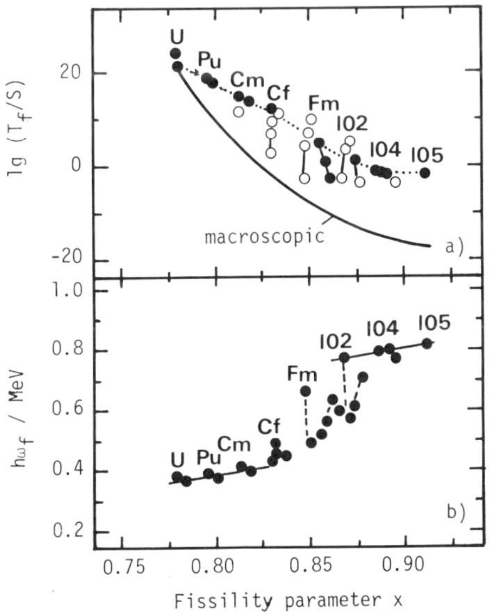

Fig. 4. The upper part shows the fission halflives of all known even-even isotopes together with the halflives expected for nuclei with macroscopic fission barriers[30]. The barrier curvature parameter is fitted to ^{232}U and kept constant for all isotopes. The lower part of the figure[31] shows the curvature parameter $\hbar\omega_f$, obtained from fission halflives, experimental shell correction energies, and the macroscopic barriers of Figure 2. Diamonds belong to nuclei with a double-humped barrier and points are nuclei with a predicted single-humped barrier. The straight lines are a fits to the $\hbar\omega_f$ for the two groups.

In Figure 4b the curvature $\hbar\omega$ of the fission barrier is plotted as a function of the fissility parameter x. $\hbar\omega$ was extracted from a Hill-Wheeler barrier transmission formula using the experimental values for the spontaneous fission halflives and the fission barriers[31]. For elements 104 and 106 high values of $\hbar\omega$ point to a narrow barrier the width of which is about 50% of the values obtained for liquid drop fission barriers near uranium. Moreover, constant halflives and fission barriers point to a barrier which is nearly independent of the fissility x of the nucleus.

Recent calculations of the fission halflives[3] of elements 106 to 110 give increasing values with increasing neutron number up to N = 162, the isotones with the highest shell effects. However, in case of $^{260}106$ the experimental partial fission halflife is 7 ms and still 175 times longer than the calculated values (0.04 ms). From experiments to produce $^{264}108$, Oganessian et al.[18] concluded that this nucleus is an α emitter with a partial spontaneous fission halflife larger than 1 ms. Also this value is 3000 times longer than the calculated value of 0.3 μs.

The fission halflives are further increased in case of odd and odd-odd nuclei by the additional specialization energy. The isotopes $^{253}104$ and $^{257}105$ have partial fission halflives of 2.7 s and 8.2 s, respectively. Only α decay was observed for $^{259}106$[33], $^{261}106$, $^{263}106$[27], and $^{265}108$. The main decay mode of $^{262}107$ and $^{266}109$ is α emission.

Calculated Q_α values and partial α halflives[3] are in good agreement with the experimental data which reflects the generally lower influence of smoothly varying shell structure effects on Q_α values.

THE LIMITATION IN HEAVY ELEMENT PRODUCTION

In the experiments performed to produce highly fissionable isotopes, a three-fold limitation of the production process has been observed.

1. A thermal limitation in the exit channel by fission losses in the evaporation cascade, Figure 1.
2. A dynamical limitation in the entrance channel by friction processes during passage of the fusion barrier.
3. A structural limitation by the nuclear structure-dependent disappearance of the shell stabilization with excitation energy.

The production of still heavier elements between Z = 110 and 114 is probably not a question of the ground-state stability of these elements, but a question of navigating between the three-fold limitation of the production process.

The Thermal Limitation

This limitation has to be overcome, if the production of heavy elements using actinide targets is to be continued. Figure 1 compares the largest 4n cross sections using ^{249}Bk and ^{249}Cf targets to produce isotopes of elements 100-106, E* $\sim$ 45 MeV, with 1n cross sections for cold fusion reactions, E* $\sim$ 10 MeV. For actinide-based reactions the two other limitations do not apply as long as the reaction aims at deformed isotopes for elements Z < 112 (N < 165). As the analysis of mass excesses and fission halflives points to nearly constant fission barriers, the average Γ_m/Γ_f values in the evaporation cascade may, in contrast to the trend observed up to element 106, stay fairly large and the cross sections may decrease more slowly than observed for Z < 106. This speculation makes reactions such as ^{254}Es(^{23}Na,4n)^{273}110, ^{249}Cf(^{26}Mg,4n) ^{271}110, or ^{235}U(^{40}Ar,4n)^{271}110 possible candidates for proceeding beyond 109. The production cross sections will

certainly be small. Estimates are uncertain and range between 30 and 0.01 pb. Only the detection of decay chains will allow us to identify the EVRs, since transfer reactions with actinide targets produce fission activities with much larger cross sections than do Pb- and Bi-based cold fusion reactions. The thermal limitation that since 1974 has prevented the actinide-based production of elements heavier than Z = 106 may be overcome by using recoil spectrometers and improved detection methods. Compared to the Berkeley 106 experiment[27], the sensitivity of the techniques used to detect $^{266}109$ was improved by at least two orders of magnitude[37]. At least heavier isotopes of the element Z < 108 could be produced.

The Dynamical Limitation

The entrance channel limitation reduces the cross section for production of element 110 by cold fusion to the level of a few pb, Figure 1. The Coulomb repulsion between the two collision partners during the fusion process finally prevents the formation of a fused monosystem. Figure 5 demonstrates that nature has set a limit to combining two nuclei with atomic number Z_p and Z_T. Systems that were fused successfully are separated from combinations leading to no EVRs. The line of separation is characterized by a well-defined ratio of Coulomb and nuclear forces acting between the two amalgamating collision partners. Beyond that line open points indicate the different failures to produce superheavy elements. For $Z_p+Z_T > 120$ the very first step in the formation of an EVR, the overcoming of the Coulomb barrier, already becomes impossible for any combination of Z_p and Z_T. In the range $120 > Z_p+Z_T > 80$ fusion is hindered, not every combination of Z_p and Z_T is successful, e.g., the isotope ^{244}Fm was made in the reaction $^{206}Pb(^{40}Ar,2n)^{244}Fm$, but could not be made out of the nearly symmetric partners ^{116}Pd and ^{136}Xe[38].

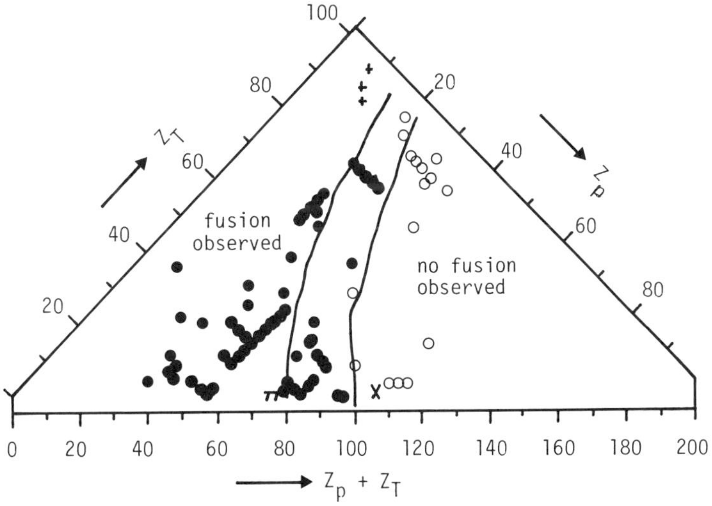

Fig. 5. Elements Z_p and Z_T fused to element Z_p+Z_T. Each combination is a point in the triangle. Successful fusion reaction (in GSI experiments over the years detected by EVRs) are indicated by full points, failures by open points. Between $x_{mean}^{UCD} = 0.72$ and 0.85, the production goes down below the 100-pb level, and $p(B_B)$ reaches values of 5×10^{-5}. Between Z = 90 and Z = 120, more and more asymmetric combinations are necessary to produce surviving EVRs[38]. Beyond Z > 120 no fusion is possible.

The ratio of disruptive Coulomb forces and attractive surface tension forces governs the amalgamation of two nuclei into one. For a monosystem this ratio is given by the fissility parameter x. For a two-touching sphere configuration, Bass[39] defined a corresponding parameter making use of the proximity force. Taking into account that the proton and neutron ratio between the two partners is equilibrated very quickly (10^{-22} s), and that the nuclear system at the Coulomb barrier for those systems of importance here is more compact than the two-touching sphere configuration, a modified parameter x_{mean}^{UCD} describing the ratio of Coulomb and nuclear forces has been defined and applied to organize the vast amount of data[38,40]:

$$x_{mean}^{UCD} = 2x(\kappa^2+\kappa+\kappa^{-1}+\kappa^{-2})^{-\frac{1}{2}} \text{ with } x = (Z^2/A^2)/(Z^2/A)_{crit}$$

$$(Z^2/A)_{crit} = 50.88[1-1.78(N-Z/N+Z)^2] \text{ and } \kappa = (A_1/A_2)^{1/3}$$

The fusion probability for a barrier which is passed by tunneling, is at the fusion barrier by definition 0.5. Fusion barriers are calculated using standard potentials which allow to reproduce fusion barriers for all the light systems. A potential introduced by Bass[41] gives these standard barriers for all possible systems. The fusion probability at the standard barrier $p(B_B)$ beyond a threshold value of x_{thr} = 0.72 becomes much smaller than the WKB-value of 0.5. The energy value at which the fusion probability actually becomes 0.5 is called the dynamical barrier, or 'extra-extra-push' barrier[32]. $p(B_B)$ characterizes the hindrance in the entrance

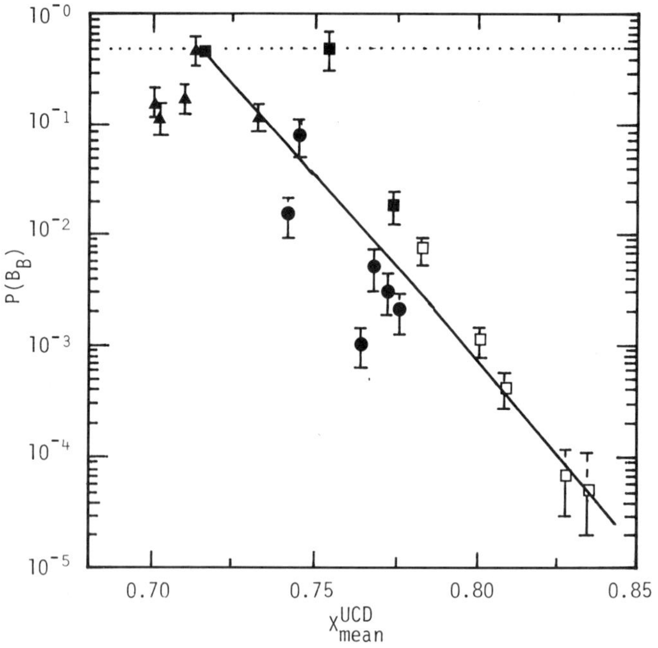

Fig. 6. The fusion probability at the Bass barrier $p(B_B)$ characterizing the dynamical hindrance in the entrance channel for all systems where EVRs have been detected[7]. Triangles are from Reference 48, circles are from Reference 49, solid squares are from Reference 31, and open squares are from cross-section values of heavy element production[13,33,37].

channel at the energy of the Bass barrier. The value of $p(B_B)$ may be derived directly from the cross-section value at the barrier energy B. Figure 6 shows $p(B_B)$ as a function of x_{mean}^{UCD} for all systems investigated beyond $x_{mean}^{UCD} = 0.71$. In this presentation the cross-section values known for heavy-element production beyond Z = 104 have been included. The scattering of the experimental values shows the importance of nuclear structure effects, e.g., $^{208}Pb(^{48}Ca,xn)^{256-x}102$ ($x_{mean}^{UCD} = 0.751$) compared to $^{123}Sb(^{86}Kr,xn)^{209-x}Fr$ ($x_{mean}^{UCD} = 0.736$). They modify $p(B_B)$ by factors larger than 10. A linear fit to all the data points gives for $p(B_B)$ an exponential dependence from $x_{mean}^{UCD}-x_{thr}$ with a slope parameter $d[\ln p(E)]/dx = -71$, equivalent to a factor of two for a change of 0.01 in x_{mean}^{UCD}:

$$p(B_B) = 0.5 \exp[-71(x_{mean}^{UCD}-x_{thr})] .$$

The constant slope points to a strongly increasing fluctuation of the dynamical barrier B for systems with dynamical hindrance.

The heavy element production via cold fusion is highly hindered, e.g., the reaction $^{209}Bi(^{58}Fe,1n)^{266}109$ shows a fusion probability which is 5×10^{-4} instead of 0.5. This entrance channel limitation nearly compensates the smaller losses in the deexcitation cascade of 1n-reactions compared to the actinide based 4n-reactions. If the ^{58}Fe beams are replaced by ^{64}Ni beams, the cross sections may decrease only marginally because the decrease may be partly compensated by the higher number of neutrons in ^{64}Ni, which will lead into a more stabilized N-Z = 52 compound system. The best cold fusion reaction for discovering element 110 is $^{298}Pb(^{64}Ni,1n)^{271}110$ at an energy near the unshifted fusion barrier. The negative result of this experiment has been described in the previous section.

The Structural Limitation

The fission barrier protects the compound systems against immediate disintegration. The fission barrier is made up of two contributions, a smoothly varying macroscopic liquid drop part and a shell correction, mainly of the ground-state mass. Near element 106 the macroscopic fission barriers fall below 1 MeV. As Figure 2b shows, the stability beyond is governed by the shell correction of the ground-state masses alone. The fluctuations of the level density around a continuously increasing Fermi-gas level density determine the shell corrections, which strongly depend on the excitation energy and the deformation of the nuclear system. The study of the temperature dependence of the shell contributions to the fission barrier makes it possible to predict which excitation energy a shell-stabilized compound system may take in order to survive the deexcitation process.

Shell-stabilized, highly fissionable nuclei are found among the actinides. Using these nuclei produced by different nuclear reactions, one can study the contribution of shell effects to EVR formation at excitation energies varying from 10 to 60 MeV. We differentiate between shell-stabilized deformed and spherical nuclei.

The experimental values for Γ_n/Γ_f obtained for three deformed nuclei are presented in Figure 7. The values given refer to an angular momentum of zero. The Γ_n/Γ_f values depend only weakly on the energy E* in the range

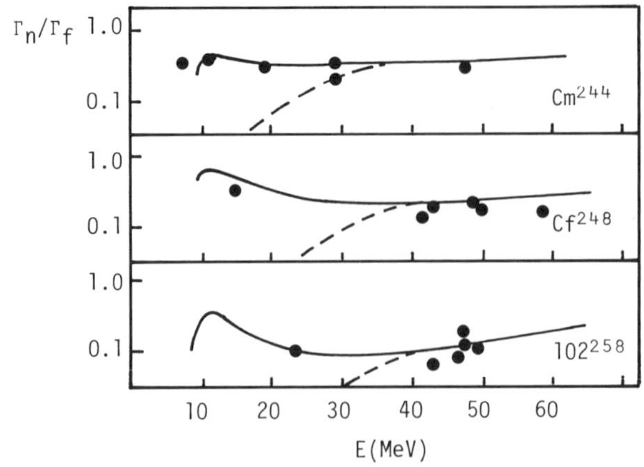

Fig. 7. $\Gamma_n/\Gamma_f(E^*)$ for the three deformed compound systems ^{244}Cm, ^{248}Cf, and $^{258}102$. Experimental values are taken from pairs of excitation functions[44,45]. The lines indicate two calculations, full line with shell effects, dotted line without shell effects.

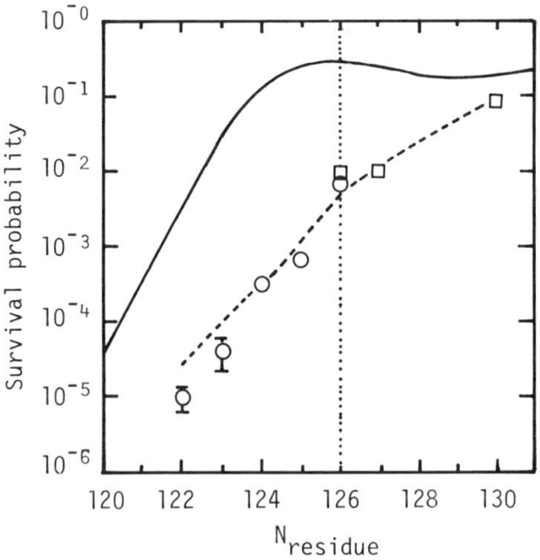

Fig. 8. The survival probability $w(E^*)$ for spherical EVRs of thorium produced by 4n reactions as a function of the neutron number of the EVR[46,47]. Circles are from (^{40}Ar+Hf), squares are from (^{48}Ca+Yb) reactions. The lines give two calculations using different damping energies for shell effects[42], full line E_D = 18 MeV, dashed line E_D = 6 MeV.

investigated. The few points in the range 10-20 MeV are even larger than at higher energies. The results of two calculations are given for comparison. The full line shows the expectation for Γ_n/Γ_f with shell effects entering into level densities and the fission barriers. An exponential damping of the shell-effects with E_D = 18 MeV, which follows from microscopic models, is used in the calculation[42-45]. The damping of shell effects leads to the shallow minimum of $\Gamma_n/\Gamma_f(E^*)$, which becomes deeper for systems dominantly shell stabilized. The hatched line is calculated without any shell effects, and obviously the observed increase of Γ_n/Γ_f in the low energy range cannot be explained. The measured cross sections must include shell effects for all the deformed nuclei analyzed. There would be no heavy elements made by fusion, unless Γ_n/Γ_f were kept large at low excitation energies by shell effects.

A very different behavior was found in experiments near ^{216}Th. Here nuclei are spherical, highly fissionable, and again their barriers are partly shell stabilized. They may be produced within a range of excitation energies between 20 and 50 MeV. These nuclei are the best approximation to the shell-stabilized spherical superheavy nuclei near N = 184, Z = 114. Figure 8 shows measurements of the survival probability for thorium isotopes produced by 4n reactions[46,47]. Two calculations of survival probabilities are shown. The measured cross sections are smaller by about two orders of magnitude, compared to an evaporation cascade calculation that included the ground-state shell effects in the fission barriers and a damping of shell effects with E_D = 18 MeV[42,43]. A calculation with a reduced value of E_D = 6 MeV is given for comparison. It reproduces the trend of the experimental data fairly well.

We know there is additional ground-state stability of spherical nuclei due to shell effects, which must be observable somewhere in the production cross sections at low energies. The analysis of the data shown in Figure 8 is compatible with shell effects becoming important at energies below 15 MeV (E_D = 6 Mev). At higher energies the spherical compound systems behave as if no shell stabilization existed, whereas in deformed compound systems the shell stabilization is of importance up to 40-50 MeV (E_D = 18 MeV).

Outlook Beyond Element 109

In Table 3 we summarize our findings on limitations in three types of reactions. The cold fusion reaction ^{208}Pb+^{64}Ni, which is entrance channel limited, the actinide based reaction, ^{254}Es+^{23}Na, which is limited by the fission losses during deexcitation, and the reaction ^{248}Cm+^{48}Ca leading into the island of spherical superheavy nuclei which is limited by the extraordinary sensitivity of shell stabilized spherical nuclei against intrinsic excitation. All three systems have predicted fission barriers which are higher than the until now heaviest isotope ^{266}109.

The recent failure to produce element 116 via ^{248}Cm(^{48}Ca,2n)$^{294}116^{178}$ with a cross-section limit of 240 pb is plausible[29]. 3 orders of magnitude higher sensitivity is demanded, a scope far out of reach for nowadays techniques. The extrapolations far from our fixpoint ^{266}109 certainly are not very accurate. Future studies on the limitations of fusion may allow to find better strategies for the time consuming search experiments. The outcome of future experiments attempting to go beyond Z = 109 is open. Only experiments can decide which of the limitations, the dynamical of cold fusion reactions or the thermal of actinide-based reactions, may be more easily overcome. The three-fold limitation will stop our search for man-made elements either at Z = 109 or at a few atomic numbers more. In any case, the game is limited and we should accept the study of the limitations to be a message in itself. The idea of superheavy elements - elements shell stabilized, not existing in a world of macroscopic nuclear drops - has already been realized. The isotope ^{266}109 is already completely shell stabilized. For ^{260}106, the spontaneous fission halflife is increased by 15 orders, and without nuclear structure effects the isotope would never have been detected. Entering the island of ε_4 stabilized α-emitters at Z = 106 we produced elements that correspond to what superheavy elements are to be. They are not spherical nuclei, but deformed sausage-like entities. They live not billions of years, but milliseconds. We are reluctant to accept the idea that our dream of superheavy elements may have already been fulfilled.

Table 3. Cross Section Estimate for SHE-Production

Reaction	Fission barrier/MeV	Excitation energy/MeV	Dynamical limitation $p(B_B)$	Deexcitation losses	Nuclear structure limitation	σ_{EVR}/pb	Efficiency SHIP
$^{208}Pb(^{64}Ni,n)^{271}110^{161}$	6.7	15	$5 \cdot 10^{-3}$	$5 \cdot 10^{-3}$	-	7	0.2
$^{254}Es(^{23}Na,5n)^{272}110^{162}$	7.6	51	0.5	$8 \cdot 10^{-10}$	-	32	0.02
$^{248}Ca(^{48}Ca,2n)^{294}116^{178}$	8.4	30	$3 \cdot 10^{-4}$	$2 \cdot 10^{-6}$	$2 \cdot 10^{-2}$	0.2	0.12

ACKNOWLEDGEMENTS

This lecture reports on the work of about 30 scientists during the years 1981-1985. I am indebted to all of them, especially I want to thank my collaborators G. Müzenberg, S. Hofmann, F. P. Heßberger, W. Reisdorf, and K. -H. Schmidt for their steady involvement in the experiment. The help of S. Hofmann in preparing this manuscript is gratefully acknowledged.

REFERENCES

1. E. O. Fiset and J. R. Nix, Nucl.Phys., A193:647 (1972).
2. K. Takahashi et al., At.Data and Nucl.Data Tables, 12:101 (1973).
3. G. A. Leander et al., Proceedings of the 7th International Conf. on Atomic Masses and Fundamental Constants, AMCO-7, Darmstadt-Seeheim, p.466 (1984) and private communication.
4. W. D. Myers, W. J. Swiatecki, Nucl.Phys., 81:1 (1966).
5. S. G. Nilsson et al., Nucl.Phys., A131:1 (1969).
6. Y. T. Oganessian, Lect.Notes Phys., 33:221 (1974).
7. P. Armbruster, Ann.Rev.of Nucl.and Part.Science, 35:135 (1985).
8. G. Münzenberg et al., Nucl.Instr.Meth., 161:65 (1979).
9. Y. T. Oganessian et al., JETP Lett., 20:265 (1974).
10. J. Randrup et al., Phys.Rev., C13:229 (1976).
11. S. Hofmann et al., Nucl.Instr.Meth., 223:312 (1984).
12. F. P. Heßberger et al., GSI Annual Report 1981, p.66 (1982); Z.Phys.
13. G. Münzenberg et al., Z.Phys., A300:107 (1981).
14. G. Münzenberg et al., Z.Phys., A309:89 (1982).
15. Y. T. Oganessian et al., Radiochemica Acta, 37:113 (1984).
16. A. G. Demin et al., Z.Phys., A315:197 (1984).
17. G. Münzenberg et al., Z.Phys., A317:235 (1984).
18. Y. T. Oganessian et al., Z.Phys., A319:215 (1984).
19. W. D. Myers, "Droplet Model of Atomic Nuclei," IFI/Plenum, New York (1973).
20. H. v. Groote et al., At.Data and Nucl.Data Tables, 17:418 (1976).
21. P. A. Seeger and W. M. Howard, Nucl.Phys., A238:49 (1975).
22. S. Liran and N. Zeldes, At.Data and Nucl.Data Tables, 17:431 (1976).
23. P. Møller and J. R. Nix, At.Data and Nucl.Data Tables, 26:165 (1981).
24. P. Armbruster, "The Int. School of Physics 'Enrico Fermi'," Varenna (1984).
25. S. Cwiok et al., Nucl.Phys., A410:254 (1983).
26. Y. T. Oganessian et al., JINR P 7-12054, Dubna (1978).
27. A. Ghiorso et al., Phys.Rev.Lett., 33:1490 (1974).
28. K. H. Schmidt et al., Z.Phys., A315:159 (1984).
29. P. Armbruster et al., Phys.Rev.Lett., 54:406 (1985).
30. M. Dahlinger et al., Nucl.Phys., A376:94 (1982).
31. F. P. Heßberger, Thesis, TH Darmstadt (1984).
32. S. Bjørnholm et al., Nucl.Phys., A391:471 (1982).
33. G. Müzenberg et al., Z.Phys.
34. Y. T. Oganessian et al., Pis'ma Zh.Eksp.Theor.Fiz., 20:580 (1974).
35. Y. T. Oganessian et al., Nucl.Phys., A273:505 (1976).
36. Y. T. Oganessian, "Int. School-Seminar on Heávy Ion Physics," Alushta, JINR D7-83-644, p.55, Dubna (1983).
37. G. Münzenberg et al., Z.Phys., A315:145 (1984).
38. H. Gäggeler et al., Z.Phys., A316:291 (1984).
39. R. Bass, Nucl.Phys., A231:45 (1974).
40. P. Armbruster, "Proc. Int. Conf. on Nuclear Physics," Florence, Italy, P. Blasi and R. A. Ricci, eds., p.343, Bologna, Tipografia Compositori (1983).
41. R. Bass, "Proc. Symp. on Deep-Inelastic and Fusion Reactions with Heavy Ions," Lect.Notes Phys., 117:281 (1980).
42. A. V. Ignatyuk et al., Sov.J.Nucl.Phys., 21:255 (1975).

NATO ASI Series

Advanced Science Institutes Series

A series presenting the results of activities sponsored by the NATO Science Committee, which aims at the dissemination of advanced scientific and technological knowledge, with a view to strengthening links between scientific communities.

The series is published by an international board of publishers in conjunction with the NATO Scientific Affairs Division

A Life Sciences **B Physics**	Plenum Publishing Corporation New York and London
C Mathematical and Physical Sciences	D. Reidel Publishing Company Dordrecht, Boston, and Lancaster
D Behavioral and Social Sciences **E Engineering and Materials Sciences**	Martinus Nijhoff Publishers The Hague, Boston, and Lancaster
F Computer and Systems Sciences **G Ecological Sciences**	Springer-Verlag Berlin, Heidelberg, New York, and Tokyo

Recent Volumes in this Series

Volume 138—Topological Properties and Global Structure of Space-Time
edited by Peter G. Bergmann and Venzo De Sabbata

Volume 139—New Vistas in Nuclear Dynamics
edited by P. J. Brussaard and J. H. Koch

Volume 140—Lattice Gauge Theory: A Challenge in Large-Scale Computing
edited by B. Bunk, K. H. Mütter, and K. Schilling

Volume 141—Fundamental Problems of Gauge Field Theory
edited by G. Velo and A. S. Wightman

Volume 142—New Vistas in Electro-Nuclear Physics
edited by E. L. Tomusiak, H. S. Caplan, and E. T. Dressler

Volume 143—Atoms in Unusual Situations
edited by Jean Pierre Briand

Volume 144—Fundamental Aspects of Quantum Theory
edited by Vittorio Gorini and Alberto Frigerio

Series B: Physics

Atoms in
Unusual Situations

ACKNOWLEDGEMENTS

The scientific contents of this school has been organized by:

. J.P. DESCLAUX and I. LINDGREN (Theory),
. S. HAROCHE and S. LIBERMAN (Rydberg atoms and laser physics),
. E. FABRE (Plasma physics),
. and J.P. BRIAND (Physics of ions),

under the recommendations of the following International Advisory Board:

. P. ARMBRUSTER (Federal Republic of Germany),
. A. DALGARNO (United States),
. D. KLEPPNER (United States),
. I. LINDGREN (Sweden),
. N.J. PEACOCK (United Kingdom),
. E. ZAVATTINI (Switzerland).

The organization of this school whose aim was to merge people who did not not know each other needed special efforts. I would like to thank all the participants for their oecumenical spirit and their help in organizing the afternoon sessions. These sessions also needed some specific material supports and I would like to thank M. Tavernier, P. Indelicato for their constant help in the organization of the afternoon sessions.

This NATO Advanced Study Institute could not have been organized without the permanent assistance of Misses M.F. Hanseler and A. De Corte whom I would like to thank on behalf of all the participants for their kindness and devotion.

I would like also to thank Mrs Ravignot and Mr Dubois for their help in the preparation of the school and express my gratitude to A. Simionovici for his great assistance in the preparation of this book.

Jean Pierre Briand
Paris

PREFACE

Atomic Physics is certainly the oldest field in which Quantum Mechanics has been used and has provided the most significant proofs of this new theory. Most of the basic concepts, except those more recently developed in field quantization, have been understood for quite a time. Atomic Physics began to serve as a basis for other fields such as molecular, solid state or nuclear physics. A renewal of interest in Atomic Physics began in the sixties, after the discovery of Quantum Electrodynamics, and later when it provided some basic tests of fundamental questions like parity violation, time reversal or Dirac theory.

More recently the development of new technologies led to the exploration of very extreme cases in which the most secrete aspects of atoms have been observed.

- Rydberg states where the atoms are so big that they can be described by classical theories;
- Heavy or super-heavy ions or exotic atoms where unknown QED or relativistic effects can be observed (very heavy hydrogenlike or heliumlike ions, positron production in very violent collisions...);
- Huge external perturbations as those appearing in super-dense plasmas or ultra-high fields.

The aim of this school was to gather atomic physicists from all over the world working in all these areas of Atomic Physics.

This NATO Advanced Study Institute which was held in Cargèse (Corsica) in June 1985 gathered about 70 physicists, working in fields such as plasma physics, laser physics, accelerator based physics and theoretical physics. The morning sessions were dedicated to general lectures in each of these fields. They provided elementary basic knowledge in theory and experimental techniques in each of the considered fields. The lectures were in most cases prepared for physicists outside the considered topic, for instance lectures in plasma physics and associated techniques for those working in theory, laser or accelerator based physics. The afternoon sessions were devoted to symposia, where all the participants having some knowledge in a given topic, tried through poster sessions, panel sessions, oral presentations or informal discussions to teach the others their own techniques, results and language.

We provide in this book all the basic lectures presented in the morning sessions and some summaries or conclusions extracted from the afternoon sessions.

Proceedings of a NATO Advanced Study Institute on
Atoms in Unusual Situations,
held June 13–26, 1985,
in Cargèse, Corsica, France

Library of Congress Cataloging in Publication Data

NATO Advanced Study Institute on Atoms in Unusual Situations (1985: Cargèse, Corsica)
Atoms in unusual situations.

(NATO ASI series. Series B, Physics; vol. 143)
"Proceedings of a NATO Advanced Study Institute on Atoms in Unusual Situations, held June 13–26, 1985, in Cargèse, Corsica, France"—T.p. verso.
Includes bibliographies and index.
1. Atoms—Congresses. 2. Quantum electrodynamics—Congresses. I. Briand, Jean Pierre. II. Title. III. Series: NATO ASI series. Series B, Physics; v. 143.
QC170.N37 1985 539.7 86-22495
ISBN 0-306-42399-5

A Division of Plenum Publishing Corporation
233 Spring Street, New York, N.Y. 10013

Printed in the United States of America

ME: NATO Advanced Institute on Atoms in Unusual Situations (1985 : Cargèse, Corsica)

Atoms in Unusual Situations

Edited by

Jean Pierre Briand

Université Pierre et Marie Curie
Paris, France

Plenum Press
New York and London
Published in cooperation with NATO Scientific Affairs Division

43. K. H. Schmidt et al., Z.Phys., A308:215 (1982).
44. A. S. Iljinov and E. A. Cherepanov, JINR P-7-84-68 Dubna (1984).
45. H. Delagrange et al., Phys.Rev.Lett., 39:867 (1977).
46. K. H. Schmidt et al., "Proc. Symp. Phys. Chem of Fission, Jülich 1979," Vienna, IAEA:1 (1980).
47. C. C. Sahm et al., Nucl.Phys., A441:316 (1985).
48. J. G. Keller, Thesis, TH Darmstadt (1984).
49. C. C. Sahm et al., Z.Phys., A319:113 (1984); C -C. Sahm, Thesis, TH Darmstadt (1984).

ATOMIC PHYSICS THROUGH ASTROPHYSICS

Alexander Dalgarno

Harvard-Smithsonian Center for Astrophysics
60 Garden Street
Cambridge, Massachusetts 02138

INTRODUCTION

Astronomical environments encompass an extreme range of physical conditions of temperature, density, pressure and radiation fields and unusual situations abound. In my lecture, I will describe some of the objects found in the Universe and discuss the atomic processes that occur.

THE EARLY UNIVERSE

In the earliest moments of the Universe the temperature was very high and the physics was determined by interactions among the fundamental particles. As the expansion proceeded and the Universe cooled to below 10^9K, the protons and neutrons combined in a sequence of nuclear reactions to produce a large fractional abundance of ^{4}He and trace amounts of ^{2}H, ^{3}He and ^{7}Li. Nucleosynthesis then ceased as the temperature fell below 10^8K. After about 10^5 years, the continued expansion reduced the temperature to 4000K. The supply of photons energetic enough to photoionize atomic hydrogen become negligible and neutral atomic hydrogen formed by radiative recombination,

$$H^+ + e \rightarrow H' + h\nu \ , \qquad (1)$$

could survive. The thermal contact that had been maintained between matter and radiation by Thomson scattering of electrons and photons was lost and matter and radiation evolved independently.

With the formation of neutral atomic hydrogen a diverse array of atomic processes occurred. Negative hydrogen ions were formed by radiative attachment

$$H + e \rightarrow H^- + h\nu \qquad (2)$$

and molecular hydrogen by associative detachment

$$H + H^- \rightarrow H_2 + e. \qquad (3)$$

Molecular hydrogen was formed also by radiative association

$$H + H^+ \rightarrow H_2^+ + h\nu \qquad (4)$$

followed by charge transfer and chemical reaction

$$H_2^+ + H \rightarrow H_2 + H^+ \qquad (5)$$

Similar reactions led to the formation of HD. Because HD has a permanent dipole moment radiative association in a free-bound transition in the ground electronic state

$$H + D \rightarrow HD + h\nu \qquad (6)$$

caused some increase in the abundance of HD as did reactions initiated by

$$H + D^+ \rightarrow HD^+ + h\nu \qquad (7)$$

$$H^+ + D \rightarrow HD^+ + h\nu \ . \qquad (8)$$

However the major source of HD was the reaction

$$D^+ + H_2 \rightarrow HD + H^+ \qquad (9)$$

and it led to a substantial enhancement of the ratio of HD to H_2 over that of D to H.

The reaction

$$H^- + H^+ \rightarrow H_2^+ + e \qquad (10)$$

is a weak source of H_2^+ molecular ions, but the main channel in the collision of H^- and H^+ is mutual neutralization

$$H^- + H^+ \rightarrow H + H \qquad (11)$$

Other neutralization reactions

$$H^- + H_2^+ \rightarrow H + H + H \qquad (12a)$$

$$\rightarrow H_2 + H \qquad (12b)$$

may occur.

The molecular ions are destroyed by dissociative recombination

$$H_2^+ + e \rightarrow H + H \ . \qquad (13)$$

The rate of (10) is sensitive to the vibrational level of the molecular ion (Giusti-Suzor, Bardsley and Derkits 1983) and processes such as

$$H + H_2^+(v) \rightarrow H + H_2^+(v') \qquad (14)$$

which modify the vibrational population must be included in a complete account of this earliest phase of atomic physics.

The atoms and molecules are embedded in the background radiation field and photodetachment

$$H^- + h\nu \rightarrow H + e \qquad (15)$$

and photodissociation

$$H_2^+ + h\nu \rightarrow H + H^+ \quad (16)$$

limited the abundance of H_2 that survived.

Values of the rate coefficients of reactions (1) through (16) have been summarised by Dalgarno (1983). The hydrogen molecules may be destroyed also by collision-induced dissociation

$$H + H_2 \rightarrow H + H + H \ . \quad (17)$$

In laboratory studies, the accepted dissociation mechanism is a complex process of multiple excitations of the vibrational levels of the ground electronic state and dissociation is controlled by transitions from the high vibrational levels into the continuum. At low densities the excited levels may be depopulated by radiation and the rate coefficient is a function of density (Dalgarno and Roberge 1979). In the low density limit collision-induced dissociation occurs only by transitions from the v=0 level. Calculations have been carried out by Blais and Truhlar (1982, 1983) for the direct dissociation process and estimates of the density-dependent rate coefficients have been presented by Roberge and Dalgarno (1982) and Shull and Lepp (1983). However another mechanism may be more effective. Osherov and Ashakov (1977) have suggested that a transition may occur from the $^2A_+$ state of H_3 formed by the approach of H and $H_2(X^1\Sigma_g^+)$ to the $^2A_-$ state which then separates to $H + H_2(b^3\Sigma_u^+)$. The $b^3\Sigma_u^+$ state is repulsive and dissociation occurs. This dissociation mechanism may modify the behavior of a shocked cosmic gas and affect a wide range of astronomical phenomena (for examples see Struck-Marcell 1982, Vishniac, Ostriker and Bertschinger 1985).

The presence of He at a level of about 25% by mass introduces some further processes. The He^+ ions can be removed by radiative recombination and by radiative charge transfer

$$He^+ + H \rightarrow He + H^+ + h\nu \ . \quad (18)$$

The molecular ion HeH^+ can be produced by the radiative association processes

$$He^+ + H \rightarrow HeH^+ + h\nu \quad (19)$$

$$H^+ + He \rightarrow HeH^+ + h\nu \quad (20)$$

and by chemical reaction with H_2^+ ions in vibrational levels $v \geq 3$

$$H_2^+(v \geq 3) + He \rightarrow HeH^+ + H \ . \quad (21)$$

The ion is destroyed by the reverse process

$$HeH^+ + H \rightarrow He + H_2^+(v<3) \quad (22)$$

but the usually rapid process of dissociative recombination is probably very slow. Mutual neutralization

$$HeH^+ + H^- \rightarrow He + H + H \quad (23)$$

will remove HeH^+ ions.

The various processes involving helium had only a minor influence in the early Universe. They may lead to detectable HeH^+ in planetary nebulae and embedded X-ray sources (Roberge and Dalgarno 1982).

The atomic and molecular physics of the early Universe has been explored most recently by Lepp and Shull (1984) who included the formation of LiH by radiative association

$$Li + H \rightarrow LiH + h\nu \qquad (24)$$

in a vibrational free-bound transition. Fig. 1 reproduces their results as a function of the redshift Z.

As the radiation temperature diminishes, the population of He^{++}, He^{+} and H^{+} ions cannot be maintained by photoionization and they are converted into neutral atoms. Some H_2 is formed by the sequence of (4) and (5) at a red shift of 500 where H_2^+ no longer undergoes photodissociation and by (2) and (3) at Z=100 where H^- no longer undergoes photodetachment. The relict fractional molecular abundance is about 3×10^{-6} and the relict fractional ionization is 1×10^{-4}. The HD abundance was larger than that indicated by an order of magnitude (Lepp, Dalgarno and Shull 1985). The HD and LiH molecules are potentially important as low temperature coolants, present in the gas out of which the first stellar or galactic objects formed. Because of radiative stabilization, they also survive longer than H_2 against collision-induced dissociation.

The importance of the molecules in the collapse of pregalactic and intergalactic clouds has been emphasized repeatedly. Without molecules the temperature is determined by impact excitation of atomic hydrogen and helium and cannot be driven much below 10^4K so that only in very massive clouds can the gravitational energy exceed the thermal energy. With the presence of molecules, temperatures of 100K can occur.

As the collapse proceeds and the density exceeds $10^8 cm^{-3}$, three-body recombination processes of the kind

$$H + H + H \rightarrow H_2 + H \qquad (25)$$

$$H + H + H_2 \rightarrow H_2 + H_2 \qquad (26)$$

convert the cloud to a fully molecular gas (Palla, Salpeter and Stahler 1983). Later in the collapse, the cloud becomes optically thick to line radiation, the temperature rises and the molecules are dissociated. Ionization is assumed to occur initially through

$$H + H \rightarrow H + H^+ + e \qquad (27)$$

with a rate coefficient of $1\times10^{-14} T^{1/2} \exp(-158,000/T)$ $cm^3 s^{-1}$ (Palla et al. 1983). Other mechanisms may produce electrons earlier. The mechanism in which a metastable atom is first excited

$$H + H \rightarrow H + H(2s) \qquad (28)$$

$$H(2s) + H \rightarrow H^+ + e + H \qquad (29)$$

has an effective lower energy threshold as do associative ionization

$$H + H \rightarrow H_2^+ + e \qquad (30)$$

and polar ionization

$$H + H \rightarrow H^+ + H^- \qquad (31)$$

followed by

$$H + H^- \rightarrow H_2 + e \ . \qquad (32)$$

However their efficiencies are unknown. Because the reverse processes of dissociative recombination and mutual neutralization preferentially populate excited states of hydrogen, detailed balance arguments are not useful.

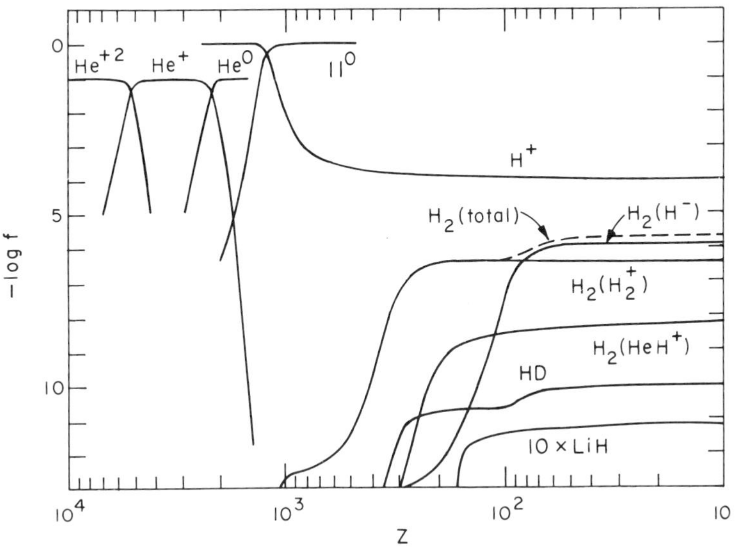

Fig. 1. Fractional abundances f as a function of the red shift Z (from Lepp and Shull 1984).

PHOTOIONIZED REGIONS

Photoionized regions include H II regions, diffuse nebulae, quasars, Seyfert galaxies and active galactic nuclei. Ionization is caused by some central source or sources of photons.

Emission nebulae have electron densities n_e between 10 cm^{-3} and 10^4 cm^{-3} and are strong emitters of HI recombination lines and N II and O II forbidden lines. Planetary nebulae are denser with n_e in the range from 10^3 cm^{-3} to as high as 10^7 cm^{-3}. The spectra of planetary nebulae consist of H I, He I and He II recombination lines together with emission lines of more highly ionized species such as O III, Ne III and Ne V.

H II regions are designated as diffuse, compact and ultra-compact. Diffuse H II regions are similar to emission nebulae but less dense. Compact H II regions are small with diameters of less than 0.15 pc. They are heavily reddened and the ionizing source if obscured. The electron densities exceed 5×10^3 cm^{-3}. Compact H II regions are identified by continuum emission in the infrared and radio.

The central stars in diffuse nebulae and H II regions are O or early B-type stars with effective temperature T exceeding 25,000 K. In planetary nebulae the central star is highly evolved and rapidly approaching the white dwarf stage. Its effective temperature may exceed 200,000 K. The ionizing source in a compact H II region is probably an OB star at an early stage of its evolution.

The kinetic temperatures of H II regions and nebulae occupy a relatively narrow range between 4,000 K and 15,000 K.

Seyfert galaxies are the most common type of active galactic nuclei and quasars and quasi-stellar objects are probably a rare extreme type, so luminous that the galaxies in which they reside are difficult to discern. Active galactic nuclei are characterized by wide emission lines from an extended range of ionization stages. The allowed lines of H I, He I and H II are very broad with widths corresponding to velocities between 5×10^3 km s^{-1} and 3×10^4 km s^{-1}. Forbidden lines of O III, Ne III, N II and S II are still broad but less so with widths of about 500 km s^{-1}. In some types of Seyfert galaxies and radio galaxies the allowed and forbidden lines have the same narrower widths.

Seyfert galaxies have a structureless continuum emanating from a small unresolved core. Conversely it appears that the intense luminous core of a QSO may be surrounded with a faint luminosity.

The narrow line regions are probably stratified with densities in the range 10^3 cm^{-3} to 10^7 cm^{-3} and temperatures of 10,000-20,000 K. In the broad line regions, the density is of the order of 10^9 cm^{-3}. The temperature is uncertain. The existence of Fe II has been advanced as evidence that T does not exceed about 40,000 K.

The large widths indicate material rapidly flowing out from a central core of such high luminosity that it is unlikely to be provided by nuclear burning as in normal galaxies. The energy source is almost certainly gravitational. A black hole of mass 10^8 M_O surrounded by an accretion disk of matter spiralling inwards and releasing energy in the form of photons with a hard spectrum is a plausible model.

The electrons produced by photoionization

$$H + h\nu \rightarrow H^+ + e \tag{33}$$

are energetic and heat the gas. Because photoionization cross sections decrease rapidly with increasing photon frequencies at photon energies well above the threshold, the cross section for atomic hydrogen at a given high frequency is orders of magnitude smaller than those for heavier elements. Ionization of heavy elements is enhanced by inner shell absorptions followed by Auger ionization as in

$$O(1s^2 2s^2 2p^4, {}^3P) + h\nu \rightarrow O^+(1s2s^2 2p^4, {}^4P) + e \tag{34}$$

$$O^+(1s2s^2 2p^4, {}^4P) \rightarrow O^{2+}(1s^2 2s^2 2p^2, {}^3P) + e \quad . \tag{35}$$

Thus in cosmic plasmas created by non-thermal or very hot ionizing sources, multiply-charged ions coexist with neutral hydrogen (and neutral helium).

The electrons are removed by radiative recombination

$$X^{(m+1)+} + e \rightarrow X^{m+} + h\nu \ , \tag{36}$$

by dielectronic recombination

$$X^{(m+1)+} + e \rightarrow (X^{m+})^* \rightarrow X^{m+} + h\nu \tag{37}$$

in which $(X^{m+})^*$ represents a resonance state of X^{m+} lying above the ionization threshold, and by charge transfer

$$X^{(m+1)+} + H \rightarrow X^{m+} + H^{+} \tag{38}$$

$$X^{(m+1)+} + He \rightarrow X^{m+} + He^{+} . \tag{39}$$

Radiative recombination populates a broad distribution of levels and transitions out of them give rise to lines ranging from the radio to the ultraviolet. They provide a powerful diagnostic probe of the environment in which they are found. The low densities permit detectable emission from very high n levels. Fig. 2 is the spectrum of a planetary nebulae showing the n = 77→76 transitions in neutral hydrogen and helium and in the n = 122-121 transition of ionized helium. Because the frequencies depend upon the mass through the Rydberg

$$R_m = \frac{109737.31}{(1 + m/M)} \text{ cm}^{-1} , \tag{40}$$

hydrogen and helium are readily distinguished. Recombination lines of carbon have also been identified in C II regions produced by early B-type stars and in the mostly neutral shells around H II regions and evidence of heavier elements has been obtained.

The lines are broadened by electron impact. The widths increase with principal quantum number n approximately as

$$\delta \sim 4.7(n/100)^{4.4}(10^4/T)^{0.1} \; n_e \nu \text{ Hz} \tag{41}$$

for frequency ν. For large n the broadening transfer power into the wings and the line merges into the background continuum and becomes undetectable. Nevertheless emission from level n = 390 of hydrogen has been identified. Still higher n lines may have been seen in carbon. A line at 26.131 MHz observed in absorption in the direction of the supernova remnant Cas A has been attributed to the 632-631 transition of carbon (Blake, Crutcher and Watson 1980). Several other lines have been seen towards Cas A (Ershov et al. 1984, Konovalenko 1984), one of which may be the 733-732 transition.

The level populations are affected by collisions of electrons and protons and by the absorption of radiation. The high n level populations are modified by absorption of the 2.7 °K blackbody background radiation.

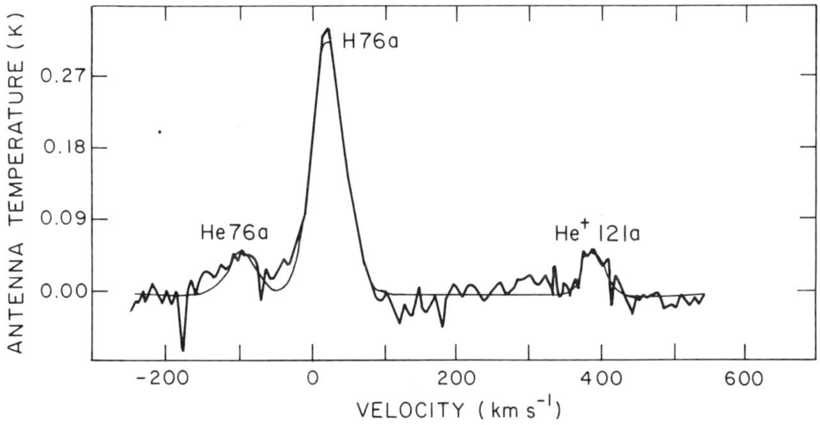

Fig. 2. Recombination lines of NGC 7027 (Terzian 1980).

However it appears that for most H II regions in our galaxy the lines observed at radio wavelengths are produced in regions where neither stimulated emission nor impact broadening is significant and the assumption of local thermodynamic equilibrium is an acceptable approximation. The approximation must become invalid as n decreases as indeed is indicated by data on millimetre lines from between 63 and 40 (Wilson and Pauls 1984, Hoang-Binh, Enermaz and Linke 1985).

Radio recombination lines have been observed from distant galaxies. For them, local thermodynamic equilibrium is not appropriate. It seems that substantial amplification is occurring as a result of stimulated emission by a background continuum source at the nucleus of the galaxy.

Shaver (1980) has discussed the possibility of detecting recombination lines from quasars. The observation would place interesting limits on the variation with time of the fundamental constants.

Recombination lines in the infrared and optical region have long been important diagnostic probes. Fig. 3 is a reproduction of the spectrum of the planetary n bula NGC 7027 (Smith, Larson and Fink 1981) showing recombination lines from levels up to n = 25. Because the theoretical description is so precise, departures from it may be attributed to reddening and the extinction due to dust can be derived (cf. Seaton 1979).

The line ratios in QSO's differ from standard recombination theory and from observations of nebulae (cf. Baldwin 1977). In addition to being optically thick in the Lyman continuum

$$H(1s) + h\nu \rightarrow H^+ + e \qquad (42)$$

it seems that QSO's are often optically thick in the Balmer continuum

$$H(2s) + h\nu \rightarrow H^+ + e \qquad (43)$$

(cf. Kwan 1984).

The recombination spectrum has discontinuities at the wavelengths

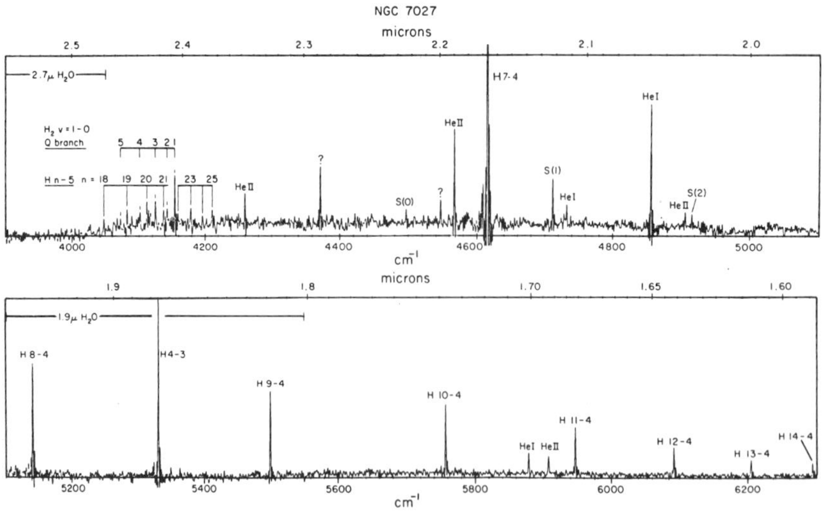

Fig. 3. The infrared spectrum of NGC 7027 (Smith, Larson and Fink 1981).

corresponding to the ionization thresholds of the excited states. The strongest jump is the Balmer jump at 364.6 nm. Between the discontinuities the intensity varies as $\exp(h\nu/kT)$ so that T can be determined from the spectral shape of the continuum. Fig. 4 is the spectrum of the extended shell around the nova CP Puppis (Williams 1982). The derived temperature is 800 K, much less than the temperatures of gaseous nebulae. It appears that after its initial ionization during the nova outburst, the shell has been cooling and recombining.

The emission lines in the spectrum in Fig. 4 are unusual in astrophysical plasmas. They are due to radiative recombination and they appear in the CP Puppis shell because of the low temperatures which enhances them compared to lines from levels populated by electron impact.

Dielectronic recombination was introduced into astrophysics by Burgess (1964) in connection with the ionization balance of the solar corona. At high temperatures, dielectronic recombination is stabilized by radiative transitions of the core electrons. For helium-like ions, the process may be represented by

$$X^{(m+1)}(1s^2) + e \rightarrow X^{m+}(1sn\ell 2s)$$
$$\rightarrow X^{m+}(1s2s) + h\nu. \qquad (44)$$

The lines lie close in wavelength of the resonance lines $1sn\ell \rightarrow 1s^2$ of $X^{(m+1)+}$. They are called satellite lines and they have considerable utility in the interpretation of high temperature plasmas.

At the temperatures characteristic of nebulae, the accessible resonance states are more readily stabilized by radiative transitions of the captured electron. Thus dielectronic recombination of O^{4+} which has a ground state $(1s^2 2s^2, {}^1S)$ can proceed through the resonance state $O^{3+}(1s^2 2p^2({}^1D)3p, {}^2F^o)$ which lies 486 cm^{-1} above the ionization threshold of O^{3+}. The ${}^2F^o$ state then radiates to the $(1s^2 2s 2p^2, {}^2D)$ state as the 3p electron is transferred to a 2s orbital. Nussbaumer and Storey (1983) have presented a table of dielectronic recombination coefficients of astrophysically important ions at nebular temperatures.

The C III line at 229.7 nm observed in planetary nebulae is excited by dielectronic recombination of C IV into the resonance 2p4d ${}^1F^o$ state which cascades to the upper $2p^2\ {}^1D$ state of the 229.7 nm transition.

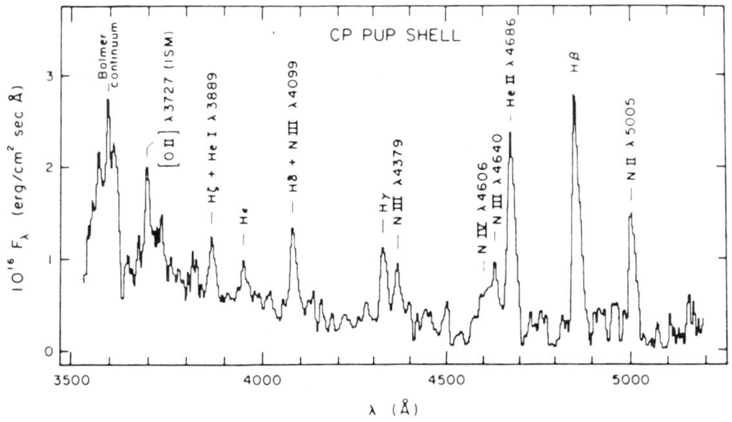

Fig. 4. The spectrum of the CP Puppis shell (Williams 1982).

Charge transfer processes (38) and (39) play a major role in determining the ionization structure of cosmic plasmas created by high frequency photons. The product ions are usually excited and the resulting emission is a useful indicator of a high excitation source. Most of the lines appear in the ultraviolet (Shields, Dalgarno and Sternberg 1983) but charge transfer of O^{3+} with H leads to several visible lines. Charge transfer into the $(2p3p,{}^1P)$ state leads to the emission of a line at 559.2 nm. Fig. 5 reproduces the spectrum of NGC 3918 obtained by Clegg and Walsh (1985) in which the 559.2 nm line is clearly present. Charge transfer of O^{3+} with H also populates the $(2p3p)^3D$ level with high efficiency producing lines at 381.1, 377.4, 376.0, 375.7 and 376.0 nm. The $(2p3p)^3D$ levels are also populated by the Bowen fluorescence mechanism. There is a close coincidence between the 1s-2p resonance line of He^+ at 30.3783 nm and the $(2p^2)^3P_2$ - $(2p3d)^3P_2{}^o$ resonance line of O^{2+} at 30.3799 nm. Thus absorption of the He^+ resonance line followed by cascading also populates the $(2p3p)^3D$ levels. The contributions of the charge transfer mechanism and the Bowen mechanism can be distinguished by using the 559.2 nm line which is due to charge transfer and the several lines which are due only to the pumping mechanism. For planetary nebulae (Dalgarno and Sternberg 1982, Clegg and Walsh 1985, Likkel and Aller 1985), charge transfer is the major contributor to the lines at 377.4 and 375.7 nm.

Lines at 377.4, 379.1 and 375.4-376.0 nm have been identified in the spectra of the optical counterparts of three X-ray bursters (Canizares, McClintock and Grindlay 1979) and have been used to support the view that the Bowen mechanism is operating. The alternative interpretation that charge transfer is responsible is an interesting possibility (Sternberg and Dalgarno 1985).

Radiative, dielectronic and charge transfer recombination all contribute to emission from photoionized plasmas but the major source is electron impact excitation. Analysis of the line intensities yields the density and temperature of the emitting material and the element abundances. The distribution of ionization stages provides evidence of the nature of the ioni-

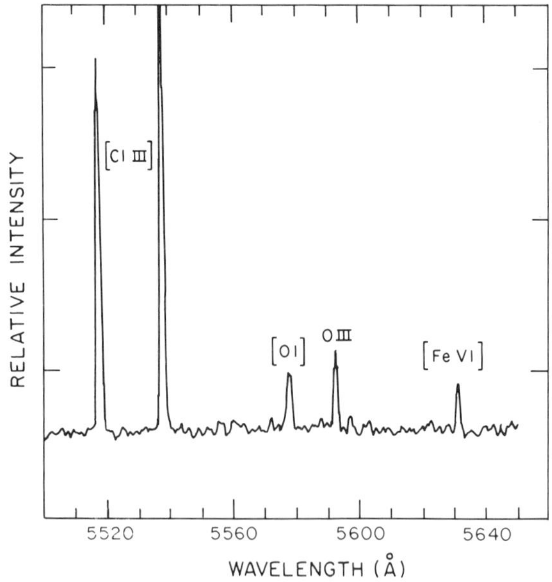

Fig. 5. The spectrum of NGC 3918 (from Clegg and Walsh 1985).

zation source. The highly-ionized Fe X present in active galactic nuclei indicates a high frequency source.

Electron impact excitation also cools the plasma. Because of the large number of excited states accessible at low energies, it is difficult in a photoionized nebula to drive the temperature beyond 20,000 K, except close to an intense source of X-rays which strips the atoms of their electrons and slows the cooling. However in a high temperature plasma, collisions become the major mode of ionization.

COLLISIONALLY-IONIZED PLASMAS

Collisionally-ionized regions include solar and stellar coronae, the hot component of the interstellar gas, supernova remnants, interstellar bubbles and probably the galactic halo. Ionization is created mainly by electron impacts

$$e + X^{m+} \rightarrow e + X^{(m+1)+} + e \qquad (45)$$

and lost by radiative and dielectronic recombination. So long as the density is low, both the ionization rate and the recombination rate are functions only of temperature. The steady-state equilibrium is called coronal equilibrium. In coronal equilibrium, the ionization distribution of any chosen element is a direct measure of the temperature. Fig. 6 illustrates the oxygen ion distribution calculated by Shapiro and Moore (1976).

For those ionization stages for which charge transfer to hydrogen or helium is rapid and for which the product ion is in its ground state the reverse process of charge transfer ionization

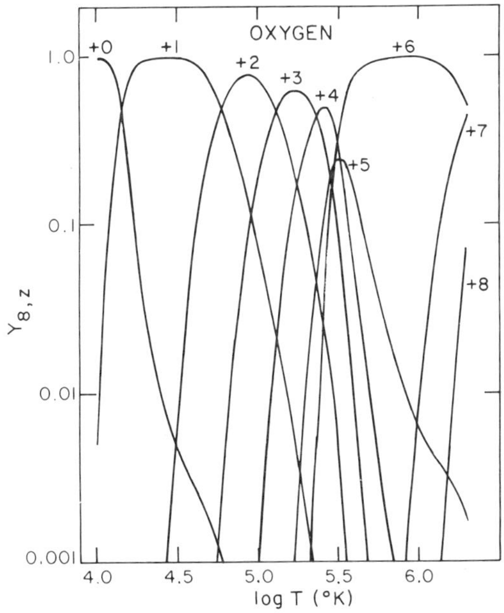

Fig. 6. The oxygen ion distribution in coronal equilibrium as a function of temperature T (from Shapiro and Moore 1976).

$$H^+ + X^{m+} \rightarrow H + X^{(m+1)+} \tag{46}$$

is a source of ionization that is more efficient than electron impact. Examples include

$$H^+ + Si^+ \rightarrow H + Si^{2+} \tag{47}$$

$$He^+ + Si^{2+} \rightarrow He + Si^{3+} \tag{48}$$

$$H^+ + Fe^+ \rightarrow H + Fe^{2+} \quad . \tag{49}$$

From diagrams such as Fig. 6, the temperature of the solar corona has been derived to be of the order of $1\text{-}2\times10^6$K. The density can be inferred from the intensities of lines emitted from the 2^1P, 2^3P and 2^3S states of helium-like ions. Because electron impacts couple the long-lived 2^3S state to the 2^3P state, the ratio of the $2^3S\text{-}1^1S$ line intensity to the $2^3P\text{-}1^1S$ line intensities is a useful measure of density. Fig. 7 illustrates the Ne IX lines seen in an active region (McKenzie and Landecker 1982). The electron density is of the order 10^{10} cm^{-3}.

Temperatures can also be obtained from the relative intensities of emission lines. For example, the comparison of model spectra (Raymond and Smith 1977, Mewe and Gronenchild 1981) with measurements of X-ray lines from Fe XVII and Fe XXI from the visual binary RS CVn Sigma Corona Borealis yield a temperature of 6.92×10^6K (Agrawal, Markert and Riegler 1985).

The existence of a hot component of the interstellar gas was established by the observation of the 103.5 nm line of OVI in absorption towards hot stars. Fig. 6 indicates a temperature near 3×10^5K. If OVI is widely distributed, its density is very small, of the order of $3\times10^{-8}cm^{-3}$ in the plane. Coronal equilibrium then implies a total density of $3\times10^{-3}cm^{-3}$. Alternatively the OVI is produced in conduction fronts in the interfaces between hot interstellar gas and cool clouds.

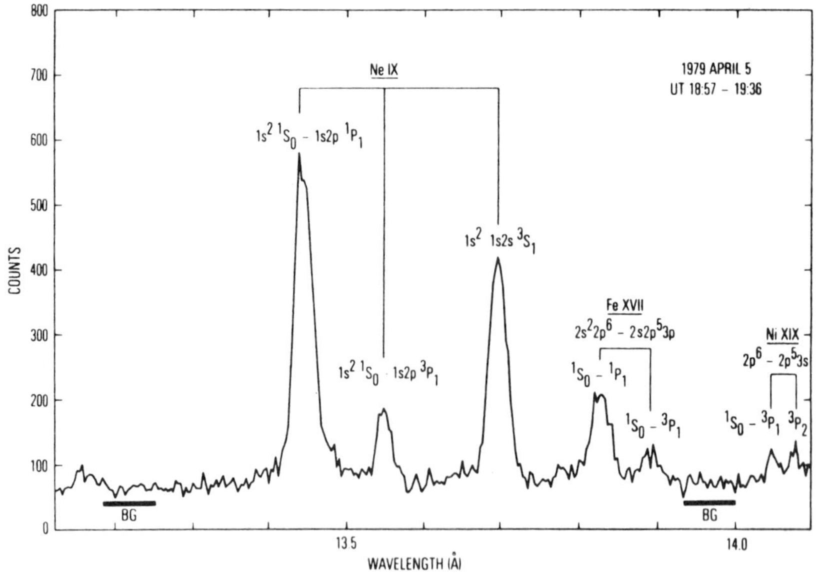

Fig. 7. The X-ray spectrum of an active region showing the Ne IX lines (from McKenzie and Landecker 1982).

The source of energy for the hot component of the interstellar gas is commonly assumed to be supernovae. In a supernova explosion, a star ejects several solar masses of material into the interstellar medium with a velocity of about 10^4km s^{-1}. The ejected material sweeps up the ambient gas and collides with it. A shock moves outwards from the impact surface into the interstellar gas and a reverse shock moves into the ejecta. The blast wave drives the gas into a thin shell inside of which there is a hot cavity of low density gas. Because of the low density of less than 1 cm^{-3}, the cavity cools slowly and it is likely to be affected by other supernovae before it has reached equilibrium. Thus the hot component of the interstellar gas may consist of a series of overlapping supernova cavities. The conduction fronts of clouds overrun by the blast wave may be the site of the OVI ions.

The hot ionized cavity is produced by collisions in the shocked gas. In shocks the directed energy of motion is converted into random thermal energy. If the velocity if large enough the temperature obtained is high, for remnants of the order of 10^6 K. Ionization is caused by the hot thermal electrons as in the solar corona. As the gas cools, it recombines and emits ultraviolet and X-ray photons which ionize the gas ahead of and behind the shock. Recombination occurs by radiative and dielectronic recombination. Because the fraction of neutral hydrogen is small, charge transfer recombination is ignored. However it may play a role. At temperatures of 10^6 K, the charge transfer rate coefficient for highly-stripped ions approaches $10^{-7}cm^3s^{-1}$ so that charge transfer recombination is important for neutral fractions as small as 10^{-6}. In equilibrium such neutral fractions are obtained at $5x10^5$K.

The temperatures of supernova remnants may be obtained by procedures similar to those developed for the study of the solar corona. Analysis of the intensities of the helium-like lines in the Puppis A SNR, shown in Fig. 8, yields temperatures of $2.2x10^6$K. Comparison with solar coronal data indicates that ionization equilibrium has not been obtained in the remnant. The observations suggest the presence of several solar masses of neon and oxygen and Canizares and Winkler (1981) have argued that Puppis A is the result of massive star explosion.

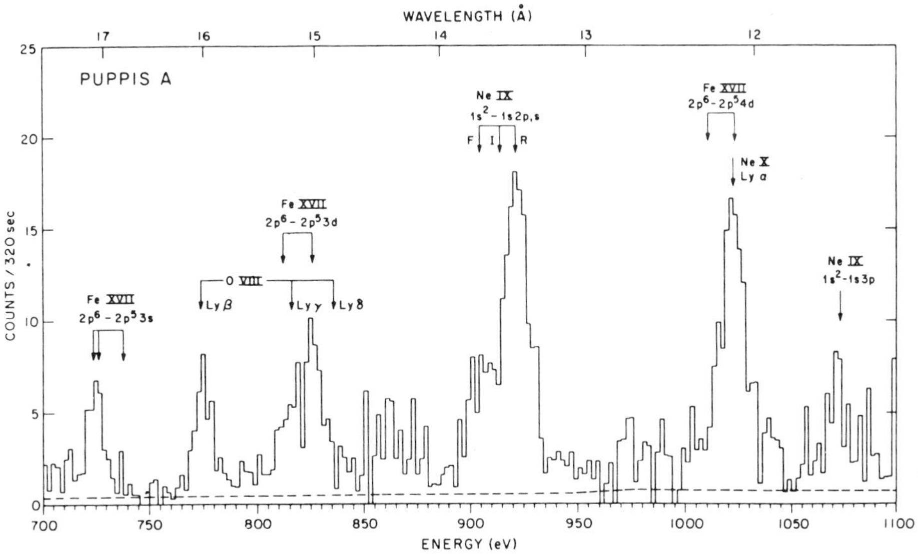

Fig. 8. The spectrum of Puppis A (from Winkler et al. 1981).

Several young supernova remnants contain fast moving knots with velocities of order 1,000-2,000km s^{-1}, which are characterized by strong forbidden lines of oxygen and some heavier elements and by a total absence of hydrogen and helium recombination lines (Dopita and Tuohy 1984, Blair, Kirschner and Winkler 1983). They are formed presumably of material from the core of the precursor star. Attempts have been made to interpret the observational data using the model of a reverse radiative shock propagating in a gas of heavy elements (Dopita, Binette and Tuohy 1984, Hamilton and Sarazin 1984). The models have met with limited success. It may be possible to resolve some of the discrepancies by including charge transfer processes between heavy elements such as

$$O^{++} + O \rightarrow O^{+} + O^{+} \tag{50}$$

$$S^{++} + O \rightarrow S^{+} + O^{+} \tag{51}$$

which may have large rate coefficients (cf. Johnson and Strobel 1982).

Interstellar bubbles are cavities in the interstellar gas produced not by supernova explosions but by strong stellar winds from early-type stars. The dynamics and conditions are similar to those of a supernova explosion except that the winds are steady and maintained throughout most of the lifetimes of the stars.

The galactic corona or halo consists of gas extending above and below the galactic plane with a scale height measured in kpc, rather than the 0.1 kpc that characterizes the distribution in the disk. The corona is detected by absorption towards hot stars in the Large and Small Magellanic Clouds (cf. Fitzpatrick and Savage 1983). An extensive range of ionization stages are found suggesting a wide range of temperatures if the ionization is the consequence of collisions in a hot gas. Temperatures as high as 3×10^5K are required to produce the observed N^{4+} ions. The corresponding total density at a height of 2 kpc is 5×10^{-5} cm^{-3}, so low that the steady-state analysis used to derive the value is not strictly valid. The coronal gas may be the hot component of the interstellar gas. At 10^6 K, the thermal scale height is 10 kpc. The hot gas flows away from the galactic plane, cools and forms clouds which fall back on to the plane in a fountain effect (cf. Shapiro and Field 1976). Other interpretations are possible in which a warm gas is ionized by galactic and extragalactic ultraviolet and X-ray photons.

In the galactic disk, there appears to be in addition to the hot component a warm phase at about 10^4 K with a density of about 0.15 cm^{-3} which may be the product of evaporating cold clouds embedded in the hot component. The ionization in it is mostly due to photoionization by the interstellar radiation field. There exist cold diffuse clouds with densities of 10-200 cm^{-3} and temperatures of 80-20 K, and giant complexes of molecular clouds with densities 10^3-10^6 cm^{-3} and temperatures 20-50 K. The diffuse clouds have fractional ionizations of about 10^{-4} produced by photoionization of those elements with ionization potentials less than that of atomic hydrogen. The dense clouds have a fractional ionization of 10^{-6} or less which is caused by cosmic ray impacts. The diffuse clouds and the outer edges of the molecular clouds cool by impact excitation of the fine-structure levels of C^+ and O. Recently the C^+ emission at 158 μ has been observed in several gas-rich galaxies (Crawford et al. 1985). In the inner regions cooling occurs by excitation of molecular energy levels.

A significant part of the heating of a dense molecular cloud is caused by cosmic rays which eject energetic electrons in primary ionization events. The energetic electrons slow down by causing ionization and excitation but eventually by elastic collisions with the thermal ambient electrons. The elastic collisions heat the cloud.

The material surrounding an X-ray source is similarly heated by the photoelectrons, creating neutral atoms and ions beyond the hot ionized zone that immediately surrounds it, a warm mostly neutral gas in which emission from neutral atoms and molecules controls the cooling (Lepp and McCray 1983). Emission from the low-lying metastable states of neutral atoms may be an indicator of an embedded X-ray source.

REFERENCES

Agrawal, P. C., Markert, T. H., and Riegler, G. R., 1985, High spectral resolution observations of the coronal X-ray emission from the RS CVn binary Sigma Corona Borealis, MNRAS, 213:761.

Baldwin, J. A., 1977, The Lyα/Hβ intensity ratio in the spectra of QSOs, MNRAS, 178:67P.

Blair, W. P., Kirschner, R. P., and Winkler, P. F., 1983, The extraordinary extragalactic supernova remnant in x-ray and optical investigations, Ap. J., 272:84.

Blais, N. C., and Truhlar, D. G., 1982, High-energy collision-induced dissociation of H_2 by H, Ap. J. Lett. 258:L79.

Blais, N. C., and Truhlar, D. C., 1983, Third body efficiencies for collision-induced dissociation of diatomics. Rate coefficients for $H + H_2 \rightarrow 3H$, J. Chem. Phys., 78:2388.

Blake, D. H., Crutcher, R. M., and Watson, W. D., 1980, Identification of the anomalous 26.131 MHz nitrogen line observed towards Cas A, Nature, 287:707.

Burgess, A., 1964, Dielectronic recombination and the temperature of the solar corona, Ap. J., 139:776.

Canizares, C. R., McClintock, J. E., and Grindlay, J. E., 1979, A spectroscopic study of the optical counterparts of three x-ray bursters, Ap. J., 234:556.

Canizares, C. R., and Winkler, P. F., 1981 Evidence for elemental enrichment of Puppis A by a type II supernova, Ap. J. Lett., 246:L33.

Clegg, R. E. S., and Walsh, J. R., 1985, Charge exchange of O^{3+} with H in planetary nebulae, MNRAS, in press.

Crawford, M. K., Genzel, R., Townes, C. H., and Watson, D. M., 1985, Far-infrared spectroscopy of galaxies: The 158 micron C^+ line and the energy balance of molecular clouds, Ap. J., 291:755.

Dalgarno, A., and Roberge, W. G., 1979, Collision-induced dissociation of interstellar molecules, Ap. J. Lett., 233:L25.

Dalgarno, A., 1983, "Physics of Ion-Ion and Electron-Ion Collisions," Plenum, New York.

Dopita, M. A., Binette, L., and Tuohy, I. R., 1984, Radiative shock wave theory. III. The nature of the optical emission in young supernova remnants, Ap. J., 282:142.

Dopita, M. A., and Tuohy, I. R., 1984, Spectrophotometry of young supernova remnants, Ap. J., 282:135.

Ershov, A. A., Ilyasov, Yu. P., Leklit., E. E., Smirnov, G. T., Solodkov, V. T., and Sorochenko, R. L., 1984, Low-frequency (42,57.84 MHz) excited-carbon lines towards Cassiopeia A, Sov. Astron. Lett. 10:348.

Fitzpatrick, G. L., and Savage, B. D., 1983, Ultraviolet interstellar absorption toward HD 5980 in the small magellanic cloud, Ap. J., 267:93.

Hamilton, A. J. S., and Sarazin, C. L., 1984, Heating and cooling in reverse shocks into pure heavy-element supernova ejecta, Ap.J., 287:282.

Hoang-Binh, D., Encrenaz, P., and Linke, P. A., 1985, Observations of radio recombination lines in the millimetre-wave spectrum of Orion A, Astron. Ap., 146:L19.

Johnson, R. E., and Strobel, D. F., 1982, Charge exchange in the Io Torus and exosphere, J. Geophys. Rev. 87:10, 385.

Konovalenko, A. A., 1984, Decameter-wavelength carbon recombination lines towards Cassiopeia A, Sov. Astron. Lett. 10:353.
Kwan, J., 1984, Photoionization models and diagnosis of physical properties of the broad-line emission gas in quasars and Seyfert nuclei, Ap. J., 283:70.
Lepp, S., Dalgarno, A., and Shull, M. J., 1985, Star formation in the early universe, in preparation.
Lepp, S., and McCray, R., 1983, X-ray sources in molecular clouds, Ap. J., 269:560.
Lepp, S., and Shull, M. J., 1984, Molecules in the early universe, Ap. J. 280, 465.
Likkel, L., and Aller, L. H., 1985, Observations of the Bowen fluorescent mechanism in planetary nebulae, in press.
McKenzie, D. L., and Landecker, P. B., 1982, X-ray lines of helium-like oxygen and neon in the solar corona, Ap. J., 259:372.
Nussbaumer, H., and Storey, P. J., 1983, Dielectronic recombination at low temperatures, Astron. Ap. 126:75.
Osherov, V. I., and Ushakov, V. G., 1977, The two-path dissociation $H + H_2 \rightarrow H + H + H$, Sov. Phys. Dokl., 22:499.
Palla, F., Salpeter, F. E., and Stahler, S. W., 1983, Primordial star formation: The role of molecular hydrogen, Ap. J., 271:632.
Roberge, W. G., and Dalgarno, A., 1982, Collision-induced dissociation of H_2 and CO molecules, Ap. J., 255:176.
Roberge, W. G., and Dalgarno, A., 1982, The formation and destruction of HeH^+ in astrophysical plasmas, Ap. J., 255:489.
Seaton, M. J., 1979, Extinction of NGC 7027, MNRAS, 187:785.
Shapiro, P. R., and Field, G. B., 1976, Consequences of a new hot component of the interstellar medium, Ap. J., 205:762.
Shapiro, P. R., and Moore, R. T., 1976, Time-dependent radiative cooling of a hot, diffuse cosmic gas, and the emergent x-ray spectrum, Ap. J., 207:460.
Shields, G. A., Dalgarno, A., and Sternberg, A., 1983, Line emission from charge transfer with atomic hydrogen at thermal energies, Phys. Rev., A28:2137.
Shull, M. J., and Lepp, S., 1983, The kinetic theory of H_2 dissociation, Ap. J., 270:578.
Smith, H. A., Larson, H. P., and Fink, U., 1981, Molecular hydrogen and the 2 micron spectrum of NGC 7027, Ap. J., 244:835.
Sternberg, A., and Dalgarno, A., 1985, Charge transfer and the Bowen mechanism, in preparation.
Struck-Marcell, C., 1982, Gas cloud collisions in protogalaxies, I. Numerical simulations, Ap. J., 259:116.
Terzian, Y., in: "Radio recombination lines", P. A. Shaver, ed., Reidel, Holland (1980).
Vishniac, E. T., Ostriker, J. P., and Bertschinger, E., 1985, Explosions in the early universe, Ap. J., 291:399.
Williams, R. E., 1982, Spectroscopic analysis of the extended shells around the novae CP Puppis and T. Pyxidis, Ap. J., 261:170.
Wilson, T. L., and Pauls, T., 1984, Radio continuum and recombination line observations of Orion A, 138:225.
Winkler, P. F., Canizares, C. R., Clark, G. W., Markert, T. H., Kalaba, K., and Schnopper, H. W., 1981, A survey of x-ray line emission from the supernova remnant Puppis A, Ap. J. Lett., 246:L27.

STUDY OF ISOTOPES FAR FROM THE STABILITY LINE

Sylvain Liberman

Laboratoire Aimé Cotton - C.N.R.S. II
Bâtiment 505 - Campus d'Orsay
91405 Orsay (France)

INTRODUCTION

Some of the nuclear properties of atoms may be revealed through their hyperfine structure. It is the case for the nuclear spins which appear in hyperfine structure, in particular in Landé factors g_1 of hyperfine sub-levels. It is also the case for nuclear moments such as the nuclear magnetic dipole moments μ_N which are responsible for the appearance of hyperfine magnetic constants $A_{\alpha J}$ of atomic levels (αJ), as well as the nuclear electric quadrupole moments Q which are involved in hyperfine electric constants $B_{\alpha J}$ of atomic levels (1). In the latter case, in some way, Q is a measure of the extent to which the nuclear charge distribution deviates from the spherical symmetry. Moreover its value may be expressed in terms of a deformation parameter β, through the formula :

$$Q = 3(5\pi)^{-\frac{1}{2}} Z R_o^2 \beta$$

where R_o is the radius of the spherical nucleus and Z the nuclear charge. Systematic studies of hyperfine structures of long series of isotopes of the same element should therefore provide precise typical nuclear quantities, the variation of which, all along the isotope series, should lead to the emergence of new ideas for establishing more refined nuclei models. Particular attention should be paid to isotope shift measurements which could give, in proper conditions, the precise evolution of the nuclear shape as it is reflected by the nuclear radius of the charge distribution, the mean square value of which is given by :

$$< r^2 > = \left[1 + (5/4\pi)\beta^2\right] < r^2 >_{sph.}$$

with $r = R_o\{1 + \beta(5/16\pi)^{\frac{1}{2}}(3\cos^2\theta - 1) - \beta^2/4\pi\}$

As a matter of fact the total isotope shift quantity represents a combination of three terms : i) the normal mass shift (or Bohr term) which is simply the energy correction arising from the fact that the nucleus has a finite mass, it is therefore easy to evaluate exactly ; ii) the specific mass shift resulting from the action of a dielectronic operator in the hamiltonian, which is not easy to calculate ; iii) the volume shift or field shift which contains the relevant information (2). Fortunately, there are few arguments in favor of an almost negligible specific mass shift in alkali atoms, at least for the heaviest ones (3), so the subject will be limited to these atoms. In any case, hyperfine structure measurements as well as isotope shift measurements require high resolution techniques in the optical range which are nowadays usually provided by means of tunable lasers,, and more specifically by dye lasers.

It must be added that hyperfine interactions resulting from a coupling of nuclear and electronic quantities need time to occur; for instance the nuclear magnetic moment μ_N couples with the magnetic field created on the nucleus by the orbiting electron. The study is therefore restricted to those states which are long lived enough, in particular to nuclear ground states or metastable states (isomers). Of course, other limitations are experimental ones.

2. EXPERIMENTAL METHOD

2.1. Principle

Fig. 1(a) gives a scheme of the experimental set-up: the relevant isotopes of the species under study, which are produced through adequate nuclear reactions as will be seen in the following section, are evaporated from the target and form an atomic beam. They are illuminated by the light of a tunable laser beam propagating perpendicularly to the direction of the atomic beam. Then they enter into a six-pole magnet in which they experience the inhomogeneous magnetic field which rules between the pole pieces. After this magnetic selection, the atoms are counted by a detector placed behind a mass spectrometer. The way this detection system works may be understood through the part b of fig. 1 which corresponds to the case of the D_1 line of the stable isotope of sodium, taken as an example. In the absence of resonant optical interaction all the atoms are statistically distributed between all the hyperfine Zeeman sublevels of the ground state $^2S_{1/2}$ which is split into two hyperfine sublevels of total angular moments F = 2 and F = 1. When subjected to a magnetic field the F = 2 hyperfine sublevel separates into 5 Zeeman sublevels, whereas the F = 1 hyperfine sublevel separates into 3 Zeeman sublevels.

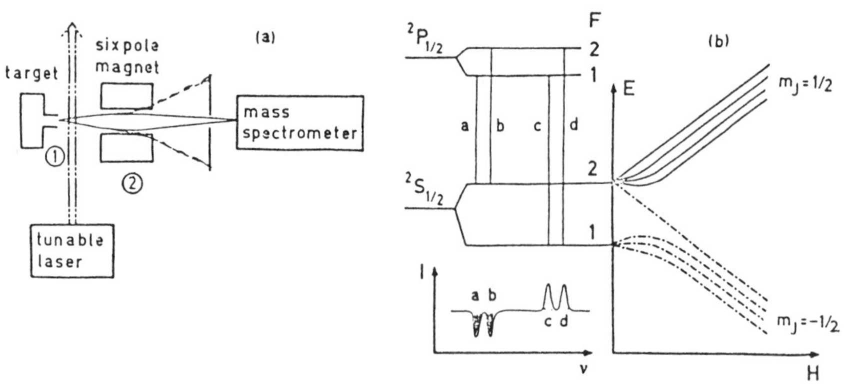

Figure 1

In a region of high magnetic field the 4 Zeeman sublevels of F = 2 corresponding to $m_j = +\frac{1}{2}$ behave identically with their energy increasing linearly with the magnetic field. In the same conditions, the 3 Zeeman sublevels of F = 1 plus the one of F = 2 corresponding to $m_j = -\frac{1}{2}$ have their energy which decreases with the magnetic field increasing. So in normal conditions half of the atoms (those with $m_j = -\frac{1}{2}$) are defocused toward the regions of high magnetic field, and the other half (those with m_j =

+ 1/2) are focused onto the symmetry axis of the system where the magnetic field is zero. In the presence of an optical interaction with the laser light properly tuned to resonance with either one of the 4 possible hyperfine transitions $^2S_{1/2} \rightarrow {}^2P_{1/2}$, an optical pumping takes place which depletes one of the two levels of $^2S_{1/2}$ to the benefit of the other. In effect taking the transition labelled a on the figure as an example, the absorption of one photon excites the level $^2P_{1/2}$ (F = 1) which deexcites spontaneously toward $^2S_{1/2}$ (F = 2) or $^2S_{1/2}$ (F = 1). In the first case the atom may be excited again by the laser light and the process occurs until the atom deexcites toward $^2S_{1/2}$ (F = 1) which happens after a few cycles. If the optical interaction depletes $^2S_{1/2}$ (F = 2) it increases $^2S_{1/2}$ (F = 1) which corresponds to an increasing of the defocused atoms ; whereas it is the contrary in the other case. Optical resonances therefore appear as negative or positive signals on a constant mean value depending on the defocusing or focusing character of the interaction. This magnetic detection procedure has revealed extremely sensitive (2) (4). It is limited in use to the only atoms which are paramagnetic (non zero J value) in their ground state.

If one is interested in measuring the nuclear quadrupole moment of alkali isotopes, then one has to study the structure of the D_2 line which connects the ground state to the $^2P_{3/2}$ excited state, owing to the fact that the electric quadrupole interaction only affects those levels with J value larger than ½. The principle is exactly the same as for the D_1 line except that the structure is a little more complicated, with hyperfine sublevels which do not all exhibit an optical pumping for ordinary selection rule reasons. Increasing of the possibilities may however arise by playing with the polarisation of the exciting light, as can be seen on fig. 2. In addition to this, in order to increase the accuracy of hyperfine internal measurements, it is possible to combine the optical pumping with double resonance RF techniques as it is described in the fig. 3.

Still using RF techniques together with optical pumping with polarized light for atoms in weak magnetic fields, it is also possible to determine unambiguously the nuclear spin of isotopes and isomers (5).

2.2. Production of isotopes and isomers

All the studied species are unstable ones and produced by nuclear reactions. The main reactions which have been used were initiated by the impinging of accelerated protons onto specific targets. The involved nuclear reactions are spallation reactions, fission or fragmentation reactions.

Spallation reactions are known to be a means of forming neutron deficient nuclei. They have been used to form the lightest isotopes of alkali atoms. Namely, for sodium isotopes aluminium target were used according to the reaction :

$$^{27}Al(p,\ 3p\ x\cdot n)^{25-x}\ Na$$

which means that an aluminium nucleus hit by a high energy proton, it induces the evaporation of 3 protons plus x neutrons (x = 0 to 5) and forms a sodium nucleus of atomic mass 25 - x.

The same type of reactions may occur for the other alkali atoms, using different atomic targets. The lighter isotopes of rubidium for instance have been obtained through the spallation of a niobium target according to :

$$^{93}Nb(p,\ 5p\ x\cdot n)^{89-x}\ Rb.$$

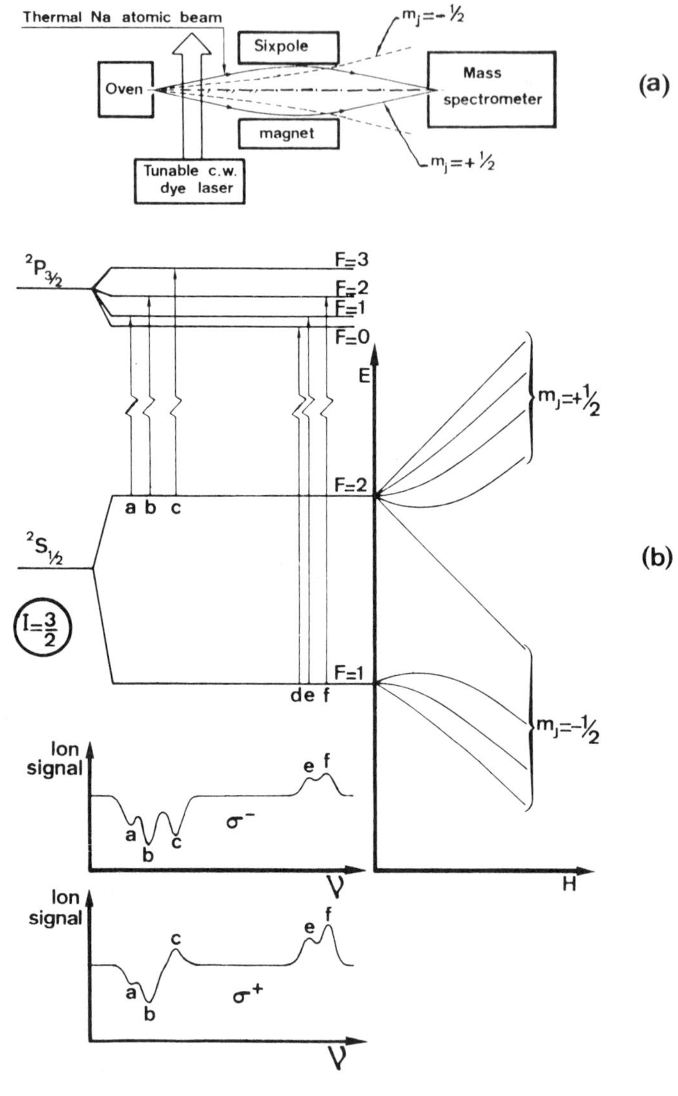

Figure 2. Na D_2 line, experimental detection system.

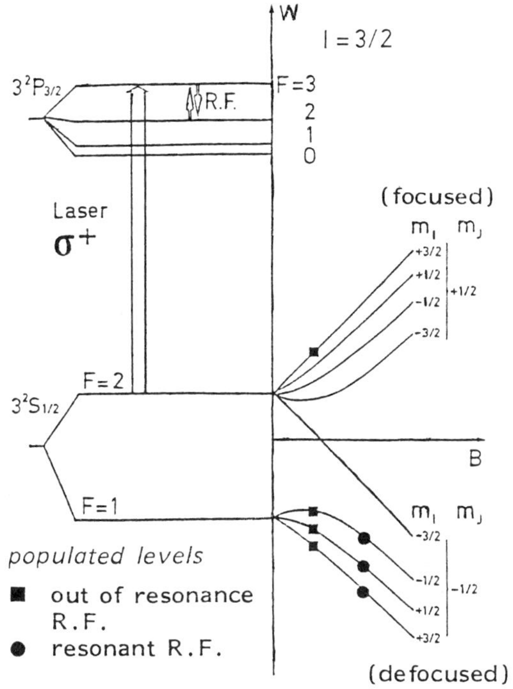

Figure 3. Na D_2 line, double resonance R.F. technique.

For cesium isotopes a lanthanum target gives the reaction :

$^{139}La(p,\ 3p\ x{\cdot}n)^{137-x}\ Cs$.

For what concerns the heavier isotopes of alkali atoms, fission reactions or fragmentation reactions have been used with uranium targets : it is generally established that these reactions give relatively larger cross-sections of formation of neutron rich nuclei (6).

As a conclusion of this section, it is to be emphasized that most of the isotopes of interest are rather short-lived ones, and consequently their study has to be done on line behind the accelerators used to produce them. In addition, all the experimental results which are reported here have been obtained at CERN in Geneva, mainly on the ISOLDE facility.

3. EXPERIMENTAL RESULTS

3.1. Sodium isotopes (7)

The experimental results obtained on both the hyperfine structure and the relative isotope shift of sodium isotopes from mass number 20 up to 31, are displayed on fig. 4. If one gets rid of the expected mass dependence of the isotope shift by multiplying it by $(\delta M/M^2)^{-1}$, one should obtain a constant value with respect to the number of mass A. The graph of fig. 5 shows that this is not the case, which is the experimental evidence

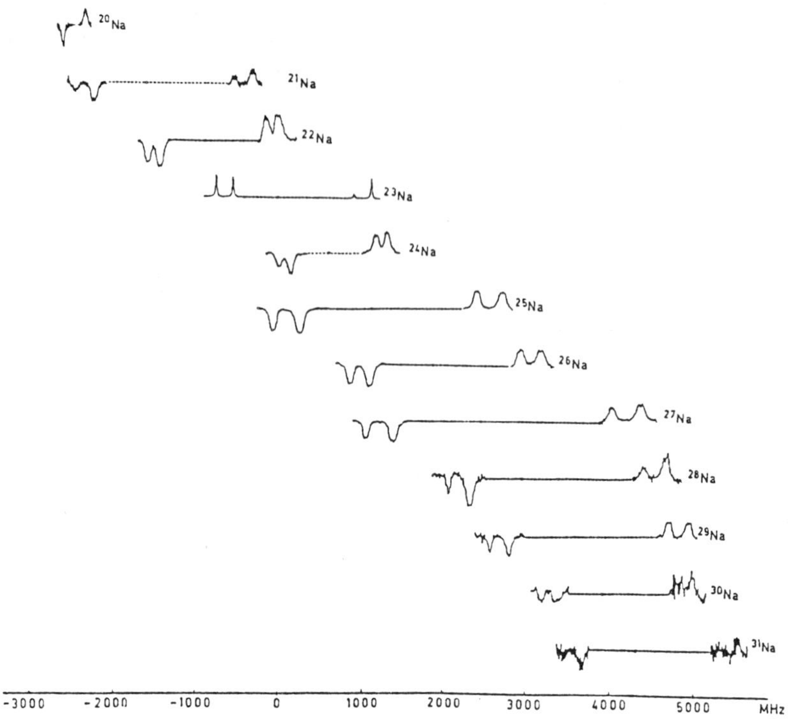

Figure 4. Experimental relative positions for Na isotope resonances

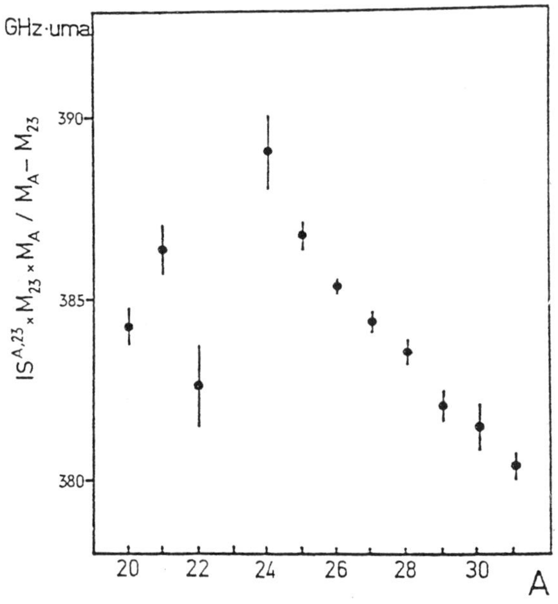

Figure 5. Observation of a volume effect

for a volume effect in the isotope shift of such a light element as the sodium atom.

Figure 3 gives the principle of a double resonance experiment performed in the excited state $^2P_{3/2}$ of sodium which has permitted to measure with good accuracy the frequency interval F = 3 - F = 2 of 6 isotopes as it is reported in fig. 6. Nuclear quadrupole moments can be deduced from these measurements, knowing precisely the magnetic hyperfine contribution from the study on the D_1 line.

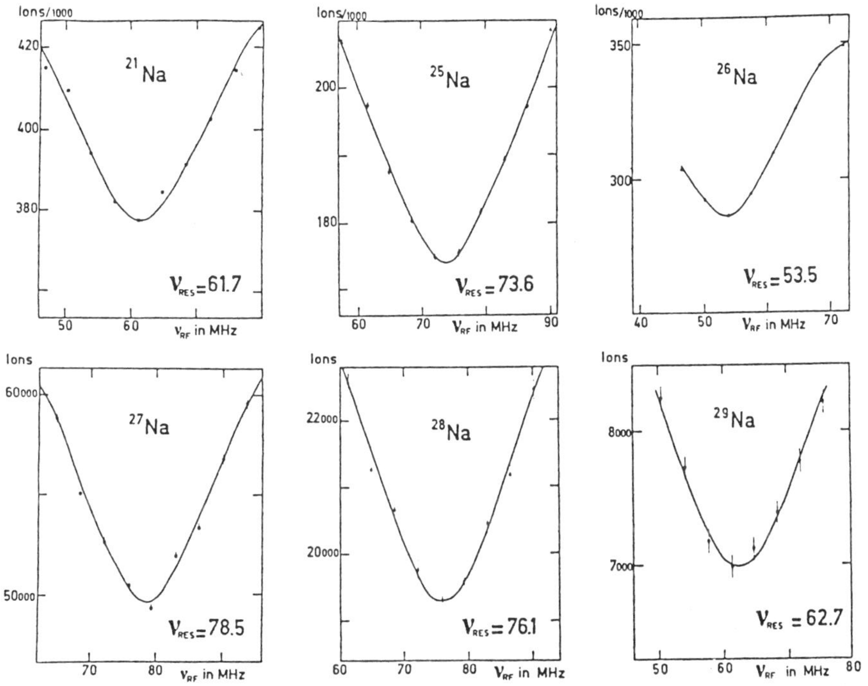

Figure 6

3.2. Potassium isotopes (8)

A schematic of the experimental results obtained for 10 isotopes of potassium is given in fig. 7. Notice that isotopes of mass number 39, 40 and 41, which are natural ones, have been studied by means of the usual fluorescence detection of the laser induced optical resonance of atoms in an atomic beam.

The results expressed as the variation of the mean square radius of the electric charges, are displayed on the fig. 8. The only characteristic feature appearing there concerns the slightly different behaviour between odd and even isotopes which is well known as the so-called odd-even staggering, and for which there is still no simple nuclear explanation.

3.3. Rubidium isotopes (1)

The experimental results obtained for rubidium atoms are shown in fig. 9, which summarizes the hyperfine structures and isotope shifts for 23 isotopes of mass numbers from 76 to 98, and 6 isomers noted with the letter m which means that it corresponds to an excited state of the nucleus

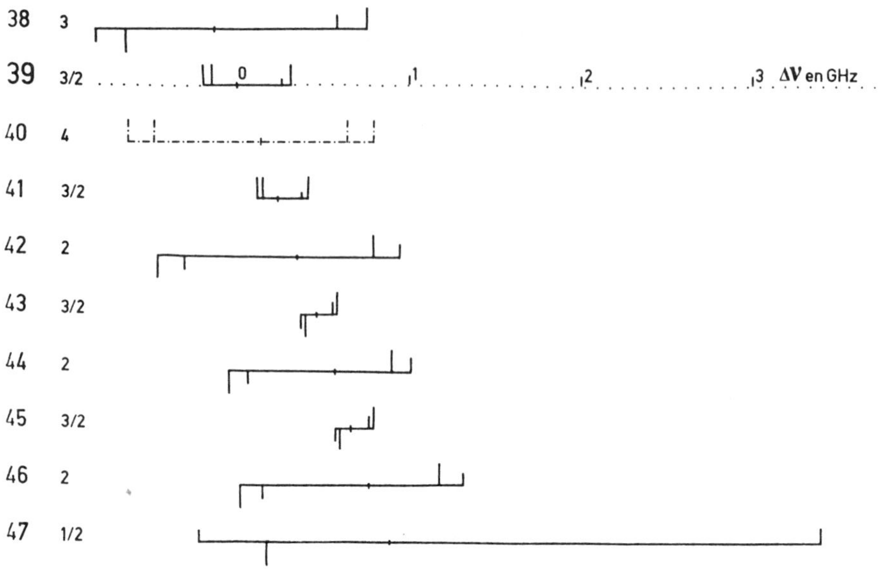

Figure 7. F=3 - F=2 frequencies in potassium isotopes.

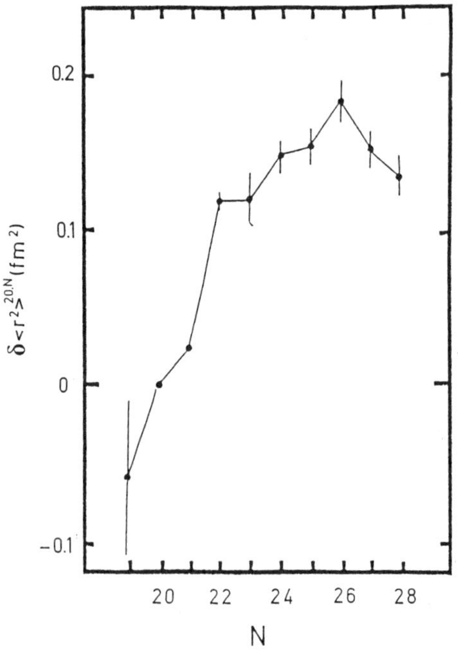

Figure 8. Variation of mean square radius of the electric charges in K isotopes.

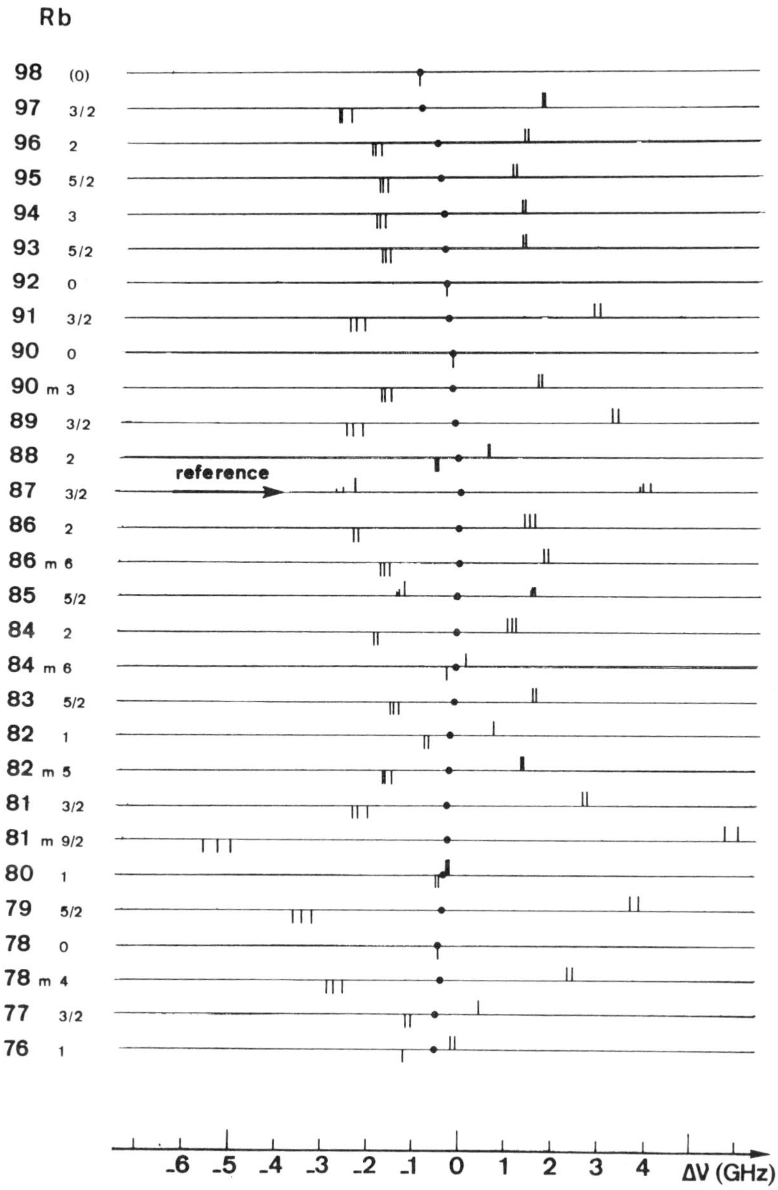

Figure 9. Hyperfine structures and isotope shifts in Rb atoms.

which is metastable with respect to the electric dipole interaction. Assuming an almost negligible specific mass shift it is rather easy to deduce the volume shift expressed in terms of $\delta < r^2 >$, as it is reported in fig. 10 (full dots for isotopes, circles for isomers). The isotope of mass

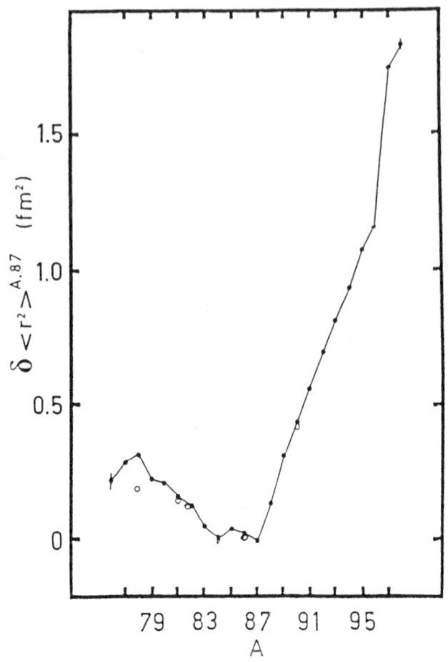

Figure 10. Variation of charge mean square radius in Rb isotopes.

number 87, which is one of the two stable isotopes, corresponds to a magic number of neutrons (N = 50) and to a minimum size of its volume. Accordingly, adding a neutron to the nucleus makes the size to increase rather rapidly with a large deformation occuring between masses 96 and 97. On the other side and quite surprisingly, removing one neutron from the nucleus also makes its size to increase although more slowly. No doubt that such striking features will stimulate new theoretical investigations.

3.4. Cesium isotopes (10)

Same as for rubidium, quite a long series of cesium isotopes has been studied, namely from mass number 118 up to 145, including 28 isotopes and 8 isomers. The experimental observations are reported on fig. 11. Assuming again in that case a negligible specific mass shift, the volume effect can be visualized through the variation of $\delta < r^2 >$ with respect to N (number of neutrons), as in the fig. 12. Three zones of specific behaviour can be clearly identified : i) the zone of isotopes of mass number larger than 137 (which corresponds to the magic number of neutrons N = 82), where the nuclear size increases almost linearly and rather rapidly ; ii) the zone of isotopes located between neutron number N = 68 and N = 82 where the nuclear size increases rather slowly and according to oscillations provoked by the normal odd even staggering ; iii) the zone of lighter isotopes corresponding to neutron numbers N = 63 to 67, in which a curious phenomenon of anomalous isotope shift has been observed. It must be noticed that the same behaviour has already been observed in the isotope shift study of mer-

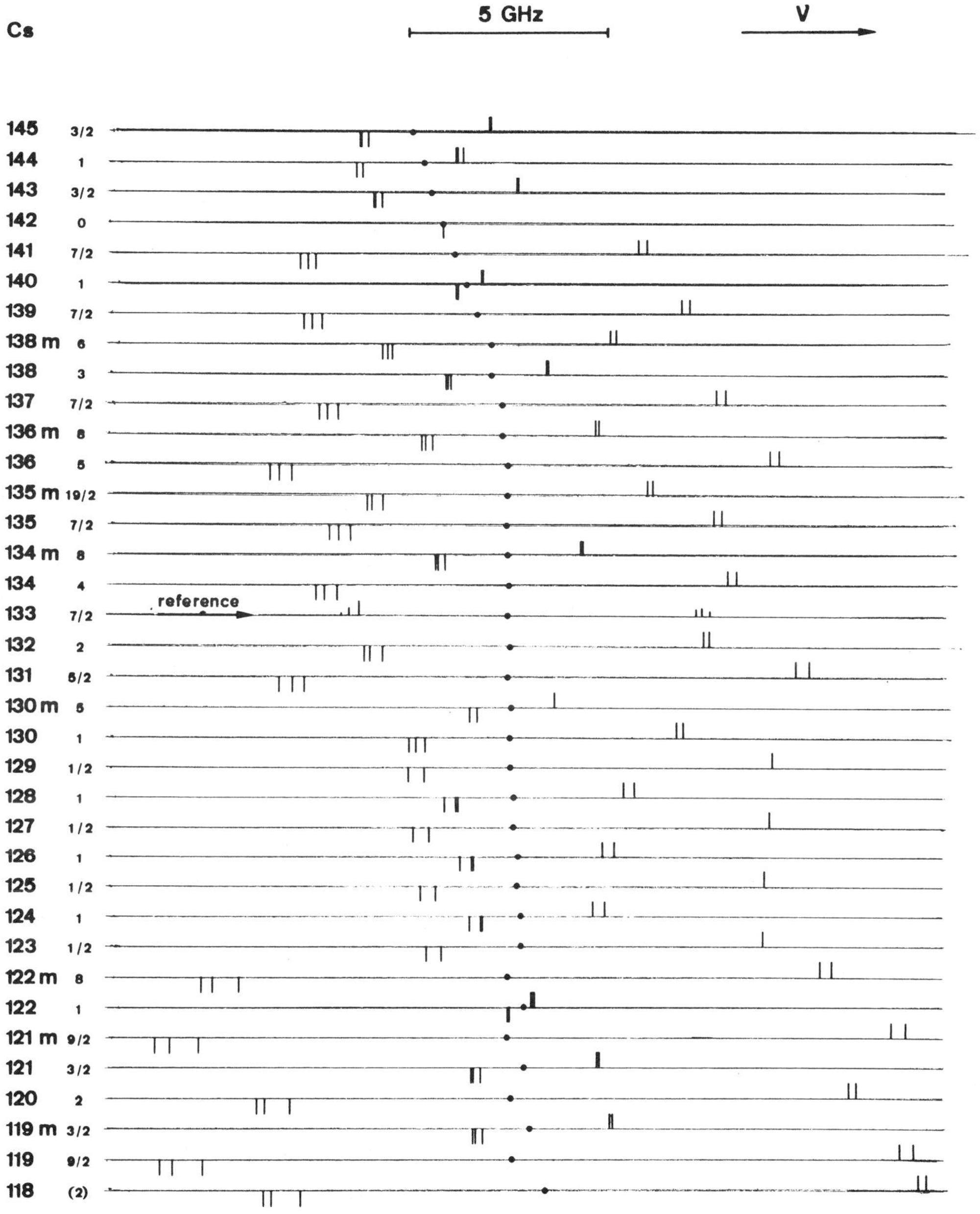

Figure 11. Hyperfine structure and isotope shifts in Cs atoms.

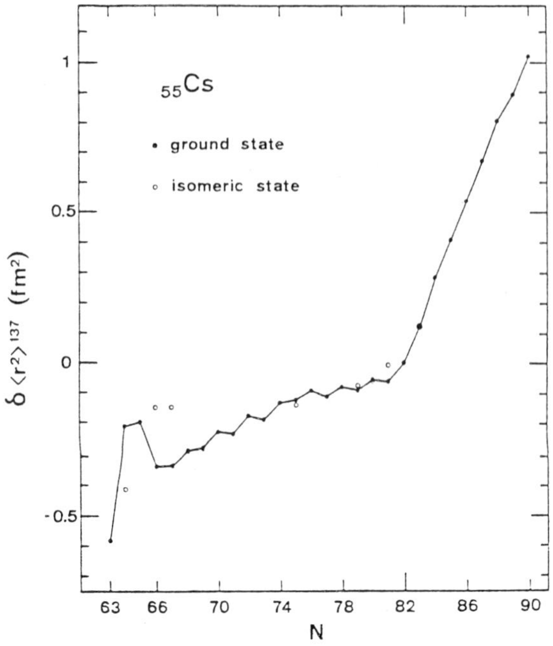

Figure 12. Volume effect in Cs atoms.

cury atoms (11) (12). It seems that two shapes of the nucleus can co-exist, one corresponding to the smallest deformations and to the lowest spin values, whereas the other corresponds to larger deformations and to higher nuclear spin values.

The case of cesium is well suited to perform a significant comparison between the values of the deformation parameter β obtained by means of the isotope shift measurements and those obtained by means of the nuclear electric quadrupole moment, at least for isotopes and isomers in the region of $N < 68$; as can be seen in fig. 13 the agreement is quite good.

3.5. Francium spectroscopy

Francium is known for rather a long time since it has been discovered in 1939 by Marguerite Perey (13). It has been established that its isotope of mass number 223 ($T_{\frac{1}{2}}$ = 22 min.) is an α-decaying daughter of ^{227}Ac ($T_{\frac{1}{2}}$ = 21,6 y) for an amount of 1,4 % (the other 98,6 % decaying by β^- emission toward ^{227}Th). So, assuming that a permanent regime has been reached, this element is naturally present in the earth crust, but in quite a small abundance since from an amount of 3 tons of uranium oxide one can get 1 g of radium from which $0{,}6.10^{-3}$ g of actinium can be extracted and then $1{,}3.10^{-11}$ g of francium, which in all corresponds to less than 28 g on earth at each time. This scarcity explains the reason why no optical line was known for this element up to now. In addition, the expected chemical aggressiveness of francium prevents to use any experimental set-up involving a cell container. An experimental arrangement involving an atomic

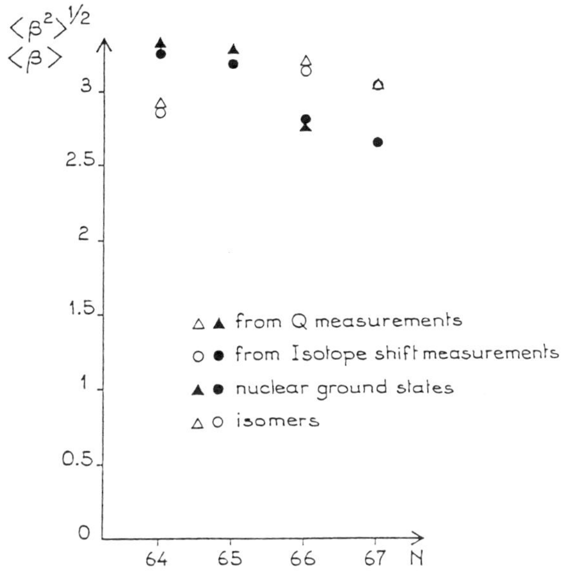

Figure 13. Comparisons of the deformation parameter β in Cs atoms.

beam machine is much better adapted to the study of this element.

On the other hand, noticeable production of several isotope species of francium can be achieved by means of usual nuclear reactions (spallation, fragmentation). So the main experimental principles utilized to investigate the other alkali atoms could be used to study the francium atom, and in a very first step to find one of its resonance line. This has been done with a specifically designed broadband tunable dye laser (14), which has permitted to locate the D_2 line and to measure its wavelength : $\lambda(D_2)$ = 718,169 nm. Further experiments have then permitted to identify the other D_1 resonance line, and the two next lines located in the blue range as it is summarized in fig. 14 (15).

Once the wavelengths of the lines have been measured it is possible to perform high resolution spectroscopy the same way as it was done for the other alkali atoms and to determine both the hyperfine structure and the isotope shift for all available isotopes. An example of such a high resolution recording of the structure of the D_2 line of ^{209}Fr is given in fig. 15. A general scheme of all recorded high resolution structure of francium isotopes from mass number 207 to 228 is displayed in fig. 16. It shows in particular the lack of recordings for the 6 isotopes 214 to 219, due to their rather short lifetime (16).

Another remark which can be done by inspecting the isotope shift measurements concerns the odd-even staggering : in order to better visualize the effect, one usually plots the quantity $Y = \frac{1}{2}(IS_{A+2} - IS_A)/(IS_{A+1} - IS_A)$ versus N number of neutrons, where IS_A means the measure of isotope shift of isotope A (with respect to an arbitrary origin). Fig. 17 shows in effect

^{212}Fr

D_1	$\sigma =$ 12 237.4093(20)cm^{-1}
D_2	$\sigma =$ 13 924.2984(20)cm^{-1}
D'_1	$\sigma =$ 23 113.0506(20)cm^{-1}
D'_2	$\sigma =$ 23 658.3918(20)cm^{-1}

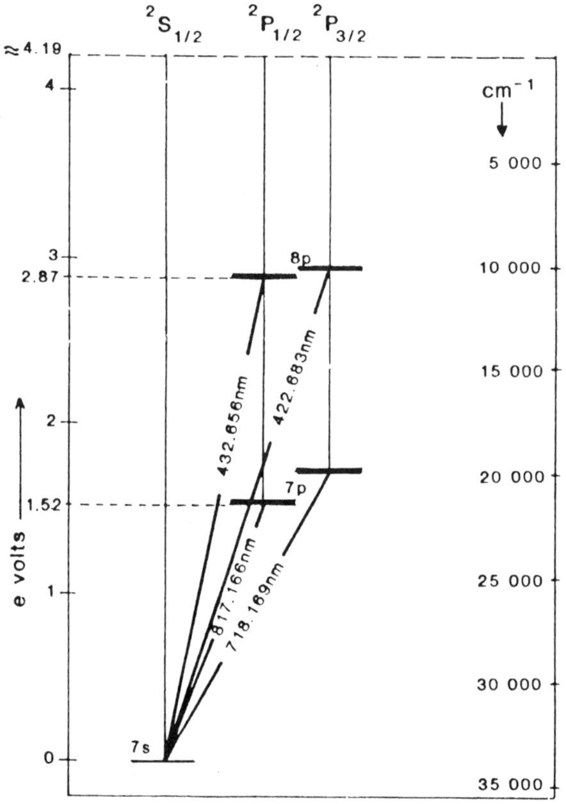

Figure 14. Resonance lines in Fr atoms.

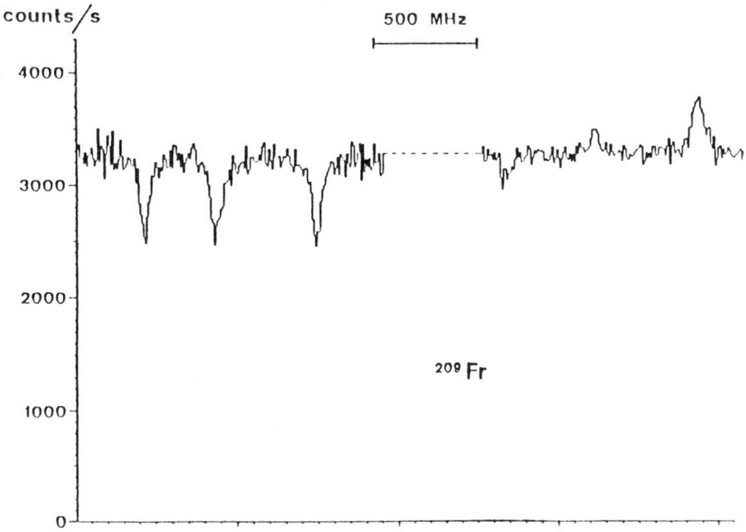

Figure 15. High resolution recording of the D_2 line in ^{209}Fr atoms.

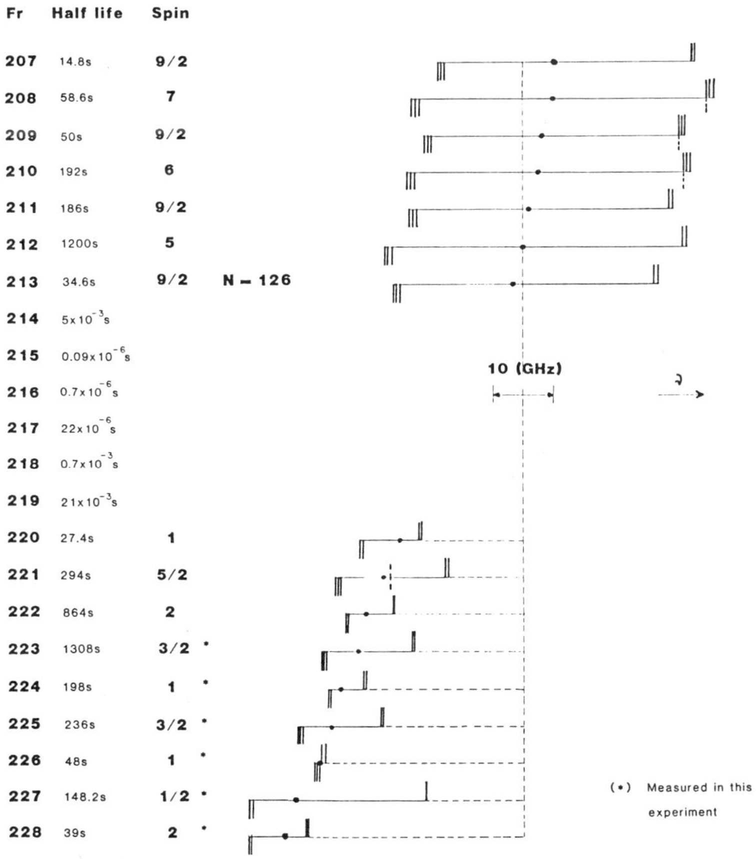

Figure 16. High resolution recordings of Fr atoms structure.

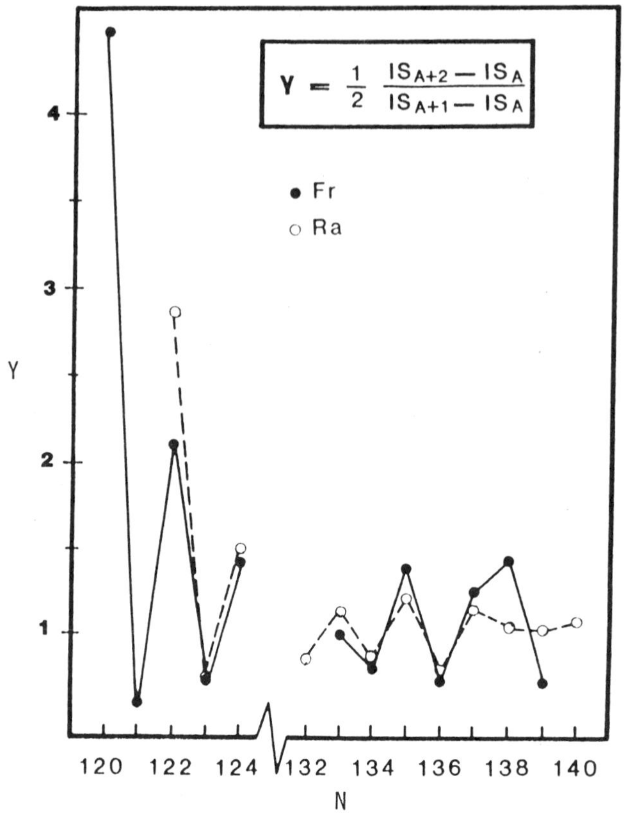

Figure 17

a clear oscillation of the quantity Y. The striking feature is that the oscillation has changed its phase for isotopes corresponding to N = 133 to 137 as compared to the lighter ones (N = 120 to 124) and the phase becomes normal again for isotopes corresponding to N = 138 and 139. Such a phenomenon has already been observed for radium isotopes (see figure 17), and it has been interpreted as the possible indication of a static octupolar deformation. Some more theoretical investigations are obviously needed to state that point more firmly.

V. CONCLUDING REMARKS

Except for the case of francium for which atomic spectroscopy experiments must be pursued, at least to have an experimental precise determination of the ionization limit, the study of alkali series of isotopes has almost arrived to an end. It has provided people with a significant amount of experimental results which are of interest for both community of atomic and nuclear physicists, and should therefore stimulate theoretical activity in both domains. It is to be emphasized that such problems as the well known odd-even staggering effect, the hyperfine anomaly, the appearance of two possible shapes for some light isotopes and isomers, the behaviour of nuclei shapes on both sides of magic numbers, have not yet been solved on a theoretical point of view.

More investigations are obviously needed for the other elements of the periodic table. But it involves several major changes in the experimental techniques, mainly due to the fact that the magnetic selection used for alkali atoms is no longer suitable for non paramagnetic atoms in their ground state. Collinear laser atomic beam spectroscopy has proven to have a few advantages which will make this technique to be thoroughly used in the near future.

Acknowledgements

I would like to gratefully acknowledge all my colleagues who have participated in the experiments described above and more particularly those of the Laboratoire Aimé Cotton and of the Laboratoire René Bernas.

REFERENCES

1. H. Kopferman, "Nuclear Moments", Academic Press (1958).
2. P. Jacquinot and R. Klapisch, Rep. Prog. Phys. 42, 773 (1979).
3. J. Bauche and R.J. Champeau, Adv. Atom. Molec. Phys. 12, 39 (1976).
4. P. Jacquinot, "High resolution laser spectroscopy", ed. K. Shimoda Springer-Verlag (Berlin) p. 51 (1976).
5. P. Gumbar, thèse 3ème cycle, Paris VI (1981) (unpublished).
J.M. Serre, Thèse 3ème cycle, Paris VI (1981) (unpublished).
6. H.L. Ravn, "Zinal Workshop on the Isolde Program, Appendix" (1984).
7. G. Haber et al., Phys. Rev. C 18, 2342 (1978).
8. F. Touchard et al., Physics Lett. B.
9. C. Thibault et al., Phys. Rev. C 23, 2720 (1981).
10. C. Thibault et al., Nucl. Phys. A 367, 1 (1981).
11. Dabkiewiez et al., Europ. Phys. Soc. Conf. on Trends in Physics, York (1978).
12. Dabkiewiez et al., J. Phys. Soc. Japan 44, 503 (1978).
13. M. Perey, Comptes Rendus Ac. Sc. 208, 97 (1939).
14. S. Liberman et al., Comptes Rendus Ac. Sc. 286, 253 (1978).
N. Bendali et al., Comptes Rendus Ac. Sc. 299, 1157 (1984).
15. H.T. Duong et al., submitted to Europhysics Letters.
16. A. Coc et al., to be published in Phys. Lett. B.

PART I

Rydberg atoms

AN INTRODUCTION TO RYDBERG ATOMS

Daniel Kleppner

Department of Physics
Massachusetts Institute of Technology
Cambridge, Massachusetts 02139 U.S.A.

INTRODUCTION

If a single valence electron of any atom is promoted to a state of high principal quantum number n, the electron experiences an essentially Coulombic potential and behaves in many respects like a highly excited electron in hydrogen. An atom in such a state is known as a Rydberg atom. How this title originated does not seem to be known, but presumably the term is associated with the fact that the spectra of many "single-electron" atoms are accurately described by Rydberg's hydrogen-like formula. (Sometimes the term "highly excited atom" is used, though this term is more properly applied to atoms in core-excited states whose energies are vastly higher than those of Rydberg atoms.) How large n must be for an atom to qualify as a Rydberg atom is vague, but a reasonable working definition is $n > 10$.

Bohr's correspondence principle requires that states with increasingly high quantum numbers assume increasingly classical properties, which might be taken to imply that Rydberg atom research essentially represents a regression from the quantum world to the well-trod world of classical dynamics. However, Rydberg atom research provides a means for changing the scale of atomic interactions by many orders of magnitude, and whenever such a change of scale occurs in physics one can look forward to new discoveries and unexpected phenomena. These introductory lectures summarize some of the discoveries and surprises that have emerged from Rydberg atom research since its modern phase began slightly over ten years ago. However, our starting point is much earlier - approximately one century ago - when Rydberg atoms first played a visible role in the development of physics.

1. SOME EARLY AND NOT-SO-EARLY HISTORY

In 1885, J.J. Balmer published his soon to be famous empirical law for the wavelengths of the visible and near uv spectrum of hydrogen [1]: $1/\lambda = R(1/2^2-1/n^2)$. He based his fit on the five known visible lines, plus four uv lines reported by the astronomers Huggins and Vogel. In a last minute addendum, Balmer compared his predictions with additional measurements reported by Huggins for n up to 16. The agreement was remarkable. Empirical formulas are always suspect, but an empirical formula with predictive power cannot easily be dismissed. The success of Balmer's formula suggested that some underlying law must be at work.

The riddle of Balmer's formula was unraveled by Niels Bohr in 1913 [2]. In Section 2 of this paper, immediately following his calculation of the Rydberg constant from the fundamental constants, Bohr speculates why very high lying states of hydrogen were never seen in laboratory emission spectra, notwithstanding that levels up to n = 33 had been detected in astrophysical spectra. The reason lies in the fact that the atom's radius scales as n^2, and its area as n^4. At laboratory pressures, high-n atoms would collide before they could radiate. Bohr adds: "according to the theory the necessary conditions for the appearance of a great number of lines is therefore a very small density of the gas: for simultaneously to obtain an intensity sufficient for observation the space filled by the gas must be very great." Interstellar space exactly fits Bohr's specifications: the density is low, and there's plenty of it. In 1965, Rydberg atoms were detected by radio astronomers who were searching for the radiation generated when electrons and protons recombined in the interstellar environment [3]. Figure 1 shows one of the early observations: the 109α line (n=110→109). Nearby peaks due to He and C are also visible (they are shifted by the reduced mass effect), as well as the 137 β line (n=139→137). Radio astronomers still hold the record for high n: n > 350 has been observed in hydrogen, and
n > 700 in carbon! However, studying the properties Rydberg atoms requires efficient techniques for creating them in the laboratory, and these techniques only started to emerge in the early 1970's, chiefly due to the advent of tunable lasers.

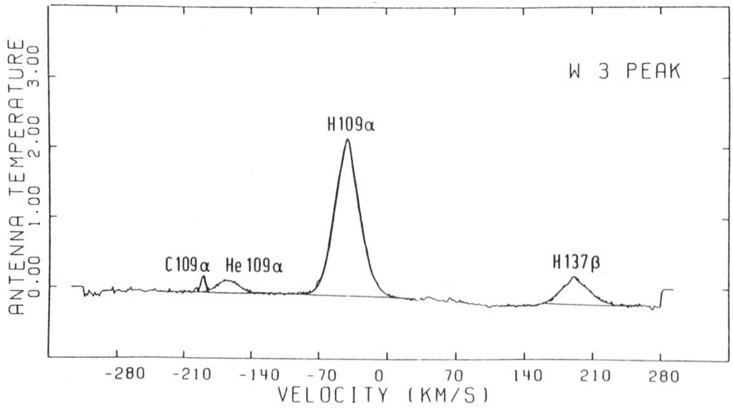

Fig. 1. Radio recombination lines. (Courtesy of P.J. Metzger.)

2. SCALING LAWS FOR RYDBERG ATOMS

Atomic units are the most natural units for Rydberg atoms. In these: $h = m = e = 1$, $c = 1/\alpha = 137$. Energy = $\alpha^2 mc^2 = 27.2$ eV (the Hartree); distance = $a_o = 0.529\times10^{-8}$ cm; electric field = $e/a_o^2 = 5.14\times10^9$ V/cm.

The unusual and sometimes bizare properties of Rydberg atoms are revealed by the scaling laws that govern their properties. These are summarized in Table 1.

Table 1

Scaling Laws for Rydberg States

quantity	scaling law	n~1,2	n~30	unit
atomic radius (dipole moment)	$(a_o)n^2$	1	10^3	Å
binding energy	$(hcR)n^{-3}$	10	10^{-2}	eV
transition frequency, $\Delta n=1$	$(2cR)n^{-2}$	10^{15}	10^{11}	Hz
geometrical cross section	$(a_o^4)n^4$	10^{-16}	10^{-10}	cm^2
linear Stark effect	$(ea_oF)n^2$	1	10^3	MHz/(V/cm)
ionizing field	$(\frac{1}{16}\frac{e}{a_o^2})n^{-4}$	10^9	10^3	V/cm
diamagnetic interacton	$(\frac{\alpha^2B^2a_o^6}{8e^2})n^4$	10^3	10^9	Hz/tesla^2
radiative lifetime	(low-ℓ) $\sim n^3$ (high-ℓ) $\sim n^5$	10^{-9}	10^{-5} 10^{-2}	s s

3. THE WORLDS OF RYDBERG ATOM RESEARCH

A series of review articles on Rydberg atom research is provided in the book <u>Rydberg states of atoms and molecules</u>, [3]. Although this field continues to move relatively rapidly, this book still provides a reasonably comprehensive overview. The following summary of major topics, each followed by a few lines of explanation and examples, is intended to suggest the flavor of the research. However, it does not start to do justice to the depth and range of the research, nor to the excitement of many of the discoveries. Reference 3 provides further details of many of these.

Spectroscopy

*Electron core interactions. Rydberg electrons can serve as ultra sensitive probes of the ionic cores of atoms. For example, microwave spectroscopy of Rydberg levels can measure an ion's polarizability to a precision of 1 in 10^6 or better, and yield static and frequency-dependent value for the dipolar and quadrupolar interactions. The accuracy of such measurements far exceeds that of any previous technique.

*Access to 2-electron systems. Rydberg spectroscopy of alkaline earth atoms can be interpreted with clarity and precision using the multichannel quantum defect theory. Such research provides lucid pictures of perturbed series, configuration interactions and autoionization. "Planetary atoms", created by exiting one of the inner electrons in a Rydberg atom, provides a new approach to the study of highly correlated systems.

*Photoionization and multiphoton processes. The n^2 scaling law for transition moments among Rydberg states makes it possible to observe numerous photoionization and multiphoton processes. The list includes blackbody photoionization, and 20-photon transitions between Rydberg states.

*Frequency standards. The energy of Rydberg levels with respect to the ground state can be predicted by quantum defect theory with such accuracy these levels can serve as secondary optical frequency references. Millimeter or microwave spectroscopy between Rydberg levels is expected to provide a value for the Rydberg constant that rivals the best optical measurement currently possible.

Collisions

*Collisions with simple atoms. Broadening and shifting of Rydberg absorption lines due to rare gas atoms was the subject of a classic work on electron scattering by Fermi in the early 1930's. Recent collisional studies include angular momentum and state charging collisions, energy transfer and associative conjugation. The cross sections are often enormous, and the collisional processes are sometimes dramatic, as in the sudden transformation of a gas of Rydberg atoms into a plasma.

*Collisions with molecules. Rydberg atoms can be employed as a source of ultra low energy electrons for studying electron attachment and other electron molecule collision processes. The broad spectrum of rydberg energies has proven to be well suited to studying the resonant excitation of internal molecular modes, as in electronic-rotational energy transfer.

*"Assisted" collisions. Cross sections for inelastic collision processes generally increase as the energy defect decreases. However, microwave photons can be added to a Rydberg atom collision complex to make up the energy difference, greatly enhancing the collision rate. Alternatively, the energy levels can be manipulated with electric fields to

achieve a zero-defect condition [4]. Fig. 2 shows the cross section for resonant state-changing collisions in sodium ns+ns → np+(n-1)p. The cross section can exceed 10^9 Å^2, a figure which is enormous even compared to the huge geometrical cross section of Rydberg atoms.

*Collisions with surface. Fast ions passing through a foil can pick up free electrons, emerging in high Rydberg states, or even in low energy continuum states in which the free electrons follow close to the projectile.

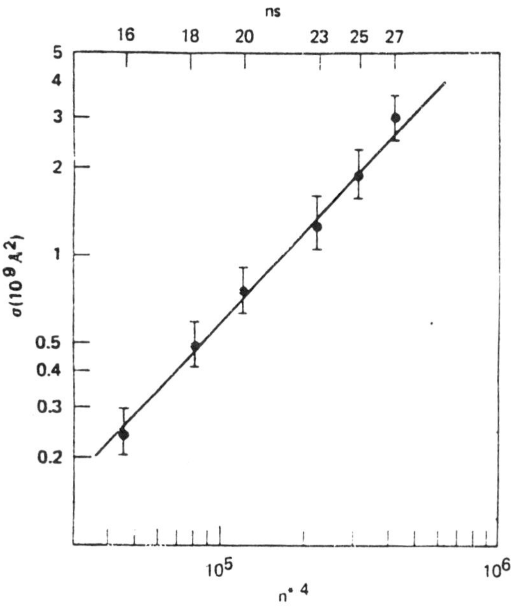

Fig. 2. Resonant cross section for ns+ns → np+(n-1)p, plotted vs. n^{*4}. (n^* is the effective quantum number of the s state.) The actual values of n are plotted above, for reference. (From Safinya et al., ref. 4.)

Atoms in Strong Fields

*Electric fields. The theory of hydrogen in strong electric fields has been developed in great detail. Rydberg state studies with hydrogen have confirmed both the general features of these theories and to many numerically sensitive predictions. In non-hydrogenic atoms such studies have led to a deeper insight on the role of symmetry and the structure of the continuum. Ionization in a static field provides a critical test of theory and a powerful experimental technique for detecting Rydberg atoms and identifying their states.

*Magnetic fields. As discussed by Dr. Gay elsewhere in these lectures, the structure of the hydrogen atom in a magnetic field of arbitrary strength poses a serious problem for atomic theory. The equations are non-separable but studies with Rydberg atoms have pointed to a hidden symmetry that provides an important key to the general understanding. The appearance of chaotic solutions to the classical equations of motion suggests the possibility of observing chaos in a small quantum system.

Rydberg atoms and radiation

Dr. Haroche discusses this subject in the following lectures. Topics in this area include

*"Traditional" processes. Spontaneous radiation, stimulated absorption and emission, blackbody radiative transfer - including transfer to the continuum - have all been observed with precision using Rydberg atoms. The shift in energy levels due to the blackbody field - far too small to see in low-lying levels - has been observed in Rydberg levels.

*Coherence effects. Superradiance on a single transition or among a cascade of transitions can be realized in relatively small systems of Rydberg atoms. The evolution of superradiance from its start up and the dynamics of damped and underdamped motion have been displayed. Photon statistics can be probed down to the single photon level, and coherence effects on the atomic levels can be witnessed.

*"Cavity quantum electrodynamics". The effects of cavities and conductors on the vacuum can be studied using single Rydberg atoms. Spontaneous emission has been enhanced and inhibited, and maser action has been realized with a single atom.

4. THE CENTRAL EXPERIMENTAL TECHNIQUES

The majority of Rydberg atom research employs laser excitation from a ground state or metastable state of atoms in an atomic beam or a gas cell. Rydberg atoms can also be created by charge transfer to fast ions as they pass through matter. The frequency selectivity of laser light makes it possible to populate well resolved Rydberg states for values of n over 200; most importantly, the high intensity of laser light can compensate for the relatively small transition moment connecting low-lying and Rydberg levels. Stepwise excitation by light from two or three lasers is often used. Multiphoton absorption has been employed with hydrogen and other atoms with relatively high ionization energy. Two-photon excitation from a single laser is suitable for some alkali metal atoms.

Rydberg atoms are most commonly detected by field ionization. The required field decreases as n^4: about 5 kV/cm is needed for n = 30. Field ionization is close to 100% efficient, generally free from background, and state specific. These properties make possible single atom experiments. An alternative detection technique, widely employed for studies in gas cells, employs a space charge limited diode. The Rydberg atom collisionally ionizes, and generates a huge current pulse due to its neutralizing effect on the space charge.

5. RYDBERG ATOMS IN ELECTRIC FIELDS - Hydrogenic and non-hydrogenic behavior

In an electric field hydrogen has a particularly simple structure: the energy levels form fan-like patterns, as shown in Fig. 3a. (The drawing is actually for lithium, $|m|=1$.) The many degeneracies (i.e. the sharp level crossings) manifest the high symmetry of the Coulombic potential. The experimental energy-level plot in Fig. 3b confirms this picture.

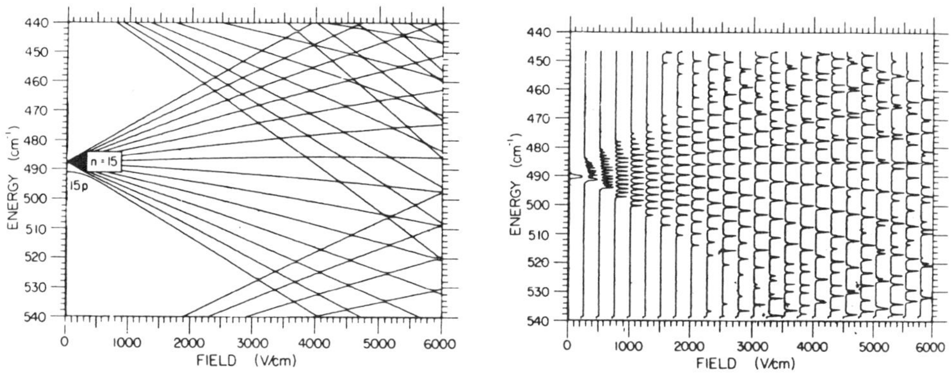

Fig. 3a (left). Calculated structure for n=15, $|m|=1$ levels of lithium in an electric field. This structure is hydrogenic in its general feature. 3b (right). Measured structure. (From ref. 5).

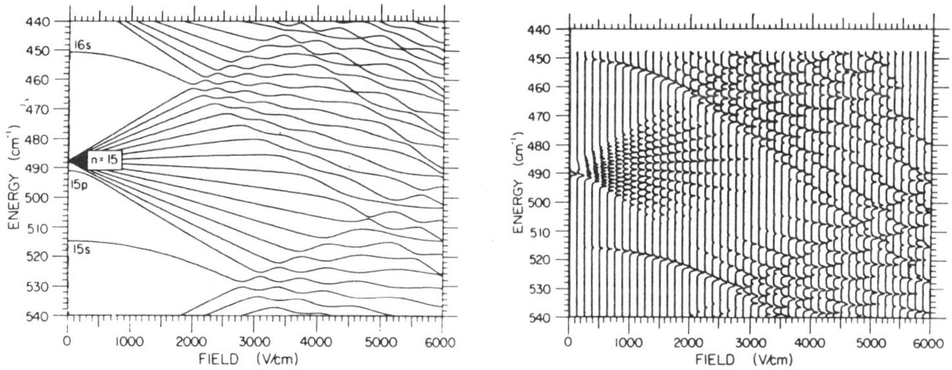

Fig. 4. Calculated (left) and measured (right) Stark structure for n = 15, $|m|=0$ level of lithium in an electric field. In contrast to Fig. 3, the levels display strong repulsion and rapid variation (from ref. 5).

In contrast, the Stark structure of the $|m| = 0$ levels of lithium, Fig. 4, reveals strong repulsions and a basic confusion of levels. The essential difference between the $|m| = 0$ and 1 manifolds arises from the presence of an s-state in the former. This s-state is non hydrogenic due to the effect of the core. The energy of a Rydberg electron, as explained below, is $-1/2(n-\delta_\ell)^2$, where the quantum defect, δ_ℓ, depends strongly on the core penetration. In lithium, $\delta_0 \sim 0.3$, whereas for higher value of ℓ, $\delta_\ell = 0$. Hence the $|m| = 1$ manifold is hydrogenic, whereas the $m = 0$ manifold is not.

Quantum defects play a major role in determining not only the structure but also the dynamics of Rydberg atoms. We digress briefly to discuss their physical significance.

6. QUANTUM DEFECTS

The energy levels of many Rydberg atoms are described accurately by $W_n = -1/2(n-\delta_\ell)^2$ (atomic units), where δ_ℓ, the quantum defect, is a constant or slowly carrying function of the angular momentum, ℓ. Often $n-\delta_\ell$ is defined as the effective quantum number, n^*. The energy level shift due to the quantum defect is approximately

$$\delta W = -\delta_\ell/n^3.$$

Note that to the same approximation, the spacing between adjacent Rydberg levels is

$$\Delta W = -1/n^3.$$

In spectroscopic measurements of Rydberg states ΔW provides a natural unit of energy, for it can be measured by the slowly varying separation between adjacent Rydberg excitation lines. In terms of this energy unit, the shift due to the quantum defect is identical to the quantum defect.

The origin of the quantum defect can be understood as follows. A Rydberg atom consists of a nucleus of charge Z, surrounded by a tight core of Z-1 electrons, and a single electron at a large distance. Rydberg electrons with high angular momentum never penetrate the core. Such states are hydrogenic: for these, $\delta_\ell \sim 0$. However, low-ℓ states - and this generally means $\ell \leq \ell_{core}$ - are perturbed. Classically, the penetration causes a precession of the elliptical orbits: quantum mechanically, it introduces a phase shift in the radial wave function. The quantum defect can be understood as follows.

The Rydberg electron spends most of its time outside the core in a region of essentially pure $-1/r$ potential. For distance less than the core radius r_c, the potential can be written $-Z(r)/r$, where $Z(0) = Z$, and $Z(r_c) = 1$.

The energy levels for hydrogen can be obtained by a WKB calculation. The effective radial potential is $V = -1/r + (\ell+1/2)^2/2r^2$. The energy is found from the phase integral.

$$\Phi_o = \int_a^b \sqrt{2(E-V)}\,dr = (n+\tfrac{1}{2})\pi$$

where a and b are the inner and outer turning points respectively. Substituting $E = -1/2\nu^2$ yields

$$\Phi_o = (\nu+\tfrac{1}{2})\pi$$

from which the Bohr result follows immediately: $\nu = n$.

Let V' be the true potential. For $r>r_c$, $V'=-1/r$. The phase integral is

$$\Phi' = \int_{a'}^{b} \sqrt{2(E-V')}\,dr = (n+\tfrac{1}{2})\pi .$$

where a' is the inner turning point. The phase integral can be rewritten as follows:

$$\Phi' = \int_{a'}^{r_c} \sqrt{2(E-V')}\,dr - \int_{a}^{r_c} \sqrt{2(E-V)}\,dr + \int_{a}^{r_c} \sqrt{2(E-V)}\,dr + \int_{r_c}^{b} \sqrt{2(E-V)}\,dr$$

where $V' = V$ for $r > r_c$. We can write this result as

$$\Phi' = \Phi_0 + \Delta\Phi(\ell, E)$$

where $\Delta\Phi(\ell,E)$ is given by the first two integrals. Note that these integrals extend only to the edge of the core. In this region $E \ll |V'|$. If E is neglected compared to V or V', then $\Delta\Phi(\ell)$ is a constant, and

$$\Phi_0 = (\nu+\tfrac{1}{2})\pi = (n+\tfrac{1}{2})\pi - \Delta\Phi(\ell)$$

$$E = \frac{-1}{2\nu^2} = \frac{-1}{2(n-\delta_\ell)^2}$$

where $\delta_\ell = \Delta\Phi(\ell)/\pi$. Thus, the quantum defect is a direct measure of the phase introduced by the core potential. Alternatively, evaluating the integral can be expanded as a power series of E/V'. In this case the quantum defect is found to be a slowly varying function of the energy.

Any central force short range perturbation - that is, perturbation of the form r^{-n}, where $n>2$ - gives rise to level shifts which can be expressed in terms of quantum defects. For example, polarization of the core by the Rydberg electron gives rise to an interaction of the form V

$$V_{pol} = -\frac{1}{2}\,\alpha\,\frac{1}{r^4}$$

where α is the dipolar core polarizability. It can be shown [6] that the corresponding quantum defect is given by

$$\delta_\ell \sim \frac{3}{4}\,\frac{\alpha}{\ell^5}$$

A quadrupole interaction gives rise to a term which varies as $1/\ell^9$. In a similar spirit, fine and hyperfine interactions, and other short range perturbations such as the Lamb shift, can be described by a quantum defect formalism.

7. FIELD IONIZATION

Field ionization - the process by which a Rydberg atom spontaneously ionizes in a static electric field - provides both an invaluable experimental technique and a problem of considerable physical interest in its own right. Field ionization in hydrogen is discussed by Damburg and Kolosov in ref. 3, and general aspects of field ionization are discussed in a number of other articles in the same volume.

A simple energy argument provides the framework for understanding some general features of field ionization. In an electric field -F along the z axis the potential for hydrogen can be written

$$V = -\frac{1}{r} - Fz.$$

This potential has a saddle point on the z axis with a maximum

$$V_{sp} = -2F^{1/2}.$$

The electron is free to escape if its total energy W exceeds V_{sp}; otherwise it is bound. Thus, one can expect a threshold energy for field ionization given by

$$W_{th} = -2F^{1/2}.$$

Alternatively, if the field is slowly increased, ionization can occur at a field

$$F_{th} = W^2/4$$

where it is understood that W is the energy of the state in the field F. These arguments hold strictly for zero-angular momentum states. The "magnetic" quantum number |m| remains good in an electric field, and it is easy to calculate a correction to W_{th} due to the centrifugal barrier for states with |m|>0.

This threshold behavior is illustrated in Fig. 5, which displays the Stark structure of lithium, n=19. Lithium was excited in an electric field by pulsed lasers, as in Figs. 2 and 3, and the Rydberg atoms were ionized by applying a large field pulse. However, the ionization pulse and the collection channel were delayed a few microseconds. Atoms which spontaneously ionized within that time were lost. The signal can be observed to abruptly disappear at the threshold field.

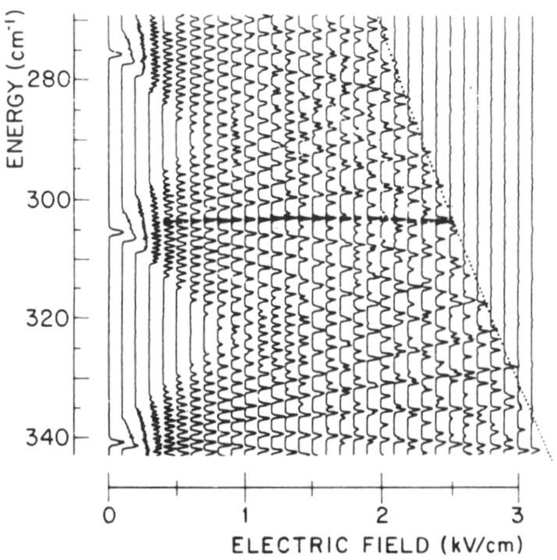

Fig. 5. "Threshold" ionization in lithium, |m| = 1. The atoms are excited by pulsed lasers, and detected by pulsed field ionization following a 3 μs delay. If the atoms spontaneously ionize during the delay period, signal is lost. The onset of ionization is accurately given by the threshold condition (dashed line). From ref. 7.

Consider what happens in the complementary experiment to the one described above. Here the atoms are excited as before, but ions are only collected if the atom spontaneously ionizes. (No pulsed ionizing field is applied, and the ions are collected immediately after laser excitation.) The results are shown in Fig. 6. It is evident that well defined states exist in the region $W > W_{th}$, though at sufficiently high fields many of the levels broaden into a continuum.

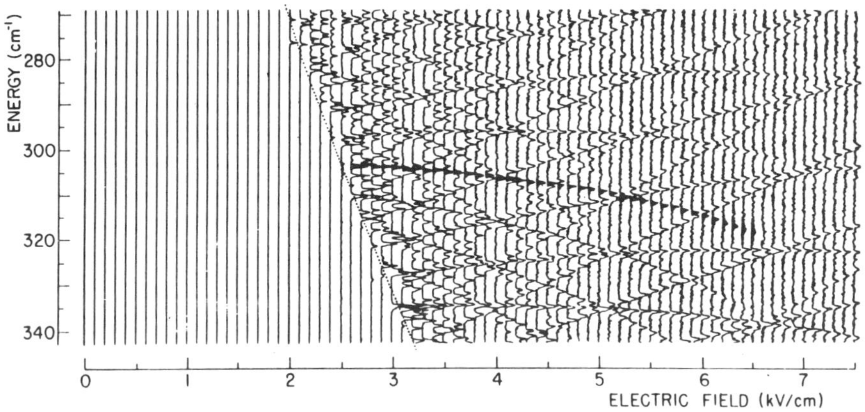

Fig. 6. Autoionizing states of lithium above the ionization threshold. One level has been darkened for clarity. Compare with Fig. 5. From reference 7.

The existence of well defined states with energy greater than $W > W_{th}$ is not really inconsistent with the energy arguments, for such arguments can only predict whether or not an electron can escape, not whether or not it will escape. In hydrogen, long-lived states exist far above the saddle point threshold. Hydrogen first ionizes by tunneling of the electron through the potential barrier at the saddlepoint. The ionization rate increases rapidly with the applied field, causing the level to broaden rapidly once the ionization rate becomes appreciable. Furthermore, the lowest-lying levels ionize first, in contrast to the saddlepoint model where the highest lying levels are first to ionize. This behavior is suggested schematically in Fig. 7a.

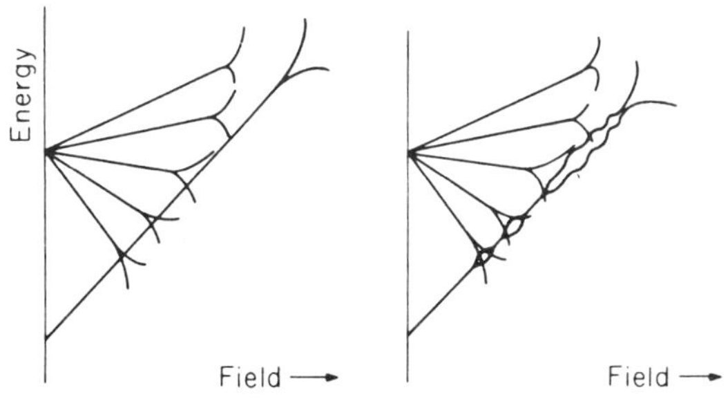

Fig. 7. Sketch of Stark levels of hydrogen in strong electric field (7a, left) and of an alkali-metal atoms (7b, right). Hydrogenic Stark levels do not interact; in an alkali, the level can mix, causing otherwise stable levels to decay. Level widths indicate the decay rate.

Frequent degeneracies are a conspicuous feature of the hydrogenic Stark spectrum. These are seen in the "sharp" level crossings in Fig. 3, and shown schematically by the discrete level which overlies, but does not interact with, the ionizing level in 7a. The absence of interactions between the Stark level of hydrogen is a manifestation of the high symmetry of the Coulomb potential. (Dr. Gay discusses the role of symmetry in the structure of Rydberg atoms elsewhere in this volume.) This symmetry is broken by the core perturbation of alkali atoms, causing the levels to mix. This is manifest in the repulsions between the levels in Fig. 4. The same effect can also mix essentially stable levels with rapidly ionizing levels, as in Fig. 7b. This is the situation in Fig. 6; the observed levels are not actually stable, but they are relatively long lived.

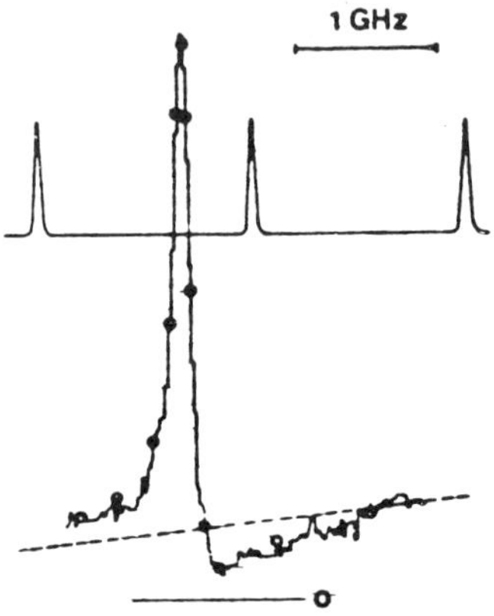

Fig. 8. High resolution study of the lineshape for Rb in an electric field of 158 V/cm in the positive energy regime: excitation energy close to 33.6 cm^{-1}. The lineshape is the characteristic Fano profile for an autoionizing resonance: the dots represent a fit to the Fano theory. From reference 8.

In summary, two separate processes give rise to field ionization. The underlying process is tunneling, and in hydrogen this is the only important process. Other atoms can ionize by the mixing of a stable level with a degenerate tunneling level due to the perturbation of the core. Such a process is essentially identical to autoionization, and in fact the photoionization line shape in this region is given by the familiar Fano profile, as shown in Fig. 8.

8. A "SAMPLER" OF RYDBERG ATOM RESEARCH

The remainder of this introduction to Rydberg atom research consists of brief accounts of a few of the many studies in this area of atomic physics.

a) A "sieve" for Rydberg atoms

The principal quantum number of a Rydberg atom, n, is large by definition, and thus the characteristic dimension of such an atom, n^2a_o, is enormous. The atom can achieve macroscopic dimensions. In principle one should be able to sort such atoms by passing them through a sieve, much as pebbles are graded in a stoneyard.

Such an experiment has been carried out for atoms in the range n = 25 to 65, using an array of electroformed slits 2μm wide [9]. Whether or not an atom passes through the slit depends on the likelihood that it will ionize in the field set up by the electron's image charge in the slit walls. Using a simple static ionization model, it can be shown that the electron will be stripped from the Rydberg atom if it approaches the wall by a distance less than 3.5 n^2a_o. From this it follows that the transmission through a slit should decrease linearly as n^2.

To demonstrate this effect, a broad beam of Rydberg atoms was directed at an array of slits etched in a gold foil, and the transmitted intensity monitored, as shown in Fig. 9a. Results are shown in Fig. 9b. The transmitted intensity decreased linearly with n^2 as expected, though the atoms appeared to ionize at a distance of 4.5 n^2a_o, rather than $3.5n^2a_o$. The slight discrepancy may be due to strong electric fields near the slit walls. As Fig. 9b shows, by tilting the foil 20° from perpendicular to the beam, the effective transmission width is decreased. Thus, the atoms can sense the orientation of the grid.

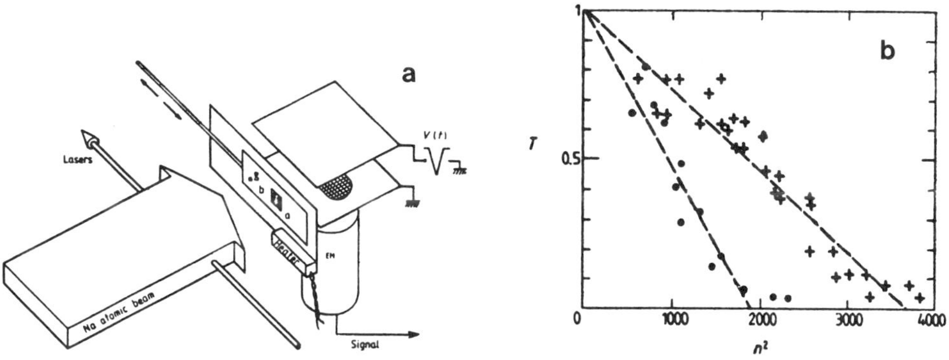

Fig. 9. a(left) Experimental set-up for studying transmission of sodium Rydberg atoms through 2 μm slits. A laser-excited atomic beam impringes on an array of slits in a gold foil mounted on a heated plate. The transmission is monitored by field ionization. For normalization, the signal through a large aperture is also monitored using a shutter.
b (right) Transmission vs n^2. Crosses are for normal incidence; dots are for a 20% tilt to the plate. From reference 9.

b) <u>Very high-n Rydberg atoms</u>

As techniques have been refined, higher and higher n-regions have become accessible to experimental study. Fig. 10a shows the set-up by which Rydberg states in barium with values of n as large as 290 have been observed [10]. These atoms are 10 μm in diameter; they are so weakly bound that they spontaneous ionize in a field of 45 mV/cm. Merely to resolve adjacent terms requires that the electric field be reduced to a few mV/cm.

The atoms are stepwise excited in an atomic beam by two cw lasers with a linewidth of approximately 1 MHz. They ionize by collisions and the ions are detected with an rf quadrupole mass selector. A spectrum in the region of n = 250 is shown in Fig. 10b. In 10c Stark structures in the vicinity of n = 78 is displayed. The irregular distribution of intensity arises from interference of the $82\,^1S_0$ and $80\,^{1,3}D_2$ states with the hydrogen-like n = 78 manifold.

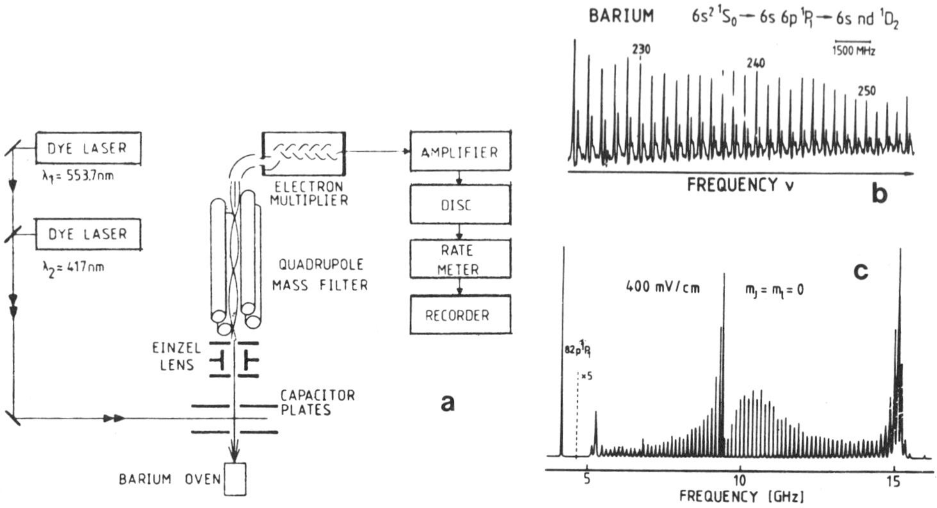

Fig. 10. a(left). Experimental set-up for studying Rydberg states of barium with values of n up to 270. b(upper right). Excitation spectrum in the vicinity of n = 250. c(lower right). Stark structure in the vicinity of n = 78 in a 400mV/cm field. The irregular pattern is due to the interference of S and D states into an essentially hydrogenic manifold. From reference 10.

c) <u>Theory of non-hydrogenic Stark structure</u>

The distinction between Rydberg states of alkali metal atoms and those of hydrogen are manifested in a small set of empirical parameters - the quantum defects. In principle, it should be possible to predict the Stark structure analytically from the theory for hydrogen plus knowledge of the quantum defects, avoiding the tedious process of diagnolizing large energy matrices. (The energy level plots in Section 5 are the results of the latter approach.) Such an analytical theory has now been developed [11]; it can yield values for the positions, strengths and widths of the spectral lines in high electric fields. Essentially, the short range interactions which give rise to the quantum defects have been expressed directly in the natural representation for electric field problems - the parabolic basis. The success of this approach can be seen by comparing the analytically calculated Stark map in Fig. 11 with the numerically calculated and experimentally measured spectra in Fig. 4.

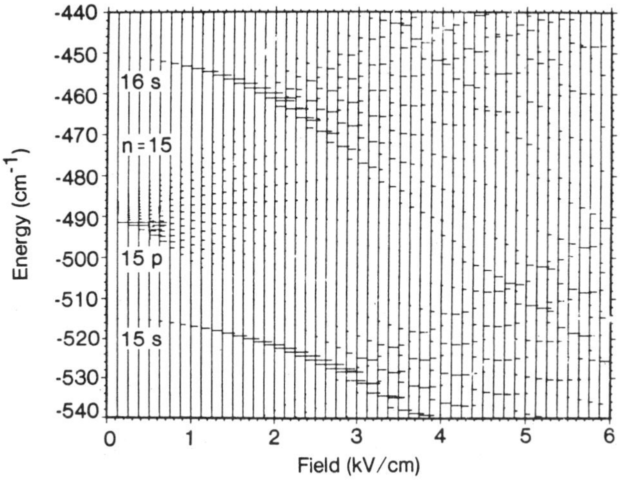

Fig. 11. Analytically calculated Stark spectrum of lithium, m = 0. The length of each line is proportional to the oscillator strength to that level. Compare with Fig. 4. From ref. 11.

d) Energy level shift by blackbody radiation

The vacuum can induce atomic transitions by spontaneous emission and it can shift energy levels, such as in the Lamb shift. Thermal radiation fields induce transitions by absorption and stimulated emission. They can also induce energy level shifts. These are conceptually related to the shifts caused by applied radiation fields (for instance the Bloch-Siegert effect in magnetic resonance) that have long been known to affect high precision spectroscopic measurements. As is the case for most types of atom-field interactions, the shifts are vastly increased for Rydberg states. The Rydberg levels all tend to shift together, however, and thus they can only be observed by measuring the optical energy difference between a Rydberg state and a low-lying state.

Such an experiment has been carried out by ultra precise optical two-photon spectroscopy of the 5S-36S transition in rubidium [12]. Fig. 12 shows the experimental set-up. The Ramsey separated oscillating field method was employed. The time delay was achieved by allowing the atomic beam to traverse the laser beam twice, using a folded cavity. Fig. 11b shows the Ramsey interference pattern for the transition, along with a theoretical fit. The linewidth is 40 kHz. Blackbody radiation from a separately heated source was passed through a chopper and focused on the beam. By using the side of the interference fringe as a frequency discriminator, level shifts as small as a few hundred Hz could be detected. Fig. 11c shows the shift as a function of temperature, along with the theoretically predicted value. Considering the minute size of the effect, typically 2×10^{-12}, the agreement is excellent.

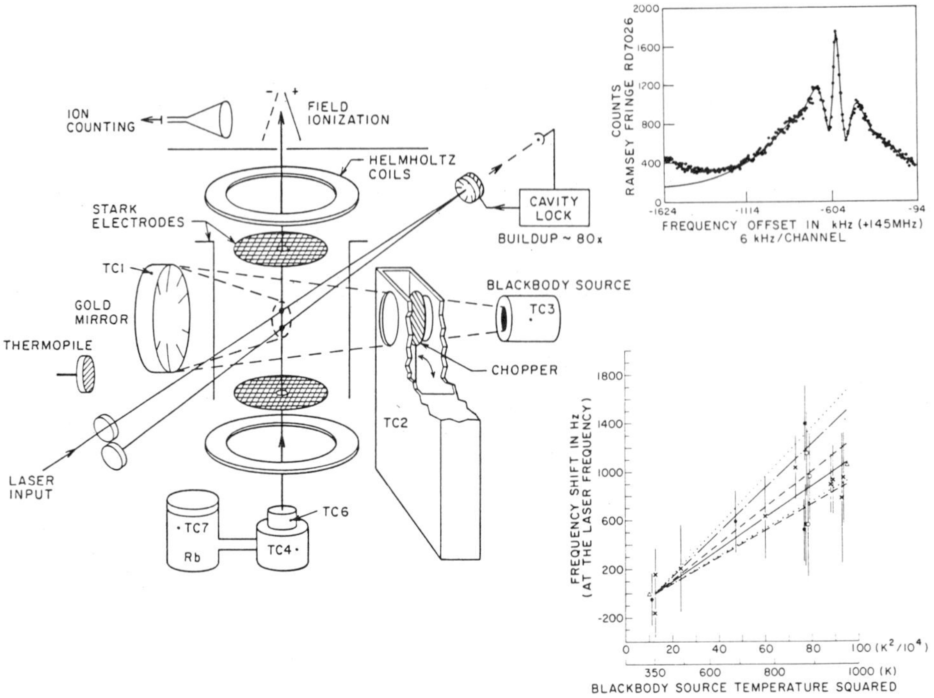

Fig. 12 (left) Apparatus for measuring blackbody-induced energy level shifts in Rydberg states of Rb. The atomic beam is directed vertically upward. b. (upper right) Ramsey pattern for the two photon optical transition. The interference fringe is approximately 40 kHz. The solid curve is a theoretical fit. c. (lower right) Blackbody-induced shift vs. temperature^2. Note that the maximum shift is only 1 kHz. Solid line is theoretical value. From ref. 12.

e) Circular states

A few optical photons generally suffice to populate almost any level of any Rydberg atom. However, populating high angular momentum states of a Rydberg atom require many photons, since each photon carries only a single unit of angular momentum. The state with $m = \ell = n-1$ is of particular interest because of its long lifetime (as Table 1 shows, the lifetimes increase as n^5, rather than n^3), large magnetic moment, and relatively small Stark effect. This state, called the "circular" state, has only a single allowed channel for dipole emission.

Figure 13 illustrates the scheme by which a circular state has been populated [13]. The lowest-lying member of a low $|m|$ Stark manifold is populated by conventional means. A series of microwave transitions transfers population to the circular state as the electric field is reduced. The transfer process can be regarded as a single multiphoton transition. Due to small shifts arising from the second order Stark effect, the transitions are forced to go along the single route to the circular state, so that the process is both state specific and highly efficient. Fig. 12b shows field ionization data for the transfer of population in the n = 19 state of lithium. As the process proceeds and the value of m decreases, the ionization peak moves to later times in the time-ramped field ionizer. 99% of the population is transferred.

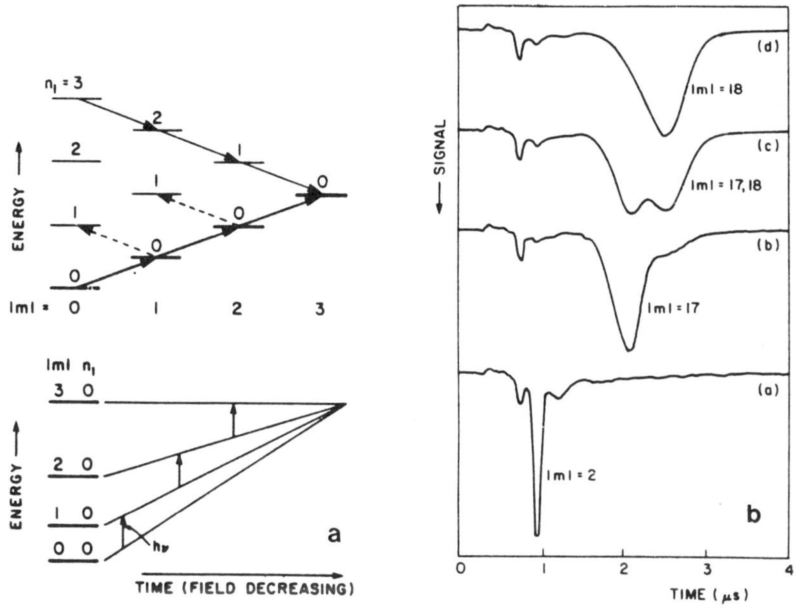

Fig. 13 a. (left) Method for populating the circular state, illustrated for n = 4. An m = 0 Stark state is populated in an electric field, and successive microwave transitions carry the system to m = 3. Because of the second order Stark effect, the transitions occur at slightly different times in the decreasing field (below), forcing the atoms to follow the desired path. b. (right) Field ionization profiles showing the population of the m = 18 level of lithium, n = 19 (curve d). The other curves show intermediate stages of the process, terminated by turning off the microwaves prematurely. From ref. 13.

f. Planetary atoms - a farewell to Rydberg states

Coulombic states are the essence of Rydberg atoms, for the potential energy varies essentially as 1/r in most of the space the electron samples. If two electrons are highly excited, however, the electron-electron interaction is comparable to the electron-core interaction and the system loses all vestiges of simple coulombic behavior. The physics of such atoms, which have come to be called "planetary atoms", represent a new subfield of atomic physics. Dr. Rau discusses elsewhere in this volume some of the unifying principles and new understandings that are emerging in this area.

Just as the alkali-metal atoms are the workhorse of Rydberg atom research, the alkaline-earth atoms are becoming the workhorse for planetary atom research. Fig. 14 shows one approach to the creation of a planetary atom. An nd Rydberg state of Ba is prepared by conventional stepwise excitation from the $6s^2$ ground state. The second 6s electron is then excited to a 7s state by two-photon or by stepwise excitation, as shown in Fig. 14a. As this process occurs, the nd electron of the original core state converts to an n'd state: the 6s-7s transition is essentially a two-electron transition. The two-electron nature is revealed by the data in Fig. 14b. If the nd electron were removed - that is, the initial state were Ba^+ 6s, the 6s-7s transition would be a single line, as shown in the top spectrum. As the value of n is decreased, the interaction between the two electrons becomes more pronounced, and the single line splits and acquires structure.

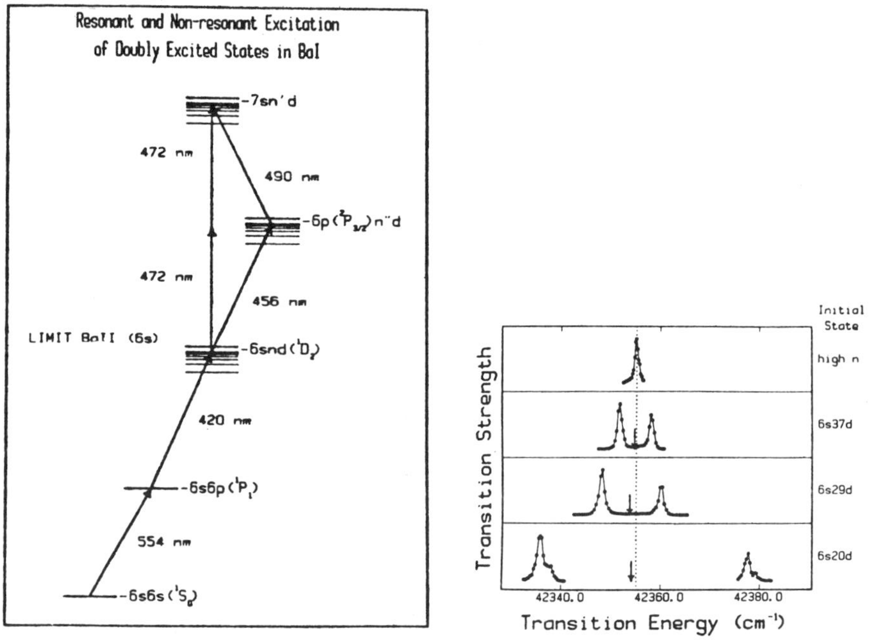

Fig. 14 a. (left) Excitation scheme for populating a doubly excited 7s n'd of barium. b. (right) Evolution of the 6s nd - 7s n'd spectral line as a function of the principal quantum number of the outer electron. From ref. 14.

Although this experiment does not actually achieve the planetary regime - the 7s electron is so tightly bound that it is still in an essentially single-electron state - it reveals the onset of planetary behavior, and points the way to an intriguing new area of atomic physics.

REFERENCES

1. J.J. Balmer, Annalen der Physik und Chemie N.F. 25, 80 (1885). Translated in Atomic Spectra. W.R. Hindermarsh, Pergamon Press, Oxford (1967). This paperback presents in translation many of the classic papers in atomic physics, with commentary.

2. N. Bohr, Philosophical Magazine 26, 1 (1913).

3. Rydberg states of atoms and molecules, eds. R.F. Stebbings and F.B. Dunning, Cambridge University Press, Cambridge (1983).

4. K.A. Safinya, J.F. Delpech, F. Gounand, W. Sandner, and T.F. Gallagher, Phys. Rev. Lett. 47, 6 (1981).

5. M.L. Zimmerman, M.G. Littman, M.M. Kash and D. Kleppner, Phys. Rev. A, 20, 6 (1979).

6. R.R. Freeman and D. Kleppner, Phys. Rev. A 14, 1614 (1976).

7. M.G. Littman, M.M. Kash and D. Kleppner, Phys. Rev. Lett. 41, 103 (1978).

8. S. Feneuille, S. Liberman and A. Taleb, Phys. Rev. Lett. 42, 1402 (1979).

9. C. Fabre, M. Gross, J.M. Raimond and S. Haroche, J. Phys. B 16, L071 (1983).

10. H. Rinnberg, J. Neukammer, G. Jonsson, H. Hieronymus, A. Kong and K. Vietzke, Phys. Rev. Lett. 55, 382 (1985).

11. D. Harmin, Phys. Rev. A 30, 2413 (1984).

12. L. Holberg and J.L. Hall, Phys. Rev. Lett. 53, 230 (1984).

13. R.G. Hulet and D. Kleppner, Phys. Rev. Lett. 51, 1430 (1984).

14. R.M. Jopson, R.R. Freeman, W.E. Cooke and J. Bokor, Phys. Rev. A 29, 3154 (1984).

RYDBERG ATOMS RADIATING IN FREE-SPACE OR IN CAVITIES : NEW SYSTEMS TO TEST ELECTRODYNAMICS AND QUANTUM OPTICS AT AN UNUSUAL SCALE

Serge Haroche

Ecole Normale Supérieure, Laboratoire de Physique
24, rue Lhomond
75231 Paris Cedex 05 - France

Rydberg atoms, characterized by a very strong coupling to the micro-wave and millimeter wave part of the electromagnetic spectrum, exhibit unusual radiative properties in free space or in resonant cavities. The rates of spontaneous emission between Rydberg levels, although small in absolute value, are very large when compared to those of ordinary atoms or molecules radiating in the same frequency range. The rates of transitions induced by external fields impinging on Rydberg atoms are also very large. In particular, blackbody radiation, even at low temperature, has dramatic effects on the Rydberg state lifetimes and thermal radiation dependent energy shifts of these states are observable. Amplification of radiation by Rydberg systems in free space or in cavities leads to the realization of new types of maser devices which operate with very low thresholds. Collective systems of Rydberg atoms radiating in cavities are in fact quasi ideal examples of superradiant sources and their statistical properties (atomic and field fluctuations) are very interesting to study as examples of macroscopic quantum systems or sources of non classical "squeezed" states of radiation.

The Rydberg maser effects are particularly spectacular when the excited atoms are prepared in a resonator of very high Q (superconducting cavity). Maser action can then be observed down to the case when a single atom at a time is present in the resonator. This limiting situation corresponds to the domain of "cavity quantum electrodynamics", a new field of experimental quantum optics in which several experiments have been recently performed or are in progress. Studying the way an isolated atom interacts with a single mode or with a limited number of modes of the electromagnetic field in a cavity provides indeed new and interesting insights in Q.E.D. It allows us to analyze in a realistic situation how the field boundary conditions around an atomic system affect its radiative properties (spontaneous emission enhancement or inhibition depending upon the cavity characteristics, cavity induced alterations of Lambshifts...).

Several review articles have been recently devoted to various aspects of Rydberg atom radiative properties in free space or in cavities[1-5]. In this course, we recall the main results already discussed in these papers and discuss some of the most recent developments in this field. In a first section, we briefly analyze the main radiative properties of Rydberg atoms in free space and mention some simple experiments in which these properties have been investigated. We devote a second section to the problem of a single

Rydberg atom in a cavity, analyzing the various cavity Q.E.D. effects which have been already observed or are presently being searched for. In a third section, we discuss the collective behaviour of Rydberg atoms in cavities and describe the interesting statistical properties of the atomic and field observables and the possible applications of these systems.

RYDBERG ATOMS IN FREE SPACE

Orders of magnitude of electric dipole transitions in Rydberg atoms[6,7]

The transitions connecting a Rydberg level (principal quantum number n ranging from about 10 to a few hundred) to other states of an atomic system can be roughly divided in three groups : (i) those which couple this level to nearby excited states, whose frequencies $\nu_{Rydberg-Rydberg}$ are of the order of R/n^3 (R : Rydberg constant); (ii) those which connect it to much less excited strongly bound states, whose frequencies $\nu_{Rydberg-bound}$ are of the order of R and (iii) those which couple it to the continuum whose frequencies $\nu_{Rydberg-free}$ are of the order of R/n^2. For a typical n value around 30, frequencies (i), (ii) and (iii) fall in the millimeter wave, near U.V. and far infrared domains respectively.

The strength of the coupling of the Rydberg systems with the radiation field is measured by the value of the atomic electric dipole matrix element between the corresponding states (higher order multipole transitions -usually much weaker than allowed dipole transitions- will not be considered here). Transitions (i) connecting states having orbitals of comparable sizes ($\sim a_o n^2$ where a_o is the Bohr radius) and angular momentum eigenvalues ℓ's differing by ± 1 unit have electric dipole moment $\langle n\ell|D|n'\ell\pm 1\rangle$ of the order of $a_o n^2$ which for $n \sim 30$ are about three orders of magnitude larger than the electric dipole matrix elements of ordinary low excited atoms or molecules. The giant size of these dipole moments is the main cause for the unusual radiative behaviour of these systems. Transitions (ii) on the other hand correspond to much weaker couplings. Furthermore, due to angular momentum conservation laws, only low angular momentum Rydberg states (ℓ = 0, 1, 2...) are concerned by these transitions. For these states, the matrix elements $\langle n\ell|D|$ bound$\rangle$ are of the order of $n^{-3/2}$ and the corresponding transition probabilities scale as n^{-3}. High angular momentum states and in particular the circular Rydberg states ($\ell = n-1$) are not coupled by electric dipole emission to ground or low excited states ($\langle n, \ell \sim n|D|$ bound$\rangle = 0$). As for transitions of type (iii), their strength depends upon the overlapping of weakly bound Rydberg orbitals and continuum states above the ionization limit. Photoionization processes corresponding to atomic excitation from a Rydberg state to a continuum band just at the ionization limit have transition rates scaling as $n^{11/3}$, i.e. increasing roughly as the "area" of the Rydberg orbitals[6]. Photoionization processes well above threshold, corresponding to the absorption of optical photon have on the other hand much smaller rates, decreasing as n^{-3} [6]. The above discussion is of course only qualitative and is intended to give a rough estimate of the orders of magnitudes of quantities which have been tabulated precisely with the help of numerical calculation of hydrogen-like wave function integrals[8,9,10].

Spontaneous emission rates

The partial spontaneous emission rate $\Gamma^{(sp)}_{n\ell\to n'\ell'}$ between two $|n\ell\rangle$ and $|n'\ell'\rangle$ levels is readily expressed in terms of the corresponding electric dipole matrix element $\langle n\ell|D|n'\ell'\rangle$ and transition frequency $\omega_{n\ell-n'\ell'}$:

$$\Gamma^{(sp)}_{n\ell\to n'\ell'} = \frac{\max(\ell,\ell')}{2\ell+1} \frac{|\langle n\ell|D|n'\ell'\rangle|^2 \omega^3_{n\ell-n'\ell'}}{3\pi\varepsilon_o \hbar c^3} \qquad (1)$$

For transitions between nearby states, $|\langle n\ell|D|n'\ell'\rangle|^2$ scales as n^4 and $\omega^3_{n\ell-n'\ell'}$ as n^{-9}. $\Gamma^{(sp)}_{n\ell\to n'\ell'}$ then varies as n^{-5}, with a typical order of magnitude 100 s^{-1} for $n \sim n' \sim 30$. This appears to be a small absolute rate, but is indeed quite large compared to the emission rates of ordinary microwave transition with dipoles n^2 time smaller (Γ is then currently in the 10^{-3} to 10^{-9} s^{-1} range).

For type (ii) transitions ($n \to n' \sim 1$), $|\langle n\ell|D|\text{ bound}\rangle|^2$ scales as n^{-3} whereas ω is roughly n independent. $\Gamma^{(sp)}_{n\ell\to\text{bound}}$ then varies as n^{-3}. In order to estimate the total spontaneous decay rate $\Gamma^{(sp)}_{T,n\ell}$ of a given Rydberg level, one has again to distinguish between low ($\ell \ll n$) and high ($\ell \sim n$) angular momentum states. For the former ones, a large number of transitions towards lower lying states are allowed and contribute to the level spontaneous decay. $\Gamma^{(sp)}_T$ is then dominated by the highest frequencies, with an overall rate scaling as n^{-3} :

$$\Gamma^{(sp)}_{T,n,\ell\ll n} \sim C\, n^{-3} \qquad (2)$$

with C of the order of 10^9 s^{-1}. For $n \sim 30$, $\ell = 0, 1, 2...$, these, Rydberg levels have a life time in the 10 to 100 μs range.

Large angular momentum states on the other hand are coupled only to lower states with $\ell' = \ell \pm 1$, i.e. $n' \sim n$. Only low frequency channels (type i transitions) are allowed and the life time of these states is much longer. One then has :

$$\Gamma^{(sp)}_{T,n\ell\sim n} \sim A\, n^{-5} \qquad (3)$$

with again $A \sim 10^9$ s^{-1}. The corresponding decay time falls in the 10^{-2} sec range for $n \sim 30$.

Formulas (2) and (3) -as most of the results presented here- have very simple interpretation in terms of classical orbits. The radiative power of a "classical" Rydberg electron is proportional to the square of its acceleration. For low angular momentum orbits (eccentric ones), the acceleration is maximum at the perihelion and is almost independent of the orbit energy. These levels thus mainly emit high energy photons at a rate proportional to the rate of passage at the perihelion $1/T_o$ (T_o is the orbit period). From the third Kepler law, T_o scales as $(a_o n^2)^{+3/2}$. One thus finds $\Gamma \sim n^{-3}$, in agreement with equ. (2).

For high ℓ "circular" orbits on the other hand, the acceleration is almost constant and proportional to n^{-4} (inverse square of orbit size). The emitted power thus scales as n^{-8}. The total energy to be emitted in one quantum jump being R/n^3, the characteristic time of emission scales as $n^{-3}/n^{-8} = n^5$ and thus the spontaneous emission rate varies as n^{-5} in agreement with equ. (3).

Observation of spontaneous emission in Rydberg states are numerous either in astrophysical[11] or in laboratory environment. In the lab, it is much easier to prepare and study the decay of low angular momentum states [4,12,13,14]. A large numer of such states in various alkalis have been studied and their decay monitored either by detecting their fluorescence or by counting as a function of time the number of Rydberg atoms remaining in the initially prepared level (using the field ionization method). Similar experiments are now in progress for high angular (circular) states[15]. In order to extract from these measurements the spontaneous rates, one has to make negligible the competing process of blackbody field induced transitions that we now study.

Blackbody induced rates in Rydberg levels[4]

We have so far considered the coupling of Rydberg atoms to the empty modes of the electromagnetic field (vacuum). At room temperature and even below, the low frequency modes of the field which are strongly coupled to Rydberg transitions are thermally excited (blackbody radiation). As a result, absorption and induced emission by resonant blackbody photons are important processes in Rydberg levels. Energy level shifts produced by off-resonant blackbody field components are also measurable quantities in these states. In this subsection, we describe the former effect and leave the discussion of blackbody induced shifts for next paragraph.

Let us recall briefly the main features of the blackbody radiation in equilibrium at temperature T in a large multimode cavity having a frequency mode distribution per unit volume $\mathcal{N}(\omega) = \omega^2/\pi^2 c^3$. The probability distribution $P(n_\phi)$ of having n_ϕ photons in a mode of frequency ω is given by the Bose-Einstein law :

$$P(n_\phi) = (1 - \beta)\,\beta^{n_\phi} \qquad \text{with } \beta = \exp - \hbar\omega/k_B T \tag{4}$$

and the average number of photons in the mode is :

$$\overline{n_\phi} = [\exp \hbar\omega/k_B T - 1]^{-1} \tag{5}$$

which, for $k_B T >> \hbar\omega$ reduces to the Rayleigh-Jeans limit :

$$\overline{n_\phi} \sim k_B T / \hbar\omega \tag{6}$$

The energy spectral density of this blackbody field is (Planck law) :

$$u(\omega) = \overline{n_\phi}\,\hbar\omega\,\mathcal{N}(\omega) = \frac{\hbar}{\pi^2 c^3}\,\frac{\omega^3}{\exp\frac{\hbar\omega}{k_B T} - 1} \tag{7}$$

which, upon integration over ω, yields the Stefan law for the total blackbody energy density :

$$u_{Total}(T) = \frac{3\sigma}{c}\,T^4 \qquad \text{with } \sigma = \frac{2\pi^5}{15}\,\frac{k_B^4}{c^2 h^3} \tag{8}$$

The Planck distribution peaks at a frequency ω^{BB}_{max} proportional to T ($\omega^{BB}_{max} \sim 10^{14}$ Hz at room temperature) and the Rayleigh Jeans limit corresponds to the low frequency tail ($\omega < \omega^{BB}_{max}$) of this distribution. For usual Rydberg transitions (n $\sim$ 10 to infinity), the resonant components of the blackbody field (T $\lesssim$ 1000 K) correspond to this low frequency tail ($\omega_{Rydberg} \lesssim \omega^{BB}_{max}$). Accordingly, most of the blackbody field energy is carried by field components whose frequencies are higher than the characteristic Rydberg frequencies. This property will be relevant for the discussion in the next subsection.

Given two atomic levels $|e>$ and $|g>$ ($E_e > E_g$), it is well known that the blackbody absorption rate $W^{BB}_{g\to e}$ and induced emission rate $W^{BB}_{e\to g}$ between them are :

$$W^{BB}_{g\to e} = W^{BB}_{e\to g} = \bar{n}_\phi \, \Gamma_{e\to g} \qquad (9)$$

where $\Gamma_{e\to g}$ is the partial spontaneous emission rate between these levels and $\bar{n}_\phi$ is the average number of blackbody photons per mode at the transition frequency. The total (spontaneous plus induced) rate of radiative transition between these levels in the presence of a thermal field is thus :

$$W^{total}_{e\to g} = (\bar{n}_\phi+1)\,\Gamma_{e\to g}; \quad W^{total}_{g\to e} = \bar{n}_\phi \, \Gamma_{e\to g} \qquad (10)$$

The above results, valid for non degenerate states can be easily modified to account for level degeneracies. They can be simply derived from the coupling of the atomic system with an ensemble of harmonic oscillator modes in thermal equilibrium[1]. Eqs. (10) are of course a central and well known result of thermal radiation theory (the rates in these equations are the ones which are required to put the atomic system in thermal equilibrium at the temperature of the field).

Clearly, blackbody fields (around room temperature) do not affect the high frequency optical transitions connecting Rydberg levels to bound states (because $\bar{n}_\phi \sim 0$ for these frequencies). On the other hand, the microwave emission transitions towards nearby Rydberg levels are speeded up by a factor $\bar{n}_\phi \sim k_B T/\hbar\omega$. In addition, significant absorption processes towards more excited states and towards the continuum (blackbody photoionization) are induced by the thermal fields. In order to discuss qualitatively the sensitivity of Rydberg atoms to these processes, it is once more convenient to distinguish between low and high angular momentum states.

Since high ℓ states are coupled only to low frequency transitions, they are particularly sensitive to blackbody fields as soon as $\bar{n}_\phi \gtrsim 1$, i.e. $k_B T \gtrsim \hbar\omega$ which corresponds to T $\gtrsim$ a few °K for n $\sim$ 30. In order to study pure spontaneous radiative effects on these levels, it is thus required to carefully cool the atomic surroundings at a temperature of the order of a liquid helium bath. In low ℓ states, on the other hand, blackbody effects on low frequency transitions compete with spontaneous decay on high frequency channels. The n-scaling law of low frequency blackbody transition rates is easily obtained from equations (9) and (1) :

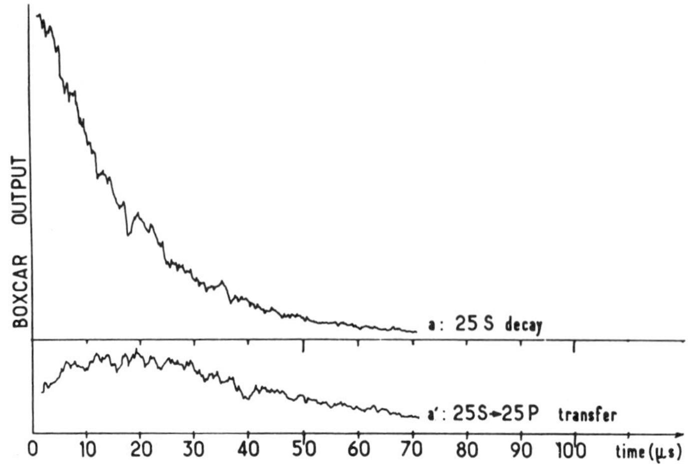

Figure 1 : Field ionization signals following the pulsed excitation of the 25S level in Na and showing : a) the decay of the 25S state, b) the 25S → 25P blackbody induced transfer at room temperature (from ref.[20]).

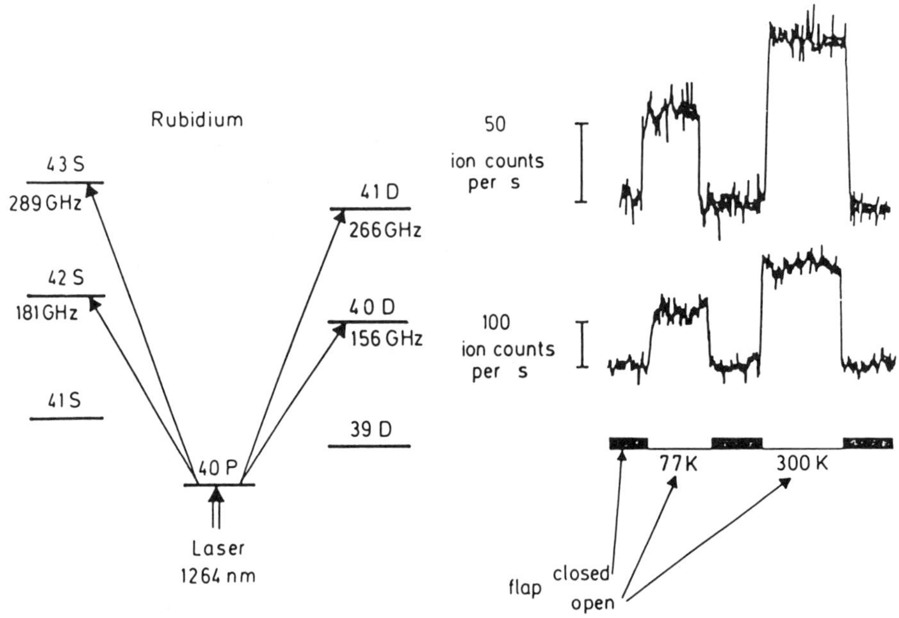

Figure 2 : Blackbody induced transition in Rubidium Rydberg levels. Left : Rydberg states relevant for the interaction with the blackbody radiation. The p → d transition probabilities are about a hundred times smaller than the p → s ones. Right : field ionization signal of the $43S_{1/2}$ (upper trace) and $42S_{1/2}$ level (lower trace) at various temperatures (77K and 300K when flaps are open; 4K when flaps are closed) (from ref.[5]).

$$W^{BB}_{e\to g} \sim \bar{n}_\phi \, \Gamma_{g\to e} \sim \frac{k_B T}{\hbar\omega} \times \omega^3 \times |< n\ell|D|n'\ell' >|^2 \sim n^{-2} \qquad (11)$$

whereas $\Gamma^{(sp)}_{T,e}$ scales as n^{-3}. The ratio $W^{BB}_{e\to g} / \Gamma^{(sp)}_{T,e}$ is thus roughly proportional to n and blackbody related effects will necessarily become significant above a given n value (practically around $n \sim 20$ at room temperature for alkali atoms). Many experiments have been performed to study these blackbody induced rates on low ℓ Rydberg states[4,5,16-20]. In one kind of experiments, one excites a Rydberg level by a short pulse of light and monitor the subsequent evolution of this level along with the one of adjacent states populated by blackbody induced transfers. Figure 1 presents the decay of the 25S state of Na prepared in this way (trace a) with the corresponding evolution of the 25p state population reached by 25s → 25p blackbody absorption at ∿ 230 GHz. This experiment[20] was performed at T=300K. Figure 2 puts in evidence the temperature dependance of these blackbody transfers as reported in ref.[5] : the laser excitation is this time continuous. CW diode lasers were used in order to prepare by stepwise process the 40p level of Rubidium. The steady state population of the 42S and 43S levels reached by blackbody absorption were monitored by field ionization of these Rydberg states (lower and upper traces on the right side of the figure). The experiment is performed in a time sequence. The atomic beam is placed in a liquid Helium cooled shield (4°K) in which a window allows to inject external blackbody radiation either at room temperature (300K) or at liquid nitrogen temperature (77K). Using appropriate flaps, one can alternate the beam surrounding temperature and thus compare the effects of the 4K, 77K and 300K radiation on the atoms. Note the increase in the ion counts from 77K to 300K.

From such experiments performed at various temperatures, one can deduce by extrapolating at 0K the spontaneous emission rates of the Rydberg transition. One can also study the sensitivity of the various transitions to field components at different frequencies, in order to assess the possibility of developing performant Rydberg atom detectors for millimeter wave radiation.

Blackbody induced shifts in Rydberg atoms[4,21,22]

The off-resonant components of the blackbody field induce virtual transition in Rydberg states whose effect is to slightly shift their energies. These shifts are thus temperature dependent. At room temperature and above, the main part of the blackbody spectrum has its frequency very large compared to the revolution frequency of the Rydberg electron. It can then be shown[23] that the main effect of the blackbody field is to superimpose to the electron motion around its orbit a fast random vibration of small amplitude, induced by the thermal field fluctuations. The energy shift of the Rydberg state corresponds to the extra potential and kinetic energies associated to this vibration. The change in potential energy can be considered as a temperature dependent modification of the quantum electrodynamical shift of the level, since it has an interpretation very close to the Welton picture of the Lamb shift[24]. At T=0K, the Lamb shift is indeed due to the change in the electronic average potential energy produced by the vacuum field induced electronic vibration. At a finite temperature, the vibration produced by the added thermal fluctuations combine its effect with the vacuum field ones. The change in potential energy produced by this vibration is merely proportional to $< \Delta V >$, the average value of the Laplacian of the hydrogenic potential seen by the Rydberg electron. Since ΔV is proportional to $\delta(\vec{r})$, this term affects only S states. It is furthermore proportional to $1/n^3$ and is very small (at most a few Hz in Rydberg levels). Much more important is the change in kinetic energy of the Rydberg electron due to the thermal field induced vibration. This change is very easy to compute "classically", in the limit where the electron is quasi-free (blackbody frequencies very high compared to Rydberg frequencies). Let us call x_ω the vibration amplitude due to a component $\mathcal{E}(\omega)$ of the blackbody field at frequency ω, along direction Ox. Since $\omega \gg \omega_{\text{Rydberg}}$, one can compute x_ω as if the electron were free :

$$x_\omega = \frac{q\,\mathcal{E}(\omega)}{m\omega^2}\, e^{-i\omega t} \tag{12}$$

The corresponding kinetic energy is then :

$$\mathcal{T}(\omega) = \frac{1}{2}\, m\omega^2\, x^2_\omega = \frac{q^2\,\mathcal{E}^2(\omega)}{2m\omega^2} \tag{13}$$

$\mathcal{E}^2(\omega)$ is related to the average number of photons in the mode by :

$$\varepsilon_o\, \mathcal{E}^2(\omega)\, \mathcal{V} = (\, \bar{n}_\phi + \frac{1}{2}\,)\, \hbar\omega \tag{14}$$

where $\mathcal{V}$ is the volume of the cavity. Hence :

$$\mathcal{T}(\omega) = \frac{q^2\hbar}{2m\varepsilon_o \mathcal{V}\,\omega}\,(\, \bar{n}_\phi + \frac{1}{2}\,) \tag{15}$$

and the total kinetic energy is obtained by adding independently the contributions of all the field modes :

$$\mathcal{T} = \int \mathcal{V}\, \mathcal{T}(\omega)\, \mathcal{N}(\omega)\, d\omega \tag{16}$$

The contribution of the 1/2 term of equ.(15) to the integral (16) is temperature independent and corresponds to a term in the free-electron mass renormalization. It can be omitted in this calculation. The other term is readily estimated by making the variable change $u = \hbar\omega/k_B T$. One gets :

$$\mathcal{T} = (\, \frac{k_B T}{\hbar}\,)^2 \int \frac{u\,du}{e^u - 1} = \frac{\pi}{6}\, \frac{\alpha^3}{R}\, (\, k_B\, T\,)^2 \tag{17}$$

All Rydberg levels thus experience a positive energy shift, independent of the level, proportional to the square of the absolute temperature. This shift is equal to 2.4 kHz at room temperature (T=300K). This shift cannot be measured on transitions between Rydberg states shifted by exactly the same amount. It can on the other hand be observed on optical transitions connecting Rydberg levels to strongly bound states, because these states experience a much smaller shift under the blackbody irradiation, as can be shown by the following simple argument. Let us consider as an example the case of Hydrogen ground state. The characteristic electron frequency of this state ($\nu \sim R/h$) is very high compared to the average blackbody frequencies. Hence an approximation opposite to the one made above is now justified : each component of the thermal field can be considered as a quasi-static field producing a quadratic Stark shift of the level :

$$\Delta E(\omega) = -\frac{1}{2}\, \alpha(0)\, \varepsilon_o\, \mathcal{E}^2(\omega) \tag{18}$$

where $\alpha(0) = 18\,\pi a_o^3$ is the static polarizability of hydrogen 1s state. Summing over the field modes, one now gets :

$$\Delta E = -9\,\pi a_o^3 \int (\, \bar{n}_\phi + \frac{1}{2}\,)\, \hbar\omega\, \mathcal{N}(\omega)\, d\omega \tag{19}$$

The 1/2 term in the integral above is a contribution to the ordinary Lamb shift of the ground state (corresponding to the low frequency components of the vacuum field) and can be omitted. The $\overline{n}_\phi$ contribution yields an integral which is nothing but the blackbody energy density at temperature T. With the help of equ.(8), we get :

$$\Delta E = - 9\pi a_0^3 \frac{3\sigma}{c} T^4 \qquad (20)$$

corresponding to -0.025Hz at room temperature.[25] The shift -at least for T not too high- can be neglected compared to the Rydberg state ones and the shift of the Rydberg to bound states optical transitions is almost entirely due to the Rydberg level. Let us note however that the ground state shift, which has a sign opposite to the excited ones, will eventually become significant and even dominant at high temperatures ($T \sim 10^4$K). Then the quasistatic approximation made above is no longer valid and a more detailed calculation has to be done. An exact calculation of these shifts -for the ground states as well as the Rydberg ones- can of course be performed whatever the temperature is by usual perturbation methods[22]. For the domain of temperature we have restricted our above analysis, these exact calculations give result in good agreement with our classical and simple pictures which have the merit of giving a very illuminating interpretation of these shifts.

The temperature dependence of the frequencies of transitions connecting Rydberg states to ground state in Rubidium has recently been observed[26] in very high precision two photon spectroscopy experiments having a resolution of 1 part in 10^{13} ! It is remarkable that Rydberg state spectroscopy has now reached a precision such that one has to worry about the temperature of the atomic environment (even when it is as low as room temperature) to account for the measured frequencies !

Collective emission of Rydberg atoms in free space[27]

Let us now turn to the description of collective radiative effects involving Rydberg atoms. If a large enough number of atoms is initially prepared in a Rydberg level within a small volume, the field radiated by the atoms reacts back on them and a collective radiative de-excitation process becomes observable. This is of course a phenomenon common to all excited atomic systems, known as superfluorescence[28,29,30]. In the case of Rydberg atoms, the unusually strong atom-field coupling makes the threshold of these collective effects to correspond to a relatively small number of radiators. In order to observe these effects, one initially prepares a sample of atoms in a given $n\ell$ state and monitors the population transfer towards lower $n'\ell'$ levels as a function of time[27]. The rate of this transfer becomes much faster than the ordinary spontaneous emission rate when the number N of atoms in the sample is larger than a well defined threshold. The detailed analyzis of the emission process for a sample of atoms radiating in free space is rather intricate[29,30]. It depends upon the actual size and shape of the medium and upon the precise structure of its energy levels. The emission is triggered in an initial phase by vacuum and/or thermal field fluctuations[31] which build up an initial macroscopic atomic polarization in the system. This polarization then acts as a source for the field which propagates across the sample and escapes through a diffraction pattern depending upon the medium size and shape. The superfluorescent emission can occur simultaneously or successively on several competing or cascading transitions along the Rydberg energy scale. The analysis of the effect thus requires to take into account the multilevel configuration of the Rydberg system. Superfluorescence has been observed on a large number of millimetre wave and infrared transitions[32]. We will not describe here the theory of the phenomenon in any detail because, as we will show in the last section below, much simpler and entirely calculable collective effects of the same kind occur when the atomic sample radiates in a cavity resonant with a Rydberg transition. Let us only derive here some very basic orders of magnitude considerations in order to estimate the free space superfluorescence threshold and the characteristic evolution time of the phenomenon (as we are interested here only in rough estimates, we do not bother to keep track of numerical factors such as 2π etc.) The electric field produced at frequency $\nu = c/\lambda$ by an atomic dipole $\langle D \rangle_{n\ell,\, n'\ell'}$ at distance r has a long range radiated part decreasing as r^{-1} (for $r > \lambda$) :

$$E_{rad} \sim \frac{1}{4\pi\varepsilon_o} \langle D \rangle_{n\ell \to n'\ell'} \frac{1}{\lambda^2 r} \tag{21}$$

and a short range electrostatic part varying as r^{-3} :

$$E_{stat} \sim \frac{1}{4\pi\varepsilon_o} \langle D \rangle_{n\ell \to n'\ell'} \frac{1}{r^3} \tag{22}$$

The collective emission of the sample (assumed to have at least one dimension $\ell \gg \lambda$) can be described to first approximation as resulting from the evolution of each atomic dipole in the long-range field coherently radiated by the N atoms of the sample. For the simple "needle" geometry (sample of length $\ell \gg \lambda$ and of transverse dimensions $a \ll \ell$), one can write :

$$E_{rad}^{(coherent)} \sim \frac{1}{4\pi\varepsilon_o} N \langle D \rangle_{n\ell,\, n'\ell'} \frac{1}{\lambda^2 \ell} \tag{23}$$

The characteristic superfluorescent emission rate T_{SR}^{-1} has a magnitude which is of the order of the Rabi nutation frequency $\dfrac{E_{rad}^{(coherent)} \cdot \langle D \rangle_{n\ell - n'\ell'}}{\hbar}$ in this field :

$$T_{SR}^{-1} \sim \frac{1}{4\pi\varepsilon_o \hbar} N \left|\langle D \rangle_{n\ell,n'\ell'}\right|^2 \frac{1}{\lambda^2 \ell} \tag{24}$$

Comparing this rate to the spontaneous emission rate on the same transition (equ. 1), one gets :

$$T_{SR}^{-1} \sim N \mu \; \Gamma_{n\ell \to n'\ell'}^{(sp)} \tag{25}$$

which means that superfluorescence corresponds to an increase of the spontaneous emission rate by a factor of N time a dimensionless geometrical factor μ which, in the case of a needle shape medium, is of the order of λ/ℓ.

Superfluorescence will be observable if T_{SR}^{-1} is larger than the reciprocal T_{rel}^{-1} of the shortest relaxation time of the atomic dipoles. Usually, for $n \sim 30$, T_{rel} is of the order of a few microseconds (it corresponds to the spontaneous emission time along the Rydberg to bound transitions or to the transit time across the millimetre dimension of the sample). Since $\Gamma_{n\ell \to n'\ell'}^{(sp)}$ is typically of the order of 10^{-2} sec in this range of n values, the threshold corresponds to $N\,{}^{\lambda}/\ell \sim 10^4$, which for $\lambda/\ell \sim 10^{-1}$, yields $N \sim 10^5$. This typical threshold is only a rough estimate which can widely change depending upon the size and shape of the medium and the transition under consideration.

We have so far neglected in this qualitative analysis the effect of the short range atom-atom coupling, mediated by the E_{stat} field contribution. This term describes the Van der Waals Rydberg-Rydberg interaction which strongly depends upon the atom distribution throughout the sample and can be analyzed as a cause of inhomogeneous broadening of the system. A rough order of magnitude estimate leads to replace $1/r^3$ by N/V (V : volume of the sample) which yields an "average" value :

$$< E_{stat} > = \frac{1}{4\pi\varepsilon_o} \frac{N}{V} < D >_{n\ell-n'\ell'} \qquad (26)$$

The inhomogeneous broadening due to this field tends to dephase the atomic dipole at a rate :

$$\Delta_{inhom} = \frac{< D >_{n\ell-n'\ell'} E_{stat}}{\hbar} \sim \frac{1}{4\pi\varepsilon_o \hbar} |< D >_{n\ell,n'\ell'}|^2 \frac{N}{V} \qquad (27)$$

In the competing process between superfluorescence and Van der Waals dephasing, the former will win if :

$$\Delta_{inhom} \lesssim T_R^{-1} \qquad (28)$$

This is a relation independent of the number of atoms which means that the sample volume V must not be too small if superfluorescence is to be observed. In the case of a needle-shaped volume, one can proceed a bit further in this simple qualitative analyzis. One then has $V = \ell a^2$ and equ.(28) then simply leads to :

$$a \gtrsim \lambda \qquad (29)$$

(a more detailed calculation yields for a needle to the somewhat less restrictive result $a \gtrsim 0.3\ \lambda$ 33). In other words, the transverse dimensions of a needle-shaped superfluorescent medium should not be smaller than a fraction of the emission wavelength. The effect of superfluorescence quenching in samples smaller than λ which results from a competition between long range dipole ordering and short range Van der Waals dipole dephasing was predicted long ago in superfluorescence theories[34]. It has been observed in Rydberg atom millimetre wave studies[35] where it is much easier to realize the condition $a < \lambda$ than in the optical domain.

SINGLE RYDBERG ATOM IN A RESONATOR. CAVITY Q.E.D. EFFECTS[1,2,3,5]

The radiative effects discussed in the previous section resulted from the interaction of Rydberg atoms with the continuum of field modes in free space. Usually, this continuum is described as the superposition of the eigenmodes of a "virtual" cavity whose dimensions are very large compared to the wavelengths of interest, the dimension of this cavity being eventually increased to infinity at the end of the calculations. We analyze in this section effects observable when the Rydberg atom is placed in a real cavity whose size is comparable to λ, thus drastically restricting the number of modes to which the atomic system is coupled. This situation, corresponding to what we will call "cavity quantum electrodynamics", is easy to realize with Rydberg atoms since λ is then of the order of a millimeter or larger and the atom to field coupling is intrinsically very large (It is of course next to impossible to realize with ordinary atoms radiating on optical transitions for which λ is much too small. Microwave transitions on ordinary atoms are on the other hand much too weak to allow the single atom-cavity coupling to become effective during any realistic experimental time). New and interesting effects are then observable with Rydberg atoms, especially when the cavity is exactly resonant with a Rydberg transition (single photon nutation effects, spontaneous emission changes induced by the cavity...).

We discuss in this section the effects produced by the cavity on a single atomic radiator. We will see in the next paragraph that the coupling with a cavity has also very important and simplifying effects on the radiative properties of a collection of Rydberg atoms.

Rydberg atom in a resonant cavity : single photon nutation effects and spontaneous emission enhancement

A single Rydberg atom motionless in a cavity resonant with a transition connecting two adjacing states $|e>$ and $|g>$ is equivalent to a two-level spin 1/2 system coupled to a harmonic oscillator (Jaynes-Cummins model[36]). This approximation is valid as long as one restricts the system evolution to a time scale short compared to the spontaneous decays towards bound states (within this time scale -usually a few microsecond- one can indeed neglect the coupling of levels $|e>$ and $|g>$ to other atomic states).

The evolution of such a system is quite easy to describe, especially if the relaxation of the cavity field is also neglected. Calling J_+ and J_-, a^+ and a the atomic and field excitation and deexcitation operators respectively ($J_+|g> = |e>$; $J_-|e> = |g>$; $a^+|n> = \sqrt{n+1}\,|n+1>$; $a|n> = \sqrt{n}\cdot|n-1>$), one can write the atom-field interaction as :

$$H_I = \hbar\Omega\left[aJ_+ + a^+J_-\right] \tag{30}$$

where :

$$\Omega = \frac{<e|D|g>}{\hbar}\cdot\sqrt{\frac{\hbar\omega}{2\varepsilon_o \mathcal{V}}} \tag{31}$$

is the atom-field coupling constant ($\mathcal{V}$: volume of cavity, usually a few cubic millimetres for Rydberg levels around $n \sim 30$). In order to establish equ.(30), one makes as usual the electric dipole and rotating wave approximations.

If one starts from an initially excited atom in an empty cavity ($|\psi(0)> = |e, 0>$), the system oscillates under the action of H_I from state $|e, 0>$ to state $|g, 1>$ and back at rate Ω. This self-induced Rabi oscillation corresponds to the elementary exchange of a single photon between the atom and the cavity. Typical Ω values for Rydberg systems around $n \sim 30$ and millimetre size cavities are in the range 10^3 to 10^4 Hz.

In a real situation, the atom is not motionless and the cavity is not lossless. The Rabi oscillation then competes with the atom escape process which has a characteristic time :

$$T_{at} \sim \lambda / v \tag{32}$$

where v is the atomic velocity. It also competes with the photon escape process which occurs in a characteristic time :

$$T_{cav} \sim \frac{Q}{\omega} = \frac{\lambda Q}{c} \tag{33}$$

where Q is the cavity quality factor.

It is poosible to realize very good low order mode superconducting cavities and to achieve the condition $T_{cav} > \Omega^{-1}$ and $T_{cav} > T_{at}$[37]. In this case, many Rabi oscillations can occur before the atom or the photon leaves the cavity and the atom escapes from the cavity before the photon does. The interaction H_I is then decoupled after a time T_{at}, leaving the atom-cavity system in a correlated state of the form $\alpha|e,0> + \beta|g,1>$. In other words, there is a probability $|\beta|^2$ that the atom has left a photon behind in the cavity. Another atom, entering later in the cavity, will then interact with this excited mode, with a probability of increasing the field to two photons, and so on... This is the principle of a single atom continuous Rydberg maser recently realized by Meschede et al[37]. Although the single photon Rabi oscillation is not directly observed in this device, the operation of this system gives indeed an indirect proof of the existence of this basic quantum mechanical two level effect.

The other interesting limiting case of atom cavity interaction corresponds to $T_{cav} << \Omega^{-1}$ (cavity with moderate Q). Then, the atom field correlation cannot develop and the Rabi oscillation is overdamped. The "broad" cavity mode can in this case be considered as a field continuum of width ω/Q to which the two-level atomic system is coupled. The excited state $|e>$ decay process towards this continuum is then irreversible and exponential, with a rate $\Gamma_e^{(sp,cav)}$ given by a Fermi-golden rule argument

$$\Gamma_e^{(sp,cav)} = \frac{2\pi}{\hbar} |< e\, 0|H_I|g,1 >|^2 \times \frac{2}{\pi}\frac{Q}{\omega}\frac{1}{\hbar}$$

$$= \frac{3}{4\pi^2}\ \frac{Q\lambda^3}{\mathcal{V}}\ \Gamma_{e\to g}^{(sp)} \tag{34}$$

The emission rate is thus enhanced in the cavity by a large factor $(3/4\pi^2)(Q\lambda^3/\mathcal{V})$, a result first derived in a different context by Purcell[38]. This effect can be interpreted as due to the change of the field mode spectral density surrounding the atom from its free space value $\mathcal{N}(\omega)=\omega^2/\pi^2c^3$ to the cavity value $\mathcal{N}_{cav}=(2/\pi)(1/\mathcal{V})(Q/\omega)$. (In fact, the enhancement factor is larger than $\mathcal{N}_{cav}/\mathcal{N}$ by an extra factor 3, due to photon polarization effect). The spontaneous emission enhancement effect can also be viewed as resulting from a kind of collective decay of the atom radiating in phase with its images in the cavity walls (see below).

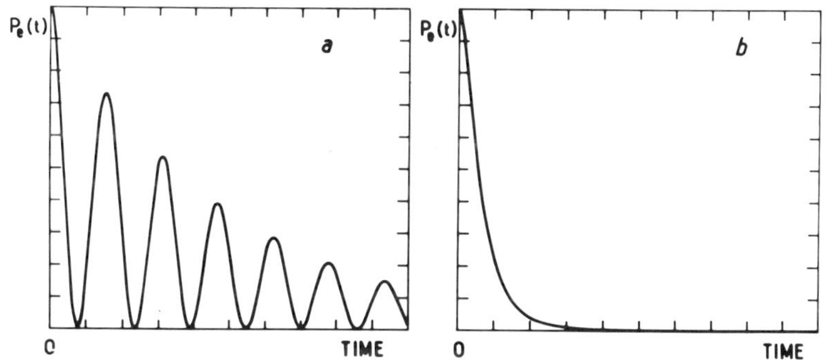

Figure 3 : The two regimes of atomic evolution in a resonant cavity a) Probability $P_e(t)$ of finding the atom in initially excited state at time t for a very good cavity ($\omega/Q = 0.2\Omega$). b) $P_e(t)$ for a moderate Q cavity ($\omega/Q = 5\Omega$). The time unit for both curves is equal to $2/\Omega$.

The above discussion is summarized on Figure 3, in which the two regimes of atomic evolution in a resonant cavity are presented as they result from a simple calculation including cavity damping processes[1,2]. Figure 3-a shows the probability $P_e(t)$ of finding the atom in its initially excited state, after a delay t following its preparation, in the case of a very good cavity ($\omega/Q = 0.2\ \Omega$). Figure 3-b presents the same quantity for a moderate Q cavity ($\omega/Q = 5\ \Omega$). The free-space transition rate on this transition is very small and, at the time scale of this figure, $P_e(t)$ basically remains equal to one in the absence of cavity around the atom. These results are of course altered by the presence of thermal photons in the cavity mode[1,2,39] so that the observation of pure spontaneous emission cavity induced effects require careful thermal shielding around $T \sim 0°K$ (in order to realize the condition $k_BT/\hbar\omega << 1$).

For the direct observation of the oscillatory regime (fig. 3-a), one must be able to control the cavity mode atom interaction time and to vary it continuously. Experiments in which this is achieved by velocity selection of the atomic beam are now in progress[40]. We show below that basically the same oscillatory regime is much easier to observe -- and has actually been observed--on collective atomic systems prepared in resonant cavities. The enhanced spontaneous emission regime in a moderate Q cavity (fig. 3-b) has recently been observed by Goy et al[41]. Figure 4-a presents

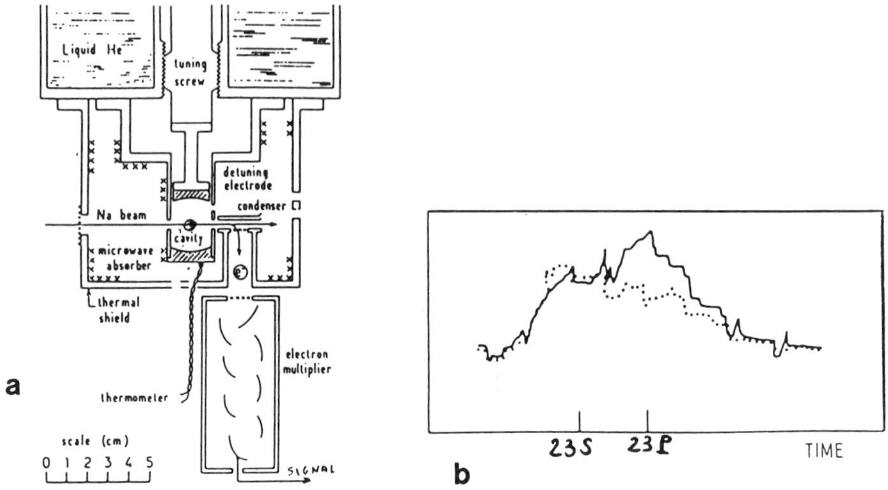

Figure 4 : Enhancement of spontaneous emission rate in a resonant cavity : a) experimental set-up; b) time resolved ionization signals showing the transfer from 23S → 23P Rydberg states when cavity is resonant (full line curve). See text for more details.

the set up used in this experiment. The cavity was of the Perot-Fabry type, operating at 147 GHz. Two superconducting Niobium mirrors in a confocal configuration are cooled to 4K by a liquid Helium bath to avoid blackbody induced effects. The cavity Q is $\sim 10^6$. Its tuning is achieved by mirror displacement. The Na atoms are excited in the $23S_{1/2}$ state by pulsed laser pumping inside the cavity (the laser beam is perpendicular to the figure plane, the region of interaction with the atomic beam is shown by the dot in the cavity center). One atom at a time is prepared on average by proper attenuation of the laser beam and the final atomic state is detected by field ionization in the condenser outside the cavity, the resulting electrons being monitored by the electron multiplier. The detection process[1,2,41] is such that Rydberg levels with different binding energies are detected at different time, the delay being an increasing function of the binding energy. The cavity induced 23S → 23P spontaneous emission transfer at 147 GHz is revealed as a change in the average time of arrival of the electrons resulting from the Rydberg atom ionization. Figure 4-b shows the signal alteration from the case where the cavity is off resonant (dotted line) to the resonant one (full line). From these signals, a cavity induced enhancement of spontaneous emission by a factor 530 was measured, in good agreement with theory.

Rydberg atoms in an off resonant cavity : spontaneous inhibition effects[42]

Let us consider now a situation where the Rydberg atom is placed in a cavity off resonant with the Rydberg atomic transition. The system evolution is then also very easy to analyze : the atom cannot radiate at all along this transition and the corresponding spontaneous emission process must be inhibited[42]. Such an effect can be observed only on circular Rydberg levels, since any cavity of real size cannot cut-off the optical modes contributing to the decay of low angular momentum levels. An experiment resulting in the observation of this inhibition effect on circular Rydberg states of Cesium has just been performed[43]. It should be noted that a quite similar spontaneous emission inhibition effect has been observed in the radiative damping of the cyclotron motion of an electron in a Penning trap[44,45]. In this case, the electrodes of the trap constituted a cavity changing the mode spectral density of the radiation field around the atoms.

Before concluding this section, let us recall a very simple classical interpretation of the spontaneous emission enhancement and inhibition effects, in the case of a simple cavity geometry. Assume that an electric dipole is radiating between two parallel conducting plates. The field in the gap between the plates results from the interference between the atomic dipole radiation and the field radiated by its electric images induced in the plates. This field has $1/r$, $1/r^2$ and $1/r^3$ contributions. It is a very simple text book problem to compute the exact field radiated back on the dipole by its successive images in the cavity walls. This calculation shows that the radiated contribution has, at the dipole location, a component $\pi/2$ out of phase with the dipole oscillation. This component induces a force on the dipole charge whose work either adds up to or subtracts from the self reaction work of the dipole on itself (depending

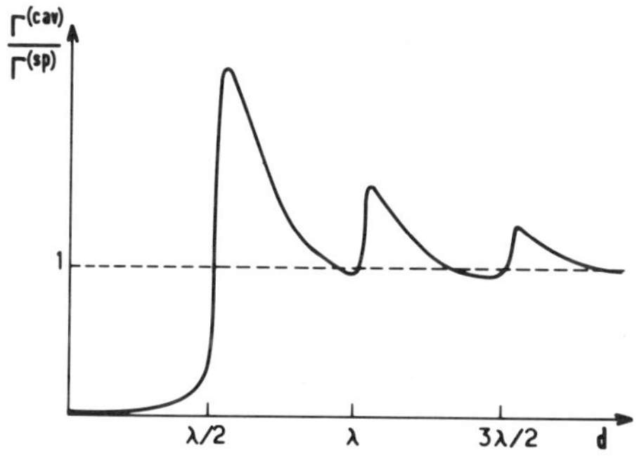

Figure 5 : Radiative damping rate of an electric dipole placed between two conducting plates parallel to the dipole as a function of the plate separation d (see text for more details).

upon the plate separation). The total atomic decay rate is obtained by algebraically adding the power dissipated by the self reaction field (responsible for free space spontaneous emission) and the one of the image dipoles. Losses in the conducting plates are accounted for by multiplying by a factor smaller than one the contribution of the successive images. The results of such a calculation are shown on Figure 5 which represents the variation of the dipole radiative damping rate as a function of plate separation d (the dipole is parallel to the plates, at equal distances from them. The plates reflectivity is 0.9, corresponding to an intrinsically moderate Q). It is clear that when d is smaller than $\lambda/2$, the dipole decay is almost totally inhibited. If $d = \lambda/2$ or $d \cong (2n+1)\,\lambda/2$, the decay rate is on the other hand enhanced (resonant cavity structure). The enhancement factor is not large in this case, because the parallel plate geometry corresponds to a multimode cavity different from the single mode cofocal Fabry-Perot structure considered above. This simple model allows us also to understand -at least qualitatively- the oscillatory regime discussed above for high Q cavity. The number of images to be considered in such a classical calculation is of course proportional to the cavity Q. If Q becomes very large, the propagation time of the radiation along the "chain" of images cannot be neglected and retardation effects in the atom image interaction process become significant, which leads to the periodic behaviour of the system.

Cavity induced energy level shifts. Van der Waals wall shifts

We have focussed our discussion so far on the modification of the atomic radiative rates due to the resonator. The presence of the cavity around the atom also affects the off-resonant modes-atom interaction processes which are responsible for the quantum electrodynamical energy level shifts. In other words, the Lamb shifts are also changed in the cavity and the position of the various atomic energy levels depend upon the presence of conducting walls around the atom. Such effects, which are in their principle quite general occur of course also in ordinary atomic systems, but here again the large size of the atom-field coupling should make them much more conspicuous on Rydberg levels. The exact calculation of these effects which depend upon the actual size and shape of the cavity will not be attempted here. A detailed calculation, in the case of a cavity made of two parallel conducting plates can be found in reference [46]. We will only give here simple physical ideas and orders of magnitude estimates. Two approaches can be used for this estimation. The first one consists in calculating the field fluctuation of the vacuum in the cavity and then to compute the perturbing effect of these fluctuations on the atomic levels, after substracting a large divergent term due to the free field fluctuations which cause the unperturbed Lamb shift. In this point of view, the atom can be considered as a kind of probe detecting the local alteration of the mode structure of the vacuum due to the cavity. As it is well known, most of the contribution to the shifts come from modes at high frequencies for which the field fluctuations at the atom location are essentially the same as in free space, unless the atom gets very close to a cavity wall (at distances d such that $d < \lambda$). In other words, the Lamb shift is expected to be modified by an amount which depends upon the atom position, strongly increasing in the vicinity of a cavity wall. This behaviour is quite different from the alteration of the atomic decay rates studied above, which are largely independent of the atom distance to the walls and can be quite important even when the atom is in the center of the cavity.

The same conclusion can be retrieved in a different way. Instead of studying the interaction of the atom with the vacuum eigenmodes of the cavity, one computes its interaction with its images in the cavity walls, as discussed above. This time, this is the component of the image field in phase with the radiating dipole which is relevant. It causes a shift of the dipole eigenfrequency with a dominant contribution proportional to $1/r^3$ where r is the atom-image distance. Since this field component varies very quickly with the atom metal separation (in fact much more quickly than the $\pi/2$ out of phase component responsible for the alteration of the damping), only the effect of the closest image practically matters and the energy shift to a good approximation depends only upon the distance of the atom to the nearest wall. This point of view closely relates the Lambshift problem to the Van der Waals interaction between the atom and a metal surface[47,48,49], and can be shown to be strictly equivalent to the vacuum fluctuation picture.

From the above qualitative analysis, it seems that the resonant mode structure of the cavity plays no role in the energy shifts, contrary to what happens in the radiative damping problem. The actual situation is a little bit more subtle. Let us assume that the cavity is nearly resonant with an atomic transition (characteristic wavelength λ). When the atom is far from the cavity walls, somewhere at its center, its distance to the nearest metal surface is of the order of λ. At such distances, retardation effects become significant in the expression of the Van der Waals interaction which varies as $1/d^4$ and no longer $1/d^3$ (This is the well known Casimir-Polder effect[50]). Clearly, the exact expression of the Van der Waals force should then depend upon the propagation characteristics of the field inside the cavity, i.e. upon the exact cavity resonant frequencies. Very small effects of this kind, in which the transition frequency between two Rydberg levels is shown to be sensitive to the cut-off frequency of a cavity made of two parallel plates have recently been predicted[51]. It is to be noted however that the interaction with the cavity responsible for these effects is exceedingly small (The shift of a Rydberg level at a distance $\lambda \sim 0.1$ mm from a metal surface is at most a few hundred Hz around $n \sim 30$), so that these cavity-structure will be very difficult to observe. The atoms in a real experiment will be distributed inside the cavity and the shifts experienced by the atoms closer to the wall will be much larger than the tiny resonant shift effects due to the retarded-cavity atom interaction far from the walls.

Let us neglect in the following these cavity-structure dependant shifts and focus on the much larger Van der Waals type atomic wall shifts (d small compared to the characteristic atomic wavelength λ). Adopting here the simple electrostatic image model, we can describe the dipole-dipole image interaction due to the metal at distance d from the atom by an effective Hamiltonian[52] :

$$W = - \frac{q^2}{4\pi\varepsilon_0} \frac{1}{16d^3} (x^2 + y^2 + 2z^2) = - \frac{q^2}{4\pi\varepsilon_0} \frac{1}{16d^3} (r^2 + z^2) \quad (35)$$

where x, y, z are the Rydberg electron displacements parallel (x,y) and perpendicular (z) to the surface and $r^2 = x^2 + y^2 + z^2$.

On atomic ground states, the above expression is only an approximation, since the metal cannot be a perfect reflector for the optical photons virtually exchanged between the atom and the surface. Corrections depending upon the electron density in the metal have to be made to equation (35) . However, this expression generally gives a good order of magnitude estimate of the atom metal coupling and its derivative with respect to d corresponds to the attractive Van der Waals force experienced by the atom near the surface[53]

$$F = - \frac{3}{16} \frac{e^2}{d^4} < r^2 + z^2 >_{\text{ground state}} \quad (36)$$

(As recalled above, even for a perfect conductor, the above equation ceases to be valid when the atom-metal distance is of the order of or larger than the characteristic atomic wavelength ~ 1000 Å for ground state).

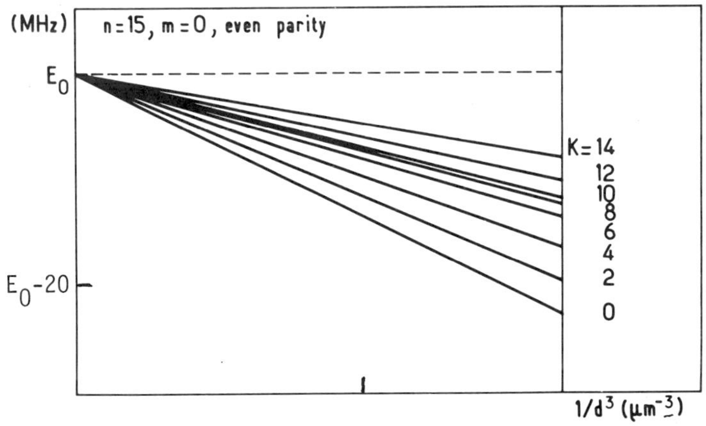

Figure 6 : Van der Waals energy levels shifts of the m=0 even parity states of the n=15 manifold of hydrogen, as a function of $1/d^3$ (d : distance from atom to metal surface).

In atomic excited states, W will usually perturb the level manifold to first order, inducing level splittings by removing the degeneracy between sublevels. Let us consider the simple case of hydrogenic Rydberg states corresponding to a given value of the principal quantum number n. The degeneracy between the n^2 angular momentum states (ℓ = 0, 1, 2... n-1; $-\ell < m < \ell$) is partially removed by W. The magnetic quantum number m corresponding to the projection of the angular momentum along Oz and the parity remain good quantum number. If we restrict ourselves for the sake of simplicity to the m=0 even parity levels, we have to diagonalize W in a n/2 or (n+1)/2 dimension subspace depending upon the parity of n. The result of this diagonalization is to split the subspace into n/2 or (n+1)/2 levels, the splittings varying linearly with the coupling strength (see Figure 6 corresponding to n=15). The shift is negative for all substates. The z^2 term in equ. (35) mixes the various ℓ states ($\Delta\ell = 0, \pm 2$ selection rule) so that ℓ is no longer a good quantum number, and it is preferable to label the levels with a new quantum number K (even K values correspond to even parity states).

An approximate value for the shift of the "center" of the n manifold (m=0) is :

$$< \Delta E_n > \sim - \frac{1}{5} \frac{e^2 a_o^2}{d^3} n^4 \qquad (37)$$

The shifts, and hence the Van der Walls forces which derive from it are thus proportional to n^4 and much larger at a given distance d than for ordinary atoms or molecules. It is interesting to notice that the Van der Waals perturbation W is very similar to the diamagnetic perturbation experienced by Rydberg levels in a large magnetic field B[54]. In this later case, the coupling is :

$$W_{magn} \propto B^2 (r^2 - z^2) \qquad (38)$$

and, apart from the sign of the z term, is identical to W. The main features of the diamagnetic Rydberg atom spectrum are indeed very similar to the one we have described above.

The Van der Waals energy shifts in Rydberg levels have two interesting features which distinguish them from ground states. First the shifts depend upon the K sublevel, which should lead to a K dependant force and hence to a kind of Stern and Gerlach type of atomic beam spatial splitting effect. This effect might however be very difficult to observe since the energy and the force depend so much upon d. More importantly, the energy shifts and the Van der Waals forces are n^4 time larger than in a ground state, so that the atomic deflection effect could be measurable at much larger distances from the metal. The force varies in fact as $(n/d)^4$, so that a n=100 Rydberg atom at 3 μm from the surface will experience the same force as a ground state atom at only 300 Å.

Let us notice here the relationship between the atom-metal Van der Waals interaction discussed here and the Rydberg-Rydberg dephasing described in the previous section in connection with the quenching of Rydberg atom superfluorescence. Both effects have a similar origin (electrostatic interaction between two Rydberg atomic dipoles or between a Rydberg dipole and its electric image) and of course the same order of magnitude. Both corresponds to highly inhomogeneous effects and are bound to produce large "broadenings" in systems where the location of the atoms with respect to each other or relative to the cavity walls are not perfectly controlled.

To summarize the above discussion, the energy shifts expected in a Rydberg atom placed inside a metallic cavity have a part which varies very quickly with the distance to the walls and a contribution experienced by the atoms even at large distances from the walls, which depend upon the cavity mode structure (relative values of the eigenmode frequencies of the cavity compared to the atom frequencies). The first contribution, of the Van der Waals electrostatic type, corresponds to shifts as large as 100 MHz for $n \sim 30$ and distances to the walls of the order of a μm. The second contribution, related to retardation effects in the Van der Waals force is at most of the order of a few kHz. Putting in evidence the straight $1/r^3$ Van der Waals atom-surface interaction by laser spectroscopy on atomic beams grazing a clean conducting mirror seems feasible. Observing the cavity mode-structure related effects on the Lambshift of microwave transitions is possible but difficult, since one should observe a small resonant change of the energy level positions superposed to a broad inhomogenous background due to the unavoidable dispersion of atomic positions inside the cavity.

In the above discussion, we have assumed that the walls of the cavity are a perfect metal. This approximation is good for usual metallic surfaces and Rydberg atoms whose characteristic frequencies are low compared to metal plasma frequencies. Departure from ideal metallic conduction can however occur if the surface is a semi-conductor and if the Rydberg electron frequencies get close to electron excitation resonances (cyclotron resonances, electron impurity bound state resonances falling in the millimeter wave domains). Each time a Rydberg transition matches one of these excitation frequencies on the surface, one should expect a resonant change in the atom-interface coupling which could be observed as a change of the atomic deflection, the atomic energy levels or the atomic lifetimes. In this respect, Rydberg atoms could be used as probe of metal or semi-conductor surface properties.

COLLECTIVE EMISSION OF RYDBERG ATOMS IN A CAVITY : RYDBERG MASERS[1, 2,20, 55,56,57]

General consideration. Orders of magnitude

We discuss now the radiative behaviour of a collection of N atoms initially prepared by a pulsed excitation in a Rydberg state inside a cavity resonant with a Rydberg-Rydberg transition. If the atoms are initially in the upper level of the transition resonant with the cavity, the situation corresponds to a transient Rydberg atom maser. This system has very interesting properties that we will review in this section. Let us start by some general order of magnitude considerations which relate Rydberg masers to the systems we have analyzed above.

At first, it is instructive to compare a Rydberg maser to a free space superfluorescent Rydberg system (see first section). Qualitatively, one expects a sample of N atoms in a resonant cavity of finesse f to behave as a collection of Nf radiators in free space (since the N atoms interact with their f images reflected back and forth in the cavity walls). In order to be specific in this discussion, we will consider the case of an open Fabry-Perot cavity of length L whose finesse is $f = Q\lambda/2L$. The characteristic radiative irreversible damping time in such a cavity is thus expected to be (see equ. 25) :

$$T_{SR}^{(cav)-1} = Nf\mu\, \Gamma^{(sp)}_{n\ell\to n'\ell'} \qquad (39)$$

where μ is a geometrical factor depending upon the geometry of the cavity mode. For a Gaussian mode in a Fabry-Perot resonator, one has[56] :

$$\mu = \frac{3}{4\pi^2}\ \frac{\lambda^2}{a^2} \qquad (40)$$

where a is the Gaussian mode waist.

The above analysis is valid as long as the f images interact quasi simultaneously with the atom sample, i.e. if :

$$\frac{c}{Lf} > \Gamma^{(sp)} N f \mu \qquad (41)$$

which means that the cavity finesse f (or Q) must not exceed a limit inversely proportional to $\sqrt{N}$. If the Q is larger than this limit, retardation effects in the atom cavity interaction are significant and according to the discussion of the 2nd section, one expects oscillations to appear in the atom cavity energy exchange process. It is interesting to notice that the transition between the irreversible damping regime and the oscillatory regime occur for cavity finesse $\sqrt{N}$ time smaller than in the single atom-cavity problem, which makes the observation of this transition much easier in the collective case.

To summarize, we expect for moderate Q cavities an irreversible cavity assisted superfluorescent process lasting a time of the order of $T_{SR}^{(cav)}$. Clearly the threshold for this collective damping is lowered by the factor f with respect to its free space value. The cavity finesse being currently of the order of 10^3-10^5 depending upon the cavity wall reflectivity, this threshold -of the order of 10^5 for free space superfluorescence- can be reduced to N=1 ! The single atom Rydberg maser case is in fact nothing but the cavity assisted spontaneous emission enhancement effect already discussed in the previous section. This remark leads us to another simple interpretation of equ. 39. Identifying $f\mu$ to $(3/4\pi^2)(O\lambda^3/\upsilon)$ where $\upsilon \sim La^2$ is the effective cavity volume, we can indeed write :

$$T_{SR}^{(cav)-} = N\,\Gamma^{(sp,\ cav)} \qquad (42)$$

and the Rydberg atom collective emission in the cavity appears -in the limit of moderate Q's- as the collective version of the single atom cavity assisted spontaneous emission enhancement effect.

The lowering of the collective emission threshold in a cavity has another important effect. We have shown above that it is impossible to get free space superfluorescence in a sample with dimensions smaller than λ, because of large Van der Waals dephasing effects. Since the threshold for collective radiation is greatly lowered in a cavity, it becomes in this case possible to prepare all the atoms in a very small volume compared to λ^3 and still avoid Van der Waals dephasing proportional to the atom density. Of course, one adds in principle the atom-metal image Van der Waals effects, but as shown above they are negligibly small as soon as the atom-wall distances is of the order of λ.

When all atoms are concentrated in such a small volume far from the cavity boundaries, they all see the same field in the cavity and are thus indiscernable with respect to the atom-mode coupling. The Rydberg maser thus realizes the ideal case of an ensemble of two level atoms symmetrically coupled to a single radiation mode, a canonical model for all superfluorescence theories[58,28,30]. As we will see below, this greatly simplifies the dynamics of the atom-field system and makes superfluorescence in a cavity a much more basic effect to study than free space superfluorescence where the atoms are not equivalently coupled to the field.

Symmetrical ensemble of two level atoms coupled to a single field mode : Quantum mechanical formalism [1,2]

We consider from now on that the Rydberg atoms are prepared in a small volume in an antinode position of the field in the resonant cavity.

The simplicity and importance of this small sample Rydberg atom maser operating in the pulsed regime lies in the fact that it is a very basic quantum mechanical system corresponding to the coupling of an angular momentum with an harmonic oscillator. The energy levels of these two sub-

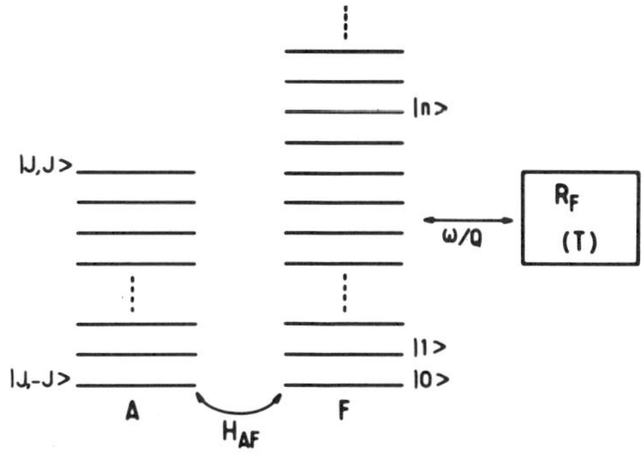

Figure 7 : Energy levels of the N two-level atom + field system (see text for more details).

systems are shown on Figure 7. The oscillator (eigenstates $|n>$ with energies $n\hbar\omega$ corresponding to n photon states) describes the cavity field mode F. The angular momentum represents the atomic medium A : all levels of the atom but the two connected by the transition resonant with the cavity can be neglected and it is well known that an ensemble of N two level atom evolving -due to a symmetrical coupling- in a superposition of states invariant by atom permutation behaves as an angular momentum $\vec{J}$ ($J=N/2$). The various degrees of excitation of the collective atomic system are the $|M>$ eigenstates (energy $M\hbar\omega_o$) of the J_z component of $\vec{J}$. The atomic (J_+, J_-) and field (a^+, a) operators which obey the standard commutation rules, respectively raise and lower the atom and field excitation by one unit. The coupling between the two systems is described by the interaction given by Equ. (30). The N atom evolution in the cavity appears thus as a mere generalization of the single atom case from a spin $J=1/2$ to a spin $J=N/2$.

Various initial states can be considered for this system at time t=0, corresponding to different experimental situations.

$$|\psi(0)> = |M = N/2;\ n = 0> \qquad (43)$$

represents the case of a pulsed Rydberg maser starting on pure spontaneous emission.

$$|\psi(0)> = \sum_M c_M |M;\ n=0> \qquad (M \sim N/2) \qquad (44)$$

describes a "superradiant" maser in which a small atomic polarization has been prepared at t=0.

$$|\psi(0)> = \sum_n C_n |M = N/2;\ n> \qquad (n << N) \qquad (45)$$

represents the case of a "triggered" maser in which the atomic system is initially fully inverted with a small coherent field impinging in the cavity at t=0. Initiation of the maser by thermal radiation can also be accounted for by describing the system initial state and subsequent evolution with a density operator instead of a wave function[1]. After time t=0, the maser evolves under the action of H_I and damping mechanisms which correspond to the dissipation of the field in the cavity walls "reservoir" within a time Q/ω. (This reservoir is represented by a box in Figure 7). Many basic features of the evolution (namely all phenomena appearing on a time scale shorter than Q/ω) can be qualitatively understood within a simplified dissipation-free model. Solving the Schrödinger equation for this model can be performed directly with a computer for N not too large ($N \lesssim 150$). For larger atomic samples, approximate calculations involving cumulants [59] provide $1/N$ expansions of the solution. The cumulant method with a density matrix approach is necessary even for small N's if field damping is to be taken into account. A variety of physical quantities can be computed as a function of time : $\langle J_z(t) \rangle$ and $\langle a^+a(t) \rangle$ describe respectively the atomic and field mean energies, $\langle J_{x,y} \rangle$ and $\langle a_1 \rangle$, $\langle a_2 \rangle$ ($a_1 = a + a^+$, $a_2 = i(a - a^+)$) represent the two $\pi/2$ out of phase components of the collective atomic dipole and electric field in the cavity, ΔJ_z, $\Delta(a^+a)$, $\Delta J_{x,y}$, $\Delta a_{1,2}$ are the fluctuations (second order moments of these quantities). One can also compute the distribution $P(M,t)$ and $P(n,t)$ of the atomic and field systems along their respective energy ladders which are quantities containing much more information than the second order moments.

Semi-classical formalism : the Bloch vector model[1,2]

The quantum mechanical model described above although simple in its principle has the drawback of leading to long numerical calculations as soon as N is large. Fortunately, it is possible to develop a semi-classical model for the Rydberg atom maser which coincides with the quantum mechanical one as $1/N \to 0$. This model furthermore provides a very simple mechanical analogy of the Rydberg maser which allows us to understand very easily most of the system features. In this point of view, the Rydberg maser is a vector evolving in an abstract space and pointing at time t in a direction $\theta(t)$, $\phi(t)$. ($\theta = \pi$ and $\theta = 0$ correspond to the fully inverted and fully deexcited systems respectively, ϕ is the phase of the collective atomic dipole). The collective atomic energy J_z corresponds to $-N/2 \cos\theta$, the atomic polarization components J_x and J_y to $N/2 \sin\theta\cos\phi$ and $N/2 \sin\theta\sin\phi$ and the electric field componant a_1 to $-\Omega^{-1}\beta\sin\phi$ where $\beta = -\dot{\theta}$ is the Bloch vector angular velocity in the abstract space. The Bloch vector is shown to obey an equation analoguous to a pendulum in a gravitational field[1,2,28,60,61]

$$\ddot{\theta} + \frac{\omega}{2Q}\dot{\theta} + N\Omega^2 \sin\theta = 0 \qquad (46)$$

In this analogy, the pendulum potential and kinetic energies become the atom and field excitation respectively. This pendulum starts on thermal and vacuum noises, which have to be introduced in the model as ad hoc fluctuations of the atomic polarization and velocity [1,2,31].

Solving the equations for an ensemble of initial conditions corresponding to a random initial tipping angle around $\theta = \pi$ (mimicking the initial atomic fluctuations ($\Delta J_x^2 = \Delta J_y^2 \neq 0$) and a random initial velocity mimicking the initial vacuum fluctuation ($\Delta a_1^2 = \Delta a_2^2 \neq 0$), one gets a statistical ensemble of pendulums position and velocity at any later time from which one can reconstruct the atomic and field fluctuations of the Rydberg masers. It can be shown by comparison of the quantum mechanical and classical calculations that they lead to the same predictions as soon as $1/N \to 0$.

Mean value and fluctuations of atomic energy in the Rydberg maser

The evolution of the atomic energy during the emission of the Rydberg maser has been experimentally studied and the results compared with the predictions of the quantum and semi-classical models outlined above[1,2,62,63]. The experiments consist in preparing a sample of Rydberg atom in a given initial state by a pulsed laser excitation and in counting, for each laser shot, the number of atoms in the upper and the lower level of the Rydberg transition at a fixed delay t after the system preparation. One then reconstructs from a large ensemble of such pulses with identical initial conditions the average atomic energy $< J_z(t) > = -N/2 < \cos\theta(t) >$ and the atomic distribution function P(M,t) which is related to the dispersion of M values at time t. The statistical analyzis is performed with the help of a computer interfaced to the experiments. More details about the set-up and the experimental procedures are given in references [61,62].

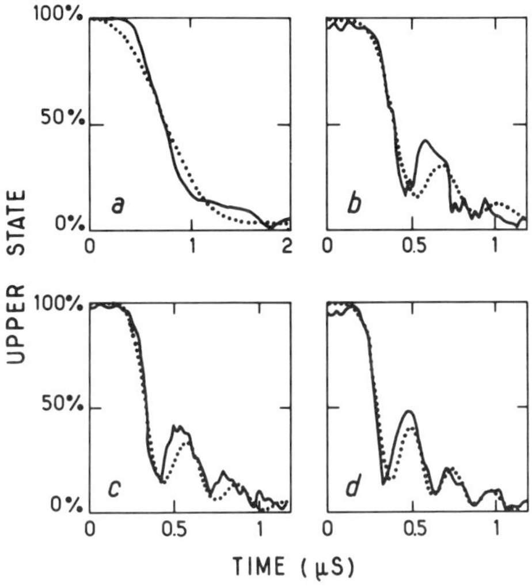

Figure 8 : Mean atomic energy evolution of Rydberg maser operating on the 36S → 35P transition in Na. Full line is experimental and dotted line is theoretical curve. Atom numbers are N=2000, N=19000, N=27000 and N=40000 respectively (see text for more details).

The mean atomic energy evolution obtained from these experiments[62] is shown in full lines on Figure 8 which represents, for increasing N values, the measured $< J_z(t) >$ values as a function of time. The maser is starting on thermal fluctuations at room temperature and operating in a cavity with a $Q = \sim 10^5$. Figures 8 -a,b,c and d correspond respectively to a ratio $\frac{Q\Omega\sqrt{N}}{\omega}$ equal to 0.8 , 2.5, 3 , and 3.5 . Cases b, c, d correspond to a damped oscillatory regime of the Bloch vector, with increasing oscillation frequencies $\Omega\sqrt{N}$. Case a corresponds to the overdamped regime ($\Omega\sqrt{N}$ smaller than the damping rate ω/Q). In the quantum mechanical picture, the oscillatory regime corresponds to a reversible exchange of energy between the atomic and field energy ladders : the atomic angular momentum first goes down ($< M >$ decreases) and the number of photon increases. Then the photons are reabsorbed and the atomic system goes up and so on... The overdamped regime on the other hand corresponds to an irreversible decay of the atomic system, the emitted photons being rapidly damped away into the cavity walls. These two regimes are the collective counterpart of the single atom Rabi oscillation and irreversible enhanced spontaneous emission discussed in previous section. The main change with respect to the single atom case is an increase of the Rabi frequency from Ω to $\Omega\sqrt{N}$ and of the irreversible decay rate from $\Gamma^{(sp)}_{cav}$ to $N\Gamma^{(sp)}_{cav}$, in full agreement with the qualitative analyzis developed above . The dotted lines on the same figure represent the results of the semi-classical model. Agreement between the theory and the experiments appear to be quite good.

The results for the P(M,t) distribution are displayed on Figures 9 and 10 corresponding respectively to the overdamped[61] and oscillatory regimes[2]. The full lines superposed to the histograms correspond to the theoretical predictions. Here again, the agreement between theory and experi-

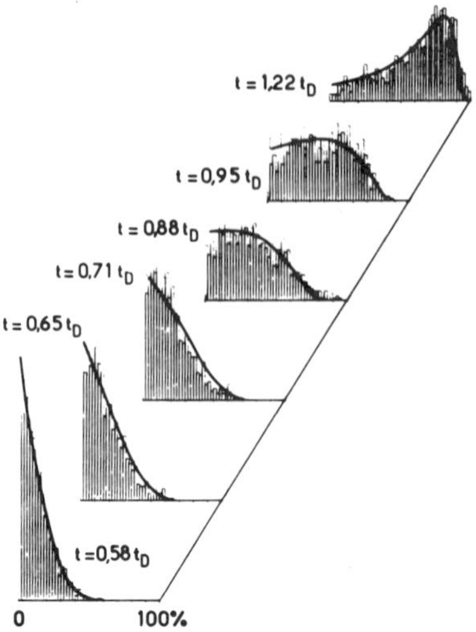

Figure 9 : P(M,t) distribution for an overdamped Rydberg maser operating on the $33S \rightarrow 32P^1/_2$ transition in Na. Histograms are experimental and full line theoretical. 0 and 100% on horizontal axis refer to M=J and M=-J respectively (percentage of deexcited atoms).

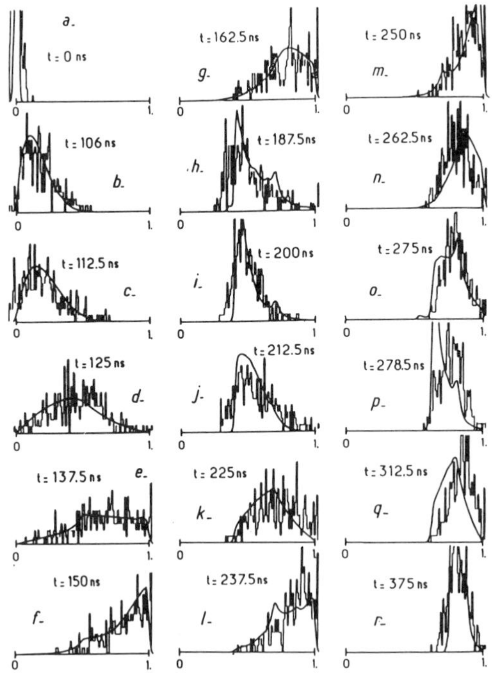

Figure 10 : P(M,t) distribution for a Rydberg maser undergoing reversible energy exchange between atoms and field. Each histogram corresponds to a given time following the maser excitation at time t=0. The full line curves are theoretical. In this experiment, the maser emits independently on two orthogonally polarized transitions and P(M,t) is the convolution product of two probability distributions associated to two identical Bloch pendulums (see ref.[2] and text for more details).

ment is satisfactory. The overdamped regime of Figure 9 corresponds to the regime of single mode superfluorescence which was developped in connection with superradiance theories[28]. The evolution of the system fluctuations displayed on Figure 9 are the first precise experimental check of this model. The fluctuations evolve from an exponential distribution at early time into a bell shaped distribution as the superradiance pulse develops, in full agreement with theory. The fluctuations in the oscillatory regime are also very intersting to analyze. It clearly appears from the successive histograms of Figure 10 that the atomic energy variance (width in M of each histogram) undergoes oscillations as the mean energy does. The J_z fluctuation is maximum around the time when the atomic energy mean value reaches its first minimum (histogram f), then decreases and goes through a relative minimum when $< J_z >$ has bounced back to its second maximum (histogram i) and so on... These results can be understood from the pendulum model. A pendulum undergoing large oscillations has a period depending upon its initial amplitude and velocity (non isochromism of large oscillations). A small initial $\Delta\theta_o$ or $\Delta\beta_o$ fluctuation (due here to thermal radiation noise) results into a large dispersion of the arrival times of the pendulum at its potential energy minimum ($\theta = 0$). This time fluctuation in turn entails large energy dispersion at any given time around this position, which explains the important energy fluctuations around $\theta \sim 0$. When the pendulum bounces back, its velocity decreases and the ΔJ_z variance is reduced since the slowest pendulum in the statistical ensemble starts to catch up the fastest ones which reduces their velocity first.

These Rydberg maser studies thus represent a complete anlyzis of the evolution of a "classical" pendulum triggered by thermal noise away from its unstable position around $\theta = \pi$. If the temperature is reduced to T=0K, similar results are expected, the thermal noise being replaced by the vacuum field fluctuations which have the same statistics. The evolution of the "macroscopic" system made of a large number N of atoms is characterized by a non linear amplification of a small initial quantum noise. The Rydberg maser is thus an interesting example of "Macroscopic quantum behaviour" in which a very small fluctuation of intrinsically quantum nature reveals itself at the "macroscopic" level by the existence of important fluctuation in the evolution of a large system.

Field fluctuations in the Rydberg maser : prediction of squeezing

The field amplitude fluctuations in the Rydberg maser are also interesting to discuss here, since they have recently attracted a lot of interest in the context of the study of "squeezed" states of the electromagnetic field[63].

Before describing these fluctuations, it is appropriate at this stage to recall the definition of these "squeezed" states. The two $\pi/2$ out of phase components a_1 and a_2 of a mode of the radiation field obeying the standard commutation rules of the position and momentum of a harmonic oscillator, their variances Δa_1 and Δa_2 satisfy the Heisenberg uncertainty relation :

$$\Delta a_1 \, \Delta a_2 \geqslant 1. \qquad (47)$$

In the vacuum field, these two fluctuations are equal and correspond to the minimum uncertainty :

$$\Delta a_1 = \Delta a_2 = 1 \qquad (48)$$

"Coherent" states of the field produced by classical current distributions and ordinary laser sources are nothing but "displaced" vacuum states in the Liouville phase space of this harmonic oscillator, for which equ. 48 still holds. The possibility of generating "non-classical" radiation states for which relation (47) would be fulfilled with $\Delta a_1 < 1$ and $\Delta a_2 > 1$ has in the past years received a lot of attention since such "squeezed" states would correspond to a reduction of the noise of a photon detector sensitive to the a_1 component below what was previously thought to be the ultimate quantum limit. Several possible sources of squeezed states, all involving non linear optical processes have been considered in the literature[64]. Rydberg masers, as shown in reference[65], are one example of such possible source. A very attractive feature of these systems as opposed to others is their great conceptual simplicity and the possibility of actually realizing in the laboratory a system very close to the theoretical model. Another advantage of Rydberg masers is also to give a very simple and illuminating interpretation of the squeezing effect in terms of the classical pendulum model.

For sake of simplicity in most of the analyzis to come, we assume T = OK (no thermal photons in the cavity) and neglect field damping (practically, we suppose that the cavity has a very large Q and restrict the discussion to a time scale shorter than Q/ω).

Obviously the version of the Rydberg maser discussed above cannot produce any squeezing. The variances Δa_1 and Δa_2 of the two field components of a "spontaneous" maser are indeed always equal (no phase preference of the emission) and Equ.47 then entails $\Delta a_1 = \Delta a_2 > 1$, as can be confirmed by a direct quantum mechanical calculation. It is however very easy to break the phase symmetry of the maser by operating it in a superradiant or triggered regime (initial states given by Eqs.44 or 45). Assuming the initial atomic polarization triggered along Oy ($< J_x > = 0$; $\phi = \pi/2$), one finds by the quantum calculation that the field component a_1 (in phase quadrature with the atomic polarization) has a variance Δa_1 which becomes smaller than 1 at some times during the system evolution. (The product $\Delta a_1 \, \Delta a_2$ remains of course always larger than one and squeezing on one field phase is obtained at the expense of an increased noise on the other). Figure 11 shows Δa_1^2 as well as the average atomic energy in the cavity as a function of time for a maser with N=100 starting with an initial atomic polarization corresponding to an angle $\theta_o = 3\pi/4$. Δa_1 first increases above one, then diminishes and reaches a minimum smaller than one when the atomic energy becomes minimum in the cavity ($\theta \sim 0$). The amount of squeezing (minimum Δa_1 value) depends upon the initial state of the system. For a given N, the squeezing around $\theta = 0$ is optimum for a small finite value of the initial atomic polarization (value of θ_o slightly different from π) or for a small but finite value of the injected field (value of β_o slightly different from 0)[65]. The optimum squeezing furthermore increases with N as $N^{-1/3}$ [59]. A minimum value of Δa_1 very close to zero (maximum possible squeezing) is obtained as soon as N becomes larger than a few thousand (Figure 12 shows the best attainable squeezing value Δa_1^2 as a function of the atom number (in logarithmic scale)).

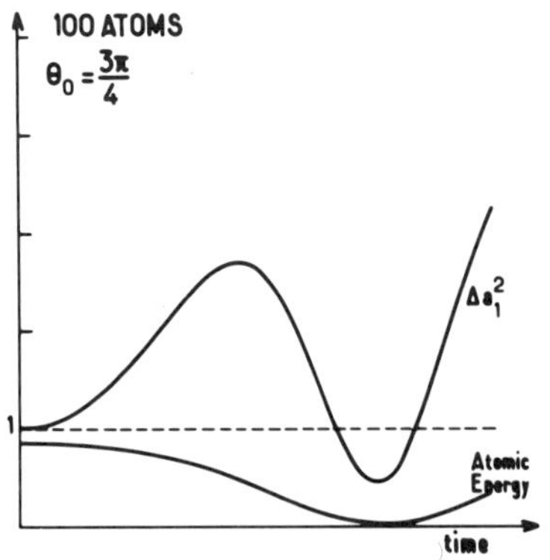

Figure 11 : Evolution of field variance Δa_1^2 for a superradiant Rydberg maser (N=100, $\theta_o = 3\pi/4$) (top curve) shown together with the average atomic energy evoution (bottom curve).

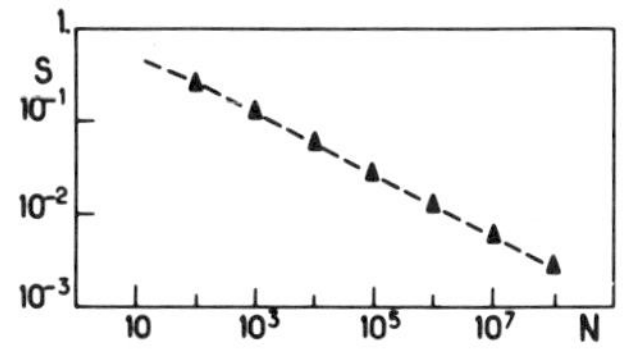

Figure 12 : Maximum attainable squeezing $S = \Delta a_1^2$ *as a function of N.*

The above results of the quantum mechanical calculation can be explained in the Bloch pendulum point of view. The requirement of triggering the maser simply means that one has to fix a phase ϕ for the pendulum so that the projection of its velocity along a given direction will not widely fluctuate from pulse to pulse. Once the phase is fixed, the pendulum motion can be simply described in the θ, β phase space. $\beta = -\dot{\theta}$ is nothing but the classically canonical conjugate quantity of θ. The area $\Delta\theta\Delta\beta$ should thus, according to Liouville theorem, be conserved during the evolution of a canonical set of identical pendulums. As discussed above, a characteristic feature of the pendulum motion is a large increase of $\Delta\theta$ when the potential energy is near minimum. Liouville theorem then entails that $\Delta\beta$ should be reduced, which corresponds to a decrease of the field amplitude noise at that time.

Detailed calculation of the Rydberg maser squeezing in the quantum mechanical as well as classical points of view can be found in refs.[59,65,66]. Taking into account dissipation of the field, or the presence of thermal radiation is possible[66] and shows that the squeezing remains significant as long as $\bar{n} \lesssim 1$ and $\omega/Q \ll \Omega\sqrt{N}$.

Observation of field amplitude fluctuations and demonstration of squeezing in Rydberg maser system has not yet been experimentally achieved because it is much more difficult to directly measure the radiation field than to detect the atom by electric field ionization. Such studies would require the use of very performant heterodyne receptors with a sensitivity approaching the quantum limit in the millimeter wave domain.

The above discussion by no way exhausts the topic of Rydberg atom radiative properties in/or outside cavities. Among the topics not covered here, let us mention the theoretical[67] and experimental[68] studies of Rydberg atom multiphoton ionization in intense microwave fields which have recently attracted some attention in connection with the problem of turbulence and evolution to chaos. In the domain of cavity-Rydberg atom physics, let us also mention the studies of coherent absorption of blackbody field in a cavity[69], which have shown that the process by which a sample of Rydberg atom absorbs the thermal field of a single mode is collective in the same sense as the superfluorescence phenomenon, the atomic sample behaving as a "Boson gas" with respect to its interaction with the field.

Let us conclude by insisting on the great variety of new effects that Rydberg atoms have allowed us to study in the domain of radiation-matter interaction processes. Some of these effects (spontaneous emission alteration in a cavity, single mode superfluorescence, maser action with a few or a single atom,...) had been considered or discussed--sometimes--long ago in theoretical papers but it is the advent of very excited state atomic physics which has opened the way to the direct observation of these effects and to the confirmation of the theory.

In the course of these very simple and elegant demonstration experiments, a lot has been learned about the possibility of using Rydberg atoms as practical tools to detect or generate small fields with interesting statistical properties and even some surprising results obtained (such as the realization that Rydberg masers were ideal sources for squeezed radiation). No doubt, further investigations of the non linear coupling of Rydberg atoms with microwave fields hold new surprises in store...

REFERENCES

1. S. Haroche in : "New Trends in Atomic Physics", Proceedings of Les Houches Summer School, session XXXVIII, G. Grynberg and R. Stora editors, North Holland, Amsterdam (1982)
2. S. Haroche and J.M. Raimond in : Advances in Atomic and Molecular Physics, 20, 347 (1985)
3. J.A.C. Gallas, G. Leuchs, H. Walther and H. Figger in : Advances in Atomic and Molecular Physics, 20, 413 (1985)
4. T.F. Gallagher in : "Rydberg States of Atoms and Molecules", R.F. Stebbins and F.B. Dunning editors, Cambridge Univ. Press, London and New York (1983)
5. P. Filipovicz, P. Meystre, G. Rempe and H. Walther in Optica Acta, published (1985)
6. C. Fabre, Annales de Physique (Paris), 7, 5 (1982)
7. C. Fabre and S. Haroche in : "Rydberg States of Atoms and Molecules", R.F. Stebbings and F.B. Dunning editors, Cambridge Univ. Press, London and New York (1983)
8. D.R. Bates and A. Damgaard, Philos. Trans. Roy. Soc. London, 242, 101 (1949)
9. J.F. Gounand, J. Phys. (Paris), 40, 457 (1979)
10. M.L. Zimmerman, M.G. Littman, M.M. Kash and D. Kleppner, Phys. Rev. A 20, 2251 (1979)
11. A. Dalgarno in : "Rydberg States of Atoms and Molecules" (same as references 4 and 7 above) and refs. in.
12. T.F. Gallagher, S.A. Edelstein and R.M. Hill, Phys. Rev. A 11, 1504 (1975)
13. M. Hugon, F. Gounand and P.R. Fournier, J. Phys. B 11, L605 (1978)
14. W.P. Spencer, A. Ganesh-Vaidyanathan and D. Kleppner, Phys. Rev. A 24, 2513 (1981)
15. R. Hulet and D. Kleppner, private communication (1985)
16. T.F. Gallagher and W.E. Cooke, Phys. Rev. Lett. 42, 835 (1979)
17. E.J. Beiting, G.F. Hildebrandt, F.G. Kellert, G.W. Foltz, K.A. Smith, F.B. Dunning and R.F. Stebbings, J. Chem. Phys. 70, 3551 (1979)
18. W.P. Spencer, A. Ganesh-Vanidyanathan and D. Kleppner, Phys. Rev. A 25, 3280 (1982)
19. H. Figger, G. Leuchs, P. Straubinger and H. Walther, Opt. Comm. 33, 37 (1980)
20. S. Haroche, C. Fabre, P. Goy, M. Gross and J.M. Raimond in : "Laser Spectroscopy IV", H. Walther and K.W. Rothe editors, Springer Verlag (1979)
21. W.E. Cooke and T.F. Gallagher, Phys. Rev. A 21, 588 (1980)
22. J.W. Farley and W.H. Wing, Phys. Rev. A 23, 2397 (1981)
23. P. Avan, C. Cohen-Tannoudji, J. Dupont-Roc and C. Fabre, J. Phys. (Paris), 37, 993 (1976)
24. T.A. Welton, Phys. Rev. 74, 1157 (1948)
25. G. Barton, Phys. Rev. A 5, 469 (1972)
26. L. Hollberg and J.L. Hall, Phys. Rev. Lett. 53, 230 (1984)
27. M. Gross, P. Goy, C. Fabre, S. Haroche and J.M. Raimond, Phys. Rev. Lett. 43, 343 (1979)
28. R. Bonifacio, P. Schwendiman and F. Haake, Phys. Rev. A 4, 302 (1971)
29. J.C. MacGillivray and M.S. Feld, Phys. Rev. A 14, 1168 (1976)
30. M. Gross and S. Haroche, Physics Reports, 93, 302 (1982)
31. F. Haake, H. King, G. Schröder, J. Haus and R.J. Glauber, Phys. Rev. A 20, 2047 (1979)
32. M. Gross, Thèse d'Etat, Paris VI (1980), unpublished
33. G. Vitrant, Thèse de 3ème cycle, Paris VI (1982), unpublished
34. R. Friedberg and S.R. Hartmann, Phys. Rev. A 10, 1728 (1974)
35. L. Moi, C. Fabre, P. Goy, M. Gross, S. Haroche and J.M. Raimond in : "Laser Spectroscopy V", A.R.W. McKellar, T. Oka and B.P. Stoicheff editors, Springer Verlag (1981)

36. E.T. Jaynes and F.W. Cummings, Proc. IEEE, 51, 89 (1963)
37. D. Meschede, H. Walther and G. Müller, Phys. Rev. Lett. 54, 551 (1985)
38. E.M. Purcell, Phys. Rev. 69, 681 (1946)
39. S. Sachdev, Phys. Rev. A 29, 2627 (1984)
40. H. Walther and G. Rempe, private communication (1985)
41. P. Goy, J.M. Raimond, M. Gross and S. Haroche, Phys. Rev. Lett. 50, 1903 (1983)
42. D. Kleppner, Phys. Rev. Lett. 47, 233 (1981)
43. R. Hulet and D. Kleppner, private communication (1985)
44. G. Gabrielse, R.S. Van Dyck Jr, P.B. Schwinberg and H. Dehmelt, Bull. Am. Phys. Soc. 29, 926 (1984)
45. G. Gabrielse and H. Dehmelt, Phys. Rev. Lett. 55, 67 (1985)
46. C.A. Lütken and F. Ravndal, Phys. Rev. A 31, 2082 (1985)
47. J.E. Lennard-Jones, Trans. Faraday Soc. 28, 334 (1932)
48. J. Bardeen, Phys. Rev. 49, 640 (1936); 58, 727 (1940)
49. W.G. Pollard and H. Margenau, Phys. Rev. 57, 557 (1940)
50. H.B. Casimir and D. Polder, 73, 360 (1948)
51. P. Dobiasch and H. Walther, published (1985)
52. C. Cohen-Tannoudji, B. Diu and F. Laloë, Quantum Mechanics, vol. II, p. 1139-1140, Wiley (1977). (Note that the expression giving W in this reference is too large by a factor 2).
53. D. Raskin and P. Kusch, Phys. Rev. 179, 712 (1969)
54. See J.C. Gay lectures in this volume
55. S. Haroche, C. Fabre, J.M. Raimond, P. Goy, M. Gross and L. Moi, J. Phys. Paris, Coll. C2, 43, 265 (1982); ibid. 43, 275 (1982)
56. L. Moi, P. Goy, M. Gross, J.M. Raimond, C. Fabre and S. Haroche, Phys. Rev. A 27, 2043 (1983)
57. P. Goy, L. Moi, M. Gross, J.M. Raimond, C. Fabre and S. Haroche, Phys. Rev. A 27, 2065 (1983)
58. R.H. Dicke, Phys. Rev. 93, 99 (1954)
59. A. Heidmann, J.M. Raimond, S. Reynaud and N. Zagury, Opt. Comm. 54, 189 (1985)
60. R. Bonifacio and L.A. Lugiato, Phys. Rev. A 11, 1507 and A 12, 587 (1975)
61. J.M. Raimond, P. Goy, M. Gross, C. Fabre and S. Haroche, Phys. Rev. Lett. 49, 1924 (1982)
62. Y. Kaluzny, P. Goy, M. Gross, J.M. Raimond and S. Haroche, Phys. Rev. Lett. 51, 1175 (1983)
63. C.M. Caves, Phys. Rev. D 23, 1693 (1981); Quantum Optics, Experimental Gravitation and Measurement Theory, edited by P. Meystre and M.O. Scully (Plenum, New York 1983)
64. D.F. Walls, Nature (London), 306, 141 (1983)
65. A. Heidmann, J.M. Raimond and S. Reynaud, Phys. Rev. Lett. 54, 326 (1985)
66. A. Heidmann, J.M. Raimond, S. Reynaud and N. Zagury, Opt. Comm. 54, 54 (1985)
67. J.G. Leopold and I.C. Percival, J. Phys. B 12, 709 (1979); R.V. Jensen, in Chaotic Behavior in Quantum Systems, edited by G. Casati (Plenum, New York 1984)
68 J.E. Bayfield and P.M. Koch, Phys. Rev. Lett. 33, 258 (1974); J.E. Bayfield, L.D. Gardner and P.M. Koch, Phys. Rev. Lett. 39, 76 (1977); P.M. Koch and D.R. Mariani, Phys. Rev. Lett. 46, 1275 (1981)
69. J.M. Raimond, P. Goy, M. Gross, C. Fabre and S. Haroche, Phys. Rev. Lett. 49, 117 (1982)

THE STRUCTURE OF RYDBERG ATOMS IN STRONG STATIC FIELDS

J.C. Gay

Laboratoire de Spectroscopie Hertzienne de l'ENS[+]-Tour 12 E01, 4 place Jussiesu, 75230 Paris Cedex 05, France

Among the numerous works carried out on Rydberg atoms for the last ten years, the class which is concerned with their properties in external fields has some singular character. As for other experiments, on Q.E.D., superradiance...the existence of scaling laws in the Rydberg ladder was an early motivation. This indeed offers a unique opportunity of experimentally studying unsolved questions at laboratory field conditions. But their singular character lies in that they aim at a complete perturbation of the atomic system itself, which,in some conditions, recalls everything but an atom. Actually, these experiments in external fields allow the deepest alterations of the atomic structure ever realized in a controlled way. This strongly contrasts with the conception which prevails in other experiments in which the atomic structure is fully respected and used as a probe of the interactions.

The purpose being different, it is not surprising that both the methods and theoretical concepts for understanding this field be also very different from the ones developed in other lectures, with the exception of the one on doubly-excited systems. Indeed these lectures on Rydberg atoms do not share any common character except that they deal with originally the same basic Coulomb-driven system.

Our purpose here is not to give extensive reviews of the state of the art which is done in Dan Kleppner's contribution. Rather we are concerned with the building of the theory and interpretation of experiments in a global way. Our reference situation will be the hydrogen atom in a magnetic field associated to a problem which has been for years a major challenge in theoretical physics. The basic reason is that this problem is a non-separable one and has to be dealt with in a non-perturbative regime where the Coulomb and external field forces do have the same magnitude. As a matter of fact, standing theoretical methods widely applied to this problem have proved mostly unefficient. A conceptual change was required which took place around 1982. The theoretical advances which followed immediately make this problem a prototype for tackling some other non-separable situation in physics, especially in atomic physics those associated with quasi-hydrogenic systems interacting with the radiation field, surfaces or other atomic systems.

The conceptual change which leads to the solution to the magnetic problem naturally follows from the necessity of solving a fundamental contradiction. On the one hand, the large amount of experimental data collected for ten years on Rydberg atoms was exhibiting very simple features even in the non-perturbative regime. These being valid for all atoms did take the force of laws. On the other hand, traditional theoretically conceptions failed to reproduce these laws, predicting instead that the phenomena should become of an unconceivable complexity. Hence the reflexion on experiments and the requirement of interpreting simple results in simple words lead to question upon the deep structure of the problem especially in the low external fields limit, in which the Coulomb field is dominant. A well-ascertained spectroscopic tradition has labelled this regime as the "inter-l mixing regime" due to the fact that l, which does not commute with the diamagnetic interaction, is no longer a good quantum number. But such a label is restrictive as it ignores the deep origin of the n^2 degeneracy of the n Coulomb shell, lying in the symmetry of the Coulomb field. Hence the first step in this conceptual change is to reintroduce what matters that is the symmetry properties of the Coulomb field and to study the way they are broken by the external field action. This is obviously doing more than a change in the labelling of a regime. Thought this way, the experimental results become coherent, organized and simple to understand.

The matter of this paper is focussed on the symmetries of the Coulomb field and how they rule the dynamics of atoms in external fields and some other questions. There are two degrees in these symmetries for the non-relativistic Coulomb problem which correspond to the symmetry group and to the dynamical group. Although this has been extensively studied for up than 60 years, from Pauli to Barut among many others, these non-standard views on the Coulomb problem are mostly ignored and hardly difficult to find in atomic physics text-books. We will thus try to reintroduce them from the first principles, showing that they lead to quite successful views on the structure of the problem and do have very strong predictive powers. Indeed these symmetry considerations are the simplest way for describing the properties of the Coulomb field in its full pecularity and to some extent those of the atom or hydrogenic ions. Moreover, dealing with Rydberg atoms in external fields, they are the obliged way for not missing the essential as, according to Fock's phrase, these experiments turn out to be some fundamental "exercises in the geometry of the Coulomb field".

But before introducing these concepts on the Coulomb symmetry and applying them to the classical and quantum magnetic problem, we will make some elementary recalls on the phenomenology and traditional approach to these questions, pointing out the need for a change.

PHENOMENOLOGICAL VIEWS ON ATOMS IN FIELDS

The basic features in the magnetic and electric field problems can be deduced through elementary considerations. The electron is submitted to the joint actions of two or three forces which may have comparable orders of magnitude. They are respectively the Coulomb, Lorentz and electric field forces, the expressions of which are well-known ($e^2 = q^2/4\pi\varepsilon_0$) :

$$\vec{F}_C = - e^2 \vec{r}/r^3 \quad (1)$$

$$\vec{F}_L = q \vec{v} \wedge \vec{B} \quad (2)$$

$$\vec{F}_E = q \vec{E} \quad (3)$$

Obviously, the relative magnitude of these tremendously depend on the degree of excitation of the electron's motion.

1. Coupling parameters in the problem

A phenomenological way of establishing the expression of the coupling parameters is as follows. We shall ignore the fact that the symmetry properties of the three forces are different and hardly complying with each others. This is actually the origin of the non-separability of this class of problems. A more refined discussion of these important aspects will be given in the following sections. Hence the coupling parameters can be expressed as the ratio of the Lorentz or electric forces to the Coulomb one. From equations (1) to (3) these are depending on the degree of excitation of the electron's motion. This can be described approximately, introducing an index n, such that r is scaling as $n^2.a_o$ (a_o the Bohr radius). In the Coulomb quantum limit, n identifies with the principal quantum number. Hence, the coupling parameter η in the magnetic problem expresses as :

$$\eta = F_C/F_L = (q\ v\ B.r^2/e^2) = B\ n^3/B_c \qquad (4)$$

where B_c is the critical magnetic field defined below. Introducing the unit of pulsation in the magnetic problem, the cyclotron frequency such that :

$$\omega_c = q.B/m \qquad (5)$$

one obtains immediately

$$\eta = B.n^3/B_c = (\hbar\omega_c/2R).n^3 \qquad (6)$$

where R is the Rydberg constant. Hence, the critical field B_c represents the value of the magnetic field for which the cyclotron frequency equals twice the Rydberg. This corresponds to the huge magnetic field strength of $2.35.10^9$ Gauss. In reduced atomic units, one defines the parameter γ :

$$\gamma = B/B_c \qquad \eta = \gamma.n^3 \qquad (7)$$

Usually, under laboratory field conditions, γ is at maximum of the order of 10^{-4}. But owing to the fact that the coupling parameter depends, through the index n, on the degree of excitation of the electron's motion, conditions such that $\eta \underline{\sim} 1$ or greater than 1 can be achieved meaning that the magnetic force can completely overwhelm the Coulomb one. This makes clear the interest of dealing with Rydberg atoms excited to high principal quantum numbers, for studying these questions. Especially equation (7) allows to state that what matters is not the absolute strength of the field compared to B_c but rather its strength compared to B_c/n^3 which looks like a critical field refering to the degree of excitation of the electron's motion.

The same analysis can be carried out for the electric field problem. The coupling parameter is then :

$$\xi = F_E/F_C = (q\ E\ r^2/\beta) = E/E_c.n^4 \qquad (8)$$

where E_c is the critical electric field on the first Bohr orbit. The magnitude of E_c is $5.14.10^9$ V/cm while the critical field on the nth Bohr orbit is only E_c/n^4.

2. Some equivalent qualitative pictures

The expression of the coupling parameters can be deduced as well from considerations based on the energy or on the Bohr semi-classical frequencies associated with each forces. For example, anticipating on some results recalled in the following sections for the Landau problem, the motion of the electron in the B field alone is somewhat equivalent to an harmonic motion with the associated classical frequency $\omega_c/2$ (Eq. (5)). Hence the magnetic contribution to the energy is of the order of $n.\hbar\omega_c$.

In the Coulomb limit, it is well known that the binding energy is of the order of $(-R/n^2)$, and the semi-classical frequency or Bohr's frequency is $2R/n^3$. Thus, η of the order of unity expresses as well the equality of the Coulomb and magnetic forces, the equality of the Bohr's semi-classical frequencies $2R/n^3 \sim \hbar\omega_c$, or the equality of the magnetic and coulombic contributions to the energy. Hence, when $\eta \simeq 1$, the total energy of the electron fulfills :

$$E \simeq E_c + E_1 \simeq -R/n^2 + n\hbar\omega_c \simeq 0$$

The transition regime in which the magnetic force becomes comparable to the Coulomb one takes place close to the zero-field threshold. Moreover, from equation (7) the field quantization of the spectrum at constant electron's energy is then $n^3.B/B_c \simeq 1$. One more approximation allows to deduce the spacing in energy close to $E \simeq 0$. Assuming the Coulomb force can be approximated with an harmonic one, the two forces problem reduces to a one harmonic force one, with an associated classical frequency $((2R/n^3)^2 + (\hbar\omega_c)^2)^{1/2}$. Hence when $\eta \simeq 1$ and $E \simeq 0$, the spacing of the energy levels should be of about $\sqrt{2}.\hbar\omega_c$, very close to the $3/2\hbar\omega_c$ value measured in 1969 on the quasi-Landau spectrum [1]. Indeed, although this model does not aim at rigour, it nicely fits the physical reality meaning at the end that the important parameters in the "atom + $\vec{B}$ field" problem do have really been taken into account.

The same kind of arguments are successful in the electric field situation, discarding (see below) any symmetry requirements. The analog of the cyclotron frequency is then the linear Stark frequency $\omega_S = 3nE/2E_c$ in atomic units. When $\xi \simeq 1$ the electric and Coulomb forces are of the same order of magnitude. This occurs for nearly zero energy of the electron and the classical Bohr frequencies are equal ($\omega_S \simeq 2R/n^3$). The quantization in $\vec{E}$ field, at constant electron's energy ($E \simeq 0$), is $n^4.E/E_c \simeq 1$. Within the harmonic approximation, the spacing of energy levels close to $E = 0$ is found to scale as $E^{3/4}$ which is one of the major feature of the so-called Stark resonances discovered in 1977 [2, 3].

3. <u>The three gross regimes of the atomic spectra in external fields - Experimental evidence</u>

From the previous phenomenological considerations, it follows that there are three types of motion in the atomic spectra when the atom experiences the action of an external field. They are discussed below in the magnetic field situation but this applies as well to the electric field case for non-hydrogenic atoms.

• If $\gamma n^3 \ll 1$ this is the low field Coulomb regime in which the magnetic force is small compared to the Coulomb one. This takes place for negative values of the electron's energy ($E < 0$). The Coulomb potential dominates the motion which is Coulomb-like. But the magnetic force, and more precisely (see below) the diamagnetic force,is responsible for the breaking of the Coulomb symmetry. How this occurs is extremely important for understanding the properties of the atom, even at higher field conditions. But the first theoretical description through classical perturbation theory, by Solovev [5], dates 1981. In such a regime, the atom still looks like an atom, although some fundamental alterations of its properties do already take place. For field values greater than B_c, this regime does no longer exist in the quantum problem.

• If $\gamma n^3 \simeq 1$ the magnetic and Coulomb forces are of comparable strengths. This is what has been called the "strong mixing regime" or quasi-Landau regime in the magnetic problem. The lack of separability of the equations in this non-perturbative regime may allow to infer that the electron's motion becomes featureless and unorganized. It was the key role of experiments to prove just the opposite, showing that, for any atom, at least a part of the spectrum was still regular. Indeed the so-called quasi-Landau resonances close to the zero-field threshold have a $3/2\hbar\omega_c$ spacing in energy and fulfills a $n^3.B/B_c = 1.56$ quantization law in field.

• If $\gamma n^3 >> 1$ the (dia)magnetic force dominates the Coulomb one. From the previous remarks the motion of the electron becomes of Landau type, that is the one of a charged particle in a magnetic field. In this regime, the role of the Coulomb force is to perturb the degenerate Landau motion. Especially, it breaks the translational invariance of the system and exerts some binding action along the $\vec{B}$ field direction. This regime does take place for positive value of the electron's energy where the atomic continua are replaced with a discrete type harmonic ladder with a spacing $\hbar\omega_c$. The quantization in field for $E >> 0$ tends to $n.B/B_c$ = Cste and thus is far different from the one obeyed at threshold ($E \simeq 0$).

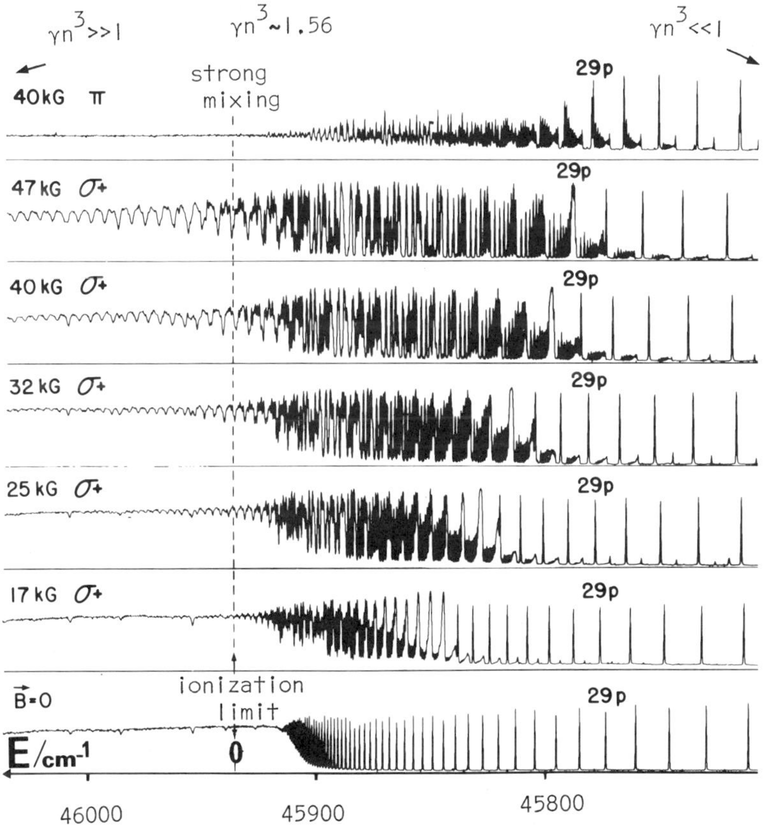

Figure 1 - Experimental spectrum on strontium atoms exhibiting from left to right the three characteristic regimes. The spacing of the resonances close to the zero-field threshold ($E \simeq 0$) is $3/2\hbar\omega_c$ (quasi-Landau regime) (courtesy Dr. F.S. Tomkins).

Hence, the atom in external fields turns out to be somewhat different from what is usually considered as an atom. In magnetic fields, its spectrum evolves between a Coulomb and a Landau signature as exemplified on the experimental plot [6] on Fig. 1. Obviously its properties should be deeply altered and will no longer look like those of an atom in some regions.

The amazing thing is that the previous considerations hold true whatever the absolute strength of the field. This is in consequence of the $\gamma.n^3$ scaling law. For fields in the microgauss or in the tesla range, the phenomena will be similar. In particular, the three gross regimes should exist once the field is non-zero. What matters indeed, is not the absolute field strength, but the quality of experimental detection. Actually, the basic features of these atoms dressed with static fields have been seen under far different experimental conditions, on fast beams in accelerators, Rydberg atoms in cells or beams or in solid state physics conditions on excitonic systems.

4. Further remarks on the phenomenological analysis

In the magnetic problem, the coupling parameter γn^3 has been introduced for the sake of simplicity through considerations on the Lorentz force. This may happen to be confusing for the reader as he may miss the physical origin of the difficulties. They are not associated with Zeeman effect but rather with the diamagnetic interaction. Anticipating on the discussion of the following sections, the Lorentz force associated with paramagnetism cancels with the Coriolis one when performing a rotation to the so-called Larmor frame with angular velocity $\vec{\omega}_C/2$. The centripetal force which remains in the rotating frame is the diamagnetic force $\vec{F}_D$:

$$\vec{F}_D = \frac{q^2B^2}{4m} \vec{\rho} \tag{9}$$

where $\vec{\rho}$ is the vector radius in cylindrical coordinates. It obviously depends on the degree of excitation of the electron's motion and makes the magnetic problem a non-separable one. Hence, the coupling parameter η should be evaluated through the comparison of the diamagnetic and Coulomb forces. One deduces immediately :

$$F_D/F_C = (q^2B^2\rho r^2/4m{\cdot}e^2) \simeq (\gamma.n^3)^2 \simeq \eta^2 \tag{10}$$

Hence the conclusions are unchanged.

The last remark concerns the way of choosing the index n in order to evaluate the coupling parameter. Our previous choice was refering to the Coulomb limit of the problem. But the conclusions do not depend on this assumption, and are still valid if one defines the index with reference to the oscillator-like Landau limit. Instead of scaling as n^2a_o (a_o the Bohr radius), the mean extension of the n orbit scales as $\sqrt{n}.a_c$, where a_c is the unit of length in the Landau problem called the cyclotron radius :

$$a_c = (\hbar/qB)^{1/2} \tag{11}$$

One deduces immediately :

$$F_L/F_C \simeq (Bn^3/B_c)^{1/2}$$

which means that the scaling parameter is once again γn^3.

5. The suitability of Rydberg atoms

Rydberg atoms can be considered a nearly perfect realization of Coulomb driven systems in 3 dimensions. The effects of the core are usually simple to account for through quantum defects corrections. State selection using single mode laser spectroscopy is possible thus allowing to populate a wide selection of energy levels ranging from n = 20 to n = 300. In addition, the use of conventional or exotic optical excitation schemes allows a further selection on the l value. The effects of the core becoming smaller as l increases, this opens the possibility of studying series having a nearly perfect coulombic behaviour. The most recent advances are those on Rydberg states of the hydrogen atom itself.

The first interest of the Rydberg system is the existence of scaling laws on the energy and on other quantities as the atomic sizes which early caught the attention of experimentalists. Though not so obvious, the second aspect is the key one for understanding the recent theoretical advances and the experiments. The symmetries of the Coulomb field rule the dynamics of Rydberg atoms.

Obviously these scaling laws provide us with a good opportunity of experimentally studying the properties of atoms in external fields, over a wide range of external fields strengths. Especially the γn^3 scaling in the magnetic field problem allows to study all the regimes at laboratory field strengths ranging from several hundredths gauss to 8 Teslas. The values of the critical electric and magnetic fields corresponding to the onset of the strong mixing regimes are given in Table 1 which illustrates the wide possibilities of realizations on Rydberg atoms.

Table 1 - Some orders of magnitude of the binding energies, Bohr frequencies and critical electric and magnetic fields according to the principal quantum number in Rydberg atoms (Mind the units).

n	1	30	100	300
$E_n = -R/n^2$	10^5 cm^{-1}	100 cm^{-1}	10 cm^{-1}	1 cm^{-1} = 30 GHz
$\omega_n = 2R/n^3$		220 GHz	6 GHz	220 MHz
$\vec{E}_n = \vec{E}_c/n^4$	5.14 GV/cm	6.30 kV/cm	51 V/cm	0.63 V/cm
$\vec{B}_n = \vec{B}_c/n^3$	$2.35.10^9$ G	87 kG	2.35 kG	87 G

Actually the suitability of such systems for the present purpose is not to prove. For the last ten years this allows to collect major experimental clues which in turn were a strong motivation for the theory [7-10].

THE NATURE OF THE PROBLEMS TO SOLVE

In this section, we present the problem to solve in a more quantitative way. We show in particular that, due to its non-separable character, the use of conventional techniques leads to numerous difficulties, implying to deal with an infinite number of coulombic states including continua. More powerful views are thus required.

1. The Two-body problem in a magnetic field

Even in the non-relativistic approximation, the simplest two-body situation in a magnetic field leads to a first difficulty which is the lack of separability of the equations in the center of mass frame [7 11 12]. This can be established for example from the classical equations of motion. Writing $\vec{r}_{12}$ the relative coordinate of the two particles, one obtains (with $j \neq i$) :

$$m_i . \ddot{\vec{r}}_i = q_i \ \dot{\vec{r}}_i \wedge \vec{B} + (q_1 q_2/4\pi\varepsilon_o) \ \vec{r}_{ij}/(r_{ij})^3 \qquad (12)$$

Writing $\vec{R}$ and $\vec{r}$ the center of mass and relative coordinates, one deduces through an integration that, when $q_1 = -q_2 = q$:

$$\vec{C} = (m_1 + m_2)\dot{\vec{R}} - q\vec{r} \wedge \vec{B} = \text{Cste} \tag{13}$$

The quantity which is conserved is not the velocity of the center of mass but mixes the center of mass and internal coordinates. This obviously holds true in the quantum hamiltonian formulation. Writing $\vec{\Pi}_i$ the momenta of the velocity and $\vec{A}$ the vector potential :

$$\vec{\Pi}_i = \vec{p}_i - q_i \vec{A}(r_i) \tag{14}$$

the expression of the hamiltonian becomes :

$$H = \frac{\Pi_1^2}{2m_1} + \frac{\Pi_2^2}{2m_2} - \frac{q^2}{4\pi\varepsilon_0}\frac{1}{r_{12}} \tag{15}$$

It is straightforward to verify that the operator $\vec{C}$ associated with equation (13) :

$$\vec{C} = \vec{\Pi} - q\vec{r} \wedge \vec{B} = (\vec{P} - q\vec{A}(\vec{R})) - q\vec{r} \wedge \vec{B} \tag{16}$$

$$[C_i, C_j] = 0$$

commutes with the hamiltonian. Hence the existence of $\vec{C}$ allows a quasi-separability of the 2-body equations. In the symmetric gauge :

$$\vec{A} = \frac{1}{2}\vec{B} \wedge \vec{r} \tag{17}$$

taking account of the conservation of $\vec{C}$, one obtains :

$$H_C = \frac{C^2}{2M} - \frac{q}{M}(\vec{C} \wedge \vec{B}).\vec{r} + \frac{P^2}{2\mu} - \frac{q}{2}\left(\frac{1}{m_1} - \frac{1}{m_2}\right)\vec{B}.\vec{L} + \frac{q^2}{8\mu}(\vec{r} \wedge \vec{B})^2 - e^2/r \tag{18}$$

where $\vec{C}$ labels the eigenvalue of the operator $\vec{C}$ ($\mu = m_1m_2/(m_1+m_2)$ is the reduced mass). The first field dependent term in (18) couples the center of mass and relative motions. $(\vec{C} \wedge \vec{B})/M$ represents the motional electric field seen in the rest frame of the reduced particle and is crossed to $\vec{B}$. The second field dependant term is the paramagnetic interaction (in which the gyromagnetic ratio is altered) responsible for Zeeman effect. The last field dependant term, proportional to $(\vec{r} \wedge \vec{B})^2$ is the diamagnetic interaction.

Hence the two-body problem in a $\vec{B}$ field is not separable in the sense that the motions of the center of mass and reduced particle are still coupled [12]. Their energies are not conserved independently. The complete solution of the problem of a one electron atom in a magnetic field, in the non-relativistic approximation, requires to solve an infinite set of one particle problems in crossed electric and magnetic fields.

Nevertheless one should remark that the approximation of infinite proton mass makes the problem tractable and is justified for most atomic physics situations. Equations (16) and (18) are yet of interest in connection with the search for magnetic trapping devices for the neutral atom as it introduces a coupling between the relative and center of mass motions [7]. Exploring the consequences of such a coupling requires to understand better the quantum one-body problem in crossed $(\vec{E}, \vec{B})$ fields, which seems now possible.

Another important exception is the one of positronium in a $\vec{B}$ field for which the infinite mass approximation does not apply (as $m_1 = m_2$). The non-relativistic treatment is thus given through (16) and (18). Note that the paramagnetic interaction disappears in this situation.

The last remark is that the previous derivation is valid for a neutral system in which $q_1 + q_2 = 0$. If this condition is not fulfilled, the components of the operator $\vec{C}$ do no longer commute. When $q_1 = q_2$, which corresponds to an important physical situation in solid state physics - that of the correlations in an electron gas - the conservative quantity just involves the coordinates of the center of mass which means that the problem is separable and retains translational invariance [13] .

2. The approximate one-body hamiltonian in a $\vec{B}$ field

Assuming the proton is infinitely massive ($\vec{C}/M \simeq 0$), the hamiltonian in the symmetric gauge becomes :

$$H = \frac{p^2}{2m} - \frac{q}{2}\vec{B}.\vec{L} + \frac{q^2B^2}{8m}(x^2 + y^2) - e^2/r \qquad (19)$$

When the magnetic field is zero, one recovers the usual Coulomb situation while in the absence of the Coulomb term, the hamiltonian coïncides with the one for the Landau problem.

Assuming both electric and magnetic fields, the general form of the one-body hamiltonian is thus ($\vec{\beta} = \vec{E}/E_c$) :

$$H = p^2/2 - 1/r + \frac{\gamma}{2}.L_z + \frac{\gamma^2}{8}.(x^2+y^2) + \vec{r}.\vec{\beta}/2 \qquad (20)$$

The field dependant terms are successively the Zeeman (paramagnetic), diamagnetic and electric field ones.

In order to clearly state the nature of the difficulties with such a hamiltonian, one will first discuss the magnetic problem ($\beta = 0$). In looking for constants of the motion, one can as well use the classical equations of the motion avoiding the choice of a gauge. Obviously the magnetic problem retains a rotational symmetry along the $\vec{B}$ field axis and spatial parity P is also a constant of the motion. Hence $[H, L_z] = 0$ and $[H, P] = 0$. Combining these two, one deduces that the parity along z axis (P_z) is also a constant of the motion. Now restricting the analysis to the spatial symmetry transformations, the so-called dynamical ones being discussed in the following sections, there are no others which leave H invariant. Especially, due to the diamagnetic interaction, $\vec{L}$ is not a constant of the motion :

$$[H, L^2] \neq 0$$

Actually, the magnetic hamiltonian when including the diamagnetic term turns out to be non-separable in any of the thirteen sets of natural coordinates in R(3). The origin of this lack of separability lies in the hardly complying symmetries of the Coulomb and diamagnetic forces. When the Coulomb force is ignored, H reduces to the hamiltonian for the Landau problem. Consequently it is also worthwhile to test if among the operators which are constant of the motion in this Landau limit, some do not have the property of commuting with the Coulomb potential. The answer is no. Indeed, anticipating on the results of the following sections, the problem is definitely non-separable even in a dynamical sense.

Hence the challenge is now to answer the fundamental questions "what are the spectrum and eigenfunctions for such a magnetic system and how to get them". How the transformation from a Coulomb to a Landau spectrum takes place ? How to describe the regularity of the observed experimental spectra in the quasi-Landau regime and what does it mean ?

3. The need for more powerful conceptual views

Computational works can hardly give the answer to such questions, especially in the strong mixing regimes. This in consequence of the huge sizes of the basis set to consider if one whishes reliable predictions. For example, the diamagnetic hamiltonian couples any Coulomb state (n_o,l_o) to any other one (n,l) including continuum states. The only selection rule is on the l value. $(x^2 + y^2)$ is the sum of a rank 0 and rank 2 tensors on the angular momentum l which leads to $\Delta l = l - l_o = 0 \pm 2$. But there are no selection rules on the principal quantum number. Moreover, as said before, $(x^2 + y^2)$ tremendously depends on the n value as n^4. This means that any truncation of the basis in computations should lead to important and uncontrolled errors, especially when dealing with states close to the zero-field threshold. In addition no structural informations on the problem, for example the existence or non-existence of a dynamical symmetry,can be obtained this way for two reasons - the first one is the uncontrolled accuracy of the computations due to the need of truncating the basis - the second one , much more subtle, is the uncontrolled modification of the symmetries in the whole problem when using the trick of truncation. Actually the weight of continuum states can be shown as large as 20 % in the calculated energies of states with $n \simeq 30$.

The only case in which standing numerical simulations can be thought convenient is the low field Coulomb limit when $2R/n^3 >> H_D$. This is the so-called inter l mixing regime, in which the Coulomb n^2 degeneracy is removed by the diamagnetic interaction (as $[H_D, L^2] \neq 0$). Being restricted to the consideration of one manifold, the problem becomes tractable and the rank of the matrices to deal with is of the order of n, for each M value. Indeed, this is still a tedious and expensive task which in addition is unlikely to provide us with some insights in the structure of the problem. The other way of conceiving this inter-l mixing regime is to remark that it is the one in which the symmetry of the Coulomb field is broken by the diamagnetic interaction. When taking proper account of the symmetry of the Coulomb field, only one analysis is required which produces the general result, valid whatever the n and M values.

The previous considerations just indicate that solving this class of non-separable problems requires to change the angle of view, introducing much more powerful concepts. Standing theoretical methods do not lead to any theoretical predictions and basically, do not lead further than done in 1939 by Schiff and Snyder [14] , but for some special case in the perturbative regime.

4. Looking for dynamical symmetries

At this stage, the example of the electric field problem is of particular interest (for $\gamma = 0$ and $\vec{\beta} \neq \vec{0}$ in equation (20)) as the system is exactly separable which is well-known for years [15 16] . The first level of thinking this feature is well-known. The origin of this lies in the separability of Schrödinger's equation in parabolic coordinates [17] . But this does not give at all any indication about the deep origin of the property. The second degree in the understanding which involves higher concepts is to make the link between the separability in spatial parabolic coordinates and the existence of a constant of the motion built on the so-called Laplace-Runge-Lenz vector $\vec{A}$. This can be straightforwardly established from the classical equations of the motion [18], for example. This property as the conservation of $\vec{A}$ in the Coulomb problem, is called a dynamical symmetry and has no simple geometrical meaning in real space.

Hence comes the hint that for tackling these non-separable problems the first question to answer is the one of the symmetry properties in a general sense, and especially those in the limiting cases. For the magnetic problem this leads to examining those of the Coulomb and Landau problems, which will be done in the following sections. The further application to the magnetic problem will provide us with the basic structural informations on the symmetries. A by-product will be a rational method for choosing proper basis sets and performing truncations in a self consistent way in the quantum approach. But also, we will show that such an analysis leads to defining a convenient set of dynamical variables in the classical problem which in turn allows to understand more in the quantum solution.

5. Some additional remarks

A few remarks are in order. In the general situation were $\vec{E}$ and $\vec{B}$ fields are applied to the atoms, there are no longer any constants of the motion. A purely numerical work is obviously impossible. But, in the low field regime, a complete analysis is easy making use of the symmetry properties of the Coulomb field [19].

It is a very common idea that, when dealing with highly excited states, semi-classical approximations are quite good and a natural tool for making any quantum prediction. This is true. But the point is that there is no obvious ways for performing such a semi-classical analysis when dealing with non-separable problems. This will be exemplified, though not discussed in details in the magnetic field situation (see reference [10] for example). The point is that when the classical motion becomes chaotic as in the magnetic problem, there are no obvious solutions for semi-classically quantizing the system.

We shall now re-introduce the basic concepts pertaining to the symmetries of the Coulomb field, showing that this is certainly the right way for understanding things on a large scale as well as for making convenient predictions. What we discuss is a global way of conceiving the atom which originally starts with Pauli's work around 1925 [19].

STRUCTURE OF THE NON-RELATIVISTIC COULOMB FIELD

We recall here some basic aspects of the Coulomb problem in classical and quantum physics. The stress is put on the structure which is the key point for understanding the dynamics of atoms in external fields and lot of other questions dealing with the perturbed Coulomb problem. There are two degrees in the symmetries of the Coulomb field. The symmetry group expresses a rotational invariance in a four-dimensional space and allows for a powerful description of the properties in term of two quasi spin $\vec{j}_1$ and $\vec{j}_2$, in a given n shell. The dynamical group allows the whole problem to be thought in terms of a pair of 2-dimensional harmonic oscillators. Both ideas are introduced through simple arguments.

1. Views from classical mechanics

The two key points in the Coulomb problem can be drawn from the classical equations of the motion. Writing $V_c = -e^2/r$ the Coulomb potential and $\vec{L} = \vec{r} \wedge \vec{p}$ the angular momentum, the hamiltonian H expresses as :

$$H = p^2/2m - e^2/r = p^2/2m + V_c \tag{21}$$

As a consequence of the isotropy of space and uniformity of time, $\vec{L}$ and H are constants of the motion. But, there is another one, with vectorial character, which is specific to the Coulomb field. This is the Laplace-Runge-Lenz vector $\vec{A}$ such that : [20] [21]

$$\begin{aligned} \vec{A} &= \vec{p} \wedge \vec{L} - me^2\vec{r}/r \\ \vec{A}.\vec{L} &= 0 \end{aligned} \tag{22}$$

For given $(\vec{L},\vec{A})$, the energy is completely defined :

$$A^2 = 2m.E.L^2 + m^2 e^4 \tag{23}$$

which means that the classical motion is degenerate.

The classical trajectories are the well-known conical ones. The existence of an axis of symmetry is associated with the conservation of the Lenz vector $\vec{A}$ (see Figure 2). Writing φ the angle $(\vec{r},\vec{A})$ equation (22) leads to :

$$\frac{1}{r} = \frac{me^2}{L^2}(1 + \frac{A}{me^2}\cos\varphi) \tag{24}$$

The existence of $\vec{A}$ is the first key feature in the Coulomb problem.

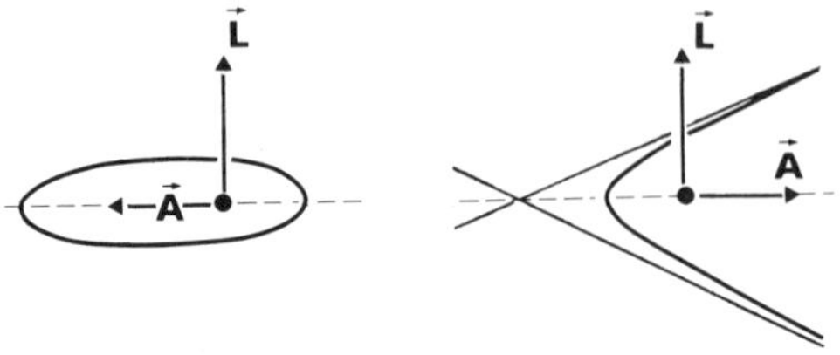

Figure 2 - The classical trajectories showing the basic elements $(\vec{L},\vec{A})$ of the Coulomb symmetry.

The second key feature in this problem involves less obvious considerations on the time and energy. From equation (21) the relationship between the time and radius vector r is of implicit type. It is yet possible to parametrize the conical trajectories in a more convenient way [20] making the substitution

$$r = \frac{e^2}{2E}(1 - \alpha\cos\xi) \tag{25}$$

where $\alpha = A/me^2$ is the eccentricity and ξ the parameter. This amounts to the usual parametrization for conical curves which for $E < 0$ leads to :

$$x = r\cos\varphi = \frac{e^2}{2E}(\cos\xi - \alpha)$$

$$y = r\sin\varphi = \frac{e^2}{2E}(1 - \alpha^2)^{1/2}\sin\xi \tag{26}$$

In the plane perpendicular to $\vec{L}$, this is the motion of a 2-dimensional harmonic oscillator with unit frequency, the parameter ξ playing the role of time. The relationship between ξ and t is (from (21) and (25)):

$$t = (me^4/8E^3)^{1/2}(\xi - \alpha\sin\xi) \tag{27}$$

$$\text{or } \frac{dt}{d\xi} = (m/2E)^{1/2}.r \tag{28}$$

The second key feature in the Coulomb problem is thus its equivalence, whatever the energy, with an harmonic motion provided one abandons the uniformity of time. The new time ξ does not flow uniformly on the Kepler ellipse as $dt/d\xi \propto r(t)$. This is as well valid for describing unbound trajectories ($E > 0$) performing the substitution $\xi \to i\xi$.

2. Traditional quantum views - Some recalls

The standard treatments of (21) make use of the separability of Schrödinger's equation in spherical and parabolic coordinates [17,22]. In spherical coordinates the description is through the set (H, L^2, L_z) of commuting operators with associated quantum numbers (n, l, m). The eigenfunctions separate into a radial and angular part such that :

$$\Psi_{nlm}(r, \theta, \varphi) = \langle r\theta\varphi | nlm\rangle = R_{nl}(r)\, Y_l^m(\theta,\varphi) \tag{29}$$

where R_{nl} is a Laguerre polynomial and Y_l^m the spherical harmonics. The bound spectrum is $E_n = -1/2n^2$ and does present a n^2 orbital degeneracy in the angular momentum quantum numbers ($0 \leqslant l \leqslant n-1$; $-l \leqslant m \leqslant l$).

The separability of Schrödinger's equation in parabolic coordinates is a feature specific to the Coulomb field. The parabolic coordinates are defined through :

$$\xi = r + z \qquad \eta = r - z \qquad \varphi = \mathrm{Arctg}(y/x) \tag{30}$$

which using the fact that L_z commutes with H leads to two equations of the type (Z_i the separation constant) :

$$\frac{d}{d\xi}\left(\xi \frac{df_1}{d\xi}\right) + \left(\frac{E\xi}{2} - \frac{m^2}{4\xi} + Z_1\right) f_1(\xi) = 0 \tag{31}$$

$$Z_1 + Z_2 = 1$$

Some tedious manipulations allow to deduce that the spectrum of the separation constants are quantized for negative energies, according to $Z_i = n_i/n$ where n_i are the parabolic quantum numbers :

$$n = n_1 + n_2 + |m| + 1 \tag{32}$$

$$-(n-1) \leqslant n_1 - n_2 \leqslant n - 1$$

The eigenfunctions are of the general form :

$$\Psi_{n_1 n_2 m} = \langle \xi\eta\varphi | n_1 n_2 m\rangle = e^{im\varphi}\, f_{n_1 m}(\xi/n) f_{n_2 m}(\eta/n) \tag{33}$$

where the $f_{n_i m}$ are hypergeometric functions [17].

With such treatments the specificity of the Coulomb field is completely hindered and any development would be by force heavily calculatory. Important questions like "is there any other coordinate systems in which the problem separate" cannot be addressed in a rational way. Around 1920, Epstein and Pauli [15 16] have yet a clear view of what was involved. This has been further developped by Fock [23], Biedenharn [24], Fronsdal [25] and Barut [26] among many others. The first step was to recognize that the separability in parabolic coordinates was linked to the conservation of an hermitic version of the Lenz vector [17] :

$$\vec{A} = (\vec{p} \wedge \vec{L} - \vec{L} \wedge \vec{p})/2 - me^2\vec{r}/r \tag{34}$$

and that the parabolic eigenfunctions are common eigenvectors of (H, L_z, A_z) with :

$$A_z\, \Psi_{n_1 n_2 m}(\xi\eta\varphi) = \frac{n_1 - n_2}{n}\, \Psi_{n_1 n_2 m}(\xi\eta\varphi) \tag{35}$$

3. The symmetry group of the non-relativistic Coulomb problem

We will introduce the symmetry group from algebraic considerations very similar to the ones in Pauli [19 24] rather than using the Fock's method [23 27]. This only requires to understand the properties of the angular momentum in 3 dimensions [22]. The treatment closely parallels the classical one. We first consider the operators $\vec{L}$ and $\vec{A}$ (eq.(34)) which commute with the hamiltonian H. It is straightforward to establish their commutation relations from the first principles, which in tensorial notations, are :

$$[L_i, L_j] = i\hbar.\varepsilon_{ijk}.L_k \tag{36}$$
$$[L_i, A_j] = i\hbar.\varepsilon_{ijk}\, A_k \tag{37}$$
$\vec{L}$ is an angular momentum in a 3-dimensional space while $\vec{A}$ is a vectorial operator, though not an angular momentum as :
$$[A_i, A_j] = -\,2i\hbar.\varepsilon_{ijk}.H.L_k \tag{38}$$
Moreover, one can show that the quantum analog of (22) and (23) holds. A closer inspection of (38) allows to be convinced that the algebra of operators becomes closed, provided one defines a scaled version of the Lenz vector :
$$\vec{a} = (-\,2mH)^{-1/2}\vec{A} \tag{39}$$
$\vec{a}$ is defined in the subspaces associated with the eigenvalue E of the Coulomb hamiltonian H and has the dimension of an angular momentum. Hence, as in the classical case, the $(\vec{L}, \vec{a})$ set of operators allows for a complete description of the Coulomb problem, and build now a (closed) Lie algebra. On obtains (with (36)) :
$$[a_i, a_j] = i.\varepsilon_{ijk}.L_k \qquad [L_i, a_j] = i\varepsilon_{ijk}.a_k \tag{40}$$
$$\vec{L}.\vec{a} = \vec{a}.\vec{L} = 0 \qquad a^2 + L^2 + 1 = -\frac{1}{2H} \tag{41}$$
Remark that the description is as well valid for positive energies extending (39) to imaginary values of $\vec{a}$.

- The SO(4) symmetry of the bound Coulomb spectrum

Equations (40) and (41) have a well-defined group theoretical meaning. They express a rotational invariance in a four-dimensional space for the bound Coulomb spectrum. This is labelled as SO(4) symmetry. Writing 1 to 3 the (xyz) spatial coordinates and 4 the extra-dimension, it turns out from the commutation relations (40) that the six operators $(\vec{L},\vec{a})$ do build the six components of the angular momentum $\vec{\Sigma}$ in the four dimensional space [24] and :
$$\begin{matrix} \Sigma_{23} = L_x & \Sigma_{31} = L_y & \Sigma_{12} = L_z \\ \Sigma_{14} = a_x & \Sigma_{24} = a_y & \Sigma_{34} = a_z \end{matrix} \tag{42}$$
Hence, equation (41) becomes $\Sigma^2 + 1 = -1/2H$ making clear the rotational invariance in the four-dimensional space for $E < 0$. The eigenfunctions of the Coulomb problem are thus the spherical harmonics in four dimensions [23 28]. For the scattering states ($E > 0$), the symmetry is SO(3,1) expressing the invariance in rotations in a four-dimensional Lorentz space. This is due to $\vec{a}$ taking imaginary values in (39) to (42).

The $\Sigma_{\alpha\beta}$ in (42) provides us with the six infinitesimal generators of the rotation group SO(4) in 4 dimensions. The first three ones are the generators of the rotation group SO(3) in real space associated with $\vec{L}$ while the last three ones from (40) do not build a 3-dimensional angular momentum. But it turns out that a description of SO(4) in terms of two independent SO(3) symmetries is possible :
$$SO(4) = SO(3) \otimes SO(3) \tag{43}$$
which allows a major simplification as the angular momentum algebra in 3 dimensions is well-known [22].

- The two "quasi-spin" description of the Coulomb symmetry

From the structure of (41) it seems wise to introduce two new operators $\vec{j}_{1,2}$ acting in a given n subspace, such that :
$$\vec{j}_{\substack{1\\2}} = (\vec{L} \pm \vec{a})/2 \tag{44}$$
which from (40) fulfills the simple equations (i=1,2) :

$$[j_{ik}, j_{il}] = i\,\varepsilon_{klm}\cdot j_{im} \qquad (45)$$
$$[\vec{j}_1, \vec{j}_2] = \vec{0}$$

Thus $\vec{j}_1$ and $\vec{j}_2$ build a system of two independant (commuting) angular momenta in a 3-dimensional space. From (44), they are constants of the motion. Consequently their spectrum and eigenfunctions are known from usual angular momentum theory. With j_i taking integer or half integer values, one deduces immediately $j_i^2 = j_i(j_i+1)$ and $j_{iz} = m_i$ $(-j_i \leqslant m_i \leqslant j_i)$. The constraints in (41) lead to (with $E = -1/2n^2$):

$$j = j_1 = j_2 = (n - 1)/2 \qquad (46)$$

Hence, the complete description of the properties of the n Coulomb shell just involves two independent angular momenta $\vec{j}_1$ and $\vec{j}_2$ in a 3-dimensional space with $j_1 = j_2 = (n - 1)/2$. This leads to the semi-classical vectorial model on Figure 3 which involves two vectors. A further complexification of (45) and (46) with $n = i/k$ allows for the description of scattering states $(E > 0)$.

From the fact that $\vec{L}$ and $\vec{a}$ are respectively axial and polar in parity operation (P) one deduces

$$P\vec{j}_{\substack{1\\2}}\cdot P^+ = \vec{j}_{\substack{2\\1}} \qquad (47)$$

Note that the fact that $\vec{j}_1$ and $\vec{j}_2$ have been recognized 3-dimensional angular momenta which provides us with their spectrum and eigenfunctions, is a key point as from (44) they are complicated differential operators depending on $\vec{r}$ and $\vec{p}$ at third order. This illustrates the strong predictive power of such an analysis based on the inner symmetries rather than on the geometrical ones in R(3).

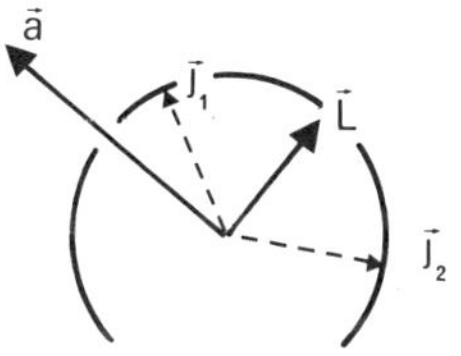

Figure 3 - The semi-classical double vectorial model of the SO(4) symmetry.

• Generalized eigenbasis for the Coulomb problem

From (45), it is now possible to define the various classes of eigenfunctions for the Coulomb problem, in a general way, without referring to a particular coordinate representation. There are two main classes, according to the usual coupling schemes of two angular momenta.

The first class of eigenfunctions is of decoupled type in which the two angular momenta $\vec{j}_1$ and $\vec{j}_2$ are quantized along two different axis $\vec{\omega}_1$ and $\vec{\omega}_2$ in space. This corresponds to $(j_1^2\ j_2^2\ j_{1\omega_1}\ j_{2\omega_2})$ in standard notations. The eigenfunctions are the direct product of the ones for each quasi-spin $|j_i m_i\rangle$, with $n = 2j+1$:

$$H|j\ m_1\ m_2\rangle = -\frac{1}{2n^2}|j\ m_1\ m_2\rangle$$
$$j_{i\omega_i}|j\ m_1\ m_2\rangle = m_i|j\ m_1\ m_2\rangle \qquad (48)$$
$$-(n - 1)/2 \leqslant m_i \leqslant (n - 1)/2$$

This corresponds to the following subgroup chain reduction of SO(4) :

$$SO(4) = SO(3) \otimes SO(3) \supset SO(2) \otimes SO(2) \qquad (49)$$

The n^2 degeneracy of the n shell is labelled through (m_1, m_2) as shown on Figure 4. From (47) these eigenfunctions usually do not have a definite parity. Convenient symmetrization may be required. Usually, they are neither eigenfunctions of the components of $\vec{L}$ or $\vec{a}$, nor of a type separable in some coordinates system in R(3).

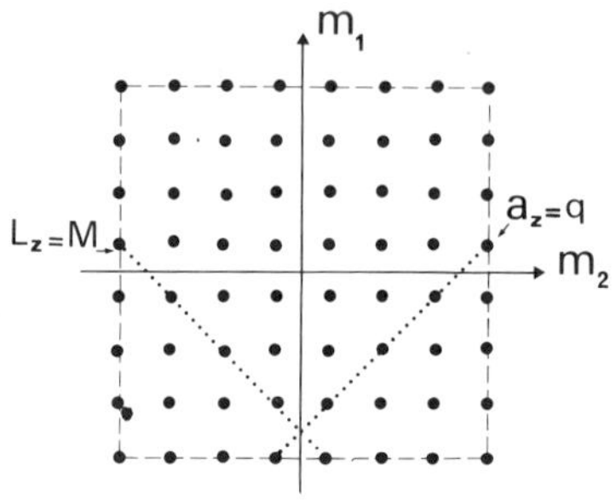

Figure 4 - The n^2 degeneracy of the n = 8 Coulomb shell as labelled through ($j_{1\omega_1} = m_1$, $j_{2\omega_2} = m_2$)

A special case is yet (j_1^2 j_2^2 j_{1z} j_{2z}) in which the two axis $\vec{\omega}_1$ and $\vec{\omega}_2$ coincide with each other. Hence from (44) and (48) the z components of $\vec{L}$ and $\vec{a}$ are defined with $L_z = m_1 + m_2$ and $a_z = m_1 - m_2$. Being a set of eigenfunctions of (H, L_z, a_z) they identify to within a phase factor with the parabolic eigenfunctions (33) :

$$\langle \xi\eta\varphi | j_1^2 j_2^2 m_1 m_2 \rangle_z = (-)^{\alpha}\, \Psi_{n_1 n_2 m}(\xi\eta\varphi) \qquad (50)$$

From (50), one solves in a general way the problem of the spatial representations of the {j_1^2 j_2^2 $j_{1\omega_1}$ $j_{2\omega_2}$} eigenfunctions. The latter are deduced through the product of two independent rotations from the (j_1^2 j_2^2 j_{1z} j_{2z}) parabolic basis. Introducing β_i the angle and $\vec{k}_i$ the axis of the rotation ($\vec{k}_i = \vec{z} \wedge \vec{\omega}_i$) one obtains :

$$|j \;\; m_i\rangle_{\omega i} = R(\vec{\omega}_i)|j \;\; m_i\rangle_z = e^{i\beta_i(\vec{k}_i \cdot \vec{j}_i)}|j \;\; m_i\rangle_z \qquad (51)$$

which means the rotated states are similar to the coherent states of the angular momentum $\vec{j}_i$ [29]. Hence :

$$|j \;\; m_1 \; m_2\rangle_{\omega_{12}} = R(\vec{\omega}_1).R(\vec{\omega}_2)|j \;\; m_1 \; m_2\rangle_z \qquad (52)$$

which upon expanding using the properties of the rotation operators [22] and (50) leads to the spatial representations in parabolic coordinates. The electronic density distributions for such generalized parabolic states are fairly anisotropic. We will further show some physical applications to this scheme.

The second class of eigenfunctions is associated with the following subgroup chains description of SO(4) :

$$SO(4) = SO(3) \otimes SO(3) \supset SO(3) \supset SO(2) \qquad (53)$$

which corresponds to the various coupling schemes of the two angular momenta $\vec{j}_1$ and $\vec{j}_2$. A well-known choice is $(j_1^2\ j_2^2\ (\vec{j}_1 + \vec{j}_2)^2 (j_1 + j_2)_z)$ which from (44) identifies with $(j_1^2\ j_2^2\ L^2\ L_z)$ where $\vec{L}$ is the angular momentum in real space. The spatial representations of these eigenfunctions in spherical coordinates thus identifies with the usual spherical eigenfunctions (29), to within a phase factor. One obtains :

$$\langle r\theta\varphi | j_1^2 j_2^2 L^2 L_z\rangle = (-)^{\beta}.\Psi_{nlm}(r,\theta,\varphi) = (-)^{\beta} R_{nl}(r).Y_l^m(\theta,\varphi) \qquad (54)$$

Mind that (54) includes both the radial and angular parts of the eigenfunctions. This scheme complies with the isotropy of space.

From (42) there are some other ways of building a 3-dimensional angular momentum from the components of $\vec{\Sigma}$. For example, the quantity $\vec{\lambda}(a_x,\ a_y,\ L_z)$ [30] fulfills

$$[\lambda_i,\ \lambda_j] = i.\varepsilon_{ijk}.\lambda_k \qquad (55)$$

and is consequently a 3-dimensional angular momentum. It corresponds to the non-standard coupling of $(\vec{j}_1,\ \vec{j}_2)$ such that $\vec{\lambda}(j_{1x}-j_{2x},\ j_{1y}-j_{2y},\ j_{1z}+j_{2z})$. Hence the $(j_1^2 j_2^2 \lambda^2 \lambda_z)$ are eigenfunctions of the Coulomb problem such that :

$$\lambda^2 | j_1^2 j_2^2 \lambda^2 \lambda_z\rangle = \lambda(\lambda+1) | j_1^2 j_2^2 \lambda^2 \lambda_z\rangle$$

$$0 \leqslant \lambda \leqslant n-1 \qquad -\lambda \leqslant \lambda_z \leqslant \lambda \qquad (56)$$

The angular momentum $\vec{\lambda}$ which plays an important role in the theory of doubly excited states [30] and diamagnetism is a pure product of the Coulomb symmetry. Though $\vec{\lambda}$ is a generator of rotations in 3 dimensions, they are not rotations in real space (R(3)). For example $e^{i\alpha\lambda_x}$ will transform a classical Kepler ellipse into another one with the same energy but different eccentricity.

In order to deduce the spatial representation of the λ eigenfunctions, one should remark that a convenient rotation in R(4) allows to transform $\vec{L}$ into $\vec{\lambda}$ as :

$$e^{i\frac{\pi}{2}a_z}.\vec{L}\ e^{-i\frac{\pi}{2}a_z} = \vec{\lambda} \qquad (57)$$

which allows to deduce the relationship between the two sets of eigenfunctions

$$| j_1^2 j_2^2 \lambda^2 \lambda_z\rangle = e^{i\frac{\pi}{2}a_z}\ | j_1^2 j_2^2 L^2 L_z\rangle \qquad (58)$$

Using $a_z = j_{1z} - j_{2z}$ allows to evaluate (58) through conventional 3-dim. angular momentum techniques and with (54) to deduce the spatial representations.

• Concluding on the SO(4) group approach

This is one of the rational way for looking at the properties of the non-relativistic Coulomb field of which the full specificity is taken into account. As a matter of fact, we have deduced new types of eigenfunctions which play an important role for dealing with the perturbed Coulomb problem. In addition the conception in terms of two independent quasi-spin $(\vec{j}_1,\ \vec{j}_2)$ is much more simpler and especially makes trivial most of the calculations of matrix elements, which are reduced to 3-dimensional angular algebra ones. A striking example is the one of the so-called parabolic and spherical basis. From the fact that they only differ through a recoupling of $\vec{j}_1$ and $\vec{j}_2$, it follows

$$|j_1^2 j_2^2 L^2 M\rangle = \sum_{m_1 m_2} \langle \frac{n-1}{2} \frac{n-1}{2} m_1 m_2 | LM\rangle | j m_1 m_2 \rangle \qquad (59)$$

where the coefficients are the usual Clebsch-Gordan coefficients.

Another remark is that the $(\vec{j}_1,\vec{j}_2)$ approach restablishes on general grounds the origin of a certain similarity [28] of the radial and angular parts of the eigenfunctions, as in (54) in a special case. More generally the analogy with a fundamental object in molecular physics, the rigid rotator, becomes obvious. They indeed share the same SO(4) symmetry group, and (52)(56) are eigenfunctions having some molecular analogs.

At last the $(\vec{j}_1,\vec{j}_2)$ approach leads to defining in a general way new irreducible tensorial sets, for dealing with relaxation processes in a given n shell or the interaction with the radiation field. This is as well valid for quasi-hydrogenic atoms or ions in the non-relativistic approximation.

The physical relevance of this description of the Coulomb problem will be made clear in the next sections especially in the crossed $(\vec{E},\vec{B})$ fields and diamagnetic situations.

4. The Coulomb dynamical group and the oscillator representation A special case

The second degree in the symmetries of the Coulomb field results of its equivalence, with an isotropic harmonic oscillator in a four-dimensional space [31 32]. This allows to build a structure, the Coulomb dynamical group, which automatically complies with the symmetry requirements and allows to interconnect all the Coulomb states. It is thus of extreme importance for dealing with the perturbed orbital Coulomb problem.

We will here consider a restricted version SO(2,2) of the full SO(4,2) Coulomb dynamical group, useful for dealing with problems in which there remains a cylindrical symmetry. This will be introduced from the first principles rather than from group theory [32] and amounts to describing the Coulomb problem as a pair of phase-coupled 2-dimensional oscillators. We will first recall some basic properties of the 2-dimensional isotropic harmonic oscillator which will be further used for solving the Landau problem.

- Properties of the 2-dimensional isotropic harmonic oscillator with unit frequency

Owing to the isotropic character, it is a degenerate system which can be thought either in terms of the uncoupled motion of two one-dimensional oscillators along the (u,v) cartesian axis or in terms of well-defined angular momentum states [32 33] in 2-dim. polar coordinatès. As for the one-dimensional oscillator, the most synthetic views result of the introduction of the set $(a_u,\ a_u^+\ ;\ a_v,\ a_v^+)$ of Fock's operators [32 33] in cartesian coordinates :

$$a_u = (u + i.p_u)/\sqrt{2} \qquad \text{with } [a_u,\ a_v^+] = \delta_{uv} \qquad n_u = a_u^+ . a_u \qquad (60)$$

where $n_{u,v}$ are the number of elementary excitations along the (u,v) axis. The spectrum and eigenfunctions follow immediately :

$$H = (a_u^+ a_u + a_v^+ a_v + 1) = (n_u + n_v + 1) = (N + 1)$$

$$|N,n_u,n_v\rangle = (n_u! n_v!)^{-1/2} (a_u^+)^{n_u} (a_v^+)^{n_v} |0,0\rangle \qquad (61)$$

where $|0,0\rangle$ is the direct product of the vacuum states for each (u,v) oscillator. Hence each N level has a (N + 1) degeneracy in (n_u,n_v). The spatial representations $\langle u,v|N,n_u n_v\rangle$ of these eigenfunctions are just the product of two one-dimensional oscillator eigenfunctions, each one involving an Hermite polynomial. The parity of the states is $(-)^N$.

But we are much more concerned with the following description which takes advantage of the isotropic character of the system. The conservation of the angular momentum L leads to introducing from (6), the new set of Fock's operators $(a_d,a_d^+ ; a_g,a_g^+)$ associated with the creation (+) or destruction of right-handed (d) or left-handed (g) elementary circular excitations [33] :

$$a_d = (a_u - ia_v)/\sqrt{2} \qquad a_g = (a_u + ia_v)/\sqrt{2} \tag{62}$$

One deduces immediately :

$$H = (a_d^+a_d + a_g^+a_g + 1) = (n_d + n_g + 1) = (N + 1)$$

$$L = (a_d^+a_d - a_g^+a_g) = n_d - n_g = M \tag{63}$$

where n_d and n_g are the numbers of right-handed and left-handed circular excitations in the system. As expected, L expresses as the difference of these occupation numbers. The eigenfunctions are then :

$$|N,n_d,n_g> = (n_d!n_g!)^{-1/2}(a_d^+)^{n_d}(a_g^+)^{n_g}|0,0> \tag{64}$$

In (ρ,φ) 2-dim. polar coordinates, the spatial representations $<\rho,\varphi|N,n_dn_g>$ of these eigenfunctions involve a generalized Laguerre function in ρ^2 [21] while the φ dependance is $\exp(iM\varphi)$. The parity of the states is $(-)^N = (-)^M$. With (n_d,n_g) non-negative integers and (63), one deduces that M has the same odd or even integer character as N and is allowed to take (N + 1) values in the range :

$$-N \leqslant M = n_d - n_g \leqslant N \tag{65}$$

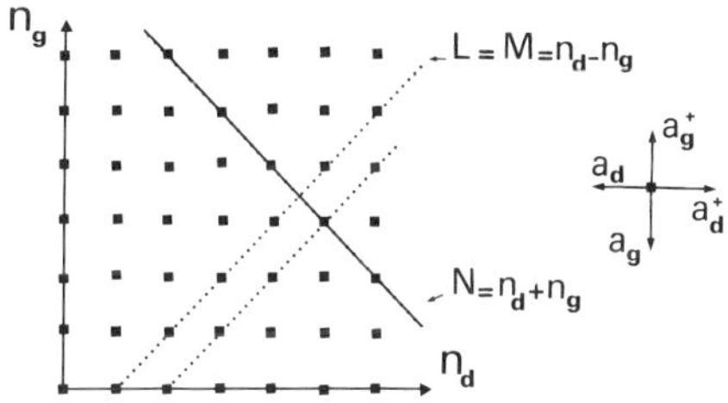

Figure 5 - The 2-dim. isotropic oscillator spectrum labelled through the (n_d,n_g) occupation numbers for circular excitations. The solid lines represent the energy (N+1) and dotted lines the angular momentum M values. The parity of the states is $(-)^N=(-)^M$. The action of the Fock operators [62] on the (n_d,n_g) ladder is shown.

An alternate labelling of the states (n_d,n_g) can be done introducing the quantum number n_ρ [21] such that :

$$N = 2n_\rho + |M| \tag{66}$$

where $n_\rho = n_g$ for $M > 0$ and $n_\rho = n_d$ for $M < 0$.

The plot on Figure 5 represents the 2-dimensional isotropic oscillator spectrum. For a given energy (N + 1), the angular momentum M values do vary by 2. There are two classes of states according to parity.

• The symmetry group of the 2-dim. isotropic oscillator

From (64) the Fock's operators allow to interconnect all the states in the spectrum. In particular they afford a complete description of the symmetries in a given energy shell (N + 1). In this respect, one is led to considering their combinations which conserve the total number N of elementary circular excitations (n_d,n_g). L as well as $a_g^+a_d$ and $a_d^+a_g$ are the simplest choices. Defining a new (vectorial type) operator $\vec{J}$ through its standard components in the rotating frame (+,-,w) as :

$$J_+ = J_u + iJ_v = a_d^+ a_g \qquad J_- = J_u - iJ_v = a_g^+ a_d \qquad J_w = \frac{1}{2}(a_d^+ a_d - a_g^+ a_g) = \frac{L}{2}$$

They are, by definition, constants of the motion $[H,\vec{J}] = 0$ and fulfill from (60)(62) : (67)

$$[J_+,J_-] = 2J_w \qquad [J_w,J_+] = J_+ \qquad [J_w,J_-] = -J_- \tag{68}$$

which proves that $\vec{J}$ is an angular momentum [22]. Hence the spectrum and eigenfunctions are known. With (63), one obtains :

$$J^2 = J_u^2 + J_v^2 + J_w^2 = j(j+1) = (H^2 - 1)/4 \tag{69}$$

$$-j \leqslant J_w = m \leqslant j$$

with j integer or half integer. Hence the identification N = 2j and M = 2m. The degeneracy is thus 2j + 1 = N + 1. Integer values of j are associated with even parity states meaning that these states have an SU(2) group structure while states with odd parity (with j half integer) are associated with spinorial type representations [32]. This makes clear the origin of some pecularities in the diagram on Figure 5. The $\vec{J}$ which are constants of the motion can as well be derived from direct considerations [32, 34].

• A special case of the dynamical group of the 2-dim. isotropic oscillator

The concept of symmetry group aims at describing the properties of the system in a given energy shell N as $[\vec{J},H] = \vec{0}$. The purpose of the dynamical group is to generalize such notion in order to describe the whole problem whatever the energy. We shall introduce this in a special case [32] where the angular momentum L = M is fixed. The structure to build should leave L invariant but not the hamiltonian H. From Figure 5 it is clear that the quadratic operators $a_d^+.a_g^+$, $a_d.a_g$ and the hamiltonian itself commute with L as they conserve the difference $(n_d - n_g)$ of the number of circular excitations. They are the simplest choice which leads to defining a new operator $\vec{S}$ through its standard components, with $[\vec{S},L] = \vec{0}$:

$$S_+ = S_u + iS_v = a_d^+ a_g^+ \quad S_- = S_u - iS_v = a_d a_g \quad S_w = \frac{1}{2}(a_d^+ a_d + a_g^+ a_g + 1) = H/2 \tag{70}$$

where S_w is half the hamiltonian (63). From their commutation relations

$$[S_+,S_w] = -S_+ \qquad [S_-,S_w] = S_- \qquad [S_+,S_-] = -2S_w \tag{71}$$

the components of $\vec{S}$ build a closed structure which is an $O_{2,1}$ algebra [32] But for signs this is similar to the commutations relations (68) of the angular momentum algebra O_3. The associated dynamical group SO(2,1) at constant L can be thought as the group of rotations in a 3-dimensional Lorentz space which leaves the metric $u^2 + v^2 - w^2$ invariant. It follows that the components of $\vec{S}$ commute with

$$S^2 = S_u^2 + S_v^2 - S_w^2 = (1 - L^2)/4 = (1 - M^2)/4 \tag{72}$$

(compare the signs in (69) and (72)). In analogy with the notations for the angular momentum $\vec{J}$, the eigenfunctions of (70)(72) are labelled as (S^2,S_w).

Figure 6 illustrates the difference between symmetry and dynamical group. The $\vec{S}$ allows to interconnect all the states whatever the energy, at constant L. Hence for each parity, a single but infinite dimensional representation of SO(2,1) produces both the spectrum and eigenfunctions. The whole structural informations on the system are included.

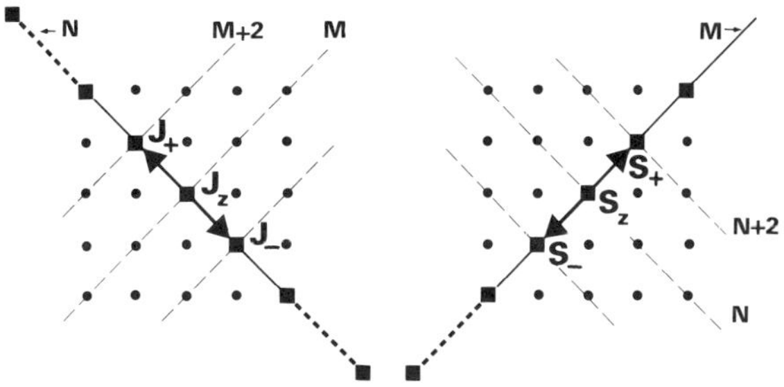

Figure 6 - The concepts of symmetry and dynamical groups for the 2-dim. isotropic oscillator. The symmetry group with the generator ($\vec{J}$) aims at describing the properties in a given (N + 1) energy shell. The dynamical group at constant L = M with the generator ($\vec{S}$) aims at describing the properties whatever the energy.

The representations of SO(2,1) are as well-known as those of SO(3)[32]. They are involved in the theory of superfluidity and parametric oscillation. SO(2,1) being non compact, the unitary irreducible representations are infinite dimensional as they should from Figures 5 and 6. The representations D_k^+ realized in the oscillator problem are specified through the index $k = (|M| + 1)/2$ and a non-negative quantum number n which identifies with the one in (66). Hence this (S^2, S_W) description fulfills :

$$S^2|n,k\rangle = k(1 - k)|n,k\rangle = \frac{1}{4}(1 - M^2)|n,k\rangle$$

$$S_W|n,k\rangle = (k + n)|n,k\rangle$$

$$S_+|n,k\rangle = (2k + n)^{1/2}(n + 1)^{1/2}|n+1,k\rangle \qquad (73)$$

From (62), (66), and (70) this is strictly equivalent to the direct deduction using the Fock's notations. For example :

$$S_+|n_d,n_g\rangle = (n_d+1)^{1/2}(n_g+1)^{1/2} \quad |n_d+1,n_g+1\rangle$$

• The dynamical group and squeezing of the 2-dim. oscillator

As an illustration of the powerful character of the dynamical group concept, we apply it to the description of the oscillator with frequency ω. But before one will deduce the explicit expression of the operators $\vec{S}$ in (70) in 2-dim. polar coordinates. This is easy from the definition (60) and (62) of the Fock's operators. One obtains [35] :

$$S_{\substack{u\\w}} = (\pm \frac{\partial^2}{\partial\rho^2} \pm \frac{1}{\rho}\frac{\partial}{\partial\rho} \mp \frac{M^2}{\rho^2} + \rho^2)/4 \tag{74}$$

$$S_v = i(1 + \rho.\partial/\partial\rho)/2 \tag{75}$$

where S_w is half the hamiltonian for the oscillator with unit frequency. The Schrödinger's equation for the oscillator with frequency ω is thus :

$$\frac{1}{2}[-\frac{\partial^2}{\partial\rho^2} - \frac{1}{\rho}\frac{\partial}{\partial\rho} + \frac{M^2}{\rho^2} + \omega^2\rho^2]\Psi = E \,.\, \Psi \tag{76}$$

From equation (74), one deduces :

$$\rho^2 = 2(S_w + S_u) \tag{77}$$

Hence (76) rewrites :

$$(S_u(\omega^2 - 1) + S_w(\omega^2 + 1) - E)\Psi = 0 \tag{78}$$

It is of general form $\vec{S}.\vec{n}$ = Cst and can be diagonalized through a Lorentz rotation around v axis with the angle $\theta = \mathrm{Log}\omega$. More generally :

$$e^{i\theta S_v} S_{\substack{w\\u}} e^{-i\theta S_v} = ch\theta.S_{\substack{w\\u}} - sh\theta.S_{\substack{u\\w}} \tag{79}$$

which is similar to the results for the SO(3) rotation group. Performing a change or tilt in the wavefunction, one obtains :

$$\tilde{\Psi} = e^{i\theta S_v}.\Psi \qquad \text{with } \theta = \mathrm{Log}\,\omega \tag{80}$$

which finally leads to the equation :

$$(2S_w - E/\omega)\tilde{\Psi} = 0$$

Hence the spectrum and eigenfunctions are given through (73). This means that one passes from the eigenfunctions of the oscillator with unit frequency to the ones of the oscillator with frequency (ω) through the hyperbolic rotation $e^{-i\theta S_v}$. The operator S_v is called for this reason "squeezing, dilatation or tilt operator".

• The Coulomb problem as a pair of 2-dim. isotropic oscillators

The simplest way of introducing this equivalence is to write down the Schrödinger's equation in semi-parabolic coordinates :

$$\mu = (r + z)^{1/2} \qquad \mu^2 + \nu^2 = 2r$$

$$\nu = (r - z)^{1/2} \qquad \mu^2 - \nu^2 = 2z$$

$$L_z = -i\hbar\partial/\partial\varphi = m \tag{81}$$

which after factorizing the $e^{im\varphi}$ dependance leads to :

$$[F(\mu) + F(\nu) + E(\mu^2 + \nu^2) + 2]\Psi = 0$$

$$\text{with } F(\mu) = \frac{1}{2}(\frac{\partial^2}{\partial\mu^2} + \frac{1}{\mu}\frac{\partial}{\partial\mu} - \frac{m^2}{\mu^2}) \tag{82}$$

owing to the multiplication of the equation with $2r = (\mu^2 + \nu^2)$ the Coulomb potential is replaced with the constant "2" term, while the energy E is the coupling constant of an harmonic potential in $(\mu^2 + \nu^2)$. Comparing to equation (76) one deduces the equivalence of the Coulomb problem with a pair of 2-dimensional isotropic oscillators at constant L = m. The frequency of the oscillators fulfills $\omega^2 = -2E$ where E is the Coulomb energy. The oscillators do share the same $L_z = m = -i\hbar\partial/\partial\varphi$ angular momentum and in that sense are "phase-coupled" in φ. μ and ν are the radius vectors of the oscillators in 2-dim. polar coordinates (cf. Fig. 7).

Hence the solutions to (82) for the eigenvalue "2" map the solutions of the Coulomb problem for the energy E.

• The SO(2,2) Coulomb dynamical group.

Introducing through (74) and (75) the two sets $(\vec{S},\vec{T})$ of SO(2,1) generators for each (μ,φ) and (ν,φ) oscillator with unit frequency, they fulfill :

$$[\vec{S},\vec{T}] = \vec{0} \qquad S^2 = T^2 = \frac{1}{4}(1 - m^2) \tag{83}$$

With $\alpha = -2E$, the equation for the Coulomb problem rewrites :

$$[(1+\alpha)(S_z + T_z) + (\alpha - 1)(S_x + T_x) - 2]\Psi = 0 \tag{84}$$

which is of the general form (78) for each independent system. It can be diagonalized through a convenient tilt. The dynamical group structure is the direct product :

$$SO(2,2) = SO(2,1) \otimes SO(2,1) \tag{85}$$

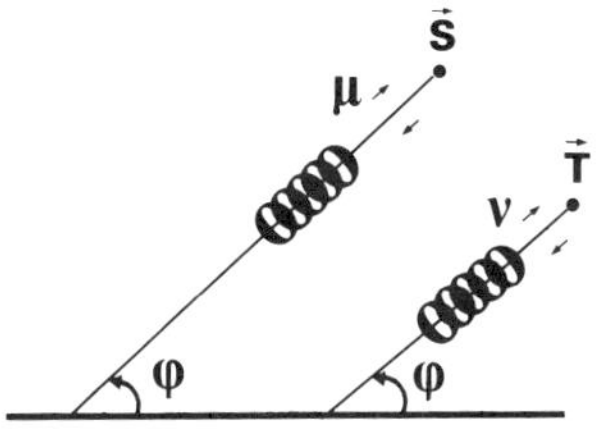

Figure 7 - The Coulomb problem as a pair of 2-dim. isotropic oscillators. The oscillators do have the same angular momentum L = m and a frequency $\omega = (-2E)^{1/2}$ where E is the Coulomb energy. μ and ν are the radius vectors for each oscillator in 2-dim. polar coordinates. To each oscillator is attached a system of generators of the dynamical group, respectively $\vec{S}$ and $\vec{T}$, allowing to describe the dynamics along μ and ν at fixed m.

Obviously the choice of the tilt angle depends on the energy E. For negative energies (E < 0), the harmonic potential is attractive and the situation is alike the one in (78). From (80) the tilt of the equations leads to :

$$\tilde{\Psi} = e^{i\theta(S_y+T_y)}.\Psi \qquad \text{with } \theta = \text{Log}\sqrt{-2E}$$
$$(S_z + T_z - 1/\sqrt{-2E}).\tilde{\Psi} = 0 \tag{86}$$

which is diagonal in the decoupled representation $(S^2\ T^2\ S_z\ T_z)$. From (73), the eigenfunctions are the direct product $|n_\mu,k\rangle \otimes |n_\nu,k\rangle$ with $k = (1+|m|)/2$. The eigenvalue of $(S_z + T_z)$ is $n_\mu + n_\nu + |m| + 1$. Hence the solution to (86) is $E_n = -1/2n^2$ as expected and the tilt angle $\theta_n = -\text{Log}n$. The solutions to the Coulomb problem for the energy E_n are thus squeezed eigenfunctions of the pair of oscillators with unit frequency :

$$|n,n_\mu n_\nu\rangle = e^{i\text{Log}n.(S_y+T_y)}|n_\mu,k\rangle \otimes |n_\nu,k\rangle \tag{87}$$

or equivalently, eigenfunctions of the pair of oscillators with frequency $\omega_n = 1/n$ (cf. Fig. 7). They do correspond to the eigenfunctions of the parabolic type while (n_μ,n_ν) identifies with the usual parabolic quantum numbers [32 35].

For the continuous spectrum ($E > 0$), the reduction of (84) into (86) is not possible. The harmonic potential is now repulsive. Using (79) and a tilt $\theta_1 = \text{Log}\sqrt{2E}$ one obtains instead, $(S_x + T_x + 1/\sqrt{2E})\hat{\Psi} = 0$ which has a continuous spectrum. From (74) S_u is half the hamiltonian of a repulsive 2-dim. oscillator at constant L_z. Hence this solves as well the problem for the Coulomb scattering states.

• Action of the $(\vec{S})$ and $(\vec{T})$ operators in the oscillator picture

The energy levels of the pair of oscillators with unit frequency can be represented at constant $L_z = m$, as on the plot on Figure 8 where n_μ and n_ν for each oscillator are the equivalent of n_ρ in (66) (and the parabolic quantum numbers). From (73), the $\vec{S}$ and $\vec{T}$ allow to interconnect the states with $\Delta n_{\mu,\nu} = \pm 1, 0$ in the ladder. For example S_+ connects the states (n_μ, n_ν) and $(n_\mu+1, n_\nu)$. For finding the operator $S_+(n \to n+1)$ which connects the states (n,m) and $(n+1,m)$ in the oscillator representation of the Coulomb problem one has to perform a convenient tilt (87). This leads to :

$$S_+(n \to n+1) = e^{i\theta_{n+1}(S_y+T_y)}\, S_+\, e^{-i\theta_n(S_y+T_y)} \qquad (88)$$

which from (79) can be further reduced. Physically this accounts for the fact that the frequencies of the equivalent oscillators associated with the n and (n+1) Coulomb shell are different.

From Figure (8) and (73), operators like $S_+.T_-$ connect states as $|n_\mu, n_\nu>$ and $|n_\mu+1, n_\nu-1)>$and preserve the $n = n_\mu + n_\nu + |m| + 1$ value. They consequently allow to describe the symmetry in a Coulomb n shell at constant L_z. From this, a correspondance can be established with the SO(4) symmetry group [35]. For example :

$$e^{i\theta_n(S_y+T_y)}(S_+T_-)\, e^{-i\theta_n(S_y+T_y)} \equiv - j_{1+}.j_{2-}$$

where (j_{1+}, j_{2-}) are the standard components of the two quasi-spin $(\vec{j}_1, \vec{j}_2)$ previously defined. Figure 8 illustrates the way they act. While $(j_{1+}.j_{2-})$ allows to move at constant (n,m) through an intermediate (n,m-1) state, $(S_+ . T_-)$allows to move at constant (n,m) through a "virtual" state (n-1,m) out of the n shell.

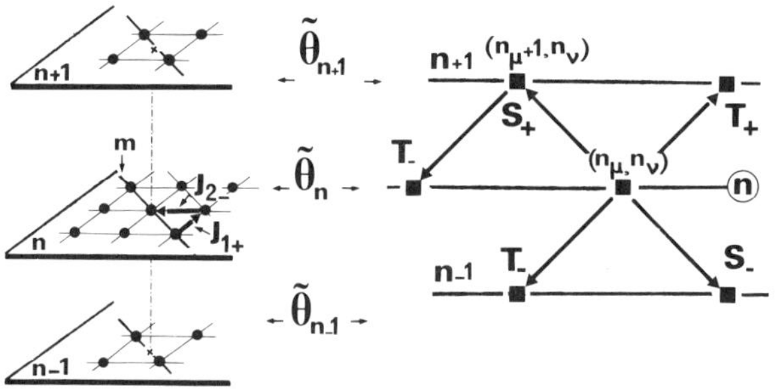

Figure 8 - Action of the (S,T) generators for the pair of oscillators with unit frequency on the oscillators energy diagram at constant $L_z = m$. A further tilt is required in order to deduce the Coulomb eigenfunctions in the oscillators representation. It physically represents the fact that the equivalent oscillators for the n and n+1 Coulomb shells have a different frequency.

• Other useful basis sets of the Coulomb dynamical group

As for the SO(4) symmetry group, an important question is the one of the reduction of SO(2,2) = SO(2,1) ⊗ SO(2,1) in term of subgroup chains. This can be addressed as well in terms of coupling of the $\vec{S}$ and $\vec{T}$ submitted to (71)(72) in analogy with the coupling of two angular momenta for the rotation group SO(3). Instead of the scheme (S^2 T^2 S_z T_z)in which the oscillators eigenstates are not coupled one can use the following ($S^2T^2V^2V_z$) with $\vec{V} = \vec{S} + \vec{T}$. From (71) the V_i build an SO(2,1) sub-group with from (74) (75) and (81)(83) :

$$V^2 = V_x^2 + V_y^2 - V_z^2 = - (\vec{r} \wedge \vec{p})^2 = - L^2 \tag{89}$$

Hence the generators commute with both L_z and L^2. This scheme complies with the isotropy of space. The generators [37 38] are those of the so-called Sturmian functions [39]. From (81)(83) one obtains :

$$V_x = (r - rp^2)/2 \qquad V_y = (i - \vec{r}.\vec{p}) \qquad V_z = (rp^2 + r)/2 \tag{90}$$

where $V_y = S_y + T_y$ identifies with the tilt operator. In this picture, one recovers immediately the idea which was expressed in (25)(29) for the classical Coulomb problem that changing the definition of time (or performing an homogeneous canonical transformation on the lagrangian) the Coulomb problem reduces to an harmonic oscillator one. One deduces :

$$\Phi = r(H-E) = (rp^2/2 - E.r - 1)$$

which from (90) is linear in V_z and V_x. From (79) it can be cast into a diagonal form.

But there is another possible SO(2,1) subgroup chain associated with the angular momentum $\vec{\lambda}(a_x, a_y, L_z)$ defined in (55) which turns out to be fundamental for dealing with the magnetic field problem. Consider the new coupling scheme (S^2 T^2 W^2 W_z) with $\vec{W}$ such that :

$$W_x = S_x - T_x \qquad W_y = S_y - T_y \qquad W_z = S_z + T_z \tag{91}$$

From (71), it is readily shown that the $\vec{W}$ build an SO(2,1) subgroup with :

$$W^2 = W_x^2 + W_y^2 - W_z^2 = - \lambda^2 \tag{92}$$

This (S^2 T^2 W^2 W_z) description thus commutes with (λ^2,L_z) and interconnects all the Coulomb states with fixed (λ,M) values. Thus it complies with the $SO(3)_{\vec{\lambda}}$ symmetry previously described and leads to defining new types of Sturmian functions [36]. This seems to have been previously unnoticed.

• Conclusions on the oscillator picture for the Coulomb problem

The inner symmetries of the Coulomb problem are thus accounted for in this approach. In addition the Coulomb bound and continuous spectra is replaced with an harmonic ladder in which calculation of matrix elements becomes oversimple. Especially the derivation of effective hamiltonian in a given n Coulomb shell is usually easy while exactly taking into account the role of continuum states. This is also true to some extents for building effective hamiltonian between two n and n+1 shells. Moreover the structure of the method is algebraic which makes it also suitable for computer types manipulations. At last the search for eventual dynamical symmetries in the perturbed Coulomb problem can be done in a rational way which also leads to a rational choice of basis sets (in a generalized sense) for tackling the question. These qualities will be exemplified in the next sections for the magnetic problem. But it is likely that they will play a key role when applied to some other questions, interactions of atoms with E.M. fields, collision processes with or without external fields or doubly excited systems.

STRUCTURE OF THE NON-RELATIVISTIC LANDAU PROBLEM

We focus here on the characters of the other limiting situation in which the Coulomb force is a small perturbation to the magnetic one. The reference situation is the one of a charged particle evolving in a magnetic field. We will recall how this Landau [40] problem can be reduced to the 2-dim. harmonic oscillator one. A more detailed treatment can be found in [10].

1. Classical mechanics of a charged particle in a $\vec{B}$ field

The solution of equation (12) with only the Lorentz force is, writing $\vec{\Pi}$ the momentum of the velocity [14] :

$$\vec{\Pi} = m\vec{v} = q\vec{r} \wedge \vec{B} + \vec{\Pi}_o \tag{93}$$

where $\vec{\Pi}_o$ is a constant vector. Hence the motion along $\vec{B}$ field is a uniform one while the motion perpendicular to $\vec{B}$ is circular and uniform at constant angular velocity $\vec{\omega}_c$ [5] . The constants of the classical motion are thus the radius of the circle Γ, the position of the center $\vec{r}_o$ and the velocity $\Pi_{o//}$ along $\vec{B}$ field. $E_\perp$ being the transverse part of the energy and (ρ,φ) 2-dim. polar coordinates, one obtains :

$$\Gamma^2 = 2E_\perp/m\omega_c^2 \qquad \vec{r}_o = \vec{\rho} + (\vec{\Pi} \wedge \vec{B})/qB^2 \tag{94}$$

Hence the transverse motion is degenerate on $\vec{r}_o$. Furthermore the quantity L_z is also a constant of the motion :

$$L_z = (\vec{r} \wedge \vec{\Pi})_z + qB\rho^2/2 = \frac{qB}{2}(r_o^2 - \Gamma^2) \tag{95}$$

which identifies in the symmetric gauge [17] with the component of the canonical angular momentum $\vec{L} = \vec{r} \wedge \vec{p}$ on the B field axis [10] . Performing a rotation to the so-called Larmor frame, the equations become [10] $m\vec{\gamma}_L = \vec{F}_D$ where $\vec{F}_D$ is the harmonic diamagnetic force [9] . Hence the trajectory in Larmor frame is an ellipse with the center at the origin. This makes clear the analogy of this problem with the one of the isotropic harmonic oscillator in two dimensions.

2. The Landau quantum spectrum

Several treatments are possible (see for example [20]) without any choice of a gauge. They are based on the commutation relations of the quantum operators associated with (93) and (94). Rather we here exploit the analogy with a 2-dimensional isotropic oscillator in Larmor frame. In the hamiltonian formulation, in the symmetric gauge, this amounts to dealing with equation (20) but without the Coulomb term. This is thus the hamiltonian of a 2-dim. oscillator with an additional $\gamma.L_z$ term :

$$H = p^2/2 + \gamma.L_z/2 + \frac{\gamma^2}{8}(x^2 + y^2) \tag{96}$$

This is readily diagonalized using the Fock approach (62)(63) for circular excitations or (ρ,φ,z) cylindrical coordinates. These are obviously the right choice as L_z is thus diagonal from (63).

Due to the paramagnetic term, the Landau spectrum differs from the 2-dim. oscillator one, and the degeneracies are different. But the eigenfunctions are the ones in (64). From (63) and (96) we obtain :

$$H = (a_d^+ a_d + \tfrac{1}{2})\hbar\omega_c + \frac{p_z^2}{2m} = (n_d + \tfrac{1}{2})\hbar\omega_c + \frac{p_z^2}{2m} \tag{97}$$

$$L_z = M\hbar = (n_d - n_g)\hbar$$

where $p_z^2/2m$ accounts for the free motion along $\vec{B}$ field. The spectrum is degenerated on the $L_z = M$ value (or on n_g). In a given n_d Landau shell,

M can take every integer value such that :

$$- \infty < M \leqslant n_d \tag{98}$$

The degeneracy is thus infinite. This is shown on Figure 9. The eigenfunctions are given through (64) at fixed n_d. Their spatial representations in cylindrical coordinates are the same as the oscillator ones. They express the cylindrical symmetry in the problem. From (94)(95) one can show [10] :

$$\Gamma^2 = (2n_d + 1)a_c^2 \qquad r_o^2 = (2n_g + 1)a_c^2 \tag{99}$$

(a_c the cyclotron radius - eq (11)) which means that both the square of the radius and the square of the distance of the center of the classical circle to the origin are defined in this representation. From which follows a semi-classical interpretation of the degeneracy shown on Figure 9. But there are lot of other possible choices of the eigenfunctions in the degeneracy for which L_z is no longer defined [10] . This one is of special interest as complying with the usual symmetry requirements in atomic physics. Especially L_z commutes with the Coulomb potential.

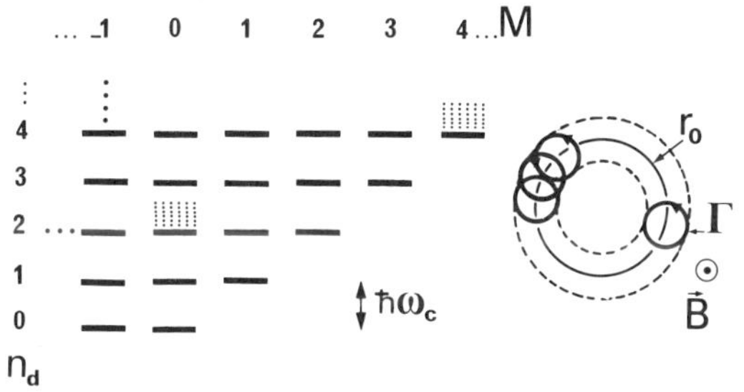

Figure 9 - The orbital Landau spectrum in the (H, L_z) representation. The labelling of the infinite degeneracy is through the $L_z = m$ integer value ($- \infty < L_z \leqslant n_d$). To each state is associated a continuum accounting for the free motion along $\vec{B}$ field. The (n_d, L_z) eigenfunctions in this representation have a cylindrical symmetry and can be represented in a semi-classical way from (99).

3. Further remarks

The essential of the physics for both the Landau and Coulomb problems is thus reduced to the knowing of the 2-dim. isotropic oscillator.

If γL_z is a constant, the treatment of the Landau hamiltonian is just analoguous to the one of the 2-dim. oscillator. Especially all the considerations on the oscillator dynamical group SO(2,1) at constant L_z apply, thus providing us with the way of interconnecting all the Landau states. This would apply to the treatment of the Landau problem perturbed by the Coulomb field. The latter breaks the translational invariance of the Landau problem [10] and is responsible for a coupling between the transverse and longitudinal motions. This will give the states the characters of resonances but for $n_d = 0$ (Landau ground state whatever $L_z \leqslant 0$) and the states with maximum $L_z = n_d$ value in each manifold.

These Landau eigenfunctions provide us with the high field asymptotic limit of the solution and have been widely used in computations or analytical works on the magnetic problem, in solid state [41] and atomic or astrophysics [42 43 44]. They have also a basic application in the treatment of electron-electron correlations in solid state physics which are responsible for the so-called fractional quantum Hall effect [13].

THE STRUCTURE OF RYDBERG ATOMS IN EXTERNAL FIELDS

In one of the previous sections we have developped some unusual views on the Coulomb problem, based on its symmetries. The first degree is associated with the symmetry group SO(4) and allows the Coulomb problem to be thought in terms of a double vectorial model involving two quasi-spin $(\vec{j}_1,\vec{j}_2)$. The second degree in the symmetries associated with the SO(2,2) dynamical group expresses to some extent that the dynamics is equivalent to the one of a pair of "phase-coupled" 2-dimensional harmonic oscillators. In this section, we apply these two key pictures to understanding the structure of atoms in external fields. We will first exemplify the powerful character of the vectorial model picture through the discussion of the low field limit of the crossed $(\vec{E},\vec{B})$ fields problem. Next this will be applied to understanding the structure of Rydberg atoms in magnetic fields.

1. Rydberg atoms in crossed $(\vec{E},\vec{B})$ fields. The low field limit

This situation provides us with a good example of the usefulness of views based on the group structure of the Coulomb problem. The experimental results amount to a direct experimental confirmation of this structure.

The conditions are those of the hamiltonian(20)with $\vec{E} \perp \vec{B}$. The external field perturbation W is assumed to be small compared to the Bohr frequency $2R/n^3$. This is thus the regime in which the external field perturbation breaks the Coulomb symmetry in a given n shell. This was first investigated by Pauli [19].

Neglecting the diamagnetic term (see § 2), W expresses as $W = q\vec{B}.\vec{L}/2m + q\vec{E}.\vec{r}$ where the first term is the paramagnetic interaction. From either the classical picture (Figure 2) or symmetry considerations, the radius vector $\vec{r}$ is proportional to $\vec{a}$ at first order (which is also a polar vector). This is the Pauli replacement [15] (in atomic units) :

$$\vec{r} \rightarrow \frac{3}{2} n \vec{a} \qquad (100)$$

This leads to introducing the so-called linear Stark frequency (in atomic units) :

$$\vec{\omega}_E = \frac{3}{2} n \vec{E} \qquad (101)$$

which obviously plays the same role as the Larmor frequency $\vec{\omega}_L = \vec{\omega}_c/2 = q\vec{B}/2m$, but is a pseudo-rotation vector. Hence, introducing the two quasi-spin $(\vec{j}_1,\vec{j}_2)$ from (44) one gets immediately :

$$W = \vec{\Omega}_1.\vec{j}_1 + \vec{\Omega}_2.\vec{j}_2 \qquad \text{with } \vec{\Omega}_{\frac{1}{2}} = \vec{\omega}_L \pm \vec{\omega}_E$$

which, as shown on Figure 10, interprets as the independant precession of $\vec{j}_1$ and $\vec{j}_2$ around the $\vec{\Omega}_1$ and $\vec{\Omega}_2$ axis. The common modulus of Ω_1 and Ω_2 is (as $\vec{\omega}_E.\vec{\omega}_L = 0$) :

$$\Omega = (\omega_E^2 + \omega_L^2)^{1/2} \tag{102}$$

From the fact (44) (46) that $\vec{j}_1$ and $\vec{j}_2$ are two commuting angular momenta the eigenfunctions are $(j_1^2 j_2^2 j_{1\Omega_1} j_{2\Omega_2})$. They are of the generalized parabolic types(52)and the spectrum is, from (48) :

$$W = (m_1 + m_2)\Omega = (j_{1\Omega 1} + j_{2\Omega 2}).\Omega = k.\Omega$$

$$-(n-1)/2 \leqslant m_i = (j_i)_{\Omega_i} \leqslant (n-1)/2 \quad \text{and} \quad -(n-1) \leqslant k \leqslant (n-1) \tag{103}$$

where $k = (m_1 + m_2)$ is an integer. The low-field crossed $(\vec{E},\vec{B})$ fields spectrum is thus analogous to the usual Zeeman or linear Stark ones as shown on Figure 10. These situations where either $\vec{E} = 0$ or $\vec{B} = 0$ are limiting cases of (103). But the eigenfunctions are of a fundamentally different type.

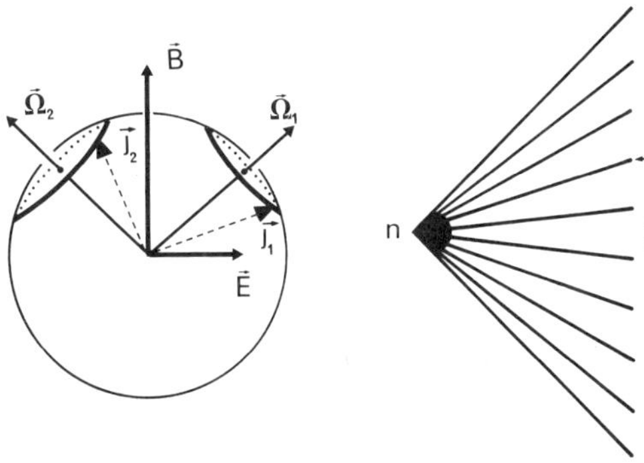

Figure 10 - The low field crossed $(\vec{E},\vec{B})$ fields situation as seen from the SO(4) vectorial model. The two quasi-spin $(\vec{j}_1,\vec{j}_2)$ are precessing around the axis $\vec{\Omega}_1 = \vec{\omega}_L \pm \vec{\omega}_E$ with the angular velocity $\Omega = (\omega_E^2 + \omega_L^2)^{1/2}$. This is the spacing of the energy levels which are labelled through the quantum number :

$k = j_{1\Omega 1} + j_{2\Omega 2} = \lambda_{\Omega_1}$. Each sublevel presents a $n-|k|$ residual degeneracy.

From(52)they express a continuous field tunability of the n atomic shell in $\vec{E}$ and $\vec{B}$ fields which opens the way to a control of the atomic properties. From(52)one passes continuously from the Zeeman to the linear Stark regime through a rotation, the angle of which being $\beta = \mathrm{Arctg}(\omega_E/\omega_L)$ and the generator $a_y = j_{1y} - j_{2y}$. This is from(52)a rotation in the four-dimensional space associated with the SO(4) symmetry.

From Figure 4 and (103), a given k sublevel does present a residual $n-|k|$ degeneracy (on $m_1 - m_2$), which is in consequence of the crossed character of the $\vec{E}$ and $\vec{B}$ fields leading to $\Omega_1 = \Omega_2 = \Omega$. This interprets in terms of the $\vec{\lambda}(a_x,a_y,L_z)$ angular momentum defined in (55). Indeed (103) rewrites as $(0 \leqslant \lambda \leqslant n-1)$:

$$W = \vec{\Omega}_1.\vec{\lambda} = \lambda_{\Omega_1}.\Omega \quad \text{with} \; -(n-1) \leqslant \lambda_{\Omega_1} = k \leqslant (n-1) \tag{104}$$

An alternate set of eigenfunctions is $(j_1^2 j_2^2 \lambda^2 \lambda_{\Omega 1})$ deduced from $(\lambda^2 \lambda_z)$ in (56) through a rotation with the generator $\lambda_y = a_y$. Hence $\vec{\lambda}$ is not an oddity but from (104) is essential for understanding the crossed $(\vec{E},\vec{B})$ fields situation in the low external fields limit.

Quantum mechanics puts the stress on the importance of measurements. In that respect the experimental demonstration of (104) allowing the direct measurement of the component $\lambda_{\Omega 1}$ of $\vec{\lambda}$ would lead to rooting $\vec{\lambda}$ in the physical reality and from (103) would amount to a direct test of the SO(4) group structure of the orbital Coulomb problem. This has been achieved recently on Rydberg states of rubidium atoms, using Doppler-free two-photon techniques and low external field conditions [45 46].

Experimental details are discussed in [45 46]. Rubidium atoms are not a perfectly coulombic system, due to the quantum defects corrections. The states which are likely to be quantized according to (104) are those with small quantum defects (with $l \geqslant 3$) which behave as an "incomplete" hydrogenic manifold with $(n^2 - 9)$ degeneracy [45]. But they can hardly be optically excited from the ground state. The nS, nP and nD series do, but have a non-hydrogenic behaviour at low fields. The experimental trick is to use the latter class of states (specifically the nS) as a probe of the hydrogenic behaviour of the former. As the energy curves of the nS states are nearly field independant, while the states in the incomplete manifold should present a linear in Ω field dependance, the two systems of curves should anticross. The tracking of the anticrossings positions, on the nS energy curve, as a function of $\vec{E}$ and $\vec{B}$ fields should provide us with a picture of the quantization (104) at nearly constant energy. This is shown on Figure 11 which displays the characteristic (E^2,B^2) dependance of (104) at constant energy W. Although Rb Rydberg atoms are not a

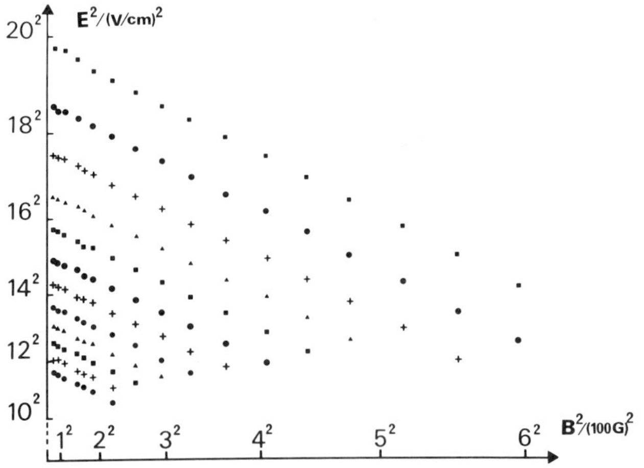

Figure 11 - The crossed $(\vec{E},\vec{B})$ fields quantization in Rubidium atoms. This is an experimental plot as a function of E^2 and B^2 of the positions of the anticrossings between the 37S state and the n = 34 incomplete manifold which displays (104) at nearly constant electron's energy. The experimentally determined values of $k = \lambda_{\Omega_1}$ are from top k = -18.3, - 19.4, - 20.4, - 21.3, - 22.3,...

perfectly coulombic system, the quantization (104) is followed up to within 1 % by most of the states [45]. (see [46] for a detailed analysis of non-coulombic corrections).

An important by-product of the plot on Figure (11) is the determination of the quantum number $k = \lambda_{\Omega 1}$, which though not an integer (due to non-coulombic corrections) varies by one unit from one sublevel to the other one. Hence the experiments allow the direct measurement of the components of the angular momentum $\vec{\lambda}$, a pure product of the Coulomb symmetry.

Consequently, the SO(4) approach is required for a full understanding of the crossed $(\vec{E},\vec{B})$ fields situation. But it also provides us with the adiabatic solution in more complicated cases involving inhomogeneous or time-depending fields. This is the case for hydrogenic ions interacting with a surface [47], the exciton spectrum in crossed fields, Rydberg atoms interacting with the radiation or microwaves fields or other atoms, or molecular Rydberg states [48].

2. The atom in a magnetic field - The low field diamagnetic behaviour

The hamiltonian is (20) with $\vec{E} = \vec{0}$. It commutes with L_z. The first question which arises is how to describe the so-called inter-l mixing regime, in which the symmetry of the Coulomb field is broken by the diamagnetic interaction. This was solved in 1981 from classical perturbation theory [5 49]. A group theoretical derivation using the Fock's method on the R(4) hypersphere appeared in 1982 [50 51]. We will present here a simpler derivation [52] based on the quasi-spin description of SO(4) [53].

• The effective diamagnetic hamiltonian in the n Coulomb shell

Conditions are such that $1/n^3 >> H_D = \gamma^2.(x^2 + y^2)/8$ or $\gamma^2 n^7 << 1$. n, L_z and parity are good quantum numbers. The problem to solve amounts to applying Wigner-Eckardt theorem in SO(4) to the diamagnetic hamiltonian. This can be done in term of the two quasi-spin $(\vec{j}_1,\vec{j}_2)$. First the invariance of $(x^2+ y^2)$ under various spatial transformations (parity and various plane reflections) leads to an expression which is quadratic and symmetrical in the two quasi-spin, of the form :

$$\rho^2 = x^2 + y^2 \rightarrow \alpha n^2 + \beta m^2 + \gamma j_{1z}.j_{2z} + \delta \vec{j}_1.\vec{j}_2$$

This requires to consider the transformations of the $\vec{j}_i$ under such operations (see(47)for example). Next, the four coefficients can be evaluated in simple cases using (48) or four "degenerated" Kepler ellipses (two circles and two lines). The exact quantum expression, valid whatever the n and m value is (with $\rho^2 = x^2 + y^2$):

$$H_D = \frac{\gamma^2}{8}\rho^2 \rightarrow \gamma^2 \frac{n^2}{16}(3n^2 + 1 - 4m^2 + 20 j_{1z}.j_{2z} - 8\vec{j}_1.\vec{j}_2) \qquad (105)$$

$$L_z = m = j_{1z} + j_{2z}$$

or using (44) and (46) :

$$\rho^2 \rightarrow \frac{n^2}{2}(n^2 + 3 + L_z^2 + 4a^2 - 5a_z^2) \qquad (106)$$

In the n shell, the diamagnetic interaction is thus responsible for a coupling of the two quasi-spin $(j_1 = j_2 = (n-1)/2)$. It is of a non-standard type and (105) cannot be diagonalized through some recoupling of $\vec{j}_1$ and $\vec{j}_2$. Nevertheless, from the fact (45) that the $\vec{j}_i$ are angular momenta, such a diagonalization is made straightforward. We will write $|n,k,m,P\rangle$ these common eigenfunctions to the Coulomb and diamagnetic hamiltonian. They have a definite parity P. k is a label [10] which replaces, for example, the l quantum number.

The diamagnetic interaction being positive, the energy shift in (105) is always positive and proportional to B^2. From Figure 4 the $n-|m|$ degeneracy of a given (n,m) subshell is removed by the diamagnetic contribution. The (n,m) subshell is diamagnetically split into several sublevels, the number of which is about $(n-|m|)/2$ (as parity is a good quantum number).

The interest of using (105) is that it affords to know, without any calculation, how such diamagnetic splitting occurs.

- Limiting symmetries in the diamagnetic manifold

Some basic structural informations on the magnetic problem can be drawn from (105). The diamagnetic manifold presents a rovibrational structure [50 53-57] as a consequence of the existence of two limiting symmetries. This can be demonstrated as follows. Consider the so-called adiabatic invariant [5] :

$$\Lambda = 4a^2 - 5\, a_z^2 \qquad (107)$$

From (106), its spectrum gives the one of the diamagnetic interaction, while in the classical picture, the constancy of Λ [49] rules the diamagnetically induced secular motion of the Lenz vector. As shown on Figure 12, there are two limiting types. For negative values of Λ, the secular motion takes place on a two-fold hyperboloid (bounded with a sphere) with as a limiting case $a \sim a_z \sim n$. The motion is thus of vibrational type along $\vec{B}$ for the states at the bottom of the diamagnetic band . In addition they are parity degenerated [56] . At the top of the band, for positive values of Λ, the Lenz vector can undergo a secular motion of rotational type on a one fold hyperboloid. In extreme conditions it is confined in the z = 0 plane perpendicular to the field as $a_z \ll a = n$. A crossover in the symmetries takes place for $\Lambda = 0$.

Such an analysis produces as well, from (105), the limiting form of the eigenfunctions. At the top of the diamagnetic manifold, the limiting symmetry is of $O(3)\vec{\lambda}$ type associated with the angular momentum $\vec{\lambda}(a_x,a_y,L_z)$ (cf (55)). Actually $a_z \ll a \simeq n$ implies that $\vec{j}_1$ and $\vec{j}_2$ are confined in the plane perpendicular to the field and correlated in opposite directions $(\vec{j}_1+\vec{j}_2 \sim \vec{0})$. Hence $L \simeq L_z$. This is summarized on Figure 13. (105) rewrites approximately :

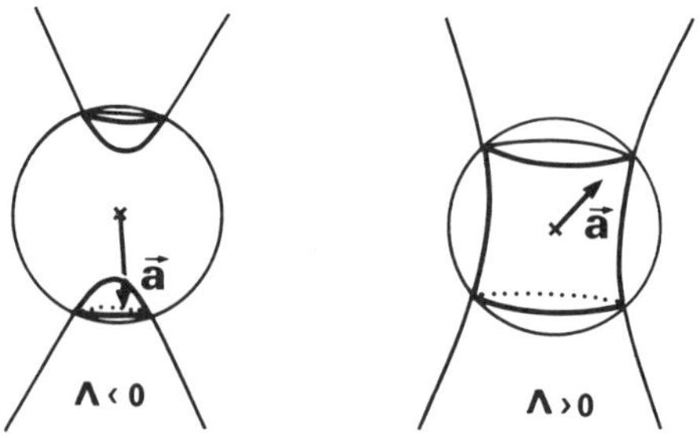

Figure 12 - The secular motion of the Lenz vector from (107). For $\Lambda > 0$ the motion is of rotational type on a one-fold hyperboloid. For $\Lambda < 0$ it is of librational type on a two-fold hyperboloid. The degenerated case for $\Lambda = 0$ corresponds to a cross-over in these symmetries (the bounding sphere represents the conservation of energy in the n shell for fixed (n,m)).

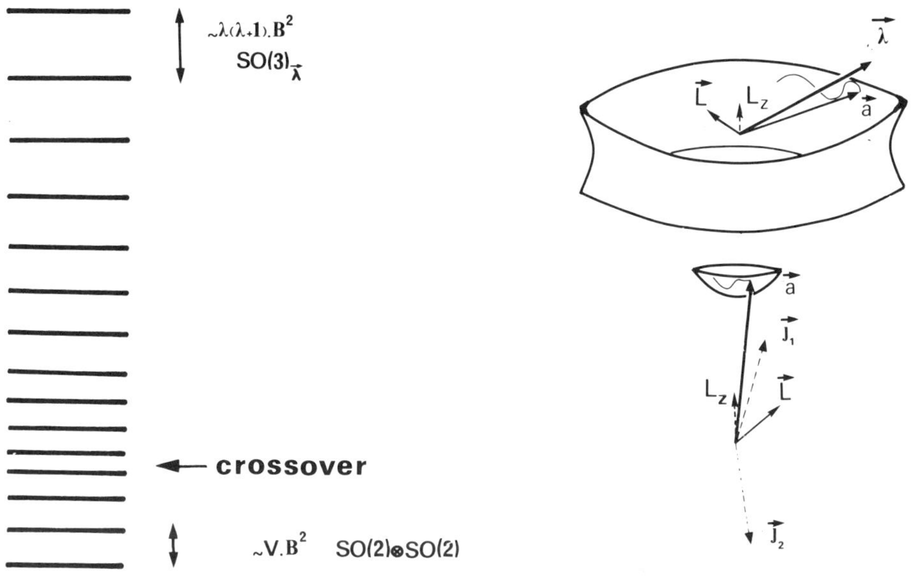

Figure 13 - The two limiting symmetries in the diamagnetic manifold and the ro-vibrational structure of the energy levels, for given (n,M) and parity (schematic).

$$H_D \sim \gamma^2 \frac{n^2}{16} (n^2 + 3 - 3m^2 + 4\lambda^2) \qquad (108)$$

which proves that the limiting eigenfunctions are $(j_1^2 j_2^2 \lambda^2 \lambda_z)$ with from (56), $\lambda^2 = \lambda(\lambda + 1)$ and $0 \leqslant \lambda \leqslant n - 1$. The maximum value of the diamagnetic shift is thus of the order of $5\gamma^2 n^4/8$.

At the bottom of the manifold, the limiting symmetry is of SO(2)⊗SO(2) type [56]. As $a \sim a_z \sim n$, $\vec{j}_1$ and $\vec{j}_2$ are along the $\vec{B}$ field axis and (105) rewrites (with $a_z = j_{1z} - j_{2z}$) :

$$H_D \sim \gamma^2 \frac{n^2}{16} (n^2 + L_z^2 - a_z^2) \qquad (109)$$

which is diagonal in the $(j_1^2 j_2^2 L_z a_z)$ parabolic basis. With $L_z = m$, the structure close to $a_z \simeq n$ is of vibrational type and, as said before, parity degenerated [50, 56]. The frequency of the oscillation is of the order of $2\sqrt{5}.n$ [50].

• The rovibrational structure of the diamagnetic manifold

From this analysis, the structure of the diamagnetic manifold shown on Figure 13 is a ro-vibrational one [50] as in molecules. The degeneracy on parity in (109) suggests a double well behaviour at the bottom of the band. Indeed one can show [50] that the vibrational type states only exists for small $L_z = m$ values while the others are mostly dominated by the $O(3)\lambda$ type approximate symmetry. In between these two limiting symmetries, a cross-over does exist which plays an important role in the dynamics of the classical system.

This compares fairly well with direct calculations as shown on Figure 14. At the top of the band, the nodal surface plots suggest a kind of rotational motion with isodensity lines concentrated close to the plane perpendicular to the field. At the bottom, the nodal surfaces rather suggest a vibrational motion along $\vec{B}$ field. Hence the two limiting symmetries are associated with spatial localizations [59, 60, 61] either in the $z = 0$ plane or around the $\vec{B}$ field axis as well as with correlated or uncorrelated motions of the two quasi-spin.

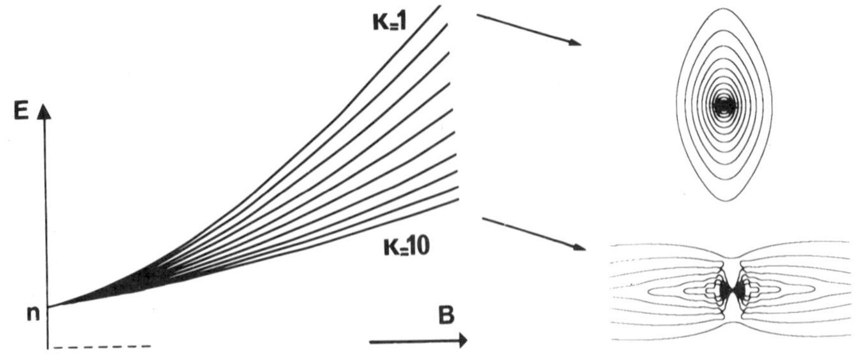

Figure 14 - Energy level structure of a diamagnetic manifold. The nodal surfaces of the eigenfunctions at the top and bottom of the manifold suggests respectively a rotational and vibrational symmetry.

The states at the top of the band with approximate O(3)λ symmetry and correlated type behaviours are of special importance as low fields precursors of quasi-Landau resonances[62,63]. The electronic density plot on Figure 15 allows to show that the mean extension along B field is scaling as $n^{3/2}$and thus is squeezed compared to the usual spherical eigenfunctions $|n,l,m>$. This manifests the tendancy of the motion to be localized in the z = 0 plane where the diamagnetic potential is a maximum. These states share some similarity with those of the oblate spherical top.

• Oscillator strengths distributions

The last question to be addressed is the one of the appearance of the optical excitation spectra which strongly depends on the parity P_z along the B field axis, that is on whether the diamagnetic eigenfunctions do have or not a node in the z = 0 plane.

Assuming the excitation is performed from a low-lying state $(n_o l_o m_o)$ to a state (n,l,m) in zero field through an electric dipolar transition, the distribution of intensities in the diamagnetic band is proportional to $|<nlm|n,k,m,P>|^2$. this quantity depends on the parity of the states,

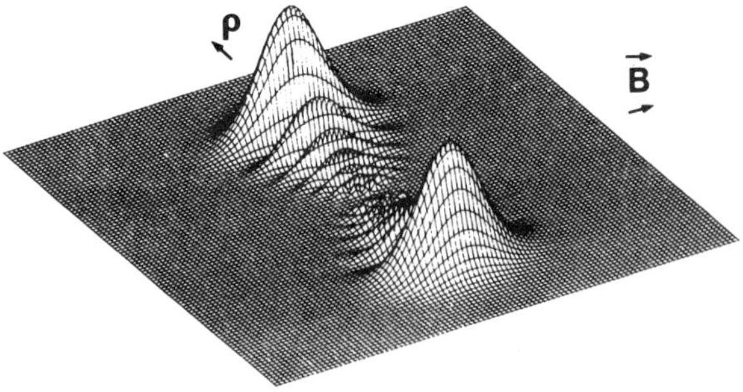

Figure 15 - Isoelectronic density curves of the eigenfunction at the top of the diamagnetic band (n=11, M=3, k=1) with even parity along z axis. The extension along $\vec{B}$ scales as $n^{3/2}$which, compared to the usual spherical eigenfunctions, manifest the localization close to the z = 0 plane.

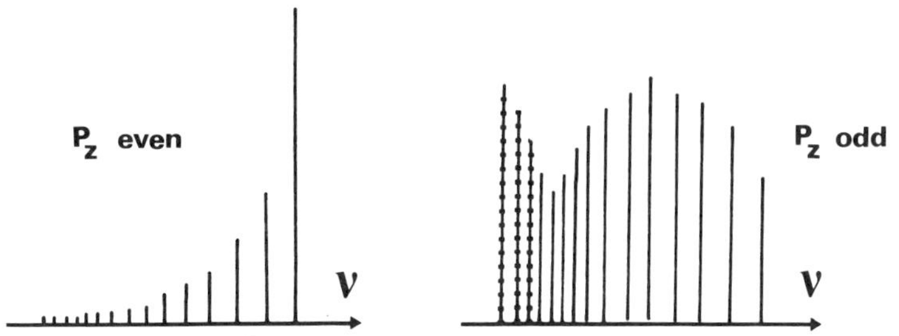

Figure 16 - Schematic representation of the intensity distributions in the diamagnetic band according to parity P_z. Solid lines are associated with the states having an approximate rotational symmetry and dotted lines with the vibrational types states which only exist for low m values.

and for given parity on the k value. The evaluation can be carried out from the limiting symmetries, using (58) and (59), or more qualitatively from Figure 13. For even parity along z axis ($P_z = (-1)^{l-m}$) the intensity distribution rapidly decreases with k and may eventually increase again at the bottom of the band. The line (k=1) at the top of the band is the dominant one. For odd P_z, due to the eigenfunctions having a node in the z = 0 plane, the k=1 line is not the dominant one and the intensity distribution is much more complicated (and not monotonic). This is schematically represented on Figure 16. According to the m value, the vibrational type states at the bottom of the band may or not exist thus altering the general appearance.

• Experimental illustrations

Experiments have been done on atomic species which do not have a perfectly hydrogenic behaviour. Quantum defect corrections (though usually small) lead to some alterations of the previous patterns especially at low fields. For example Figure 17 exhibits the diamagnetic structure of M = 3 odd parity states of Caesium [54 62] with even P_z parity along z axis. The fact that the line at top of the band is not the dominant one

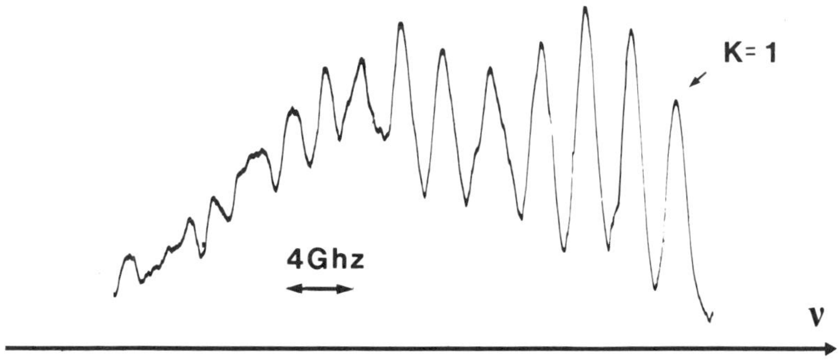

Figure 17 - The low field diamagnetic spectrum of Caesium [62] (n $\simeq$ 40 M = 3 odd parity states, even P_z) for B = 6 kG. The spacing of the lines is proportional to B^2. The intensity of the line at the top of the band is not yet dominant due to small quantum defect corrections. The intensity of the lowest energy states is enhanced due to relaxation processes in the diamagnetic band.

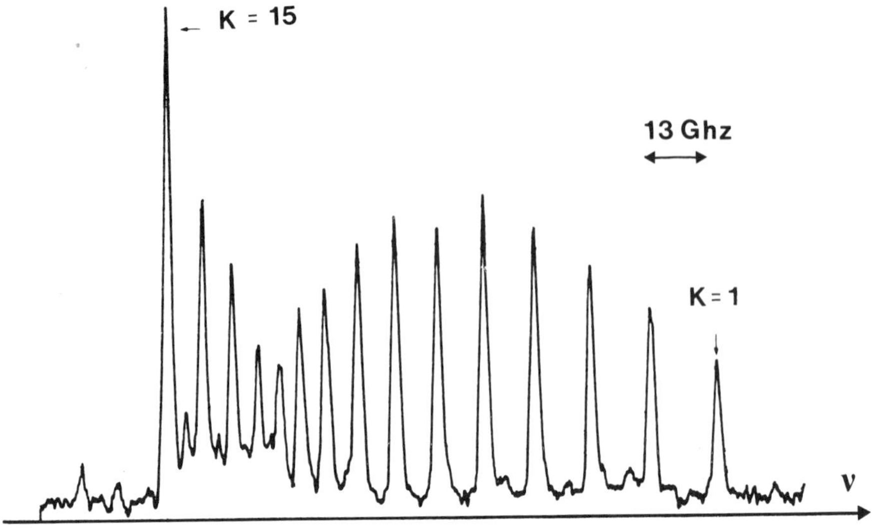

Figure 18 – The low field diamagnetic spectrum of Lithium [64] (n = 31 m = 0 odd parity, odd P_z) for B = 19.4kG. Both the rotational and vibrational types states are seen. The cross-over in the symmetries is strikingly illustrated (courtesy J. Pinard et al., lab. Aimé Cotton).

is due to the small ($\delta = 0.033$) quantum defect of the $l = 3$ states. But at higher field values the patterns will tend to the theoretical ones in Figure (16).

The plot on Figure 18 [64] exhibits nicely the diamagnetic structure of (odd, m = 0) lithium states with odd P_z parity. Both the rotational and vibrational states are seen in this situation. The cross-over in the limiting symmetries is characterized with a strong variation of the intensities and a noticeable irregularity in the spacings of the lines [54].

• Conclusions on the SO(4) analysis

This approach leads to simple understanding of a seemingly complicated situation and has very strong predictive power. Again the $(j_1^2\ j_2^2\ \lambda^2 \lambda_z)$ states play an important role [50] as limiting eigenfunctions at the top of the diamagnetic band (they do coincide only when $n \rightarrow \infty$). What we have found is that the states of the top of the band with even P_z parity do have the maximum oscillator strength in optical spectra. They also express a correlated behaviour of the two quasi-spin $(\vec{j}_1, \vec{j}_2) \rightarrow \vec{\lambda}$ and a spatial localization in the plane perpendicular to the field, where the diamagnetic potential is a maximum. Indeed experiments [62 63] have shown around 1981 that these states are the low field precursors of the quasi-Landau resonances which appear in the strong mixing regime, and that there was a kind of adiabatic continuity of the phenomena between the low and high field limit. To some extent, this is also true for all the states with approximate rotational symmetry, whatever parity, but fairly more difficult to experimentally prove as the lines are not necessarily strong. As a matter of fact, the use of other excitation techniques (discharges, electron bombardment, ...) should lead to other appearances of the diamagnetic spectrum as the selection rules will be different.

The analysis apply as well to other situations in atomic physics, involving the perturbed Coulomb problem [48 65] . For example the symmetry breaking term $(x^2 + y^2 + \alpha z^2)$ leads immediately to [48 53] :

$$V = \frac{n^2}{2}\{\frac{n^2}{4}(6+\alpha) + (2+3\alpha)/4 - 2m^2(1-\alpha) - (4-\alpha)\vec{j}_1.\vec{j}_2 + 10(1-\alpha)j_{1z}.j_{2z}\}$$

which in particular allows to solve the problem of the interaction of Rydberg atoms with surfaces, in the Van Der Waals adiabatic approximation. The irregularity of the spacings numerically found [65] thus appears as the result of a cross-over in the symmetries.

3. The atom in a magnetic field - The inter n and strong mixing regimes

In this regime in which the structure of the Coulomb field itself is perturbed or altered by the diamagnetic interaction, the question is how to extend the previous pictures and to what extent are they still valid ? An important clue was from experimental results [62 63] . They allowed to show that the positions of the dominant lines were behaving continuously in $\vec{B}$ field whatever the strength and were branched to well-defined states in the low field limit. For accounting for such an adiabatic picture while complying with the low fields symmetries and allowing the inter n mixing of the channels, the dynamical group approach is required. This allows to take into account both the symmetries at large and the important role of continua.

- The onset of the inter n mixing regime and the question of exponentially small anticrossings [66 55 59 67]

An important numerical discovery in this problem was the one of exponentially small anticrossings,between the states at the bottom of the (n + 1) band and those at the top of the (n) diamagnetic band (see Figure 19). This comes from numerical calculations performed in conventional basis for the Coulomb problem thus requiring truncations. This feature was interpreted in terms of the existence of an approximate symmetry in the magnetic problem [66] but also conjectured as the proof of unexpected features in the classical dynamics [55] . Some analytical explanations were also given [49 59 67] in term of spatial segregations of the eigenfunctions. The overlapping of the two classes of states with O(2) x O(2) and O(3)λ approximate symmetries is through the exponentially decreasing tails leading to a size of the anticrossing scaling as e^{-2n}.

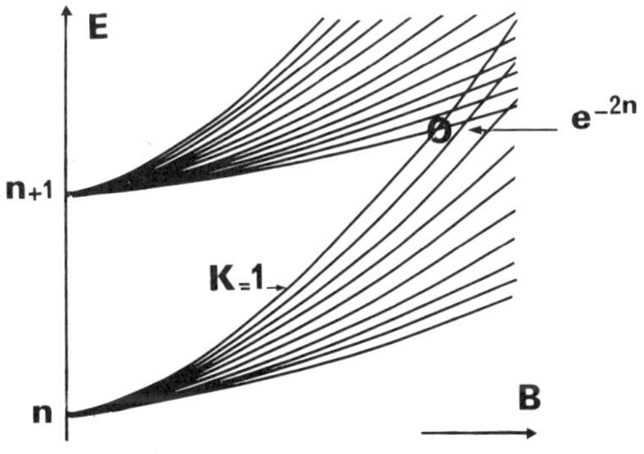

Figure 19 - The onset of the inter n mixing regime. The size of the anticrossing between the states at the top of the n manifold (with approximate O(3)λ symmetry) and those at the bottom of the (n+1) manifold (of vibrational type) is approximately scaling as e^{-2n}.

Although it turns out that the result is basically correct, the way it has been deduced is not. One can question especially on the role of truncations in such an approach (neglect of an infinite number of states including continua, effects on the real symmetry). A self consistent way of looking for dynamical symmetries is indeed to use the dynamical group approach.

• The magnetic problem from the Coulomb dynamical group

Due to the multiplication by r, the diamagnetic hamiltonian in semi-parabolic coordinates (81) writes down as :

$$H_D = \frac{\gamma^2}{8}\,\mu^2\,\nu^2(\mu^2 + \nu^2) \tag{110}$$

and the complete equation for the energy E rewrites (see (82)) as :

$$[F(\mu) + F(\nu) + E(\mu^2 + \nu^2) - H_D + 2]\,\Psi = 0 \tag{111}$$

The diamagnetic interaction acts as an anharmonic coupling between the pair of 2-dimensional harmonic oscillators at constant $L_z = m$. This important idea is illustrated on Figure 20. Indeed, it rules the whole dynamics of both the quantum and classical systems.

Introducing from (74) (75) and (83), the generators $(\vec{S},\vec{T})$ of the dynamical group SO(2,2), one readily deduces with (E the perturbed energy and $\alpha = -2E$) :

$$\{(1+\alpha)(S_z+T_z) + (\alpha-1)(S_x+T_x) - 2 + H_D\}\Psi = 0$$

$$H_D = \gamma^2(S_z+ S_x)(T_z+T_x)(S_z+T_z+S_x+T_x) \tag{112}$$

where the coupling term H_D expresses as a polynomial of degree three which is symmetrical in the $(\vec{S},\vec{T})$.

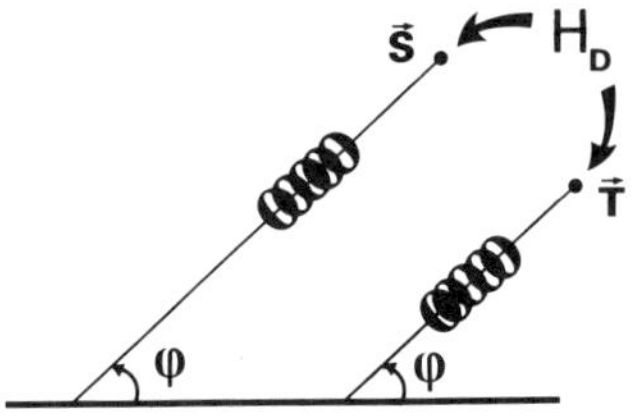

Figure 20 - The magnetic problem in the oscillator picture. The diamagnetic interaction couples the motion of the pair of 2-dim. oscillators at constant $L_z = m$ along the (μ,ν) axis.

Either (111) or (112) are the general equations for the magnetic problem, the understanding of which has been reduced to the one of the coupled motion of a pair of 2-dimensional oscillators. All the structural informations on the Coulomb field, especially its symmetries at large and the role of continua are accounted for in (112) and in the commutation relations (71)(83). From Figure 20 this also provides us with a fairly simple physical picture for the problem which will be further used.

As a general remark, (112) has to be solved for the eigenvalue "2" meaning that the energy E in the perturbed Coulomb problem will be obtained as an implicit function. This restores a certain equivalence between the energy E and the magnetic field strength γ.

• Looking for dynamical symmetries in the magnetic problem

The non linear equation (112) provides with the correct approach to this question as it naturally allows for inter n mixing of the states. Before, we will illustrate the concept of "dynamical symmetry" in a simple case for the electric field problem.

Assuming the $\vec{E}$ field along z axis, the Stark hamiltonian z.E commutes with L_z and in semi-parabolic coordinates (81), expresses as (mind the r multiplying factor) :

$$H_E = \beta(\mu^2 + \nu^2)(\mu^2 - \nu^2) = \beta(\mu^4 - \nu^4)$$

From (74)(75) and (82) one deduces : [35 36]

$$H_E = \beta\{(S_z + S_x)^4 - (T_z + T_x)^4\} \qquad (113)$$

The equation for the Stark-Coulomb problem splits into two independant equations for each $(\vec{S})$ and $(\vec{T})$ oscillators in μ and ν. In the picture of Figure 20, the two oscillators are uncoupled and the Stark effect looks like an anharmonic potential acting on each oscillator. The equations for each oscillator can be further solved for example in the (S^2, S_z) and (T^2, T_z) representations in which the n mixing naturally takes place. The subequations are obviously not diagonal in these representations. Hence the Stark dynamical symmetry expresses that the two SO(2,1) subgroups of SO(2,2) = SO(2,1)⊗SO(2,1) complies with (82) and (113) whatever the field and energy.

But this is a simple case in which the dynamics of the two oscillators is uncorrelated. One can imagine other schemes of the SO(2,1)⊂SO(2,2) types in which the oscillators dynamics is of coupled type, while a dynamical symmetry exists. For example the $(S^2\ T^2\ W^2\ W_z)$ coupling scheme of SO(2,1)$\vec{\lambda}$ type defined in (91)(92) leads [48] to building a new hierarchy of perturbing potentials to the Coulomb field which fulfills these conditions.

Indeed, SO(2,2) is as well-known as SO(4). It follows that no further reduction of (112) can be done using a convenient coupling scheme or subgroup chains reduction of SO(2,2). There is no exact dynamical symmetry in the magnetic problem. The fact that there is no approximate one holding for the whole magnetic problem can be deduced from a study of the classical motion.

• The low field limiting symmetries and their meaning in the oscillator picture

The description of the breaking of the Coulomb symmetry in the n shell only requires to perform a convenient tilt (86) on (112) with $\theta_n = -\text{Log}\,n$. The diamagnetic hamiltonian is then multiplied with a $(-2E_n)^{-2}$ factor [35] scaling as n^4, as expected. Thus, at first order in B^2, the symmetry breaking part in the diamagnetic hamiltonian is deduced from (112) by retaining the terms which conserve the n value. Figure 8 indicates how to proceed. For example, terms like $S_z.T_+.S_-$ conserve the n value while $S_z.T_z.T_+$ do not. One deduces immediately [35] :

$$(\tilde{H}_D)_n = \frac{\gamma^2}{4}(-2E_n)^{-2}(S_z + T_z)\{S^2 + T^2 + 6S_zT_z + 2S_+T_- + 2S_-T_+\} \qquad (114)$$

which using the correspondance with SO(4) [35] can be shown equivalent to (105). Using the same analysis as in section (2) with SO(4), one can conclude that they are two limiting symmetries in the diamagnetic manifold. At the top of the manifold they are of SO(2,1)$\vec{\lambda}$ type associated with the $(S^2\ T^2\ W^2\ W_z)$ coupling scheme defined through (91) while at the bottom they are of $(S^2\ T^2\ S_z\ T_z)$ types.

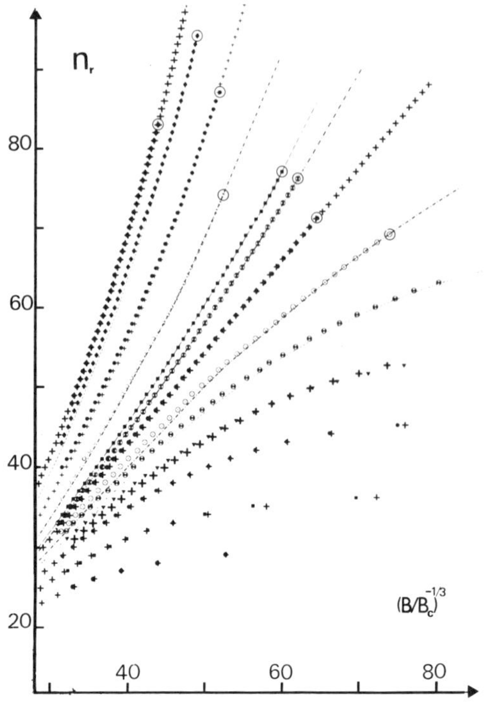

Figure 21 - Experimental plot of the positions of the dominant lines for Caesium (M=3 odd parity) states as a function of $B^{-1/3}$. Each curve corresponds to a given electronic energy scanning the B field [62]. At threshold the quasi-Landau resonances obey a $n^3.B/B_c = 1.56$ quantization (to within 5 %) and all the points are aligned. The quantum interpretation of the positions for $E < 0$ is now possible, for all the states.

• Adiabatic building up of the Landau spectrum

The low field characters of the diamagnetic problem are now expressed in a way which potentially authorizes the inter n mixing of the channels and thus is more general. The increase of the field will require to add more and more channels belonging to the oscillator ladder but pertaining to the same approximate symmetry type. This closely parallels the experimental idea of an adiabatic or continuous building up of the spectrum which is shown on the plot on Figure 21 for the dominant lines associated with $\vec{\lambda}$ type approximate symmetries [62].

From Figure 20 one deduces the following simple ideas. The states at the top of the manifold and to some extent quasi-Landau resonances are correlated states of the two oscillators at constant L_z. The relative phase of their oscillations along the (μ,ν) axis is a well-defined one and their motions are of collective types. To the contrast, the states at the bottom of the manifold are uncorrelated states of the pair of oscillators for which the relative phase of the oscillations along μ and ν is not fixed. This obviously bears some analogy with the motion of a pair of electrons [68] in doubly excited states of atoms.

• What do we learned from the dynamical group approach

At first sight it seems that we have learned only little from such an approach, no more than in the symmetry group approach. Actually, (112) and its subsequent analysis are valid whatever the n value which makes a fundamental difference with the SO(4) approach.

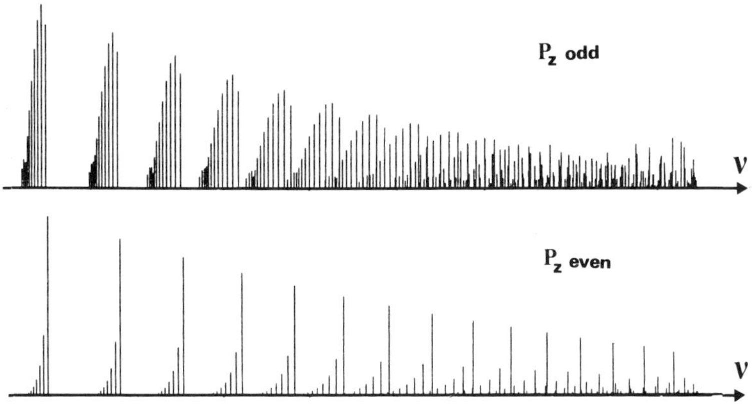

Figure 22 - The quantum simulations of the diamagnetic spectrum below threshold from the L-type Sturmian computations. Note the very different appearance according to even and odd z-parity (courtesy C.W. Clark and K.T. Taylor [71]).

Although it is not possible to reduce (112) further, the structure of the equation is algebraic which allows convenient manipulations. The evaluation of matrix elements of oscillators type does not lead to any difficulty. In addition H_D being a polynomial of degree 3, the selection rule $|\Delta n| \leqslant 3$ is fulfilled as well as $|\Delta n_\nu| \leqslant 2$ and $|\Delta n_\mu| \leqslant 2$. There is thus a finite number of non-zero matrix elements. Furthermore, Coulomb continua are exactly taken into account as well as the full symmetries. Hence this allows convenient perturbation expansions of the equations in a completely self consistent way [69]. This is in particular a convenient way for building effective hamiltonian [35 36].

This allows for rationnally choosing the basis sets for numerical simulations. The $(S^2T^2S_zT_z)$ and $(S^2T^2W^2W_z)$ ones are here of interest. They are Sturmian functions of a generalized type which are non-orthogonal but complete. The usual Sturmian set [39] defined in (89)(90) complies with the isotropy of space, as the generators commute with L^2 and L_z. It allowed the first quantum simulations of the diamagnetic spectra[70 71] as shown on Figure 22, on a Cray I computer. Although this set respects the algebraic structure of equation (112), it does not fit correctly the symmetry of the diamagnetic interaction. Instead, the $\vec{\lambda}$ type set of Sturmian functions allow such a simulation with a microcomputer [69] and only requires the mixing of 100 channels at maximum. Such a choice as well as direct perturbation expansions from (112) allow the description of the bound spectrum in the magnetic problem (for $E < 0$ and $\gamma n^3 < 1$) up to the strong mixing regime. This is in agreement with both the experimental data of Figure 21 on Caesium atoms [62] and Sturmian L-type computations [71].

4. Classical chaos and its quantum analog in the magnetic problem

Up to a certain point, close to the zero-field threshold, both calculations fail to converge, whatever the method. On the other hand the experimental results do present a continuity across threshold (see Figure 21). Of special interest is that these difficulties are associated with a peculiarity in the classical dynamics of the system which becomes completely chaotic [72 73 74 75].

It is a well-known feature of coupled oscillators systems that the dynamics can become chaotic. This has been extensively studied especially

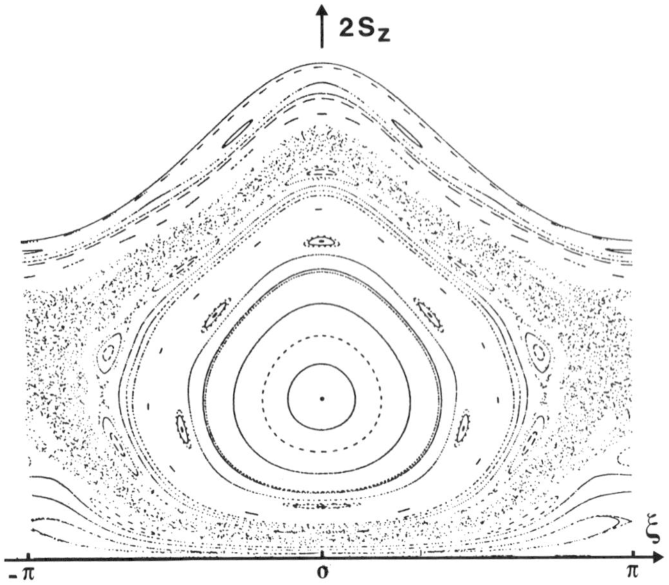

Figure 23 - Poincare mapping for the magnetic problem in the oscillator representation. The closed curves are associated with the $SO(2,1)\vec{\lambda}$ type approximate symmetry and express a correlated behaviour in the oscillator system. They are associated to quasi-Landau resonances which are fairly stable states located close to the maximum of the diamagnetic potential. The curves near the boundaries express an uncorrelated behaviour in the oscillator systems as the relative phase is not bounded. The onset of chaos is seen close to the separatrix where the cross-over in the symmetries occur. The correlated states surlive up to a point where the whole phase space becomes chaotic. (Courtesy D. Delande, unpublished [36]).

for photochemistry purposes. Hence from the coupled oscillators picture on Figure 20, it is not surprising that the magnetic problem in classical mechanics exhibits such a transition to chaos. Actually, the Stark problem will not exhibit such a feature due to the existence of a dynamical symmetry (the oscillators are not coupled). The existence of chaos is indeed a good indicator of the non-existence of a dynamical symmetry.

From (112) in the dynamical group approach, it is straightforward to deduce the equations of the classical motion in terms of the $(\vec{S},\vec{T})$ generators which play here the role of convenient classical dynamical coordinates (which especially respect the symmetries). A convenient way of characterizing the classical dynamics of a system is to perform a cut in phase space which is called a Poincare surface of section [76]. The results for the magnetic problem are shown on Figure 23 in a system of coordinates which can be thought as the energy of the oscillator along the μ axis and ξ the relative phase of the two oscillators. This plot both illustrates the existence of two limiting symmetries and the onset of chaos (that is their destruction).

The closed curves are associated with states in which the relative phase ξ of the two oscillators is well defined. This is the correlated motion associated with the $SO(2,1)\vec{\lambda}$ type approximate symmetry. These states have a tendancy to localize in the z = 0 plane perpendicular to the field (they are Ridge states in the approach of [68,77]) and give rise to quasi-Landau resonances. The other trajectories close to the boundaries are such that the relative phase ξ of the two oscillators is not bounded. They are associated with the $(S^2T^2S_zT_z)$ uncorrelated behaviour of the two oscillators (and motion of vibrational type along $\vec{B}$ field). The chaotic behaviour in the system is triggered in the region where there is a crossover between these two limiting symmetries. Further studies [36] allow to show that with increasing energy, the $\vec{\lambda}$ type states and thus quasi-Landau resonances are more stable than the others of uncorrelated type. Above a well-defined energy, but below the zero-field threshold, the whole phase-space becomes chaotic [36].

This precisely occurs in the vicinity of the point where the previous quantum calculations fail down. The question is thus to quantize the system up to the Landau limit in the region where the classical system is entirely chaotic, while the experimental data (see Figures 1 and 21) is still regular. This is certainly not the less interesting aspect of this magnetic field problem to provide us with this challenging example of the quantum analog of chaos.

CONCLUDING REMARKS

These considerations do not aim at exhausting the matter. In this magnetic problem, considered as a prototype, we have tried to state that the most simple and powerful views are drawn from unusual concepts, essentially based on the Coulomb symmetry. Considered this way, the quasi-Landau phenomenon expresses to some extent the survival of a symmetry which has its deep rooting in the structure of the Coulomb field itself. But obviously, the argument applies as well to other classes of perturbing potentials along the same lines meaning the phenomena is quite general.

The symmetry and dynamical group approach allow to understand the quantum and classical aspects of the magnetic problem on a quite large scale. All the advances for the last four years have been done incorporating the coulombic dimension of the atom this way. Thinking the dynamical properties in term of two quasi-spin or a pair of two-dim. harmonic oscillators is not so complicated. The notion of correlations which naturally follows from the picture has a dynamical character. It expresses a localization of the motion in phase-space rather than in real space. Both $\vec{\lambda}$ and $\vec{W}$ in the SO(3) and SO(2,1) coupling schemes are complicated differential operators in coordinates systems.

A consequence of these dynamical correlations is somewhat unexpected. It gives the states a greater stability close to the z = 0 plane perpendicular to the field where the diamagnetic potential is a maximum. This is known as "Ridge behaviour", early conjectured by Fano for this problem [77], which also rules the dynamics of doubly-excited systems or photochemical reactions. The fact that these ridge states are more stable than the valley ones (of uncorrelated types) is certainly not the least discovery in this problem. The analogy with other models used for interpreting the dynamics of polyatomic molecules or collective excitations in nucleus is striking.

Though the consequences of this extended tunability of the atom in external fields are not explored, one can expect the production of new classes of states with unusual properties. A simple example was shown in the low field crossed $(\vec{E},\vec{B})$ fields situation. More generally, lot of other situations involve the Coulomb field and a strong perturbation. This is true for the two-centers Coulomb problem, molecules, atoms interacting with inhomogeneous fields or electromagnetic fields and doubly excited systems. It is not unlikely that an analysis according to the previous schemes leads to far greater insights into the physics of these situations.

ACKNOWLEDGEMENTS

I would like to thank Dr. D. Delande for his essential contribution to the ideas leading this work and for his comments on the manuscript.

+ Laboratoire associé au Centre National de la Recherche Scientifique (LA18)

REFERENCES

1. W.R.S. Garton and F.S. Tomkins, Ap. J. 158 (1969) 83
2. S. Feneuille, S. Liberman, J. Pinard and P. Jacquinot, C.R. Heb. Acad. Sc. 284 (1977) 291
3. R.R. Freeman and N.P. Economou, Phys. Rev. A 20 (1979) 2356
4. A.R.P. Rau, Phys. Rev. A 16 (1977) 613
5. E.A. Solovev, JETP Lett. 34 (1981) 265
6. R.J. Fonck, D.H. Tracy, D. Wright and F.S. Tomkins, Phys. Rev. Lett. 40 (1978) 1366
7. J.C. Gay, "High-magnetic-field atomic Physics" in Progress Atomic Spectroscopy, volume C, H.J. Beyer and H. Kleinpoppen ed. (Plenum 1984)
8. D. Kleppner, M.G. Littman and M.L. Zimmerman "Rydberg atoms in strong fields" in Rydberg States of Atoms and Molecules; R.F. Stebbings and F.B. Dunning ed. (Cambridge Un. Press, 1983)
9. C.W. Clark, K.T. Lu and A.F. Starace, "Effects of magnetic and electric fields on highly excited atoms" in Progress Atomic Spectroscopy, volume C, H.J. Beyer and H. Kleinpoppen ed. (Plenum 1984)
10. J.C. Gay, "New trends in atomic diamagnetism" in Photophysics and Photochemistry in the V.U.V., S.P. Mc Glynn et al. ed. (Reidel, 1985)
11. W.E. Lamb, Phys. Rev. 85 (1952) 259
12. L.P. Gorkov and I.C Dzyaloshinskii, JETP 26 (1968) 449
13. R.B. Laughlin, Physica 126B (1984) 254
14. L.I. Schiff and H. Snyder, Phys. Rev. 55 (1939) 59
15. W. Pauli, Z. Phys. 36 (1926) 336
16. P.S. Epstein, Phys. Rev. 28 (1926) 695
17. L.D. Landau and E.M. Lifschitz, Mécanique Quantique (MIR, Moscou, 1966)
18. P.J. Redmonds, Phys. Rev. 133 (1964) 1352
19. Y. Demkov, B.S. Monozon and V. Ostrovskii, JETP 39 (1970) 775
20. L.D. Landau and E.M. Lifschitz, Mécanique (MIR, Moscou, 1966)
21. A. Durand, Mécanique Quantique (Dunod, Paris, 1970)
22. A. Messiah, Mécanique Quantique (Dunod, Paris, 1964)
23. V.A. Fok, Z. Phys. 98 (1935) 145
24. L.C. Biedenharn, Phys. Rev. A 126 (1962) 845
25. A.O. Barut and C. Fronsdal, Proc. Roy. Soc. A287 (1965) 532
26. A.O. Barut - Lectures in Theoretical Physics (Gordon and Breach, N.Y. 1967)
27. M. Bander and C. Itzykson, Rev. Mod. Phys. 38 (1966) 330
28. B.R. Judd, Angular Momentum Theory for Diatomic Molecules (N.Y. Academic Press, 1975)
29. A.M. Perelomov, Sov. Phys. Usp. 20, 9 (1977) 703
30. D.R. Herrick, J. Math. Phys. 16 (1975) 1047
31. P. Kustaanheimo and E. Stiefel, J. Ang. Math. 218 (1965) 204
32. M.J. Englefield, Group Theory and the Coulomb Problem - Wiley (N.Y. 1971)

33. C. Cohen-Tannoudji, B. Diu and F. Laloë, Mécanique Quantique (Hermann, Paris, 1973)
34. L.I. Schiff, Quantum Mechanics (McGraw Hill, N.Y. 1968)
35. D. Delande and J.C. Gay, J. Phys. B Letters 17 (1984) L335
36. D. Delande, Thèse de Doctorat d'Etat ès Sciences, Paris (1985) à paraître
37. E.U. Condon and H. Odabasi, Atomic Structure (Cambridge U. Press, 1980)
38. A.O. Barut and R. Raczka, Theory of Group Representations and Applications (Polish Scientific Pub., Warsaw, 1980)
39. M. Rotenberg, Ann. Phys. 19 (1962) 262
40. L.D. Landau, Z. Phys. 64 (1930) 629
41. H. Hasegawa in Physics of Solids in Intense Magnetic Fields, E.D. Haidemenakis ed. (Plenum, 1969)
42. J.C. Gay, Comments Atom. Mol. Phys. 9 (1980) 97
43. G. Wunner and H. Ruder, J. Physique 43, C2 (1982) 137
G. Wunner, Contribution this volume
44. S.M. Kara and M.R.C. Mc Dowell, J. Phys. B, 13 (1980) 1337
45. F. Penent, D. Delande, F. Biraben and J.C. Gay, Optics Comm. 49 (1984) 184
46. F. Penent, Thèse 3e Cycle, Paris (1984) unpublished
47. J. Andrä, private communications
48. D. Delande and J.C. Gay, unpublished
49. E.A. Solovev, JETP 82 (1982) 1762
50. D.R. Herrick, Phys. Rev. A (1982) 323
51. E.G. Kalnius, W. Miller and P. Winternitz, Siam J. Appl. Math. 30 (1976) 630
52. J.C. Gay, D. Delande, F. Biraben and F. Penent, J. Phys. B Letters 16 (1983) L693
53. J.C. Gay and D. Delande, Comments Atom. Mol. Physics 13, 6 (1983) 275
54. D. Delande, Thèse 3e Cycle (Paris, 1981) unpublished
55. D. Delande and J.C. Gay, Physics Letters A 82 (1981) 393
56. C.W. Clark, Phys. Rev. A24 (1981) 605
57. J.J. Labarthe, J. Phys. B Letters B14 (1981) L467
58. D. Delande, C. Chardonnet, F. Biraben and J.C. Gay in Colloque CNRS 334 "Atomic and Molecular Physics close to Ionization thresholds in High Fields", J.P. Connerade, J.C. Gay and S. Liberman ed., J. Physique Paris 43, C2 (1982) 97
59. C.W. Clark and K.T. Taylor, Nature 292 (1981) 437
60. D. Kleppner, M. Littman and M.L. Zimmerman, Scientific American 244 (1981) 108
61. D. Delande, F. Biraben and J.C. Gay in "New Trends in Atomic Physics" Les Houches Summer School XXXVIII, G. Grynberg and R. Stora ed. 352 (North Holland, 1984)
62. J.C. Gay, D. Delande and F. Biraben, J. Phys. B Letters B13 (1980) L720
63. J.C. Castro, M.L. Zimmerman, R.G. Hulet and D. Kleppner, Phys. Rev. Lett. 15 (1980) 1780
64. P. Cacciani, Thèse 3e Cycle, Paris (1984) unpublished
65. S. Haroche, Contribution this volume
66. M.L. Zimmerman, M.M. Kash and D. Kleppner, Phys. Rev. Lett. 45 (1980) 1092
67. T.P. Grozdanov and E.A. Solovev, J. Phys. B 15 (1982) 1195
68. A.R.P. Rau, Contribution this volume
69. D. Delande, unpublished
70. A.R. Edmonds, J. Phys. B 6 (1973) 1603
71. C.W. Clark and K.T. Taylor, J. Phys. B 13 (1980) L737
C.W. Clark and K.T. Taylor, J. Phys. B 15 (1982) 1175
72. M. Robnik, J. Phys. A 14 (1981) 3195

73. A. Harada and H. Hasegawa, J. Phys. A 16 (1983) L259
74. J.B. Delos, S.K. Knudson and D.W. Noid, Phys. Rev. A 30 (1984) 1208
75. G. Hose, H.S. Taylor and A. Tip, J. Phys. A 17 (1984) 1203
76. for example V.I. Arnold and A. Avez, Problèmes Ergodiques de la Mécanique Classique (Gauthiers-Villars, Paris, 1967)
77. U. Fano, Reports on Progress in Physics 46, 2 (1983) 97

PART II

Atoms or ions in dense plasmas or strong fields

ATOMS IN DENSE PLASMAS

Richard M. More

Lawrence Livermore National Laboratory
University of California
Livermore, CA 94550

When laser light is focussed to intensities of 10^{14} - 10^{16} Watt/cm^2 onto a cold solid, the target surface promptly rises to temperatures ~0.1 - 1 keV and produces a highly ionized plasma. Heat energy absorbed from the laser penetrates into the cold target by nonlinear electron heat conduction driven by enormous temperature gradients (~ 10^9 °K/cm), large thermoelectric and magnetic fields are generated, thermally produced x-rays are emitted and the heated material expands in high-velocity hydrodynamic flow. For all these processes the working fluid is a dense plasma of highly-charged ions. Because the densities and/or temperatures greatly exceed those available in previous laboratory plasma devices, we find many interesting new topics for scientific investigation.[1,2,3,4]

This paper covers some aspects of the theory of atomic processes in dense plasmas. Because the topic is very broad, we have selected a few general rules which give useful guidance about the typical behavior of dense plasmas. These rules are illustrated by semiclassical estimates, scaling laws and appeals to more elaborate calculations.

Included in the paper are several previously unpublished results including a new mechanism for electron-ion heat exchange (section II), and an approximate expression for oscillator-strengths of highly charged ions (section V). However the main emphasis is not upon practical formulas but rather on questions of fundamental theory, the structural ingredients which must be used in building a model for plasma events. What are the density effects and how does one represent them? Which are most important? How does one identify an incorrect theory? The general rules help to answer these questions.

Unfortunately in most cases one cannot yet directly employ experimental data to resolve theoretical questions. This circumstance has been, for many years, a basic difficulty of research on dense plasmas. For example, in

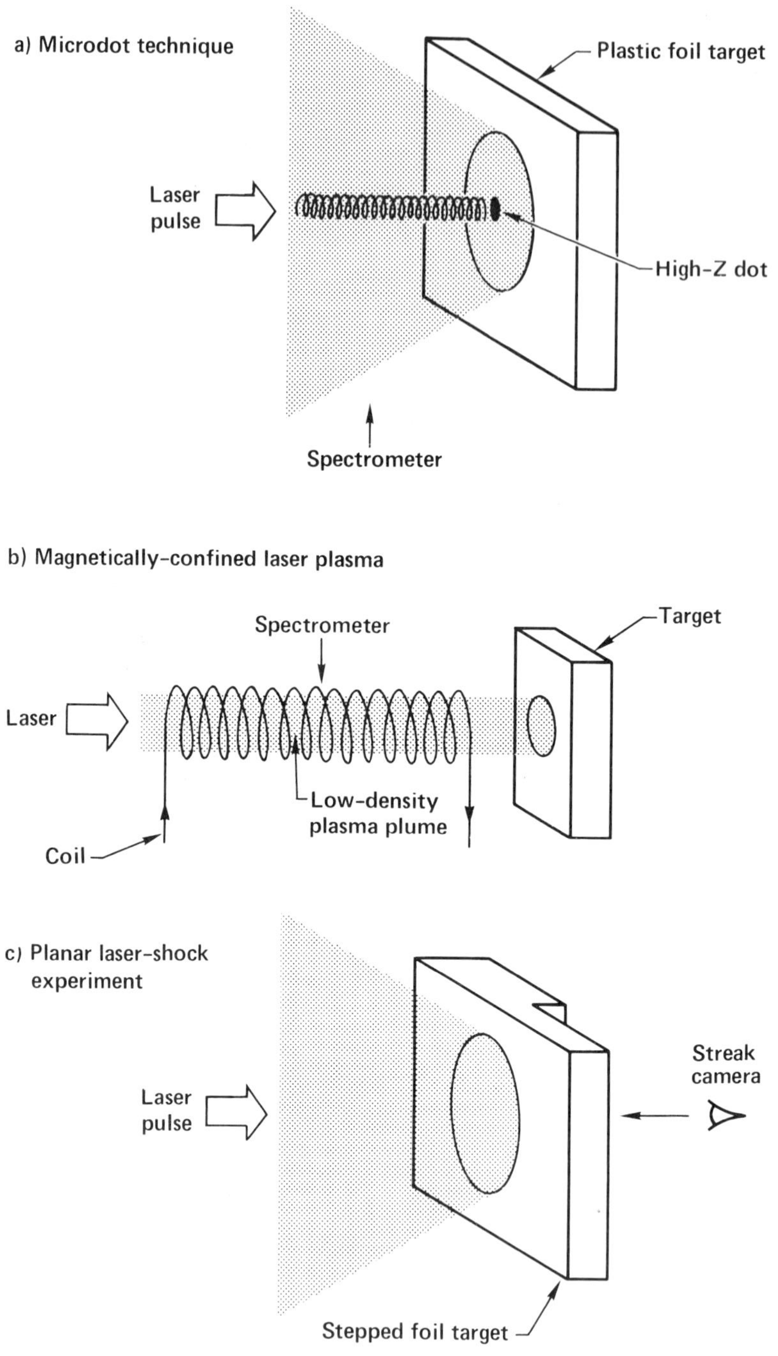

FIG. 1. Three favorable geometries for scientific experiments in laser plasmas. In each case the emitted or absorbed signal is produced by an approximately homogeneous plasma region.

laser-fusion implosion experiments, the observed x-ray signals consist of a superposition of contributions from a range of density-temperature conditions along the line of sight into the plasma. Such data can be compared with predictions of existing theory, but cannot easily be unfolded to yield accurate measurement of specific atomic processes. This difficulty is now solved, in principle, by the development of experimental geometries more favorable to scientific studies (Fig. 1).

In the plasma microdot technique (Fig. 1a) a small spot of high-Z (i.e., Z > 10) material is imbedded in a low-Z target such as plastic. Under irradiation the high-Z material expands as a plasma column, axially confined by the neighboring low-Z plasma, and strongly dominates the x-ray emission; a spectrometer observing this plasma from the side sees a unique density-temperature section of the high-Z plasma column. The technique has been applied to flow visualization (Herbst et al.[5]), plasma spectroscopy (Gauthier et al.[6]) and the verification of LASNEX hydrodynamic calculations (Rosen et al.[7]).

A similar technique (Fig. 1b) relies on a magnetic field to confine an expanding laser plasma at densities $\sim 10^{18}/cm^3$. This steady plasma has been shown, using Thomson scattering, to be remarkably homogeneous and therefore an excellent source for plasma spectroscopy experiments (Crawford and Hoffman[8]).

A third geometry yielding a very dense plasma is the planar shock-wave technique (Fig. 1c). In this case, if one knows the material equation of state and measures the shock speed (e.g., in the stepped-target geometry shown), the laws of conservation of mass, momentum and energy enable one to uniquely determine the density-temperature conditions of the hot dense plasma. With proper selection of focal spot size and pulse length, one can realize a homogeneous steady plasma at densities greater than the initial solid density.[9] Interesting recent experiments with shock-produced plasmas include absorption spectroscopy (Bradley et al.[10]), measurement of electrical conductivity (Ng et al.[11]), and hydrodynamic studies with short-wavelength light (Fabbro et al.[12]).

These new experimental techniques, a worldwide proliferation of pulsed laser research facilities, and the strong scientific interest stimulated by the recent demonstration of a laser-pumped soft x-ray laser[(7)] together signal that dense plasma research will be very active in the next few years.

I. THERMAL IONIZATION

The most obvious result of laser heating is thermal ionization produced by electron impact and/or photoelectric absorption of x-rays.

The first general rule is a well-known characterization of plasma ionization in complete thermal equilibrium LTE:

G-1.) An equilibrium plasma ionizes to a charge state $Q(Z, \rho, T)$ at which

$$I(Z,Q) \simeq \xi\ kT \tag{1}$$

The coefficient ξ is a fundamental measure of plasma density effects.

In Eq. (1), Z = nuclear charge, Q = ion (average) charge state, $I(Z,Q)$ = ionization potential. ξ is essentially the electron entropy:

$$\xi = \ln\left(\frac{2}{n_e \lambda^3}\right) = S/k - \frac{5}{2} \tag{2}$$

where $\lambda = (2\pi\hbar^2/m_e kT)^{1/2}$ = electron thermal deBroglie wavelength; n_e = electron number density = $Q\rho/AM_p$; S = entropy per free electron; k = Boltzmann constant. Equations (1) and (2) follow from the Saha equation for plasma ionization (Zel'dovich and Raizer[13]); the derivation assumes equilibrium non-degenerate plasma conditions.

Figure 2 gives contour plots of the ionization state, the ratio I/kT and $\xi = \ln\,(2/n_e\lambda^3)$. One sees $\xi \simeq 10\text{-}15$ for low-density hot plasma, but $\xi < 5$ at dense-plasma conditions. It is evident that ξ and I/kT are approximately equal. The figures are based on the Thomas-Fermi cell model, but other theories give similar results.

The ionization potentials $I(Z, Q)$ are the basic atomic data required to calculate equilibrium (LTE) ionization. There are two ways to obtain quick estimates of the ionization potentials: Thomas-Fermi theory,[3] or the Bohr model,

$$I_n = \frac{Q^2 e^2}{2 a_o n^2} \tag{3}$$

which relates the ionization potential to the ion charge Q and the principal quantum number (= n) of the outermost bound electron.

Combining Eqs. (3) and (1) we find

$$Q \simeq \left(\frac{\xi\ n^2 kT}{13.6\ \text{eV}}\right)^{1/2} \propto T^{1/2} \tag{4}$$

as a scaling law for partial ionization in LTE. Equations (1)-(4) are used to compare physical processes; for example, the free electron specific heat $\simeq 3/2\ Qk$ is less than the ionization specific heat $\simeq I(Q)\ dQ/dT \simeq 1/2\ \xi Qk$ for a partially-ionized plasma with $\xi > 3$.

The scaling of Q with $T^{1/2}$ fails (i) for fully-ionized plasmas, where $Q = Z$; (ii) at low charge states, $Q < 1$; (iii) near closed-shell configurations, where Q is relatively independent of T over a range of temperatures; and (iv) at

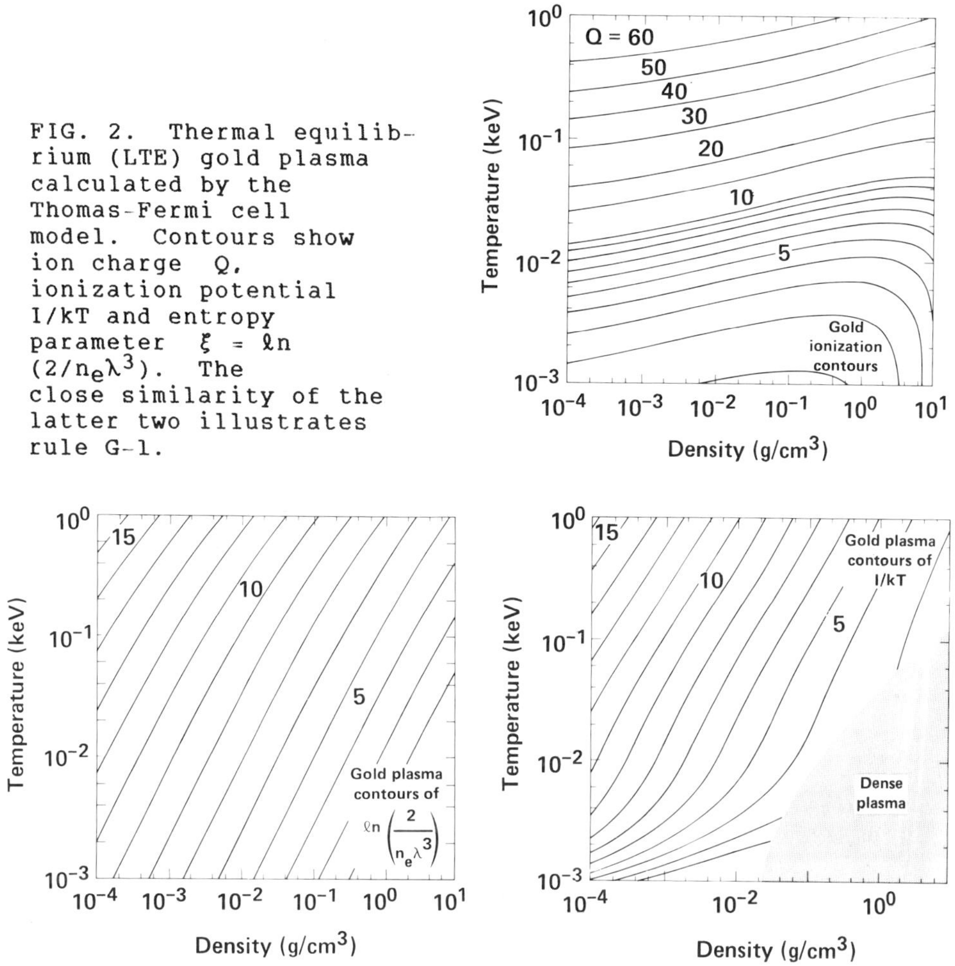

FIG. 2. Thermal equilibrium (LTE) gold plasma calculated by the Thomas-Fermi cell model. Contours show ion charge Q, ionization potential I/kT and entropy parameter $\xi = \ln(2/n_e\lambda^3)$. The close similarity of the latter two illustrates rule G-1.

very high densities where degeneracy and pressure-ionization occur. The scaling also fails at low densities where most plasmas are optically thin and depart from thermal equilibrium. The scaling $Q \propto T^{1/2}$ is illustrated in Fig. 3.

Energy Level-shifts

Even for isolated ions, the energy-levels of core and valence electrons change as the ionization state increases. These changes are simply described by an algebraic screening model often used in fusion research. Let us examine a hypothetical Bohr orbit in a many-electron atom having average orbit radius r_n, velocity v_n. The assumptions of the Bohr theory are:

$$\frac{Q_n e^2}{r_n^2} = \frac{m v_n^2}{r_n} \quad , \quad m v_n r_n = n\hbar$$

The first relation expresses Newton's law (F = m a), the second the quantization of angular momentum. The solution of these two equations is

$$r_n = a_o n^2/Q_n \tag{5}$$

$$v_n = Q_n e^2/n\hbar \tag{6}$$

The useful result of this simplified treatment is the idea that r_n and v_n depend upon an effective charge associated with the electric field at radius $r = r_n$, i.e., upon the <u>inner screening</u> charge

$$Q_n = Z - \sum_{m \leq n} \sigma_{nm} P_m \tag{7}$$

Here P_m is the population of the mth shell and σ_{nm} is a screening coefficient which tells what fraction of the charge of shell m resides inside radius r_n. In Eq. (7) the screened charge is assumed to be independent of subshell quantum numbers (j or ℓ) and a linear function of the populations.

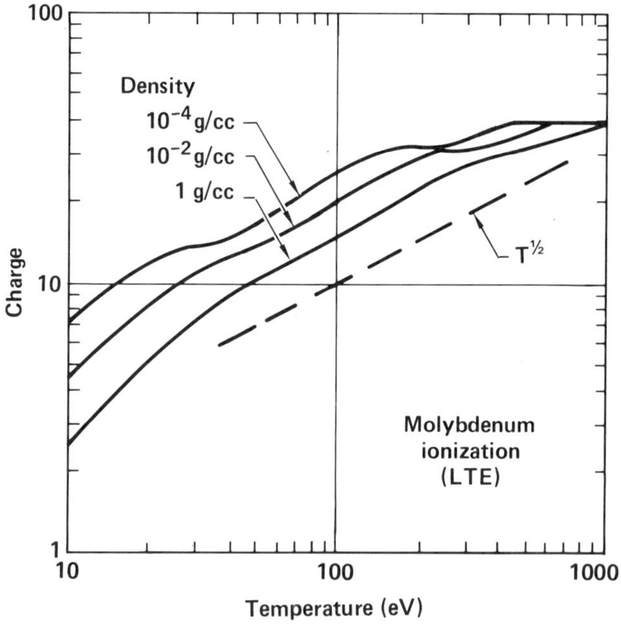

FIG. 3. Molybdenum charge state as calculated by the screened hydrogenic average-atom theory. The figure shows the scaling $Q \propto T^{1/2}$, and the tendency of Q to decrease with rising density in the nondegenerate density range. The ionization shelf structure at closed-shell configurations (Q = 40, 32, 14) disappears at high densities, illustrating rule G-7.

The energy E_n of the electron also involves the electrostatic potential $V(r)$ at radius r_n.

$$E_n = \frac{1}{2}mv_n^2 - eV(r_n)$$

The potential $V(r)$ has a contribution from <u>outer screening</u>: i.e., from electrons outside radius r_n. A useful approximation for this potential is:

$$V(r_n) = \frac{Q_n e}{r_n} - \sum_{m \geq n} \sigma_{mn} \frac{P_m e}{r_m}.$$

The first term is the potential outside a spherical core of charge + Q_n e, and the second term sums the potentials inside spherical shells of charge - $\sigma_{mn} P_m e$ located at radii $r_m > r_n$. With this model for the electrostatics,

$$E_n = -\frac{Q_n^2 e^2}{2a_o n^2} + \sum_{m \geq n} \sigma_{mn} \frac{P_m e^2}{r_m} \qquad (8)$$

The approximation based on Eqs. (7,8) is the <u>screened hydrogenic model</u> (Strömgren[14], Mayer[15], Lokke et al.[16], Zimmerman and More[17]). In this treatment, inner screening enters the calculation of Q_n, r_n and v_n, while outer screening affects only the potential $V(r_n)$ and energy E_n. Because of screening the one-electron eigenvalues E_n are strong functions of the ion charge-state (Fig. 4).

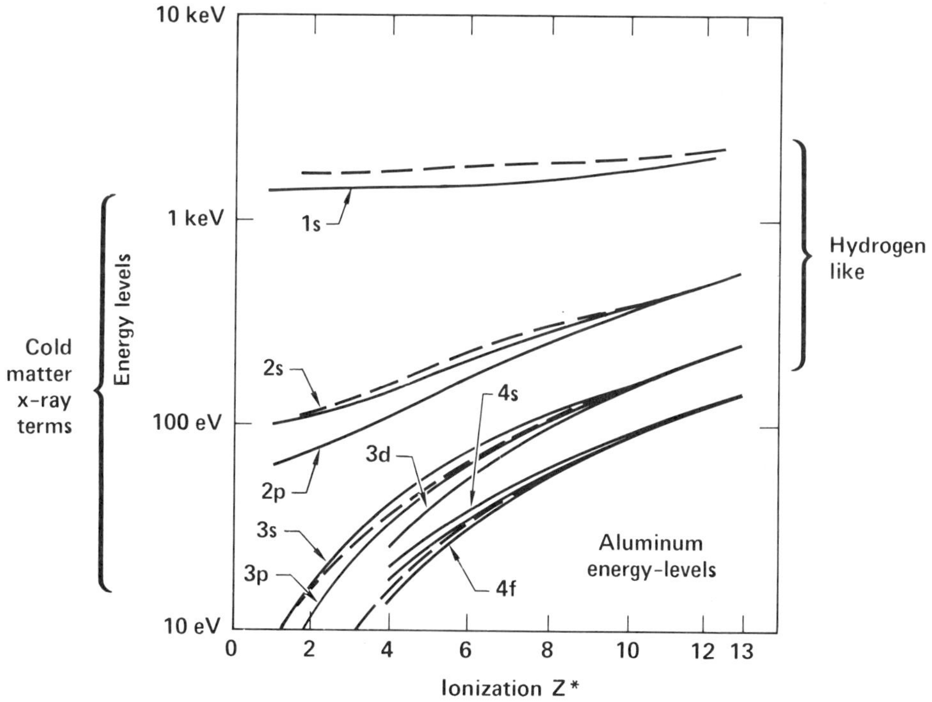

FIG. 4. Variation of Aluminum one-electron eigenvalues with ion charge. Dashed curves are the screened hydrogenic model, Eq. (8); solid curves are WKB eigenvalues for Thomas-Fermi ions. The calculations illustrate rule G-1a.

This simplified model is actually better than it appears at first sight for three reasons:

(1) One can optimize the choice of screening constants σ_{nm}, and thereby obtain a reasonable approximation to results of more accurate quantum calculations.[18]

(2) There is a Koopman's theorem which relates the one-electron energy-level E_n to a formal total ion energy of tractable functional form[17,18]

$$E_n = \frac{\partial E_{ion}}{\partial P_n} \tag{9}$$

with

$$E_{ion} = -\sum_n \left(\frac{Q_n^2 e^2}{2a_o n^2}\right) P_n = \sum_n E_n P_n - U_{ee} \tag{10}$$

Because of this result, one can employ E_{ion} as a model Hamiltonian which generates a consistent approximate statistical theory of the spectra of many-electron atoms.[3,17,18]

(3) Equations (5-8) are close cognates of more persuasive formulas obtained by simplification of the WKB theory of complex atoms. For example, the non-relativistic WKB theory yields an energy eigenvalue[18,19]

$$E_{n\ell} = -\frac{Q_n^2 e^2}{2a_o(n-\Delta_{n\ell})^2} + E_n^o$$

together with a precise specification of the inner-screened charge Q_n, the quantum defect $\Delta_{n\ell}$ and the outer-screening correction E_n^o. This WKB formula is close enough to the simpler model of Eq. (8) to give useful practical guidance in developing the model (Fig. 4).

Inner screening dominates the increase of the ionization potential I(Z,Q) with ion charge Q. On the other hand, outer screening is the reason why K-shell absorption or fluorescence energies change with removal of outer electrons.

Combining these ideas, we extend the rule G-1

G-1a.) Excitation energies are approximately proportional to $I(Z,Q) \propto kT$. Line spectra and absorption edges change with temperature to follow the maximum of the black-body spectrum.

The rule is illustrated by calculations of aluminum photoelectric absorption (Salzmann and Wendin[20]) based on Saha ionization-equilibrium calculations of the ion populations (Fig. 5).

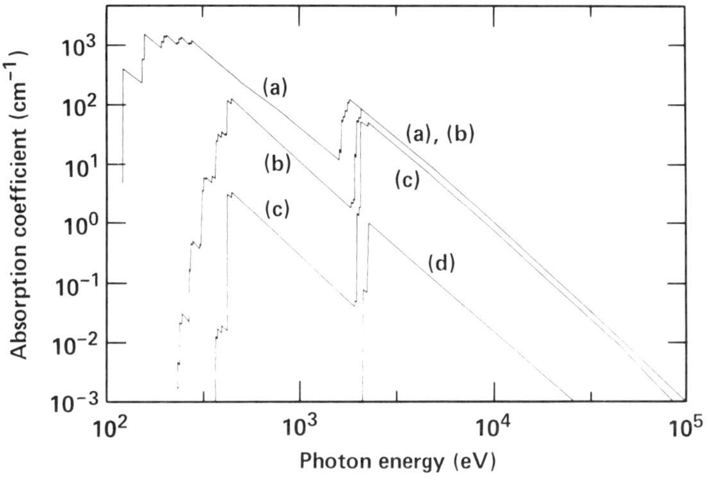

FIG. 5. Aluminum photoelectric absorption coefficient for density $n_i = 10^{21}/cm^3$ at temperatures a.) 32 eV, b.) 100 eV, c.) 320 eV, d.) 1000 eV (Salzmann and Wendin[20]). The motion of the effective absorption edge illustrates rule G-1a.

For fully-ionized plasmas, the thermal average (Rosseland) opacity has a temperature-dependence $\sim T^{-3}$ resulting from the $(h\nu)^{-3}$ dependence of the bremsstrahlung absorption cross-section.[21] For partially-ionized plasmas, the dominant photoelectric cross-section has the same frequency-dependence $(h\nu)^{-3}$, but the opacity now has a much weaker temperature- dependence $\sim T^{-1}$ due to the thermal shifts of photoelectric edges.

Shifts in energy-levels associated with ionization are much larger than any possible level-shift due directly to density itself. This point is important because the ionization state depends on density through Eq. (1). The density derivative $(\partial E_n/\partial \rho)_T$ includes a large red-shift associated with recombination while the derivative at constant charge state $(\partial E_n/\partial \rho)_Q$ is much smaller or zero.[22]

II. ION INTERACTIONS

The time-average distance between ions is obviously much greater than the ion size in low-density plasmas. Is this instantaneously true? Do ions touch during their thermal motion in the plasma?

The ion core radius R_i is related to the ionization potential I by

$$R_i(Z,Q) \propto Qe^2/I(Z,Q)$$

(this estimate is exact in Thomas-Fermi theory[3]) and for ions having thermal energies $\sim kT$ the minimum ion separation R_L is that for which

$$\frac{Q^2e^2}{R_L} \simeq kT$$

Together with Eq. (1) this implies

$$\frac{R_L}{R_i} \simeq Q\xi \gg 1 \tag{11}$$

especially for low-density plasmas where $\xi \gg 1$. Equation (11) says that ion cores are well-separated even during the closest encounters.* This in turn implies that the Coulomb law $U(R) = Q_1 Q_2 e^2/R$ is a good representation of the ion-ion potential.

For low-density (ideal) plasmas there are three length scales which describe ion spatial correlations:

$$\begin{aligned} R_L &= \text{Landau length} = Q^2e^2/kT \\ R_o &= \text{ion-sphere length} = (3/4\pi n_i)^{1/3} \\ D_i &= \text{(ion) Debye length} = (kT/4\pi Q^2e^2n_i)^{1/2} \end{aligned} \tag{12}$$

For the usual low-density plasma,[23]

$$R_L \ll R_o \ll D_i \tag{13}$$

When this inequality is satisfied, there are many ions in a sphere of radius D_i and the ion correlations are described by the Debye-Huckel theory. In this theory, a static point charge is screened within a length D given by

$$\frac{1}{D^2} = \frac{1}{D_i^2} + \frac{1}{D_e^2} = \frac{4\pi Q^2e^2n_i}{kT_i} + \frac{4\pi Qe^2n_i}{kT_e} \tag{14}$$

The first term corresponds to ion screening and the second to screening by free electrons. It is evident that ions dominate the static screening whenever $Q > 10$; this is the case for laser plasmas generated from SiO_2, Al, etc.

In order to understand the ion correlations induced by Coulomb interactions, it is useful to examine the ion number density fluctuation $\langle \delta n_i^2(k) \rangle$ as a function of wave-vector k. In the Debye-Huckel theory (Salpeter[24]):

* The conclusion fails in dense plasmas where R_L is no longer the distance of closest approach, see Eq. (18) below.

$$
\langle\delta n_i^2(k)\rangle = \begin{cases} \dfrac{n_i}{V}\,\dfrac{D^2(k^2 + 1/D_e^2)}{1 + k^2D^2} & \text{electron and ion screening} \\[2ex] \dfrac{n_i}{V}\,\dfrac{k^2D_i^2}{1 + k^2D_i^2} & \text{ions only } (T_e \to \infty) \\[2ex] \dfrac{n_i}{V} & \text{ideal gas without screening} \end{cases} \tag{15}
$$

Figure 6 plots $\langle\delta n_i^2(k)\rangle$ for a typical low-density plasma. One sees that Coulomb interactions tend to inhibit the long-wavelength charge-density fluctuations relative to those which would exist in an uncharged ideal gas. This suppression of fluctuations is the k-space image of the Debye screening itself.

We abstract a general principle from the example given in Fig. 6:

G-2.) Coulomb interactions tend to suppress (reduce) thermal fluctuations.

The argument for the general rule is the following: usually a non-interacting (ideal) many-body system has many highly degenerate states which are equally populated in thermal equilibrium. Coulomb interactions split this degeneracy; fewer low-energy states are populated more heavily and so the non-ideal gas has less thermal fluctuation.

The general rule is also illustrated by ion Stark broadening: Coulomb interactions between the plasma perturbers reduce the large microfields responsible for ion Stark wings in comparison with the non-interacting (Holtsmark) calculation (Ecker[25], Mihalas[26]). Another example: ion Doppler line-width is reduced when one includes the effect of Coulomb interactions between the ions (Dicke[27], Burgess and Lee[28]). Another example: calculations of ion charge-state distributions show that more rigorous models which include the electron-electron interactions predict less ionization fluctuation $\langle\delta Q^2\rangle$ than a non-interacting average-atom model (Green[27], More[2]).

Ion Coupling

The numerical study of ion correlations in plasmas has centered around a standard idealized model, the one-component

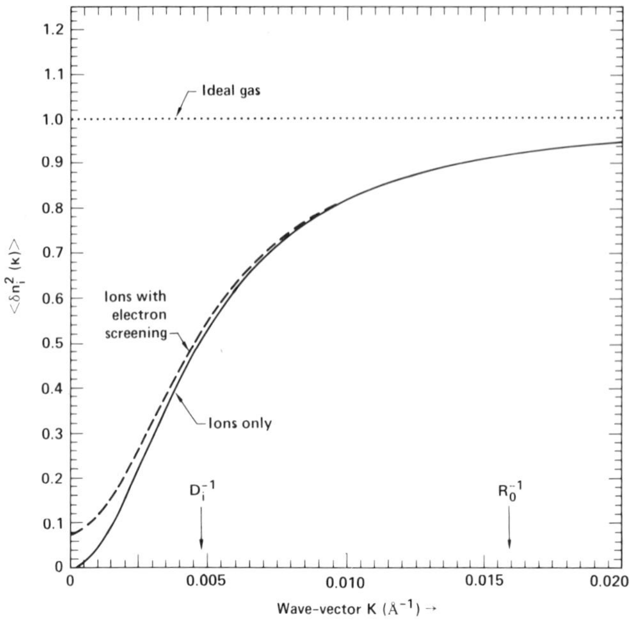

FIG. 6. Ion number-density fluctuation for a plasma with $n_i = 10^{18}/cm^3$, $Q = 10$, $T_e = 1$ keV and $T_i = .8$ keV. The large-scale charge fluctuations are strongly inhibited by Coulomb interactions, illustrating rule G-2. Electron screening reduces the effective Coulomb repulsion and permits some (neutral) density fluctuation.

plasma (Brush et al.[30], Hansen[31]). This is a model of classical point charges + Qe moving in a uniform fixed background of negative charge.

The equilibrium statistical mechanics of the one-component plasma is generated by the configurational partition function Z_c:

$$Z_c = \int d^3R_1 \int d^3R_2 \ldots \int d^3R_N \quad e^{-\sum_{i,j} \frac{Q^2e^2}{|R_i - R_j| kT}}$$

$$= R_o^{3N} \int d^3x_i \ldots \int d^3x_N \quad e^{-\Gamma \sum_{i,j} \frac{1}{|x_i - x_j|}}$$

In the second form, the integration variables are changed from ion positions $\vec{R}_i$ to scaled positions $\vec{x}_i = \vec{R}_i/R_o$, where $R_o = (3/4\pi n_i)^{1/3}$.

It follows that the configurational partition function depends essentially on only one variable, the ion coupling parameter Γ, defined by

$$\Gamma \equiv \frac{Q^2e^2}{R_o kT} \qquad (16)$$

The ion coupling parameter describes the strength of ion correlations and controls the validity of different theoretical approaches. For weak coupling ($\Gamma \ll 1$) the Debye-Huckel theory is a good approximation. In this case the average Coulomb energy per ion is

$$\Delta E_c = -\frac{1}{2}\frac{Q^2e^2}{D_i} = -\frac{\sqrt{3}}{2}\Gamma^{3/2} kT \qquad (17a)$$

In the weak coupling case ($\Gamma \ll 1$) this energy is much less than the average thermal energy ~ kT, so the Coulomb forces are weak perturbations; the same conclusion is implied by the fact that many ions ($N \propto (D_i/R_o)^3 \gg 1$) are required to screen a test charge.

For strong coupling ($\Gamma \geq 0.5$) one has a nontrivial many-body problem which is well-suited to computer simulation. In this case the screening cloud around each ion involves a modest number of particles, so that a simulation using a few hundred computational ions becomes reasonably accurate.[31-33] The Monte Carlo (MC) simulations examine many configurations and weight the contributions by a Boltzmann factor corresponding to the total ion interaction energy (computational algorithms are described in Ref. 34). Molecular Dynamics (MD) simulations simply solve the classical equations of motion to generate time histories of the ion dynamics; physical quantities are calculated as time averages.

What are the effects of strong ion correlation? The rule G-2 goes directly to the heart of the question: ions repel each other and try to avoid close encounters. This forces them into an increasingly orderly arrangement which ultimately (for $\Gamma > 170$) becomes a crystalline state.

For $\Gamma > 1$, the ion-sphere model is not a bad approximation: each ion is surrounded by a cavity filled with electrons but devoid of neighbor ions. The cavity radius is ~ R_o; numerical simulations show that the nearest neighbor distance is $\simeq 1.7\, R_o$. The energy per particle is approximately

$$\Delta E_c \simeq -\frac{9}{10}\frac{Q^2e^2}{R_o} = -\frac{9}{10}\Gamma\, kT \qquad (17b)$$

which corresponds to the electrostatic interaction of a point ion with a uniform neutralizing electron gas in the ion-sphere volume. MC simulations have given more accurate values for this Coulomb energy $\Delta E_c(\Gamma)$.

An interesting feature of the strongly coupled plasma is the breakdown of Eq. (13). R_L, originally defined as the (energetic) distance of "closest approach," becomes greater than the average ion separation R_o. Indeed,

$$\frac{R_L}{R_o} = \Gamma > 1$$

In the strongly coupled plasma the ions are forced together against their Coulomb repulsions. The lengths R_L and D_i

become meaningless, and the single distance R_0 serves as distance of closest approach, average pair-separation and screening length.

Simulations of the one-component plasma provide a variety of useful data. The Coulomb interaction energy ΔE_c enters the plasma hydrodynamic equation of state and the equilibrium ionization (ΔE_c generates the continuum lowering). The probability distribution for plasma electric fields $P(|\vec{E}|)$ can be obtained from the plasma simulations; this determines the ion Stark profile. The ion pair distribution g(r) gives the probability to find a neighbor ion a specified distance r away from one central ion; g(r) is used in calculation of quantum interference-scattering effects which arise in electrical conductivity or plasma bremsstrahlung emission.

Charge-state Variations

The one-component plasma is an idealized model; although it is very useful it also omits certain physical processes. Among these are phenomena associated with changes in the ion charge Q.

For real plasmas the time-averaged ion charge $Q = Q(\rho,T)$ is dependent upon density and temperature and this means that care must be exercised in adapting results from the OCP simulations (e.g., integrating the OCP specific heat with respect to temperature).

For temperatures much below the Fermi temperature, the ion charge depends only upon density, $Q(\rho,T) \simeq Q(\rho,0)$. Examination of Thomas-Fermi calculations for degenerate matter shows that a rather accurate description of this pressure ionization is given by[3]

$$R_i(Z,Q) \simeq R_0 \qquad (18)$$

which determines the plasma ionization state $Q(\rho)$ at low temperatures in terms of the ion core radius $R_i(Z,Q)$ and ion-sphere radius R_0. With this equation one sees that ion cores indeed come into close contact in degenerate high-density matter.

In the nondegenerate range the plasma ionization state is proportional to $T^{1/2}$ according to Eq. (4). Laughlin has pointed out that this power law implies that Γ is essentially independent of temperature in a partially-ionized plasma.[35] This means that ion thermal motion (e.g., amplitude of ion vibration with respect to the neighbors) is approximately independent of temperature; the potential between ions grows stronger as rapidly as the thermal energy does.

The ion charge also changes with time through ionization and recombination and this causes interesting new effects in the ion dynamics.

We will describe one such process, a Raman effect in electron-ion energy exchange, which is predicted theoretically for partially-ionized plasmas and caused by sudden charge fluctuations occuring in ionization and recombination.[2,36,37] The effect is a dense-plasma process, but is quantitatively important at surprisingly low plasma densities where it affects the viability of one well-known x-ray laser scheme.

In this case, the pumping efficiency of a resonant photopumped laser depends strongly on the Doppler linewidth, which is determined by the ion temperature T_i, controlled in turn by the heat exchange between electrons and ions (Hagelstein[37]).

Collisional Heating of Ions

Ordinarily the electron-ion heat exchange is calculated from the Landau-Spitzer formula,[23,38]

$$\frac{dE_i}{dt} = \frac{m_e}{M_i} (kT_e - kT_i)\, n_e \sigma_{eff} v_e \tag{19}$$

where E_i = energy per ion; m_e, M_i = electron, ion mass; kT_e, kT_i = electron, ion temperatures; n_e = electron number density; $v_e = (kT_e/m_e)^{1/2}$ = electron thermal velocity; and

$$\sigma_{eff} = 4\sqrt{2\pi} \left(\frac{Qe^2}{kT_e}\right)^2 \log \Lambda \tag{20}$$

is an effective collision cross-section. The numerical coefficient in Eq. (20) comes from a well-known formula for the electron-ion coupling time τ_{ei} (Spitzer[23]). Equations (19)-(20) describe energy transfer occuring in electron-ion collisions; a small fraction (proportional to $m_e/M_i \ll 1$) of the kinetic energy of the hotter particle is transferred in each Coulomb collision.

Ionization Heating of Ions

The Raman heat-exchange mechanism is also a consequence of electrical forces. Consider an ion of charge $+Qe$, surrounded by a neutralizing or screening cloud of electrons and ions. The electrical energy stored in this screening cloud is $\Delta E_c \sim Q^2e^2/R_s$, where R_s is the appropriate screening length.

After electron impact ionization, the screening cloud must adjust to the increased ion charge $Q' = Q + 1$. For a time short compared to the ion plasma oscillation period $(1/\omega_{pi})$ the extra unit charge is screened by electrons at a large distance $\simeq D_e$ = electron Debye length. At the radius R_o of the nearest neighbor ions this leaves an unscreened electric field which causes the neighbors to relax outward from the central ion. This outward motion is rapidly thermalized into ion kinetic energy. Thus in each ionization event the ions gain a thermal energy of order $d(\Delta E_c)/dQ \sim Qe^2/R_s$.

This reasoning gives an estimate[2]

$$\frac{dE_i}{dt} \simeq \frac{Qe^2}{R_s} \varphi\, n_e \sigma_{ii} v_e \tag{21}$$

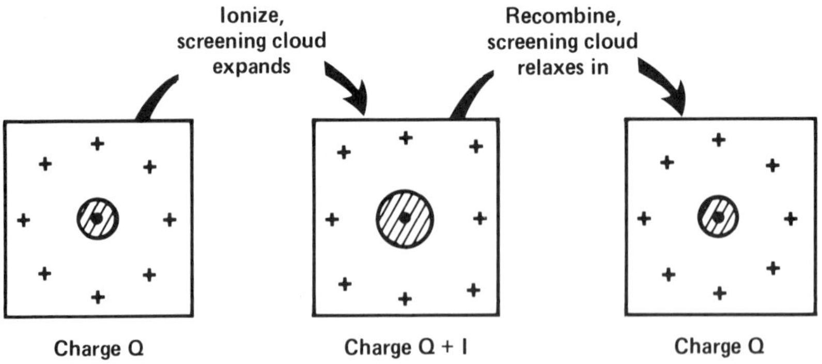

FIG. 7. Schematic representation of electron-ion energy exchange during ionization.

for the rate of ion heating associated with the ionization process. The rate of energy exchange is proportional to the rate of impact ionization $n_e \sigma_{ii} v_e$, and φ is a numerical factor intended to correct for the possibility of recombination occuring before completion of the changes in the ion screening cloud.

It is important to observe that the energy transfer is not necessarily cancelled by a subsequent recombination event. Indeed, if the time between ionization and recombination exceeds $1/\omega_{pi}$, and if T_i is sufficiently low, the screening cloud around the ion of charge Q + 1 has time to relax into a low-energy configuration; when recombination occurs, the screening cloud must again re-adjust and this transfers additional energy to the ion system.

A very similar heat-exchange mechanism is proposed by Hagelstein, who terms it <u>ion proximity heating</u>.[37]

Using Eqs. (19),(21) we can compare the Raman or proximity heating to the collisional energy-transfer: the ionization cross-section is usually much smaller than the Coulomb cross-section but the energy exchanged per ionization event ($\sim Qe^2/R_s$) can greatly exceed the small energy transfer ($\sim m_e kT_e/M_i$) in a collision.

The process described by Eq. (21) is analogous to phonon emission occuring in the absorption of visible light by impurities or color centers in transparent solids.[39] In that case absorption raises the defect center to an excited state which interacts with the crystal with an altered potential; in relaxing to the groundstate configuration in this potential the defect center releases energy as phonons. This produces a classic optical Raman effect: emission is red-shifted relative to absorption. Equation (21) describes an electronic analogue of this process.

After estimating the ion temperature rise caused by this mechanism in a rapidly ionizing plasma, we reconsider Eq. (21) to reconcile it with the principle of detailed balance.

Example: A Rapidly Ionizing Plasma

The importance of the ionization heat-exchange is illustrated by one recently proposed design for a laser-pumped soft x-ray laser.[37] In this experiment a low ion temperature (T < 10 eV) would have the favorable consequence of reducing the Doppler width of the pumping transition. In one specific proposal neon ions at a number density ~ $10^{18}/cm^3$ would be rapidly stripped to charge Q = + 8 by photoelectric absorption of kilovolt x-rays from an intense laser-plasma source. For the proposed plasma conditions one has $kT_i \simeq 1 - 5$ eV and $kT_e \simeq 60 - 200$ eV. With these assumptions, the screening lengths defined by Eqs. (12, 14) are:

$$R_o \simeq 60 \text{ Å}$$

$$D_i \simeq 10\text{-}20 \text{ Å}$$

$$D_e \simeq 200\text{-}400 \text{ Å}$$

Since $D_i \ll R_o$ one has a strongly coupled plasma despite the relatively low ion density.

In the ion-sphere description of this strongly coupled plasma, the central ion is assumed to be surrounded by an electron gas which is nearly uniform in the range 0 < r < R_o; outside this range the other ions and electrons are taken to comprise a uniform neutral plasma.

When an ion of arbitrary charge Q' is placed in this plasma, it will repel neighbor ions in order to produce a cavity containing only electrons, whose radius is such as to assure electrical neutrality. The screening energy is then $\Delta E_c = 9/10\ (Q'^2 e^2/\ R')$ with $Q' = (4\pi/3)\ R'^3\ n_e$ and constant free electron density n_e determined by the average ionization state $\langle Q' \rangle$. When we include the change of cavity radius with ionization of the central ion we find

$$\left(\frac{\partial \Delta E_c}{\partial Q'}\right) = \frac{3}{2}\frac{Q' e^2}{R_o}$$

A minimal ion temperature can be calculated by integrating these energy changes with respect to Q' from Q' = 0 to the final charge state Q' = Q. Assuming an ideal-gas ion specific heat we find:

$$kT_i \geq \frac{1}{2}\frac{Q^2 e^2}{R_o} e^{-R_o/D_e} \tag{22}$$

This mechanism alone thus predicts $kT_i \geq 7$ eV for the case considered. The actual ion temperature would be higher still because of the collisional heat transfer of Eqs. (19,20) and because of cyclic Raman heating occuring when the steady-state ionization is achieved.

The limitation indicated in Eq. (22) applies to a plasma composed of a single substance (e.g., pure neon). If this material were diluted with a low-Z impurity (e.g., hydrogen gas) the same ion heating energy could be shared with more nuclei resulting in a lower ion temperature.

A Question of Detailed Balance

The formulation of electron-ion heat exchange based on Eq. (21) suffers from one significant defect: it appears to predict a continuous heating of the ions and thus appears to predict that the ion temperature T_i continues to rise even when $T_i > T_e$. This cannot be correct: the heat exchange mechanism must move the two species toward thermal equilibrium and the net rate of heat transfer must approach zero as the temperature difference approaches zero (see rule G-10).

This formal requirement (associated with detailed balance) is certainly satisfied by the collisional heat-transfer described by Eq. (19).

In order to resolve this question for the Raman process it is useful to consider ionization and recombination occuring in the presence of a fluctuating electrostatic potential Φ produced by plasma particles near a central ion.

Ionization, Recombination in a Potential Φ

We consider an ion subject to a potential Φ generated by the neighboring charged particles. If the plasma density is sufficiently low the nearest neighbor ion is often the main source of this potential, whose magnitude is therefore $\sim Qe/R_o$; for definiteness the reader may consider this case while noting that the reasoning is independent of the source of the potential.

It is assumed that the spatial variation of the potential Φ is on a size-scale much larger than the ion radius R_i and the time variation is slow compared to electron transit times. An electron of energy gains a kinetic energy = e Φ as it approaches the central ion, and the modified impact ionization cross-section is therefore approximately $\sigma_{ii}(\epsilon_o + e\Phi)$, where $\sigma_{ii}(\epsilon_o)$ was the original cross-section for impact ionization without the potential.

After ionization there are two low-energy free electrons. One or both could be at low kinetic energies ($\epsilon < e\Phi$) where they are actually loosely bound by the plasma fluctuation. However electrons in this portion of phase-space are extremely collisional and therefore may still be regarded as free.

The ionization rate is affected by a change in the distribution function for the incident free electron and by the probability of finding the target ion in its specified charge state. Assuming the ion probabilities are determined by an equilibrium Boltzmann factor proportional to $\exp(-Qe\Phi/kT_i)$, and assuming the electron distribution is a nondegenerate Maxwellian function of the asymptotic energy ϵ_o, the result is a modified ionization rate containing the Boltzmann factors $\exp(e\Phi/kT_e - Qe\Phi/kT_i)$ and a cross-section evaluated at the shifted energy $\tilde{\epsilon}_o = \epsilon_o + e\Phi$.

The three-body recombination which is the time-reverse of this ionization also has an altered rate and the changes are slightly different. In this case the distribution functions contribute Boltzmann factors

$$\exp(2e\Phi/kT_e - (Q + 1)e\Phi/kT_i)$$

corresponding to two electrons incident upon an ion of charge Q+1. The 3-body cross-section is also evaluated at modified energies determined by energy conservation. Again there is a question about the low-energy portion of electron phase-space where one or both incident electron(s) could be formally bound to the potential fluctuation; again it seems reasonable to assume these electrons are in equilibrium with the adjacent positive-energy states.

If ion densities are determined by the LTE Saha equation containing the electron temperature, the ratio of recombination and ionization rates becomes

$$\exp[e\Phi(\frac{1}{kT_e} - \frac{1}{kT_i})] \tag{23}$$

Because this ratio is not unity for $T_i \neq T_e$, the Saha population ratios are altered when the temperatures are unequal. An effect of this character has previously been predicted.[40]

Using the corrections described we can readily form the rate of electron-ion energy exchange,

$$\frac{dE_i}{dt} = \langle e\Phi\{1 - \exp[e\Phi(\frac{1}{kT_e} - \frac{1}{kT_i})] \, n_e\sigma_{ii}v\rangle \tag{24}$$

The average is taken over configurations of the local environment and over the distribution function of the incident electron. The recombination contribution was related to the ionization rate as described above, using the LTE Saha detailed-balance condition. If we expand Eq. (24) assuming the potential $e\Phi$ to be smaller than either kT_e or kT_i, we find

$$\frac{dE_i}{dt} \simeq \langle (e\Phi)^2\rangle \, (\frac{1}{kT_i} - \frac{1}{kT_e}) \, \langle n_e\sigma_{ii}v\rangle \tag{25}$$

This expression is a generalization of Eq. (21) for the case of nearly equal electron and ion temperatures; it predicts zero heat transfer at equilibrium and changes sign with the temperature difference. In the other limit ($kT_i \ll kT_e$), Eq. (24) is qualitatively equivalent to Eq. (21) and so this more comprehensive approach resolves the difficulty associated with the requirement of detailed balance.

The discussion given here refers to the process of impact ionization and 3-body recombination. A similar coupling of electron, ions and photons arises from photoionization and radiative recombination.

III. CORE INTERACTIONS AND PRESSURE IONIZATION

The ultimate dense plasma occurs inside white dwarf stars, at densities $\sim 10^6$ g/cm^3 ; all bound electrons are liberated, leaving point nuclei imbedded in a homogeneous degenerate electron gas. The electron gas is increasingly ideal as its density rises, because the kinetic energy per electron increases as $E_f \sim \rho^{2/3}$, more rapidly than the electrostatic energy $\langle Ze^2/r \rangle \sim \rho^{1/3}$.

Although the Chandrasekhar (degenerate) gas model applies to densities far beyond the laboratory range,[21] one finds similar behavior for the outer (valence) electrons in matter at less extreme densities:

G-3.) Strong compression raises electron kinetic energies relative to potential energies (i.e., high pressure "liberates" electrons as it crushes them down into smaller volumes).

The rule is illustrated by the virial theorem:

$$3pV = 2K + U \tag{26}$$

For a cold, low-density system the pressure p is small and one has the isolated-atom result, $K \simeq -1/2\ U$. At high densities the Coulomb energy U becomes small compared to the kinetic energy K and one obtains the ideal gas law $K \simeq 3/2\ pV$ (this form is valid in degenerate or nondegenerate cases).

The virial theorem is often used in calculation of high-density plasmas so it is worth comment that Eq. (26) is valid only in certain circumstances. Equation (26) is correct for an isolated quantum system. It is also valid when applied to one unit cell of a periodic system such as a crystalline solid. In these cases there is no difficulty deciding how much kinetic or potential energy belongs to the volume V and the wave-functions obey boundary conditions which assist the required integration by parts.

In a dense plasma each atom (ion) is surrounded by a random environment of neighbor atoms without high symmetry or sharp boundaries. This situation is often described by the <u>cell model</u>: one nucleus located at the center of a spherical cavity in a uniform positive charge background (Liberman[41]). The cavity radius ($= R_0$) is set by the density (Fig. 8). The electron distribution is calculated by finite-temperature self-consistent field theory. A similar physical picture is assumed by the Thomas-Fermi or Thomas-Fermi-Dirac (TFD) cell models. Several authors have explored more elaborate prescriptions for the exterior positive charge density (Perrot[42], Cauble et al.[43]).

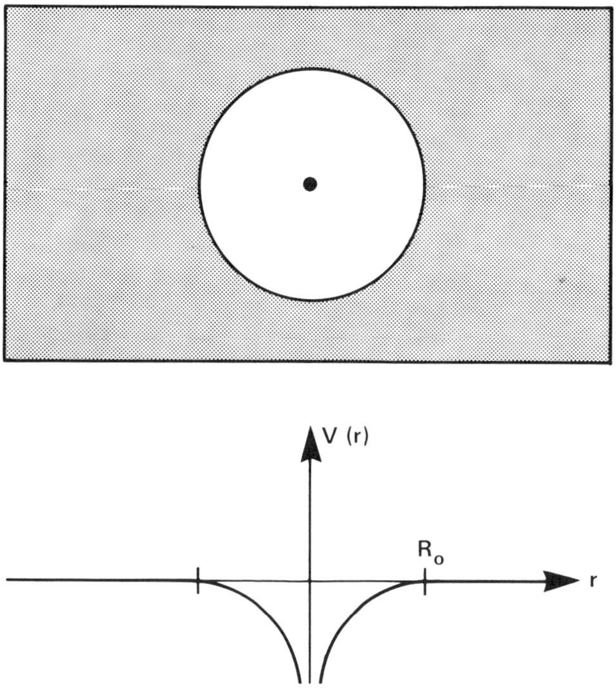

FIG. 8. Schematic representation of the spherical-cell model. A nucleus of charge Ze is located at the center of a cavity of radius R_o in a positive charge background $\rho_+(r)$. The cavity radius is fixed by the matter density. The electron distribution is calculated from a self-consistent average potential V(r) and the chemical potential is chosen so that the cell is electrically neutral.

For the spherical-cell self-consistent field theory, we can establish an extended virial theorem referring to the atomic or cavity volume V_{at}:

$$3pV_{at} = K_{(1)} + K_{(2)} + U \tag{27}$$

with

$$K_{(1)} = \frac{\hbar^2}{2m} \sum_s f_s \int_{V_{at}} |\nabla\psi_s|^2 \, d^3r$$

$$K_{(2)} = -\frac{\hbar^2}{2m} \sum_s f_s \int_{V_{at}} \psi_s \nabla^* \psi_s^2 \, d^3r$$

In Eq. (27), ψ_s = wave-function in the self-consistent potential (the sums run over all one-electron states, bound and free) and f_s = Fermi function describing the occupation

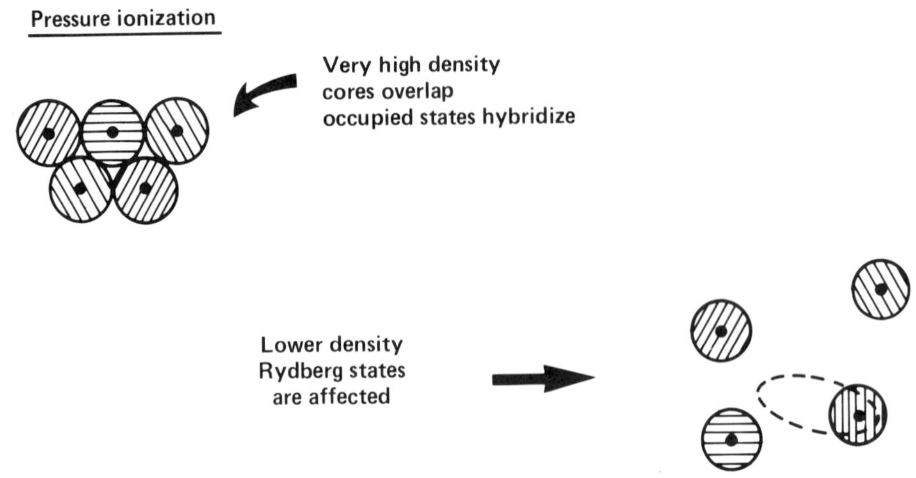

FIG. 9. Pressure ionization affects occupied or unoccupied eigenstates, depending upon the density.

of ψ_s. The pressure p appearing on the left side of the equation is equal to $P_{rr}(R_0)$, the radial element of the quantum kinetic stress or pressure tensor evaluated at the ion-sphere boundary.[44]

Pressure Ionization

The great difference between atomic structure for isolated atoms (ions) and those in the dense plasma is the importance of the continuum in the latter case. With increasing density, an increasing number of bound states are shifted into the continuum.

At high densities, pressure ionization occurs when ion cores are forced together; the outermost bound states hybridize and become propagating waves. At these densities atoms are arranged in a relatively regular close-packed structure and so ideas of solid-state band structure are directly applicable. The effective ion charge $Q(\rho)$ is approximately determined by Eq. (18) in this case.

At lower densities pressure destroys excited states which are only occasionally occupied. In this case the atoms are separated from their neighbors by random distances, and the fluctuations in the local environment probably play an important role. The low-density precursor of pressure ionization may be seen in the interactions between Rydberg atoms.

One requires a simple estimate of the density at which a given level is pressure-ionized, but this immediately raises questions and there is disagreement in the literature. We will summarize the most popular viewpoints and evaluate the formulas for a representative plasma (Table I). The reader will find further discussion in unpublished reports of Burgess,[45] Peacock,[46] and Brush and Armstrong.[47]

1.) Ion-sphere model (Unsold,[48] Carson et al.,[49] More and Zimmerman,[17] More,[18] Burgess and Lee[50]):

$$r_n \simeq 1/3\ R_o \tag{28}$$

Here r_n = orbit radius, Eq. (5); R_o = ion-sphere radius, Eq. (12). This formula follows from several lines of reasoning: it is the criterion for significant overlap of wave-functions on adjacent ions, it follows from a comparison of the binding energy and the continuum lowering, and it corresponds to a density at which the nuclear attraction for the electron is exceeded by the electric field of a neighbor ion.

2.) Debye-Huckel model (Rogers et al.,[51] Weisheit and Shore,[52] Vinogradov et al.[53]):

$$r_n \simeq D \tag{29}$$

For the case examined in Table I this criterion predicts many more bound states than does Eq. (28). Equation (29) is derived by locating the highest boundstate of the Debye-screened potential $V_{DH}(r) = (Qe/r)\exp(-r/D)$. The question is the validity of $V_{DH}(r)$ when used for this purpose.

3.) In the white-dwarf literature, one occasionally sees a mysterious criterion[54]

$$\epsilon_n \simeq E_f \tag{30}$$

This criterion must be intended to refer to degenerate plasmas. If applied to the nondegenerate plasma of Table I, Eq. (30) predicts an enormous number of bound states.

4.) Planck-Larkin method (Rogers[55]). This formula reduces or removes the contribution of bound-states obeying $|E_n| < kT$; for the plasma of Table I this translates to $n_{max} = 2$, a strange or bizarre result. However the Planck-Larkin formula is properly used only in thermodynamic calculations, and the continuum states are also modified so as to essentially replace the removed bound-electron contributions. For dense plasmas the calculations by the Planck-Larkin method often agree with Thomas-Fermi results to surprisingly high precision (see Table II).

5.) Landau-length model In ideal plasmas satisfying Eq. (13), it appears that bound states having orbit radii comparable to $R_L = Q^2e^2/kT$ would eventually be strongly perturbed by close approach of a neighbor ion. This does not mean that states having $r_n > R_L$ would not generate (broadened) emission lines.

6.) Inglis-Teller limit For sufficiently large quantum numbers the Stark effect will cause adjacent levels to overlap in energy and make unresolvable contributions to the emission or absorption spectrum (Griem[56]). This occurs at

$$E_{n+1} - E_n \simeq r_n \left(\frac{Qe^2}{R_o^2}\right) \quad \text{or} \quad r_n \simeq \frac{R_o}{\sqrt{n}} \tag{31}$$

As the table shows, the predictions of these six models span a large range.

TABLE I

Pressure ionization by various criteria for an Aluminum plasma at $T = 1$ keV, electron density $n_e = 10^{19}/cm^3$

MODEL	n_{max}
Ion sphere model	24
Debye-Huckel model	70
Fermi energy model	870
Planck-Larkin formula	2
Landau length	8
Inglis-Teller	19

At present there is no experimental evidence which would clearly select between the models. It appears that the most convincing theoretical arguments favor the ion-sphere picture.[18] Without repeating this discussion in detail, we summarize the main points:

a.) For strongly-coupled plasmas ($\Gamma > 1$) the ion-sphere picture is essentially required by our knowledge of ion correlation; in this case the Debye length has no physical interpretation. In spherical cell SCF calculations, bound-states reach zero binding energy and become resonances at a density determined approximately by Eq. (28) with the coefficient 1/3 as given.[44]

b.) For lower densities ($\Gamma < 1$) the question concerns the existence of discrete bound states having $R_o < r_n < D$, i.e., large orbits which encircle many ions. According to the Debye model, electrons in such states experience a weak spherically symmetric potential $V_{DH}(r) = (Qe/r) \exp(-r/D)$ which is known to be the average of the actual potential of the plasma. However the fluctuations from this average are very strong; the neighbor ions give rise to potentials $\sim .8\ Qe/R_o$ which greatly exceed $V_{DH}(r)$ for $r \gg R_o$. In particular, the electron mean free path for scattering by these potential fluctuations is less than the orbit circumference for hypothetical bound-states with

$R_o < r_n < D$; it is difficult to believe that a discrete quantized spectrum exists under these conditions.

Continuum lowering

The mechanism of pressure ionization is at least partially understood as continuum lowering: as the density rises the spatial average $\langle V(r) \rangle$ of the electrostatic potential becomes increasingly negative until it ultimately exceeds the free-atom binding energy of any given electron.

In the calculation of ionization balance, one generates this effect automatically if the Coulomb interaction energy per ion,

$$\Delta E_c = -a \frac{Q^2 e^2}{R_s}$$

(a = numerical coefficient, R_s = screening length) is added to the energy or free energy of the system. For low densities ($\Gamma \ll 1$) one assumes $a = 1/2$, $R_s = D_i$; for strongly coupled plasmas one has the ion-sphere form with $a = 9/10$, $R_s = R_o$.

The ionization process may be thought of as a change of Q occuring at constant free electron density n_e. Following the reasoning already presented near Eq. (22) the continuum lowering is then (for $\Gamma > 1$)

$$\left(\frac{\partial \Delta E_c}{\partial Q}\right)_{n_e = \text{const}} = \frac{3}{2} \frac{Qe^2}{R_o} \tag{32}$$

This value is also the spatial average of the ion potential $V_i(r) = Qe/r$ over the ion-sphere volume, but not the average of the complete potential

$$V(r) = \frac{Qe}{r} + \frac{1}{2} \frac{Qe}{R_o} \left(\frac{r}{R_o}\right)^2 - \frac{3}{2} \frac{Qe}{R_o}$$

which includes the contribution of a uniform free-electron gas.

Interpolation between ion-sphere and Debye-Huckel screening formulas is most often accomplished with the Stewart-Pyatt model.[57] However this approach has certain defects and a completely satisfactory theory does not yet exist.[2]

Resonances and Pressure Ionization

In quantum spherical-cell SCF calculations, the effect of high pressure is to lower the continuum successively through various one-electron boundstates.

Above a density $\rho_{n\ell}$ the boundstate of energy $E_{n\ell}$ becomes a scattering resonance or shape resonance in the partial-wave potential

$$V_\ell(r) = V(r) + \frac{\hbar^2}{2m} \frac{\ell(\ell+1)}{r^2}$$

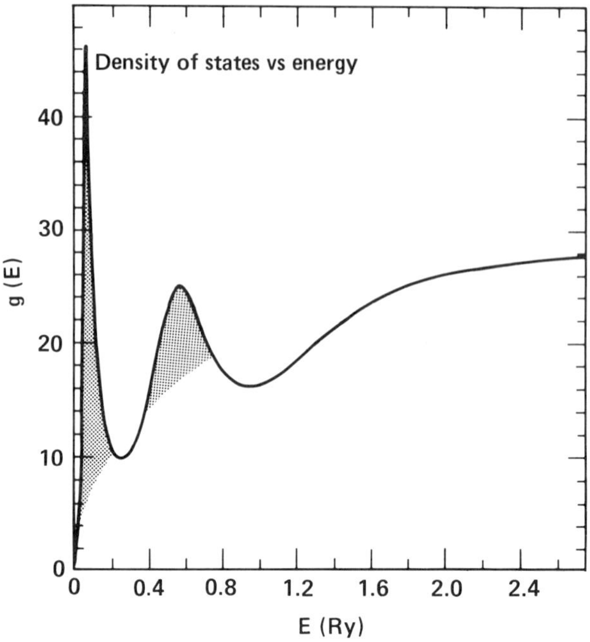

FIG. 10. Theoretical electronic density of states for aluminum calculated by the self-consistent field model (D. Liberman provided these data). Shading has been applied to enhance the contrast of 3p and 3d resonances from the $\epsilon^{1/2}$ background continuum density of states. The density is 2.7 g/cm^3 and the temperature is 50 eV.

which has an interior well and a barrier at $r \approx R_o$ associated with the centrifugal potential. According to scattering theory, the pressure-ionized state is described as a resonance state of complex energy

$$\tilde{E}_{n\ell} = E^o_{n\ell} + i\,\Gamma_{n\ell} \tag{33}$$

The real part $E^o_{n\ell}$ indicates the central energy of the resonance; the imaginary part $\Gamma_{n\ell} = \hbar / \tau_{n\ell}$ determines the resonance width or lifetime $\tau_{n\ell}$. The precise definition of the complex energy $\tilde{E}_{n\ell} = \hbar^2 Q^2_{n\ell}/2m$ is that there exists a solution $\phi_{n\ell}(r)$ of the radial Schroedinger equation which obeys the boundary conditions

$$\phi_{n\ell}(r) \rightarrow 0 \qquad r \rightarrow 0$$

$$\phi_{n\ell}(r) \rightarrow i^{\ell} N_{n\ell} e^{-iQ_{n\ell} r} \qquad r \rightarrow \infty \tag{34}$$

The normalization $N_{n\ell}$ of these wave-functions is fixed by a remarkable formula which is the analytic continuation of the boundstate normalization formula and which implies the first-order resonance perturbation theory,[58]

$$\delta\tilde{E}_{n\ell} = \int_0^{R_o} \phi^2_{n\ell}(r)\ \delta V(r)\ dr \qquad (35)$$

where $\delta V(r)$ is an arbitrary change in the potential. Equation (35) gives a unified expression for the perturbation in $E^0_{n\ell}$ and $\Gamma_{n\ell}$.

In terms of these resonance energies one can transform the well-known expression for the continuum density of states

$$g(\epsilon) = c_i V\sqrt{\epsilon} + \frac{2}{\pi} \sum_\ell (2\ell + 1) \frac{d\delta_\ell}{d\epsilon} \qquad (36)$$

to a simple and general form[44]

$$g(\epsilon) = c_1 V\sqrt{\epsilon} + \frac{b}{\sqrt{\epsilon}} + \sum_{n\ell} 2(2\ell+1)\ \mathrm{Re}\left[\frac{1}{i\pi} \frac{1}{(\epsilon-\tilde{E}_{n\ell})} \sqrt{\frac{\epsilon}{\tilde{E}_{n\ell}}}\right] \qquad (37)$$

The derivation of Eq. (37) uses a representation of the scattering matrix (S-matrix) originally proven by Regge;[59] it is rigorously correct for one-electron nonrelativistic potential-scattering theory.

When a state is pressure ionized, its binding energy approaches zero and moves to positive energies where it begins as a very narrow resonance. For such a state, the right-hand side of Eq. (37) gives a sharp resonance peak very similar to the delta-function contributions of the bound states.

Resonances at higher energies become broad and the contributions overlap and merge. Using the Born approximation for the scattering phase-shift, one can show that at high energies Eq. (36) gives

$$g(\epsilon) = c_1 V\sqrt{\epsilon} + \frac{c_1 V}{2\sqrt{\epsilon}} \Delta E + \ldots \qquad (38)$$

This corresponds simply to a continuum lowered by

$$\Delta E = \frac{1}{V} \int_0^\infty 4\pi r^2 V(r) dr \qquad (39)$$

From Eq. (37) one can calculate the average thermal population of a resonance state. The result is[44]

$$P_{n\ell} = 2(2\ell+1)\ \mathrm{Re}\left[\frac{1}{i\pi} \int_0^\infty \sqrt{\frac{\epsilon}{\tilde{E}_{n\ell}}}\ \frac{f(\epsilon)d\epsilon}{\epsilon-\tilde{E}_{n\ell}}\right] \qquad (40)$$

This is not exactly a fermi function because the resonance state is not exactly an eigenstate. Nevertheless Eq. (37) shows the various resonances make independent additive contributions to the total number of electrons in the continuum.

These resonance formulas give the most precise description of pressure ionization in the spherically symmetric cell model. A much more complex picture would emerge if we could perform calculations for models containing a cluster of atoms. Perhaps some progress along these lines can be made through perturbation theory, based for example upon Eq. (35).

IV. QUALITATIVE BEHAVIOR OF DENSE PLASMAS

Electron-ion coupling

The interaction of ions with free electrons is measured by a coupling parameter defined as

$$\Gamma_{ei} = \frac{Qe^2}{R_o kT} \tag{41}$$

This is the ratio of potential energy $\sim Qe^2 / R_o$ to kinetic energy $\sim kT$ evaluated for an electron at the ion-sphere (interatomic) boundary. Of course, electrons closer to the ion interact more strongly, but the spatial average $\langle Qe^2 / rkT \rangle$ is proportional to Γ_{ei}.

Assuming the definition Eq. (41) one can show:

$$\Gamma_{ei} \leq 0.54 \text{ all } \rho, T \tag{42}$$

for hydrogen plasma described by the Saha equation. This surprising result is established in Appendix A. It is a rigorous consequence of the Saha equation, but that theory breaks down at high plasma densities (see G-5a).

The inequality (42) can be generalized,

G-4.) The interaction of ions with free electrons never becomes large.

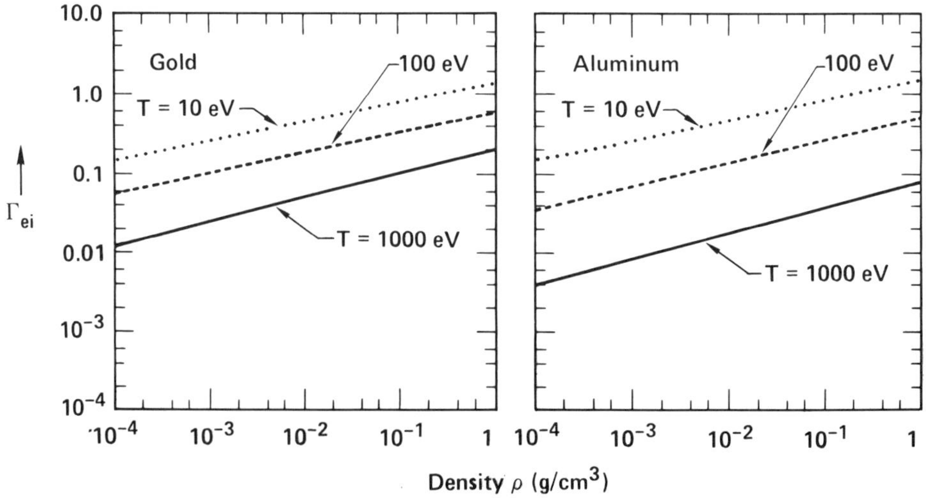

FIG. 11. Electron-ion coupling parameter Γ_{ei} for gold and aluminum plasmas. The small values illustrate rule G-4.

For materials other than hydrogen it is simplest to test the rule by examining numerical ionization data. Figure (11) shows $\Gamma_{ei}(\rho, T)$ calculated by the Thomas-Fermi cell model, accurate enough for present purposes; Γ_{ei} remains less than unity throughout the nondegenerate plasma range. At the lowest temperatures, Γ_{ei} approaches unity as one reaches solid density, but for degenerate electrons one should redefine Γ_{ei} using the Fermi energy rather than kT as a measure of electron kinetic energy; with this modified definition Γ_{ei} is again small at high pressures.

There is a logical reason for the general rule: if the attraction of ions for electrons strongly exceeds the electron kinetic energy, some electrons will recombine and the ion charge Q will decrease, reducing Γ_{ei}.

Electron-Electron Interaction

The interaction between free electrons is characterized by a coupling parameter Γ_{ee}^{free} which is smaller than Γ_{ei} by a further factor $Q^{2/3}$ (the 2/3 power reflects a corrected separation r_{12}). This means that $\Gamma_{ee}^{free} \ll 1$ and hence electron correlation effects are small, at least for free electrons.

Coulomb interactions between bound electrons are larger because the bound electrons are closer together. One can characterize these interactions by forming a bound-electron coupling parameter:

$$\Gamma_{ee}^{bound} = \frac{e^2}{r_n kT} \simeq \frac{I_n}{QkT} \sim \frac{\xi}{Q} \qquad (43)$$

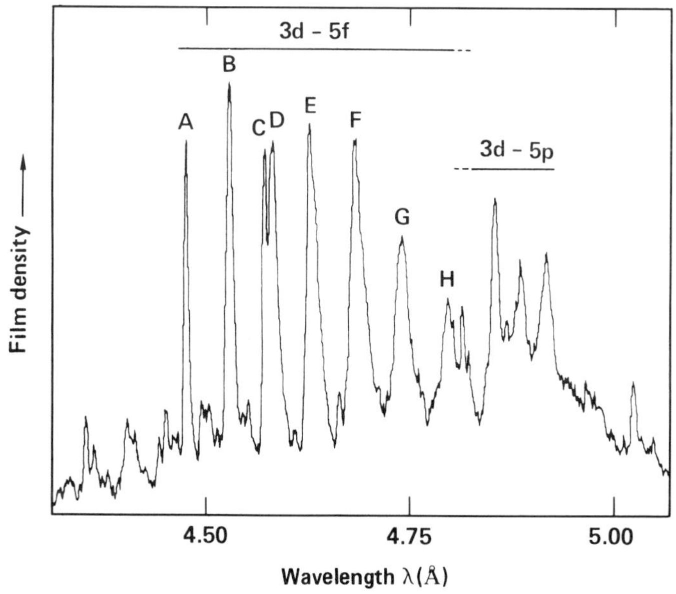

FIG. 12. Experimental spectrum of Ta plasmas illustrating sequence of UTA spectra (Andebert et al.[61]).

Evidently these interactions are often important (typical values of ξ and Q were given in Fig. 2). Electron interactions produce term splittings which are readily calculated for low-Z atoms or for ions carrying few bound electrons. For high-Z multielectron atoms the spectra become very complex and there is a nontrivial difficulty going beyond the one-electron (SCF or average-atom) approximation.

In recent years, statistical methods have been developed for the analysis of electron interactions in complex atoms. The term splitting for an arbitrary configuration is studied by moment expansion techniques by Bauche, Klapisch et al., a technique which is called the method of unresolved transition arrays (UTAs).[60] A representative recent application to analysis of laser-produced Ta spectra is given by Audebert, Gauthier et al. (Fig. 12).[61]

The distribution of configuration probabilities has been investigated through a high-temperature series expansion method (Green[62]), a technique which holds one population fixed and averages over occupations of other states (Grimaldi and Grimaldi-Lecourt[63]), and through direct numerical evaluation of configuration probabilities for a model Hamiltonian based on Eqs. (8-10).[2,3]

Figure 13 shows the complex spectrum of a gold plasma which is analyzed by identification of numerous specific transitions together with a schematized background from satellite configurations (Busquet, Pain, Bauche and Luc-Koenig[64]).

Degeneracy of Free Electrons

Qualitatively, an electron gas of density n_e is degenerate when the Fermi energy E_f,

$$E_f = \frac{\hbar^2}{2m}(3\pi^2 n_e)^{2/3} \tag{44}$$

exceeds the temperature kT ; this translates into a practical density criterion:

$$n_e \geq (1.4 \times 10^{23}/\mathrm{cm}^3)\left(\frac{kT}{10\ \mathrm{eV}}\right)^{3/2} \tag{45}$$

The criterion is usually satisfied for plasmas produced by shock compression of solid targets.

Degeneracy has a strong influence on plasma process coefficients, typically changing the form of their temperature-dependence. In many cases, the change amounts to replacement of the temperature by the Fermi temperature ($E_f = kT_f$), as in the electrical conductivity where a proportionality to $T^{3/2}$ is replaced by $T_f^{3/2}$. However, that rule is not general; the electron thermal conductivity in a non-degenerate plasma is proportional to $T^{5/2}$ and this is replaced by proportionality to $T \bullet T_f^{3/2}$.

Many theoretical treatments of plasma ionization incorrectly predict complete recombination at high densities.

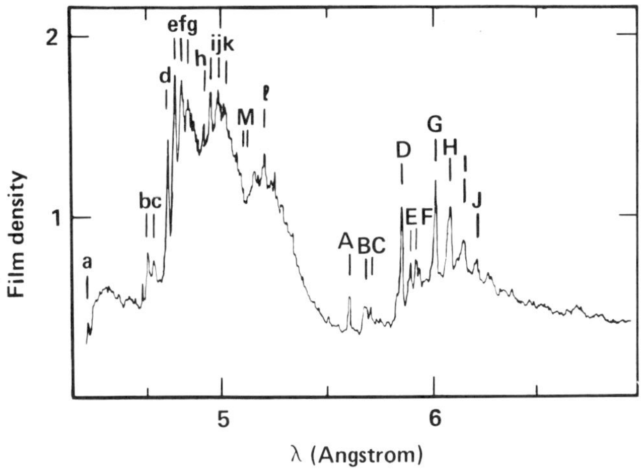

FIG. 13. Experimental spectrum of Au laser plasma illustrating complex line cluster features (Busquet et al.[64]).

For example, ionic rate equations employ an impact ionization rate proportional to $n_e\, n_i$ and a three-body recombination rate $\propto n_e^2\, n_i$; with these rates the plasma recombines at high densities and cannot reach a degenerate state.[65]

What key ingredient must a theory include in order to handle the high-density degenerate case?

Degeneracy implies full (complete) occupation of all states having energies less than the Fermi energy E_f. Without pressure ionization there are an infinite number of bound states and these will absorb all Z electrons leaving no free electrons. Only if bound states are removed from the calculation by density effects will electrons be forced into free states to produce a degenerate free-electron gas.

This appears to be a general rule:

G-5.) Degeneracy of free electrons cannot occur without pressure ionization.

The rule applies to the construction of theories, not to nature. Experience with condensed matter leaves no question that both degeneracy and pressure-ionization occur at sufficiently high density in the real world.

At present one does not have a detailed configuration theory (i.e., Saha equation or ionic rate equation) which goes over to the high-density degenerate range in a satisfactory way. This difficulty of the Saha approach is one of the main unresolved questions of dense plasma theory.

The Saha equation (or ionic rate equations) are formulated in terms of exact energy-levels of the isolated ion. That is, the Hamiltonian is assumed to be diagonal when expressed in a basis of wave-functions in which bound electrons are localized on specific atoms (ions), or alternatively, many-body basis states in which each atom (ion) has an integral number of bound electrons.

However the Hamiltonian contains additional terms which are not diagonal in terms of basis states localized on individual atoms. Wave-functions $\psi_{n\ell}(r-R_i)$ centered on one ion overlap states $\psi_{n\ell}(r-R_j)$ centered on a neighbor ion, and this overlap leads to nonzero matrix-elements of kinetic and potential energy operators.

In low-density plasmas the off-diagonal hopping terms of the Hamiltonian are exponentially small, at least for states with small quantum numbers, but they rise with density. In the dense plasma the off-diagonal terms become very important. Alternatively, the charge on one ion is not a constant of the motion (even neglecting collisional or radiative processes which are treated as perturbations).

The upshot is that the Saha equation is untenable at high densities, and with it the entire language of hydrogen-like, helium-like ions (ion stages), ionization rate, etc.

Fortunately, the average-atom model remains workable at these conditions. The interatomic transfers associated with pressure-ionization are generated by one-electron operators (e.g., kinetic energy or central-field potential energy operators) and are readily incorporated in the average atom model. For this reason the average-atom theory yields results like the usual quantum theory of solids at high densities. Even the spherical-cell SCF model is accurate enough to agree with high-pressure shockwave data and appears to be essentially correct throughout the high-density degenerate range.(*)

To summarize,

G-5a.) Pressure ionization can be described in the AA model but not in the Saha theory. Alternatively, at very high densities, the AA is essentially correct.

It is not quite clear whether (G-5a) is a general rule or simply a generalization of recent experience. Hopefully the reader will be challenged to make his own judgement on this central question in the physics of dense plasmas.

(*) The problem with the average-atom model is that it does not exactly represent the electron-electron interaction, so that it predicts an incorrect line spectrum.

Excited States

Now we return to nondegenerate plasma conditions and examine the degree to which the highly-charged ion is susceptible to thermal excitation or perturbation by plasma microfields:

G-6.) The Z* theorem: at low densities, an ion is an isolated structureless point charge.

At low densities ($n_e < 10^{15}$ /cm^3) where this rule applies, one has little interest in atomic properties.[23] Ground-state ions are spatially separated by distances large compared to their size, scatter electrons like point charges and general behave in a simple fashion.

This behavior is a consequence of rule G-1: excitation energies are proportional to the ionization potential I(Z,Q) and at low densities become large in comparison with the temperature. Excited states therefore become exponentially improbable at low density.

This argument points the other way at high density where the key parameter ξ is not large (see Fig. 2):

G-6a.) At high density, all types of ionization (recombination) become rapid; ions are highly excited and no state persists long. Many plasma perturbations become strong but their effects overlap and average.

We give only a few examples here. In the complex spectra of high-density plasmas, satellite lines corresponding to multiply-excited ions are often prominent.[64] (See Fig. 13). Experimental spectra even understate this effect because the observed spectra usually come from nonequilibrium surface regions which typically have lower excited state populations than the bulk plasma.

Second, we can compare rates for ion excitation or ionization with those for other electron-electron collisions. For example, the ratio of impact ionization to electron-electron scattering can be estimated using the Lotz formula:

$$\frac{\langle \sigma_{ii} v \rangle}{\langle \sigma_{ee} v \rangle} \sim \frac{1}{\xi^2} e^{-\xi} \qquad (46)$$

The result shows the rates become comparable when ξ becomes small. Another (related) estimate shows that the ion

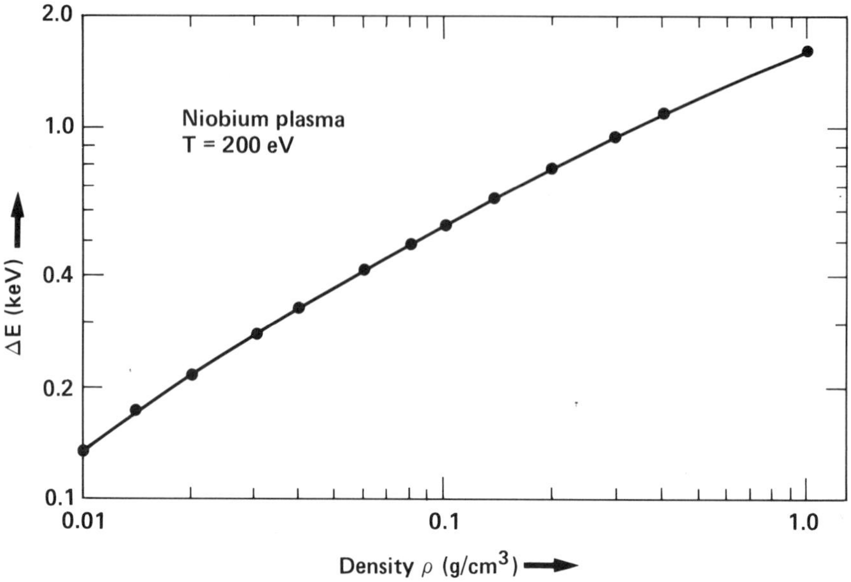

FIG. 14. Thermal average excitation energy per atom for Niobium plasma at various densities, calculated by solution of the Saha equation based on Eq. (10). As density rises the degree of excitation increases, illustrating rule G-6a.

scatters most electrons as a point charge when $\xi \gg 1$ but the quantum-mechanical form-factor or scattering amplitude becomes important when ξ becomes small.[2]

The comparison given in Eq. (46) is interesting because it bears on the possibility that a high rate of ionization might deplete the high-energy tail of the electron distribution more rapidly than electron-electron collisions can restore the Maxwellian form.

Third, one can examine the average excitation energy per atom for plasmas in thermal equilibrium (LTE). Figure (14) shows this excitation energy calculated by a Saha equation based on the screened hydrogenic model of Eqs. (8-10). The figure shows that excitation energy rises as density increases (for fixed temperature). For the case shown, the excitation energy reaches ~ 1 keV, enough to permit up to five autoionizations (for the typical ion) and the atomic internal heat capacity is comparable to the free electron specific heat.[3]

Rule G-5a enhances itself in the sense that the excited atoms (ions) existing at high densities are themselves more easily ionized, excited, polarized, etc. than groundstate ions would be.

Classical Behavior

It is a fundamental aspect of quantum statistics that all the exotic quantum phenomena (from superfluidity to the quantum hall effect) are demonstrated in cryogenic systems:

G-7.) Classical approximations become more accurate at high temperature, high density, high Z.

Our question here is a practical matter: can we determine the range of validity of Thomas-Fermi calculations, which are entirely classical except for including the exclusion principle in the form of the restriction

$$0 < f(r,p) < 1$$

for the one-electron distribution function.

The numbers given in Table II may help convince the reader that there is something to be explained: the statistical theory is in surprisingly close agreement with much more elaborate relativistic quantum-statistical calculations using both DCA and average-atom methods.

The TFD free energies have been supplemented with the Scott correction,[66] which is a constant in this case. The ACTEX (Activity expansion) theory[67] is based upon a Saha equation with elaborate summation of plasma corrections; it also employs the Planck-Larkin transformation, which helps explain the greatly increased charge state (see Table I). The INFERNO model is a spherical cell relativistic SCF theory.[41] The surprise is that the statistical model is within a few percent of the other theories.

Table II

Free energy F, entropy S and ionization state of Gold plasma at T = 630 eV and ρ = 10, 19.3, and 63 g/cm^3. F and TS are given in keV per atom.

	TFD	ACTEX	INFERNO	MAXIMUM DIFFERENCE
F	639	654	653	2.3%
TS	214	-	210	1.9%
Z*	42.8	48.9	42.6	
F	622	638	637	2.6%
TS	190	-	187	1.6%
Z*	39.6	48.3	40.9	
F	595	-	611	2.7%
TS	151	-	148	2.0%
Z*	37.9	-	38.5	

The general assertion that classical statistics becomes correct at high temperature must confront the pronounced quantum effect of K-shell ionization, visible in any quantum ionization calculation. Is there a definite mathematical theorem underlying rule G-7?

One precise statement of the general rule is obtained from the quantum density matrix $\rho(r, r')$ of a nondegenerate free electron in volume V:

$$\rho(\underline{r},\underline{r}') \equiv \langle \underline{r}|\frac{1}{Z}e^{-\beta H}|\underline{r}'\rangle$$

$$= \frac{1}{V} \exp - [\pi(\underline{r} - \underline{r}')^2/\lambda]^2 \qquad (47)$$

Here Z = Trace $[\exp(-\beta H)]$ = partition function; $H = p^2/2m$ = free electron Hamiltonian, and $\lambda = (2\pi\hbar^2/m_e kT)^{1/2}$ = thermal deBroglie wavelength.[68] This expression reduces to the diagonal (=classical) form as $T \to \infty$.

Equation (47) says that quantum effects can only occur on a small size-scale $\leq \lambda$, i.e., only for the innermost bound electrons at high temperature. The remaining electrons are therefore adequately described by the classical (statistical) theory.

A similar (inequivalent) result is the observation that the exchange energy per electron,

$$f_{exch} = \frac{1}{12} \frac{e^2}{\lambda} (n_e \lambda^3) \qquad (48)$$

becomes small as $T \to \infty$; of course the exchange energy is a quantum correction to the classical TF theory.

For dense plasmas the high thermal excitation indicated by rule G-6a tells us that the effects of various quantum states will be averaged or smoothed leaving only continuous (i.e., classical) behavior.

Figure 15 compares ionization states $Z^*(T)$ for aluminum and gold plasmas at constant density $\rho = 10^{-3}$ g/cm^3 calculated by the screened hydrogenic model (which includes quantum shell effects) and the Thomas-Fermi theory.[2] The comparison shows the TF approximation to be more accurate at high Z. Figure 16 shows a perspective drawing of $Z^*(\rho,T)$ for aluminum. In this case the figure shows the quantum ionization shelf-structure, associated with the helium-like ion, disappearing at higher densities.

Active Particles

The next "general rule" poses problems: we are not certain it is a rule, nor what would be the proof. However it summarizes a variety of situations, and has proven useful in detecting errors in complex calculations -- and that is the acid test of utility. The statement is simple enough:

G-8.) A small number of high-energy particles can easily dominate the properties of an equilibrium plasma. A small number of low-energy (inactive) particles cannot.

The obvious example is electrical: a few ions in a neutral gas can easily dominate the conductivity; a few neutrals in a strongly ionized gas have little effect.

In the calculation of thermal conduction by electrons or radiation it emerges that the net heat current is strongly dominated by a small minority of electrons (or photons) at the high-energy tail of the distribution. In these cases the reason is the rapid increase of mean free path with energy. Ionization is also dominated by the most energetic particles, as rule G-1 implies.

Other examples of active particles are the suprathermal electrons produced in laser-plasma interaction, and high energy x-rays they produce through bremsstrahlung or K-shell ionization and fluorescence. These have great importance as preheat sources, even when they carry as little as a few percent of the absorbed energy.

In thermonuclear plasmas it is not unusual to find the fusion rate dominated by collisions involving energetic ions in the hot tail of the distribution. For this reason, the processes involving these energetic ions (energy loss, escape, energy transfer to knock-on particles) assume special importance.

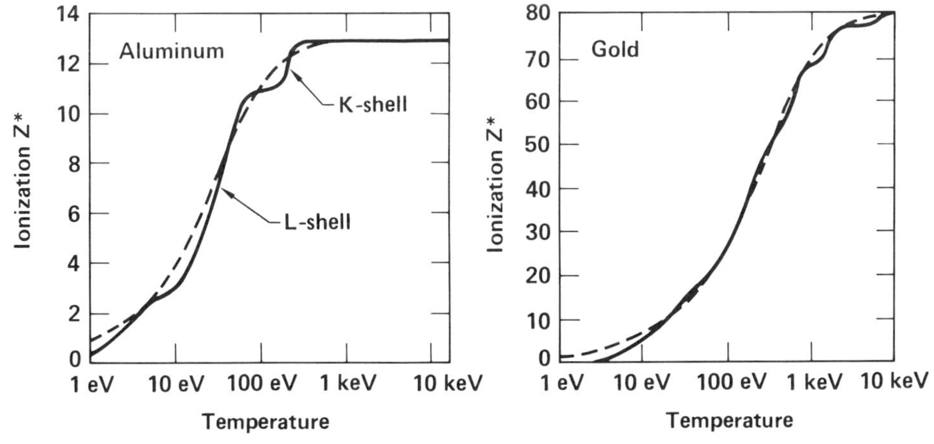

FIG. 15. Ionization state of Aluminum and Gold at density 10^{-3} g/cm^3. Solid line is the screened hydrogenic average-atom model. Dashed line is the Thomas-Fermi theory. The two approaches are in closer agreement for the larger Z, illustrating rule G-7.

A nice example of rule G-8 is given in the Appendix. There it is shown that the few bound electrons cannot dominate the stopping-power, e.g., for fast alphas resulting from thermonuclear reactions, even under the special conditions

$$v_{Bohr} \ll v_{ion} \ll v_e = (kT_e/m_e)^{1/2}$$

where the free-electron stopping is inhibited by a large factor $(v_{ion}/v_e)^3 \ll 1$. One would reach an incorrect conclusion in this case without include the process of pressure ionization.

Another (related) example is that the large photoelectric cross-section of bound electrons cannot dominate the opacity of a mostly-ionized equilibrium plasma (again, the proposition refers to the case of high temperatures at which most ions are fully stripped, so bound electrons are indeed a minority).

The rule is not perfectly general, as we have seen from interesting counterexamples given in the lectures of Dalgarno; neutral hydrogen injected into highly-ionized plasma produces distinctive and dominant radiative emission via charge-transfer.[69] Of course there would be very little neutral hydrogen in these plasmas if they were closer to thermal equilibrium.

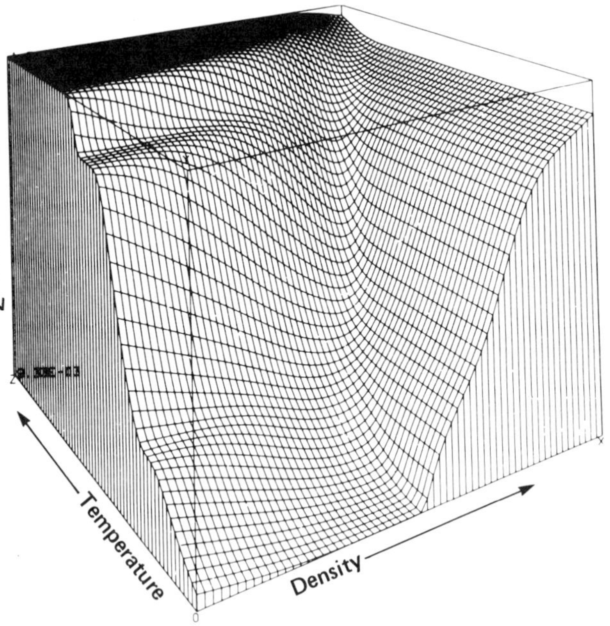

FIG. 16. Contour plot of Aluminum ionization state calculated by the screened hydrogenic model of reference 17. The density range is 10^{-4} to 10^{+4} g/cm^3; the temperature range is 10 eV to 10^4 eV. Note (1.) Pressure ionization at high densities, (2.) reduced prominence of the K-shell ionization plateau as density rises. This illustrates rule G-7.

V. INTERCONNECTION OF PLASMA PROCESSES

As a target is irradiated by laser or particle beam, the absorbed beam energy is transferred into electron, ion and photon distributions and ultimately appears as kinetic energy of hydrodynamic motion, emitted x-rays, magnetic fields, etc. These energy conversions are described by plasma process coefficients: rate coefficients which enter equations describing the macroscopic plasma hydrodynamics. For example, the specific heat and pressure measure the conversion of absorbed energy into temperature and hydrodynamic motion, the stopping power dE/dx tells how ion (or electron) beam energy is deposited, the coupling τ_{ei} describes heat exchange between electrons and ions, and the opacity κ_ν determines the x-ray mean free path $\ell_\nu = 1/\rho\kappa_\nu$.

The plasma process coefficients reflect atomic events including scattering, ionization and bound-electron transitions as they occur in the dense plasma environment. The rates are functions of plasma composition, density and temperature and/or state of nonequilibrium. The practical

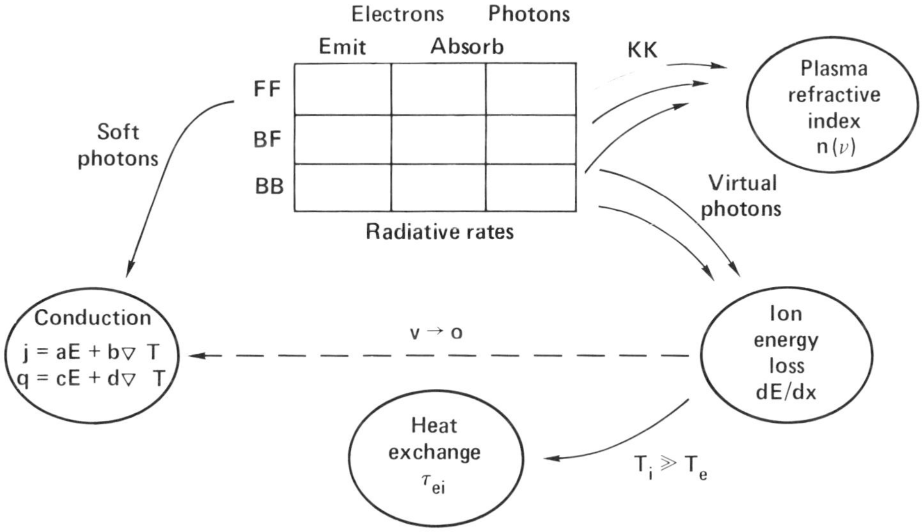

FIG. 17. Schematic illustration of rule G-11. The rectangular box represents radiation formulas corresponding to free-free (FF), bound-free (BF) and bound-bound (BB) emission or absorption processes. The absorption formulas can be written as electron or photon cross-sections. These processes are related by detailed balance and continuity in energy. The soft photon ($h\nu \to 0$) limit of the FF cross-section is equivalent to the electrical conductivity. The plasma index $n(\nu)$ is obtained by Kramers-Kronig transformation of the photon absorption cross-section, and the fast-ion stopping dE/dx equivalent to absorption of virtual photons. The heat-exchange between electrons and ions is an average of dE/dx in the special case $T_i >> T_e$ and finally the low-velocity ion stopping is given by a friction coefficient related to the electrical conductivity.

output of theory is measured by our ability to predict the process coefficients. The discussion in this section will emphasize three general rules,

---

G-9.) Observable plasma properties change continuously as bound states convert into free states.

G-10.) Detailed balance: each process rate is related by time-reversal to the rate for the inverse process.

G-11.) The plasma process coefficients are linked together by a network of exact and/or qualitative consistency relations.

The rules are illustrated throughout the following. One example of rule G-11 is the consistency of pressure $p(\rho, T)$ and energy $E(\rho, T)$; for equilibrium plasmas this is a simple thermodynamic equation, but for nonequilibrium plasmas it developes into an extension of conventional statistical mechanics (More[3], Boercker and More[70]). Another example is the Kramers-Kronig relation between plasma refractive index $n(\nu)$ and the absorption opacity κ_ν.[71,72] These are exact connections.

A qualitative consistency must exist between opacity and the charged-particle stopping-power: dE/dx is essentially a mean opacity analogous to the Planck or Rosseland means, i.e., a certain average of the photon absorption cross-section. This connection is seen by representing the electromagnetic field of a fast charged particle as a superposition of virtual photons having the frequency distribution:

$$N(h\nu) \simeq \frac{2\alpha}{\pi h\nu} Q^2 \left(\frac{c}{v}\right)^2 \log\left(\frac{av^2}{b_o h\nu}\right) \tag{49}$$

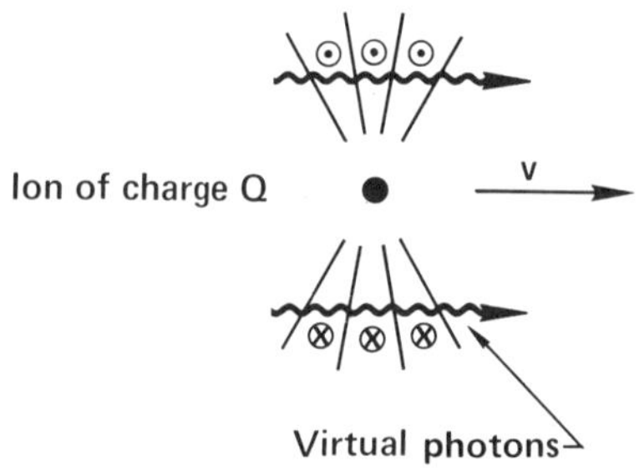

FIG. 18. A fast ion is surrounded by electric and magnetic fields which are approximately equivalent to a cloud of comoving photons.

In this well-known expression v = ion speed (assumed $\ll$ c), c = speed of light, $\alpha = e^2/\hbar c$ = fine-structure constant, a = constant ~1 and b_o = minimum impact parameter.[72]

The charged-particle energy-loss results from absorption of these virtual photons by bound or free electrons of the target plasma. Strictly speaking, virtual photons do not obey the dispersion relation $\omega = c\,k$ of free photons, but this does not matter to the extent that one uses the dipole approximation in the absorption calculation. The bound-bound absorption opacity generates an energy-loss of the form:

$$-\left(\frac{dE}{dx}\right)_{BB} = \int h\nu\, N(h\nu)\, \rho\kappa_\nu^{BB}\, dh\nu \tag{50}$$

$$= \frac{4\pi Q^2 e^4}{mv^2}\, n_I \sum_{n,m} f_{nm}^{abs}\, P_n\left(1 - \frac{P_m}{D_m}\right) \log\left(\frac{av^2}{b_o h\nu_{nm}}\right)$$

This expression looks like the bound electron contribution to the Bethe (high-velocity) stopping theory when the appropriate minimum impact parameter $b_o = \hbar/mv$ is selected.

In the Bethe theory, bound-bound (line) and bound-free (photoelectric) transitions together are represented by the expression

$$\left(\frac{dE}{dx}\right)_{bound} = \frac{4\pi Q^2 e^4}{mv^2}\, n_I \log\left(\frac{2mv^2}{\bar{I}(Z,Q)}\right) \tag{51}$$

where $\bar{I}(Z,Q)$ is a logarithmic mean of excitation and ionization energies as indicated by Eg. (50). There is also a free electron contribution, calculated by Skupsky and Deutsch,[73,74] which has a similar analytic form in which $\bar{I}$ is replaced by $\hbar\omega_{pe}$, the electron plasma frequency of free electrons, in the high-velocity case. For a practical stopping formula, one therefore requires a convenient representation of the bound-electron contribution, i.e., $\bar{I}(Z,Q)$.

This quantity has been calculated by Thomas-Fermi theory using arguments based on the inhomogeneous electron-gas model, closely related to the original Bloch theory.[3,75] The Thomas-Fermi calculations are accurately reproduced by

$$\bar{I} = a\,Z\,\frac{\exp\left[1.29x^{.72} - .18x\right]}{\sqrt{1 - x}} \qquad x = Q/Z \tag{52}$$

The coefficient a is $\simeq$ 10 eV. With this formula one obtains a simple and useful description of the high-velocity stopping power of partially-stripped nondegenerate plasmas. Figure 19 compares Eq. (52) with quantum calculations performed by E. McGuire.[76]

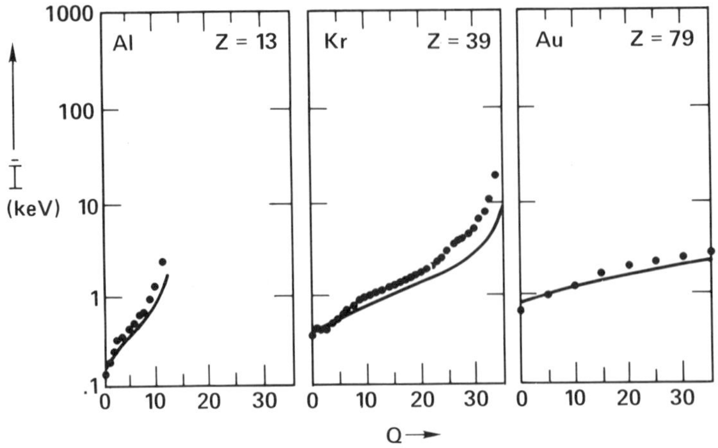

FIG. 19. The mean ionization-excitation potential $\bar{I}(Z,Q)$ for ions of Al, Kr and Au. Solid line is the TF result, Eq. (52). Dots are results from quantum generalized oscillator-strength calculations kindly provided by Dr. E. McGuire.[76]

Another example of rule G-11: the collisional electron-ion heat exchange coefficient of Eqs. (19-20) can be calculated from the stopping power because in the special case $T_i \gg T_e$ the heat transfer is the same process as fast-ion energy-loss to the cooler electrons. In this case, however, we use the energy-loss expression valid for $v_e \geq v_i$ (this is the free-electron contribution in Eq. (A-10)), and average over the ion Maxwell distribution,

$$\frac{1}{\tau_{ei}} = \frac{1}{T_i}\frac{dT_i}{dt} = \frac{1}{\frac{3}{2}n_i kT_i}\langle v_i \frac{dE_i}{dx}\rangle \tag{53}$$

With the proper choice of Coulomb logarithm, the result is the Landau-Spitzer heat-exchange time (see Eqs. (19, 20) above).

These examples illustrate the utility of the connections between plasma processes: if we encounter a dense-plasma mechanism which alters the photon absorption opacity, Eqs. (50, 53) will suggest related improvements in the formulas for stopping power and electron-ion coupling τ_{ei}.

Rule G-10 belongs to the general structure of statistical mechanics; it plays a powerful role in developing consistent descriptions of the plasma processes. For example, it relates absorption and emission cross-sections. Another typical application was given in section II.

Rule G-9 concerns the continuum and bound states near it. In general terms, the uppermost bound states have much in common with the continuum states. In a more precise sense there are three related ideas: correspondence, analyticity in energy and the continuity of bound and free states.

The correspondence principle of the old quantum theory is the idea that a classically defined quantity q_n (e.g., orbital frequency or radiation emission rate) approaches the classical value at large quantum number n:

$$\lim_{n \to \infty} q_n^{QM} = q_n^{Classical} \tag{54}$$

The principle was originally formulated for hydrogenic spectra which have infinitely many bound states, so that the limit $n \to \infty$ is defined. From the quantum viewpoint, the correspondence principle is simply the fact that the WKB approximation becomes reasonably accurate at large quantum numbers (but is not exact in general).

The analytic connection between bound and free states is illustrated by the well-known result that bound-electron eigenvalues $E_{n\ell}$ are poles of the continuum scattering or S-matrix $S_\ell(E) = e^{2i\delta_\ell(E)}$, or by the formula of Seaton[77] which relates quantum defects $\Delta_{n\ell}$ (for large quantum number n) to the low-energy scattering phase-shifts $\delta_\ell(E)$.

These are connections of negative-energy and positive-energy solutions of the Schroedinger equation, proven by extrapolation or analytic continuation in energy. They are rigorous results of the one-electron theory and in most cases apply to any potential $V(r)$, although they may require modification for long-range potentials.

The continuity principle states that observable plasma properties including the energy per particle, the pressure and opacity coefficients are continuous functions of density even at those special densities $\rho_{n\ell}$ where a bound state is pressure-ionized into the continuum. Likewise, the properties are continuous functions of nuclear charge Z.

These continuity principles have an interesting history.[44,78] As with many of the general rules, they provide a way to identify incorrect or unsatisfactory physical models. They also give powerful guidance in developing approximate formulas for bound electrons. One example is the Burgess-Merts formula for dielectronic recombination, obtained by extrapolating an impact-excitation cross-section to energy transfers where the incident electron is captured.[79] However the classic application of the continuity principle is the work of Kramers which we review and extend in this section.

Kramers' Hydrogenic Radiation Formulas

The emission of x-rays by free-free (FF), free-bound (FB) and bound-bound (BB) processes in a plasma are described by the following kinetic equations, which give the number of photons emitted per cm^3-sec-keV:

$$\frac{dN^{FF}}{d^3r\ dt\ dh\nu} = n_I \int \frac{2d^3p_o}{h^3} |v_o|\ f(\varepsilon_o)\frac{d\sigma^{Bremss}}{dh\nu}\ (n_\nu+1)[1-f(E)] \quad (55)$$

$$\frac{dN^{FB}}{d^3r\ dt\ dh\nu} = n_I \sum_n \int \frac{2d^3p_o}{h^3} |v_o|\ f(\varepsilon_o)\frac{d\sigma^{FB}}{dh\nu}\ (n_\nu+1)[1-\frac{P_n}{D_n}] \quad (56)$$

$$\frac{dN^{BB}}{d^3r\ dt\ dh\nu} = n_I \sum_n \sum_{no} P_{n_o}\ A_{n_o} \rightarrow n\ (n_\nu+1)[1-\frac{P_n}{D_n}] \quad (57)$$

In these equations, n_I = number density of target ions, $f(\epsilon_o)$ = distribution function for incident free electrons, n_ν = number of photons per mode at energy h_ν, P_n = population of bound-states in shell of principal quantum number n and $A_{n_o \rightarrow n}$ = Einstein rate coefficient, p_o = m v_o = momentum of incident electron. The degeneracy factor [1 - $f(\epsilon)$] or [1 - Pn/Dn] describes possible occupation of the final state.

The Kramers cross-sections are:[13,80]

$$\frac{d\sigma^{Bremss}}{dh\nu} = \frac{8\pi}{3\sqrt{3}}\ \frac{Z^2\alpha^3a_o^2}{h\nu}\ (\frac{e^2/a_o}{\varepsilon_o}) \quad (58)$$

$$\frac{d\sigma^{FB}}{dh\nu} = \frac{8\pi}{3\sqrt{3}}\ \frac{Z^4\alpha^3a_o^2}{n^3}\ (\frac{e^2/a_o}{\varepsilon_o})\ (\frac{e^2/a_o}{h\nu})\ \delta(\varepsilon_o - h\nu - E_n) \quad (59)$$

$$A_{n_o} \rightarrow n = \frac{8\pi^2e^2\nu^2}{mc^3}\ f^{emiss}_{n_o,n}\quad I(h\nu) \quad (60)$$

where the line profile $I(h_\nu)$ and emission oscillator-strength obey

$$\int_{-\infty}^{\infty} I(h\nu)\quad dh\nu = 1 \quad (61)$$

$$f^{emiss}_{n_o,n} = \frac{n^2}{n_o^2}\ f^{abs}_{n,n_o} = \frac{n^2}{n_o^2}\ \frac{32}{3\pi\sqrt{3}}\ \frac{n\ n_o^3}{(n_o^2 - n^2)^3} \quad (62)$$

The Kramers bremsstrahlung emission cross-section is derived from classical radiation theory: a free electron of energy $\epsilon_o = 1/2m\ v_o^2 > 0$ approaches a point nucleus of charge + Ze along a hyperbolic trajectory. The electron gains a kinetic energy $\simeq Ze^2 / r_{min}$ as it approaches its innermost radius $r_{min} \simeq a_o\ell^2/Z$ ($\hbar\ell$ is the angular

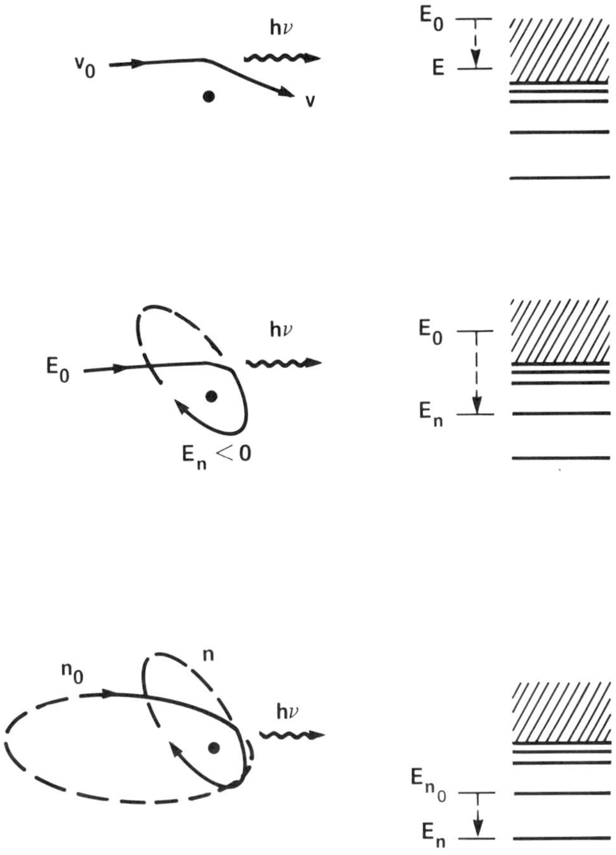

FIG. 20. Schematic representation of free-free, free-bound and bound-bound emission processes. The formulas which describe these processes are closely related because the emission occurs during the strong acceleration near the target nucleus.

momentum). For the collisions which produce the majority of the emission, r_{min} is small; Kramers' calculation is most accurate for the case where $Ze^2 / r_{min} \gg \epsilon_0$. The opposite limit is described by the Born approximation. In the Kramers case, the electron follows a hyperbolic path with strong curvature; the closest portion of the orbit is at radius $\sim r_{min}$ and is traversed at velocity

$$v_{max} \approx \sqrt{\frac{2mZe^2}{r_{min}}} \text{ giving radiation of frequency } \lesssim \frac{r_{min}^{3/2}}{\sqrt{2mZe^2}} .$$

The radiated energy spectrum is calculated by fourier analysis of the acceleration and then summed over impact parameters (angular momenta); the result is Eq. (58).[81]

We emphasize that this calculation describes the classical non-relativistic bremsstrahlung of an electron colliding with an isolated point charge. There are many physical corrections:

a.) Even for collisions with a point charge ion, the quantum theory gives a correction referred to as the Gaunt factor.[82]

b.) For collisions with a high-Z point charge there are significant relativistic corrections. These corrections can be appreciable even for $\epsilon_o \ll mc^2$ if the kinetic energy $\approx Ze^2 / r_{min}$ is relativistic.

c.) For collisions with partially-stripped ions, the emission cross-section is likely to vary between a Kramers cross-section determined by the ion charge Q for soft photons to that determined by the nuclear charge Z for hard photons produced at small radii. This bound-electron screening correction is discussed by Lamoureux and Pratt[83] and Kogan and Kukushkin.[84]

Numerical calculations of bremsstrahlung from partially stripped ions including effects (a,b,c) are reported by Feng and Pratt,[85] Lee and Pratt.[86]

d.) One dense-plasma effect is plasma degeneracy, represented by the final-state factor [1-f] in Eq. (5). In the Kramers approximation the effect of degeneracy on the net emission rate turns out to have simple analytic form.[2,87]

e.) In dense plasmas, the plasma dielectric function alters the dispersion relation of the outgoing photon. This effect is interesting because it alters the detailed-balance equations, but the results are not quantitatively large in typical laboratory plasmas.[88]

f.) Another dense-plasma effect is screening of the ion potential by the exterior plasma. This has been calculated analytically by use of the Born approximation for the scattering;[89] however the Born approximation is very inaccurate in the case of greatest practical interest (the Kramers case mentioned above).[90,91] In reality, most of the radiated photons originate at small radii where plasma screening is not large.[92]

g.) A more sophisticated calculation of the screening effect describes the environment with ion pair-correlation functions.[93] The result is interference between scattering by adjacent ions. This calculation is also limited to the Born approximation and produces its main effect for low-energy electrons (for which the Born approximation is surely invalid).

h.) In practice there is considerable interest in the possibility of significantly non-Maxwellian electron distributions in laser-plasma experiments. The predicted changes in bremsstrahlung spectrum, following from Eq. (5), are quite substantial and may help identify non-Maxwellian distributions in experimental continuum emission spectra (Lamoureux, Moller and Jaegle[94]).

Of these corrections to the Kramers' formula for bremsstrahlung, we will focus on the largest: the question of effective charges which replace the nuclear charge Z when the plasma is not fully ionized. Normally recombination and line emission are strongly dominant in a partially-ionized plasma; exactly the same question of effective charges arises for those processes. We will examine Kramers' connection of bremsstrahlung, recombination and line emission in order to derive modified Kramers formulas which apply to partially-stripped ions.

To calculate radiative recombination, Kramers extrapolates the bremsstrahlung emission cross-section of Eq. (58) to events in which $h\nu > \epsilon_o$. All transitions to energies near the quantized final energy E_n are grouped together as

$$\frac{d\sigma^{FB}}{dh\nu} = \frac{d\sigma^{Bremss}}{dh\nu} \frac{\partial E_n}{\partial n} \delta(\epsilon_o - h\nu - E_n) \tag{63}$$

For hydrogenic ions, $E_n = - Z^2 e^2/2a_o n^2$ and this equation correctly connects Eqs. (58, 59). For ions with bound electrons we employ Eq. (8) for the eigenvalue E_n, with a level spacing

$$\frac{\partial E_n}{\partial n} \simeq \frac{Q_n^2 e^2}{a_o n^3}$$

For recombination into shell n it is natural to assume the effective charge Q_n enters the extrapolated bremsstrahlung cross section; this is because the radiation occurs close to the inner turning point. With these assumptions Eq. (63) gives a nonhydrogenic cross-section (we omit the energy-conserving delta-function):

$$\sigma^{RR}(\epsilon_o) = \frac{8\pi}{3\sqrt{3}} \frac{Q_n^2 a_o^2 \alpha^3}{h\nu} \left(\frac{e^2/a_o}{\epsilon_o}\right) \frac{\partial E_n}{\partial n} \tag{64}$$

$$\propto Q_n^4 / (n^3 \epsilon_o h\nu)$$

This expression has been extensively compared to more elaborate quantum calculations and gives very satisfactory agreement. One example of these comparisons is reported by Huebner, Argo and Ohlsen.[95] To obtain good agreement, it is essential that an inner-screening effective charge be employed in Eq. (64).

Next we examine the case of line emission. In order to treat this as a continuation of radiative recombination we employ a special trick, one which has a very pleasing intuitive content. The idea is to regard the upper electron's interaction with the nuclear potential as a collision. Because the radiative transition happens during the strong acceleration occuring close to the nucleus, we can imagine that it is not important whether the asymptotic energy is positive ($\epsilon_o > 0$) or negative ($E_{n_o} < 0$).

We assume there are P_{n_o} electrons in the upper level n_o, statistically distributed over the subshell states, so that

$$P_{n_o,\ell} = 2(2\ell + 1)(P_{n_o}/2n_o^2)$$

The electrons in states of small angular momenta (elliptical orbits) approach the nucleus with the orbital frequency $\nu_{n_o} = (1/h)(\partial E_{n_o}/\partial n_o)$ giving a collision rate:

$$\frac{dN_\ell}{dt} = P_{n_o,\ell}\ \nu_{n_o} \tag{65}$$

Now we define an effective incident electron flux,

$$\Phi_{n_o} = \left(\frac{P_{n_o}}{2n_o^2}\right) \frac{2m|E_{n_o}|}{\pi^2 \hbar^2} \ \frac{1}{\hbar} \frac{\partial E_{n_o}}{\partial n_o} \tag{65}$$

If the idea of a uniform incident flux makes sense, then a geometrical calculation of the collision rate should agree with Eq. (65). With an impact parameter b determined by $mv_{n_o}b = \hbar(\ell + 1/2)$ the geometric calculation is

$$\frac{dN_\ell}{dt} = \Phi_{n_o}\ 2\pi b\ db \tag{67}$$

One readily verifies that this agrees with Eq. (65).

To summarize: the bound electron(s) in state n_o are equivalent to a uniform incident flux Φ_{n_o} in terms of the rate of close approaches to the nucleus. With this equivalence, the line emission rate is a straightforward extrapolation of the radiative recombination rate of Eq. (56),

$$\frac{dN^{BB}}{d^3r\ dt\ dh\nu} = n_I\ \Phi_{n_o} \frac{d\sigma^{BB}}{dh\nu} (n_\nu + 1)[1-P_n/D_n] \tag{68}$$

with the formal extrapolation

$$\frac{d\sigma^{BB}}{dh\nu} = \frac{d\sigma^{FB}}{dh\nu} \tag{69}$$

Again the charge Q appearing explicitly is replaced by the effective charge Q_n of the lower state on the grounds that the acceleration occurs at small radii.

This reasoning gives

$$\frac{d\sigma^{BB}}{dh\nu} = \frac{8\pi}{3\sqrt{3}} \frac{\alpha^3 a_o^2}{h\nu} \left(Q_n^2 \frac{dE_n}{dn}\right)\left(\frac{e^2/a_o}{E_{n_o}}\right) \delta(E_{n_o} - h\nu - E_n)$$

When Eq. (68) is forced back into the usual form of Eq. (60) we find an expression for non-hydrogenic oscillator strengths,[2]

$$f^{abs}_{n \to n_o} = \frac{4}{3\pi\sqrt{3}} \left(\frac{e^2/a_o}{h\nu} \right)^3 \frac{Q_n^4 \, Q_{n_o}^2}{n^5 \, n_o^3} \tag{70}$$

Equation (70) gives a remarkably simple prediction of shell-averaged oscillator-strengths for arbitrarily charged ions.

From the derivation it is obvious that Eq. (70) reduces to the hydrogenic form of Eq. (62) whenever $Q_n = Q_{n_o}$. Of course Eq. (62) is not exact for hydrogenic ions; there are relativistic and quantum corrections of the order of 50 % (Bethe and Salpeter,[96] Rose[97]).

However Eq. (70) manages to nicely reproduce the largest effect of screening (Fig. 21). In the example shown, certain oscillator-strengths change by as much as a factor ~ 10 during ionization and Eq. (70) follows this dependence to ~ 50 % accuracy.

Eq. (70) explains a semi-empirical scaling law for 1s-np transitions recently observed by Benka and Watson.[98]

The physical content of Eq. (70) has been recognized by many previous workers: the dominance of the effective charge of the innermost turning point and the factors $\partial E_n/\partial n$ which which are equivalent to wave-function normalization factors. It is also true that more accurate oscillator-strengths are readily obtained from quantum calculations. However Eq. (70) represents a significant advance for applications in which a high priority is placed upon simplicity and generality.

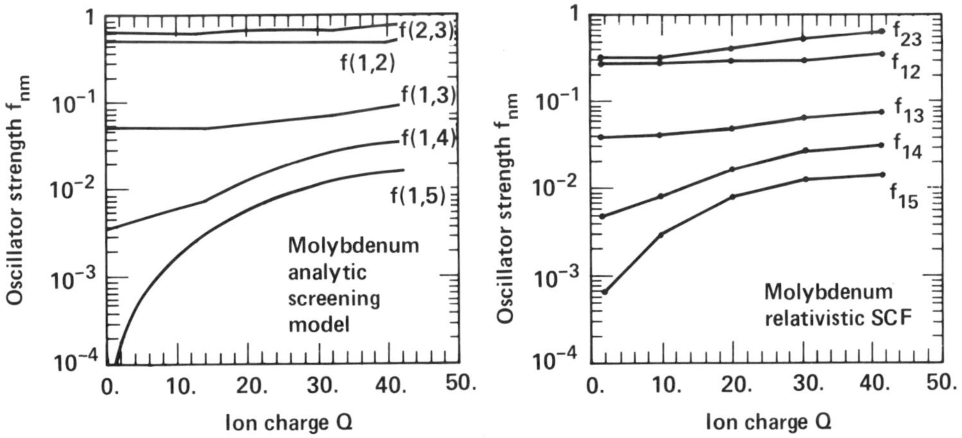

FIG. 21a, 21b. Absorption oscillator-strengths for Molybdenum ions. $f_{nn'}$ is averaged over initial states and summed over final states. The analytic screening model consists of Eq. (70) evaluated with the screening coefficients of reference 18. The relativistic SCF calculations were performed by D. A. Liberman.

Reverting to the broader perspective of this section, it is useful to reconsider Eqs. (63) and (69). These equations have shown how to convert the formulas for brems- strahlung into expressions for recombination and line emission. The point of greatest interest is that when we calculate radiative processes including various density effects, some version of these connection formulas should remain valid. This will provide powerful guidance toward the future development of a complete understanding of high-density plasma radiative phenomena.

VI. NONEQUILIBRIUM PHENOMENA

We conclude this survey with a few comments on non-equilibrium (NLTE) phenomena.

If we consider high-density nonequilibrium plasmas there are two simplifications: first, at high enough density our equations will predict LTE conditions; for each ionization or excitation process, according to rule G-10, forward and reverse rates cancel in (or near) LTE. This reduces the number of independent cross- sections and also points to reduced sensitivity to approximations in the rates.

Second, many density effects are essentially the same for LTE or non-LTE plasma conditions. For example, Coulomb interactions affect the classical motion of point-like ions, independent of the radiation spectrum or bound-electron excitation state; the resulting pair-correlation is the same in LTE or non-LTE cases. Likewise the phenomena of pressure ionization should be very similar in LTE, NLTE cases.

These remarks indicate that non-LTE plasmas are not entirely unlike the LTE plasmas considered so far.

NLTE in Laser Experiments

How is NLTE observed experimentally? The answer is indirect: experimental spectra are compared to elaborate computer simulations of laser target hydrodynamics including energy transport, ionization, and x-ray production with special treatment of nonequilibrium electron, photon and ion distributions. Calculations performed with LTE and non-LTE ionization assumptions differ dramatically.

Rosen et al.,[99] compare theoretical and experimental spectra for gold disk targets irradiated at 3.10^{14} W/cm^2 (λ = 1.06μ). The LTE calculation predicts 2-3 keV line emission which is ~ 100 times stronger than either crystal spectrometer data (confirmed by filtered XRDs) or the NLTE calculation. This enormous difference is due to the plasma ionization state: the NLTE calculation predicts lower ion charges and emission of softer photons.

The comparison of simulation and experiment is neither simple nor direct because the computer calculation inevitably involves ad hoc prescriptions for several key aspects of the physics, including laser absorption mechanisms, magnetic field generation and inhibited electron thermal conduction. The non-LTE calculations are also relatively crude, using

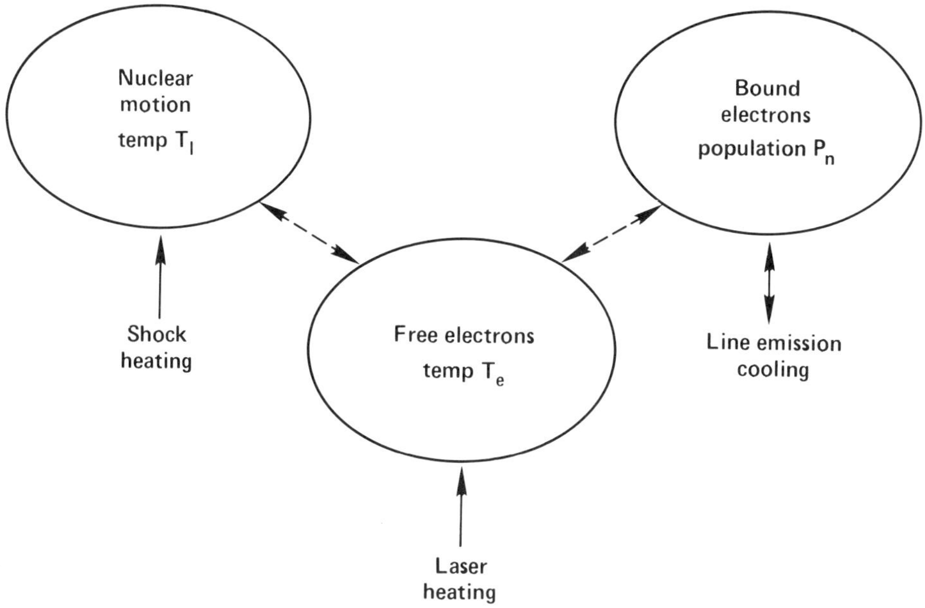

FIG. 22. Although nonequilibrium processes are relatively complicated, one achieves some degree of understanding by concentrating upon the dominant flow of thermal energy.

screened hydrogenic energies, uncorrected hydrogenic oscillator-strengths and omitting dielectronic recombination and autoionization rates. Nevertheless the strong effect of NLTE ionization cannot be missed. Without NLTE computational capability one cannot realistically describe high-Z laser interactions.

Lower-Z targets are often fully-ionized in the laser interaction region and the question of non-LTE becomes moot.

In short-wavelength irradiation one would expect reduced NLTE effects for several reasons: first, the laser absorption now occurs mainly by inverse bremsstrahlung which does not produce energetic suprathermal electrons; second, the absorption occurs at higher densities where the collisional rates are large enough to pull atomic populations toward LTE; third, the higher density also implies greater optical depth and a stronger radiation field which also brings the atoms closer to equilibrium.

Experiments on Be, CH, Ti and Au disks at shorter wavelength (λ = .53 μ) are analyzed by Mead et al.[100] These authors conclude that use of a NLTE model remains essential for mid- to high-Z targets, and large errors would be made in ionization and coronal temperatures if LTE were artificially enforced in the calculation.

Because of the complexity of the nonequilibrium processes we offer only one simple generalization which helps to interpret or guess the role of nonequilibrium in laser plasma hydrodynamics:

G-12.) Steady-state nonequilibrium plasmas can be described with heat-bath pictures which exhibit the mechanism and direction of energy flow.

In a steady ablation plasma, energy is absorbed via laser heating of the free electrons and then converted to radiation and ion expansion flow. Because up to 50 % of the absorbed energy is emitted as recombination and line radiation, one can say that this energy flows through or is processed by the bound electrons.

The bound electrons function as a thermodynamic subsystem (heat bath) collisionally coupled to a hotter bath of free electrons and radiatively coupled to the colder photon field. For optically thin plasmas most photons escape and the ambient radiation field is low, corresponding to a zero-temperature bath or heat sink.

According to LASNEX plasma simulations, the non-LTE ionization in laser plasmas is intermediate between a coronal equilibrium, corresponding to zero photon temperature, and LTE per se.

At these conditions the bound populations are described by an intermediate temperature determined by the relative strength of collisional and radiative couplings.

For heavy atoms the radiative rates (scaling with Q^4) become very large, as do the optical depths. In high-Z disk

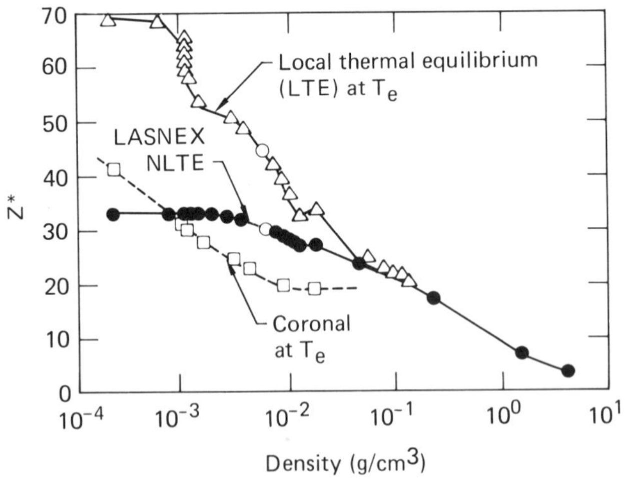

FIG. 23. LASNEX calculations of ionization for a planar-target ablation plasma (Z = 79). The NLTE charge state is significantly different from LTE or coronal models using the local free-electron temperature.

experiments one finds bound populations and even ion charge states effectively determined by the photon field which itself is cooler than the electron distribution.

These statements broadly characterise nonequilibrium target calculations. In typical ablation plasmas we do not find evidence of transient NLTE until relatively low densities are reached, at which point the charge state is frozen. At the higher densities a steady-state picture is adequate.

Reference 3 describes a quantitative development of these ideas: for plasmas with $|T_R - T_e| \ll T_e$, the NLTE steady-state is characterized by the principle of minimum entropy production, which roughly translates into a statement that the atomic populations adjust themselves to minimize the efficiency of conversion of electron thermal energy to x-rays.

Figure 21 indicates an entirely different heat-bath picture for nonequilibrium plasmas, in which electron and ion systems are coupled by slow energy exchange. During most of the laser-target interaction the ion temperature is predicted to lag behind the electron temperature by as much as 50 %. This form of nonequilibrium has less effect on the overall plasma dynamics, but the ion temperature plays a role in setting the Coulomb logarithm which enters the laser absorption, affects ion Doppler broadening, and of course enters into thermonuclear reaction rates.

These remarks underline the importance of nonequilibrium phenomena in laser-produced plasmas. The reader will find further information in specialized review articles on non-LTE atomic physics (McWhirter[101]), spectroscopic diagnostics (deMichelis and Mattioli,[102] Peacock, and Griem[103]), line formation and transport (Mihalas,[26] Kunasz[104]) and x-ray laser physics (Hagelstein,[37] Pert,[105] Jaegle[106]).

APPENDIX

A. Saha Problems

For hydrogen plasmas the Saha equation reduces to a simple (quadratic) form. Assume

n_o = number/cm^3 of neutral hydrogen atoms

n_+ = number/cm^3 of positive ions (protons)

$n_i = n_o + n_+$ = total atomic density

Because the plasma is neutral, the electron number density is $n_e = n_+$. The Saha equation is then

$$\frac{n_e n_+}{n_o} = \frac{2}{\lambda^3} \frac{1}{G_N} \exp\left(-\frac{I_o}{kT}\right) \qquad \text{(A-1)}$$

where $\lambda = (2\pi\hbar^2 / mkT)^{1/2}$

$=$ electron thermal deBroglie wavelength

$I_o = e^2 / 2a_o = 13.6$ eV = ionization potential

$\epsilon_n = -I_o / n^2$ = electron energy for principal quantum number n

$$G_N = \sum_{n=1}^{N} 2n^2 e^{-(\epsilon_n - \epsilon_1) / kT}$$

= atomic partition function (N = maximum allowed quantum number)

The average ionization state is $Q = n_e/n_i = n_e/(n_o + n_+)$, and define

$$A \equiv \frac{1}{2} n_I \lambda^3 G_N(T) \exp(I_o / kT)$$

Eq. (A-1) is then

$$1 - Q = AQ^2 \quad \text{or} \quad Q = \frac{1}{2A} (\sqrt{1+4A} - 1) \qquad \text{(A-2)}$$

Example 1. Show that $\Gamma_{ei} = Qe^2 / R_o kT < .54$ for all ρ, T.

This result illustrates rule G-4 of the text. Γ_{ei} is the electron-ion coupling parameter defined in Eq. (41). The theorem is based on the Saha equation and fails at high-density/low-temperature conditions where the Saha equation fails. The proof is very simple: one forms the quantity Γ_{ei}^3 and uses Eq. (12) and the definition of the deBroglie length λ to show:

$$\Gamma_{ei}^3 = \frac{n_i \lambda^3}{2} \left(\frac{8}{3\sqrt{\pi}}\right) Q^3 \left(\frac{I_o}{kT}\right)^{3/2}$$

The Saha equation is used to replace $n_i \lambda^3$, giving

$$\Gamma_{ei}^3 = \frac{Q(1-Q)}{G_N} \frac{8}{3\sqrt{\pi}} \left(\frac{I_o}{kT}\right)^{3/2} \exp\left(-\frac{I_o}{kT}\right) \qquad \text{(A-3)}$$

From this it can be seen that Γ_{ei}^3 never exceeds $0.154/G_N$, because

$$Q(1-Q) \leq 0.25 \text{ all } Q$$

$$x^{3/2} e^{-x} \leq 0.41 \text{ all } x \; .$$

No matter how one handles the truncation of excited states in G_N we expect $G_N > 1$ and the theorem follows.

Example 2. The Saha equation agrees with the average-atom model for strongly ionized plasmas but disagrees significantly for the nearly-neutral case.

The relative merits of average-atom and Saha theories are a perennial subject of discussion. For hydrogen we have exact analytic solutions which throw some light on the (rather subtle) difference of the two approaches.

Using the Saha equation, we form the average population $\langle P_n \rangle$ for the shell of principal quantum number n:

$$\langle P_n \rangle = 2n^2 \frac{e^{-(\epsilon_n - \epsilon_1) / kT}}{G_N} \frac{n_o}{n_i}$$

$$= 2n^2 \frac{Q}{e^{(\epsilon_n - \mu) / kT}} \qquad \text{(A-4)}$$

In the second form of this equation we employ the definition of the electron chemical potential μ appropriate to nondegenerate free electrons:

$$n_e = Qn_i = \frac{2}{\lambda^3} e^{\mu/kT} \qquad \text{(A-5)}$$

In the average-atom model, the shell populations are calculated by Fermi-Dirac statistics,

$$\bar{P}_n = \frac{2n^2}{1 + \exp(\frac{\epsilon_n - \mu}{kT})} \qquad \text{(A-6)}$$

In this equation the chemical potential μ is determined by the requirement of neutrality,

$$Q + \sum_n P_n = 1 \qquad \text{(A-7)}$$

For a given plasma density (= n_i) and temperature, the charge state Q and electron chemical potential μ of the two theories could differ.

Comparing Eqs. (A-4) and (A-6) strongly for ionized plasmas $(Q \simeq 1)$ we see that the two theories predict essentially equal chemical potentials and equal (small) boundstate populations.

In the nearly neutral case $(Q \ll 1)$ the situation is more interesting. For the Saha equation, this limit is achieved whenever $\exp((\epsilon_1 - \mu)/kT) \ll 1$; for the average-atom theory, the nearly-neutral case occurs only if $\mu \simeq \epsilon_1$. The resulting approximate forms are:

$$Q_{Saha} \simeq \frac{\exp(\epsilon_1/kT)^{1/2}}{n_i \lambda^3} \quad ; \quad Q_{AA} \simeq \frac{2}{n_i \lambda^3} e^{\epsilon_1/kT} .$$

Surprisingly these do not agree; instead

$$Q_{AA} \simeq 2\, Q^2_{Saha} \ll Q_{Saha} \qquad \text{(A-8)}$$

and in this limit the average-atom model predicts a much lower degree of ionization.

Further analysis of the equations reveals that this prediction is caused by an exaggerated binding energy for negative ions in the average-atom theory. Of course, the "exact" Saha theory altogether ignored the possibility of negative ions.

Example 3. We consider a highly ionized hydrogen plasma in complete thermal equilibrium, containing only a small density of neutral atoms. We will show that the neutrals cannot dominate the stopping power for charged particles even though bound electrons are individually much more effective at absorbing energy transfers due to their lower velocities.

This calculation illustrates rule G-8. The assumptions are

$$v_o \ll v_{ion} \ll v_e \tag{A-9}$$

where $v_o = e^2 / \hbar$ = Bohr velocity, $v_e = \sqrt{kT/m}$ = free electron thermal velocity. With these assumptions the stopping power for an ion of charge Z_1, velocity v_{ion} is a sum of contributions from bound and free electrons,

$$\frac{dE}{dx} = \frac{4\pi Z_1^2 e^4}{mv_{ion}^2} \left[n_o \log\left(\frac{2mv_{ion}^2}{\bar{I}}\right) + a\, n_e \left(\frac{v_{ion}}{v_e}\right)^3 \log\left(\frac{2kT}{\hbar\omega_p}\right) \right] \tag{A-10}$$

where $a = (1/3)\sqrt{2/\pi}$ and $\bar{I} \sim 15$ eV is the hydrogen average ionization excitation potential. The question is whether the first term can ever be dominant as a consequence of the velocity ratio in the free electron term.

Using the Saha equation, the ratio of the two contributions can be expressed

$$\frac{(dE/dx)_{Bound}}{(dE/dx)_{Free}} = a_o^3 n_+ \left(\frac{V_o}{V_{ion}}\right)^3 (3\pi^2 G_N) L \tag{A-11}$$

where L = ratio of logarithms, $a_o = \hbar^2/2m$.

This ratio is small ($\ll 1$) except at high densities where $a_o^3 n_+ \simeq 1$ but in that case the K-shell is pressure ionized according to the ion-sphere criterion.

Note that $G_N \simeq 2N^3/3 \propto n_+^{-1/2}$ so the product $n_+ G_H$ does not become large at low densities.

ACKNOWLEDGEMENT

The author is grateful to Drs. D. A. Liberman, F. J. Rogers, and E. J. McGuire for providing numerical results quoted in the text.

Work performed under the auspices of the U.S. Department of Energy by the Lawrence Livermore National Laboratory under contract number W-7405-ENG-48.

REFERENCES

1. An excellent overview is given by the series Laser Program Annual Report, Lawrence Livermore National Laboratory, UCRL-50021 (1978 to the present).
2. R. M. More, Atomic Physics in Inertial Confinement Fusion, preprint UCRL-84991, Lawrence Livermore National Laboratory (1981).
3. R. M. More, Atomic and Molecular Physics of Controlled Thermonuclear Fusion, p. 399, Ed. by C. Joachain and D. Post, Plenum Publishing Corp. (1983).
4. Proceedings of Workshop Conference on Radiative Properties of Hot Dense Plasmas, J.Q.S.R.T. 27, p. 209-385 (1982); Radiative Properties of Hot Dense Matter, Ed. by J. Davis, C. Hooper, R. Lee, A. Merts, and B. Rozsnyai, World Scientific, Singapore (1985).
5. M. J. Herbst, P. G. Burkhalter, D. Duston, M. Emery, J. Gardner, J. Grun, S. P. Obenschain, B. H. Ripin, R. R. Whitlock, J. P. Apruzese and J. Davis, Laser Interaction and Related Plasma Phenomena, Vol. 6, Ed. by H. Hora and G. Miley, p. 317, Plenum Press, New York (1984).
6. Dr. J.-C. Gauthier et al., unpublished paper presented at this meeting.
7. M. D. Rosen, P. L. Hagelstein, et al., Phys. Rev. Lett. 54, 106 (1985); D. L. Matthews, P. L. Hagelstein, et al., Phys. Rev. Lett. 54, 110 (1985).
8. E. A. Crawford and A. L. Hoffman, in Laser Interaction and Related Plasma Phenomena, Vol. 6, p. 353, Ed. by H. Hora and G. Miley, Plenum Press, New York (1984).
9. R. M. More, in Laser Interaction and Related Plasma Phenomena, Vol. 5, p. 253, Ed. by H. J. Schwarz, H. Hora, M. J. Lubin and B. Yaakobi, Plenum Press, New York (1981).
10. D. K. Bradley, J. Hares, A. Rankin and S. J. Rose, The analysis of colliding-shock experiments, Rutherford Appleton Laboratory, preprint RAL-85-020, (1985).
11. A. Ng, D. Parfeniuk, P. Celliers, L. DaSilva, University of British Columbia, Vancouver B.C., preprint "Electrical Conductivity of a Dense Plasma" (Nov. 1985).
12. R. Fabbro, B. Faral, J. Virmont, F. Cottet, J. P. Romain and H. Pepin, Phys. Fluids 28, 3414 (1985).
13. Ya. B. Zel'dovich and Yu. P. Raizer, Physics of Shock Waves and High-Temperature Hydrodynamic Phenomena, Vol. 1, Ed. W. D. Hayes and R. F. Probstein, Academic Press, New York (1966).
14. B. Strömgren, Zs. f. Ap. 4, 118 (1932).
15. H. Mayer, Los Alamos Scientific Laboratory report LA-647 (unpublished) 1947.
16. W. Lokke and W. Grasberger, Lawrence Livermore Laboratory report UCRL-52276 (1977).
17. G. B. Zimmerman and R. M. More, J.Q.S.R.T. 23, 517 (1980).
18. R. M. More, J.Q.S.R.T. 27, 345 (1982).
19. A. Sommerfeld, Atomic Structure and Spectral Lines, 3rd Ed. Methuen, London (1934).
20. D. Salzmann and G. Wendin, Phys. Rev. A18, 2695 (1978).

21. V. C. Reddish, Physics of Stellar Interiors, Edinburgh University Press, Edinburgh (1974).
22. R. M. More, Advances in Atomic and Molecular Physics, 21, 305 (1985). We note that the usual discussion concerns line shifts (not level shifts).
23. L. Spitzer, Jr., Physics of Fully Ionized Gases, 2nd Rev. Ed., Interscience Publishers, New York (1962).
24. E. E. Salpeter, J. Geophysical Research 68, 1321 (1963).
25. G. Ecker, Z. fur Phys. 148, 593 (1957).
26. D. Mihalas, Stellar Atmospheres, 2nd Ed., W. H. Freeman & Co., San Francisco, p. 294 (1978).
27. R. Dicke, Phys. Rev. 89, 472 (1953).
28. D. Burgess, D. Everett and R. Lee, J. Phys. B-12, L755 (1979).
29. J. Green, J.Q.S.R.T. 4, 639 (1964).
30. S. Brush, H. Sahlin and E. Teller, J. Chem. Phys. 45, 2102 (1966).
31. J.-P. Hansen, Phys. Rev. A8, 3096 (1973).
32. M. Baus and J.-P. Hansen, Physics Reports 59, 1 (1980).
33. H. E. DeWitt, Strongly Coupled Plasmas, p. 81, Ed. by G. Kalman and P. Carini, Plenum Publishing Corp. (1978).
34. R. W. Hockney and J. W. Eastwood, Computer Simulation Using Particles, McGraw-Hill, New York (1981).
35. R. B. Laughlin, Lawrence Livermore National Laboratory, Livermore, CA, preprint UCRL-90304, to appear in Physical Review A.
36. R. M. More, in Laser Program Annual Report, UCRL-50021-84, p. 3-68, Lawrence Livermore National Laboratory, Livermore, CA (1984).
37. P. Hagelstein, Physics of Short Wavelength Laser Design, Lawrence Livermore National Laboratory, Livermore, CA, UCRL-53100 (1981).
38. L. D. Landau, JETP 7, 203 (1937). The careful reader will observe there is a numerical mistake in the final result of this paper.
39. C. C. Klick and J. H. Schulman, Solid State Physics 5, 97 (1957);
D. Dexter and R. Knox, Excitons (Interscience, New York, NY, 1965).
40. G. Ecker and W. Kröll, Physics of Fluids 8, 354 (1965).
41. D. A. Liberman, Phys. Rev. B20, 4981 (1979);
D. A. Liberman, J.Q.S.R.T. 27, 335 (1982).
42. F. Perrot, Phys. Rev. A26, 1035 (1982).
43. R. Cauble, M. Blaha and J. Davis, Phys. Rev. A29, 3280 (1984).
44. R. M. More, Advances in Atomic and Molecular Physics, 21, 305, (1985).
45. D. Burgess, unpublished report CLM-P 567, Culham Laboratory, Abingdon, UK (1978).
46. N. J. Peacock, unpublished report CLM-P 519, Culham Laboratory, Abingdon, UK (1977).
47. S. Brush and B. H. Armstrong, Proceedings of Workshop Conference on Lowering of the Ionization Potential, JILA report 79, Univ. of Colorado, Boulder, CO (1965).
48. A. Unsold, Z. Astrophys. 24, 355 (1948).
49. T. Carson and H. Hollingsworth, Mon. Not. R. Astron. Soc. 141, 77 (1968);
T. Carson, D. Mayers and D. Stibbs, Mon. Not. R. Astron. Soc. 140, 483 (1968).

50. D. Burgess and R. L. Lee, Journal de Physique 43, Colloque C2, 413, (1982).
51. F. J. Rogers, H. C. Graboske, H. DeWitt, Physics Letters 34A, 127 (1971).
52. J. Weisheit and B. Shore, Astrophys. J., 194, 519 (1974).
53. A. Vinogradov, I. Sobelman, and E. Yukov, Sov. J. Quantum Electron. 4, 149 (1974).
54. S. Brush, H. Sahlin and E. Teller, J. Chem. Phys. 45, 2102 (1966).
55. F. J. Rogers, unpublished preprint.
56. H. R. Griem, Plasma Spectroscopy, McGraw-Hill, New York (1964).
57. J. Stewart and K. Pyatt, Astrophys. J. 144, 1203 (1966).
58. J. Humblet, Mem. Soc. R. Sci. Liege 12, 9 (1952); B. Ya. Zel'dovich, Soviet Physics JETP 12, 542 (1961); R. M. More, Phys. Rev. A 4, 1782 (1971).
59. T. Regge, Nuevo Cimento VIII, 671 (1958).
60. C. Bauche-Arnoult, J. Bauche, and M. Klapisch, Phys. Rev. A 20, 2424 (1979); 25, 2641 (1982); M. Klapisch, E. Meroz, P. Mandelbaum, A. Zigler, C. Bauche-Arnoult, and J. Bauche, ibid. 25, 2391 (1982). C. Bauche-Arnoult, J. Bauche, and M. Klapisch, Phys. Rev. A 31, 2248 (1985).
61. P. Audebert, J.-C. Gauthier, J.-P. Geindre, and C. Chenais-Popovics, C. Bauche-Arnoult, J. Bauche, M. Klapisch, E. Luc-Koenig, and J.-F. Wyart, Phys. Rev. A32, 409 (1985).
62. J. M. Green, J.Q.S.R.T. 19, 639 (1964).
63. F. Grimaldi and A. Grimaldi-Lecourt, J.Q.S.R.T. 27, 373 (1982).
64. M. Busquet, D. Pain, J. Bauche, and E. Luc-Koenig, Physica Scripta 31, 137 (1985).
65. D. Duston and J. Davis, Phys. Rev. A21, 1664 (1980). Figure 3 of this paper illustrates the (incorrect) recombination predicted by rate equations at high densities.
66. J. Scott, Philos. Mag. 43, 859 (1952); J. Schwinger, Phys. Rev. A22, 1827 (1980).
67. F. J. Rogers, Phys. Rev. A24, 1531 (1981).
68. This equation for the density matrix of a one-electron system is easily derived from the definition given.
69. A. Dalgarno, lectures presented at this meeting.
70. R. M. More, Lawrence Livermore National Laboratory report UCRL-84379, Two-Temperature Equation of State for Dense Plasmas (1980); D. A. Boercker and R. M. More, Phys. Rev. A, to be published.
71. B. J. B. Crowley and R. M. More, Proceedings of Workshop Conference on Atomic Physics for Heavy-Ion Fusion, Rutherford-Appleton Laboratory, October, 1984.
72. J. D. Jackson, Classical Electrodynamics, 2nd Ed., J. Wiley & Sons, Inc., New York, p. 724 (1975).
73. S. Skupsky, Phys. Rev. A16, 727 (1977).
74. International Workshop on Atomic Physics for Ion Driven Fusion, Journal de Physique, Colloque No. 8, Tome 44, Suppl. an FASC.II (1983); C. Deutsch, G. Maynard and H. Minoo, Journal de Physique C8, 67 (1983).
75. J. A. Harte and R. M. More, Lawrence Livermore National Laboratory, unpublished report UCRL-50021-82, "Laser Program Annual Report," p. 3-66.

76. E. J. McGuire, PRA26, 125 (1982);
E. J. McGuire, J. M. Peek and L. C. Pitchford, Phys. Rev. A26, 1318 (1982).
77. M. J. Seaton, Comptes Rendus 240, 1317 (1955);
N. F. Mott and H. S. W. Massey, Theory of Atomic Collisions, 3rd Ed., p. 68, Oxford Univ. Press, London (1965).
78. R. Peierls, Surprises in Theoretical Physics, p. 137, Princeton Univ. Press, Princeton, N.J., 1979.
79. A. Burgess, Ap. J. 139, 776 (1964);
A. Burgess, Ap. J. 141, 1588 (1965);
A. Burgess and A. S. Tworkowski, Ap. J. 205, L-105 (1976); A. Merts, R. D. Cowan, and N. H. Magee, Jr., unpublished report LA-6220-MS (1976).
80. H. A. Kramers, Philos. Mag. 271, 836 (1923).
81. L. D. Landau and E. M. Lifshitz, Classical Theory of Fields, 2nd Ed., Pergamon Press, Oxford, 1962.
82. W. J. Karzas and R. Latter, Astrophysical Journal Suppl. no. 55, Vol. VI, p. 167 (1961);
P. J. Brussard and H. C. Van de Hulst, Rev. Mod. Phys., 34, 507 (1962); I. P. Grant, Mon. Not. R. Astron. Soc. 118, 352 (1958).
83. M. Lamoureux and R. H. Pratt, Radiative Properties of Hot Dense Matter, p. 241, Ed. by J. Davis et al., World Scientific, Singapore (1985).
84. V. I. Kogan and A. B. Kukushkin, Soviet Physics JETP 60, 665 (1984).
85. J. J. Feng and R. H. Pratt, unpublished preprint.
86. C. M. Lee and R. H. Pratt, Phys. Rev. A12, 707 (1975).
87. J. P. Cox and R. T. Giuli, Principles of Stellar Structure, Vol. 1, Gordon and Breach, New York, 1968.
88. J. Dawson and C. Oberman, Physics of Fluids 6, 394 (1963).
89. B. F. Rozsnyai, J.Q.S.R.T. 22, 337 (1979).
90. J. M. Green, R. and D. Associates unpublished report RDA-TR-108600-003 (1980).
91. M. Lamoureux, I. J. Feng, R. H. Pratt and H. K. Tseng, J.Q.S.R.T. 27, 227 (1982).
92. L. Kim, R. H. Pratt and H. K. Tseng, Phys. Rev. A32, 1693 (1985).
93. S. Ichimaru, Basic Principles of Plasma Physics, W. A. Benjamin, Inc., Reading, Mass. 1973.
94. M. Lamoureux, C. Möller and P. Jaegle, Phys. Rev. A30, 429 (1984).
95. W. Huebner, M. F. Argo and L. D. Ohlsen, J.Q.S.R.T. 19, 93 (1978).
96. H. A. Bethe and E. E. Salpeter, Quantum Mechanics of One- and Two-Electron Atoms, Academic Press, Inc., New York, 1957.
97. S. J. Rose, Rutherford-Appleton Laboratory, unpublished preprint RL-82-114, "The Effect of Relativity on the Oscillator Strengths of Hydrogen-Like Ions," Dec. 1982.
98. O. Benka and R. Watson, Phys. Rev. A29, 2255 (1984).
99. M. D. Rosen, D. W. Phillion, V. C. Rupert et al., Phys. Fluids 22, 2020 (1979).
100. W. C. Mead, E. M. Campbell et al., Phys. Fluids 26, 2316 (1983).
101. R. W. P. McWhirter in Plasma Diagnostic Techniques, ed. by R. Huddlestone and S. Leonard, Academic Press, New York, 1965.

102. C. DeMichelis and M. Mattioli, Rep. Prog. Phys. 47, 1233 (1984).
103. H. R. Griem, Handbook of Plasma Physics, eds. M. N. Rosenbluth and R. Z. Sagdeev, Vol. 1, p. 73, North-Holland, 1983.
104. P. Kunasz, in Radiative Properties of Hot Dense Matter, ed. by J. Davis, C. Hooper, R. Lee, A. Merts, B. Rozsnyai, p. 3, World Scientific, Singapore, 1985.
105. G. J. Pert, preprint, University of Hull, "SUV and X-ray Lasers," to be published.
106. P. Jaegle, lectures presented at this meeting.

ATOMS AND IONS IN VERY HIGH FIELDS

K. Burnett

Spectroscopy Group
Blackett Laboratory
Imperial College London SW7 2BZ

INTRODUCTION

This lecture will take a very quick tour of how d.c. and a.c. electric fields affect atoms and ions along with atomic and molecular processes. Some of the topics I'll discuss are the subject of great interest and speculation at the present time. I can only give the most basic of pictures of what is going on.

My discussion will be qualitative throughout even when discussing sophisticated aspects of theories. I can only apologise for my naiveté in advance to those aware of the more technical aspects of the theory.

The one theoretical construct I shall make much use of is that of a dressed state.[1] This has proved to be of immense utility in much of the physics.

Static Electric Fields

The effect of d.c. electric fields on atoms has been studied since the beginning of atomic theory.[2] We all know that electric fields can shift the positions of atomic levels and for sufficiently high fields destroy bound states altogether. To precise the effect of an electric field is to turn all bound states into resonances.[3]

For many practical purposes we can classify states as effectively bound (or free) when they are well below (or above) the classical barrier for escape.

A more careful analysis includes tunneling and reflection at the classical barrier. An analysis of this type employing the JWKB approximation was given by Lanczos.[4]

For hydrogen a direct computational treatment scheme has been established using complex co-ordinate techniques. 5

This reduces the problem to finding the complex energies ie resonance energies and widths by numerical integration of the transformed Hamiltonian. The transformed Hamiltonian has square integrable complex eigenfunctions at the effective complex energies.[6]

This direct and computationally efficient method can as yet only be applied to strictly single electron problems. For hydrogen it is also possible to calculate the position and widths of high lying Stark states using sophisticated forms of perturbation theory.[7]

For many electron atoms the theoretical problem is fierce. For some states it is possible to combine the methods of Multichannel quantum defect(MCQDT) theory with the treatment of the single electrons in a d.c. field.[8] This development is indeed important since it deals with bound states and autoionizing resonances in the presence of d.c. fields in one fell swoop.

You may be wondering why I have discussed situations where resonances are still reasonably well defined. This is partly because most of spectroscopy is done with fitting lines or bumps, when there are no features the spectroscopists go home. It is also due to the fact that to completely destroy the energy levels of hydrogen (say) needs fields $\sim 10^9$V/cm. This is not attainable at least using d.c. set ups. Most precise experiments have therefore been done with Rydberg states.[9] For an easily attainable field $\sim 10^3$V/cm one gets a field comparable with the coulomb one at $n \sim 33$.

This doesn't mean that the subject is uninteresting - it just means that precise spectroscopy cannot be done (as yet) for high d.c. fields. The theory is, therefore, highly developed for hydrogen and for Rydberg states of many electron atoms.

One can of course make estimates of the survival of bound states in d.c. fields using JWKB or even classical methods.[9]

Even though intense d.c. fields $\left(E \sim \frac{e^2}{4\pi\varepsilon_0 a_0{}^2}\right)$ cannot be produced effectively static fields are often of great importance. In the theory of resonant charge transfer eg one is concerned with the decay of the bound state of one atom into another state.[10] Although the problem is really time dependent it can often be considered effectively static: because the timescale for tunnelling through the coulomb barrier is often short enough. An almost identical problem arises in the resonant charge transfer from an excited atom to a metal surface. In the latter case it is the image charge due to the atomic nucleus that provides the large field. This resonant decay into the metal continuum can produce resonances with widths $\sim$eV.

The time varying fields produced by ions in a plasma have been discussed elsewhere in these lectures.[12] I'd like to mention one situation that seems most interesting to me.

At present the field of plasma chemistry is developing most rapidly.[13] As yet we do not have a detailed knowledge of how plasmas enhance chemical reaction rates. The plasma will obviously heat the reactants, increasing reaction rates. The plasma will also provide extra ion-molecule reaction pathways for the process. It is interesting to consider what effect electric fields have on activated complexes; does it have a relation to heterogeneous catalysis and surface produced fields?

Even larger pulsed electric fields can be produced in heavy ion collisions. In this manner one can produce d.c. fields that are sufficiently large to produce electron- positron pair production.[14]

A.C. Electric Fields produced by Lasers

In the a.c. stark effect the Hamiltonian is time dependent. It is possible to convert this rigorously to a time independent problem using Floquet[15] space methods. This method is exact and has been used along with numerical techniques to describe important cases ie hydrogen, where a precise and complete numerical approach is warranted.

The Floquet space methods are elegant, if somewhat lacking in physical appeal. Luckily these methods are somewhat humbler and easily understood relative to those of dressed states.[1] For the precise relationship between the two techniques one should consult the reference given.[16]

In the dressed state method one takes the quantum theory of the radiation field seriously and considers combined states of atom plus radiation in the Schrodinger picture. For a two level atom in a radiation field we get a set of two level dressed states. Each pair of near degenerate states has the form : ground state plus N photons, excited state plus N-1 photons. In the presence of the $\underline{E}.\underline{d}$ (time independent) interaction these pairs of levels interact and produce shifted 'dressed' states.[17]

We have in essence converted the time dependent problem into a time-dependent one. This same procedure can be applied to a problem where many states are involved although the dressed state manifold becomes correspondingly more complex.

Let's consider how it would apply to an example of a negative ion with one bound state and a continuum in the presence of laser field.[18] Let us suppose the photon's energy is larger than the ionization potential.

The dressed state $|g+N\hbar w>$ is now degenerate(and coupled by $\underline{E}.\underline{d}$ interaction to) the nearly continuum states with one less photon $|e+(N-1)\hbar w>$. This problem is now a standard one described by Fano.[19] We would expect this problem to be very similar to a scattering problem with a resonance due to an autoionizing state imbedded in the continuum. In fact this problem has been called laser-induced autoionization.[20]

The role played by configuration interaction (CI) in the conventional case is now played by the electric field-dipole coupling. So we can vary the effective CI and we can also vary the position of the resonance by changing the photon's frequency. If we used floquet method combined with the dilation transformation this decaying state would appear as a complex quasi-energy of the dilated floquet Hamiltonian.

As is well known, the Fano problem is an idealisation of the more common situation where one has a series of autoionizing resonances.[21] This situation can be handled using MCQDT. The same applies here. The laser will induce a series of autoionizing resonances. It can be dealt with using MCQDT[21] by handling the laser field on the same basis as the CI interaction.

The open and closed channels are now dressed atomic channels. The interchannel coupling is induced by the laser. This coupling can be parameterized in the same way as the interchannel couplings in MCQDT since it can be shown to be localised in the core.[22]

Because of the freedom to vary the coupling position and strength a great range of dynamic phenomena can be investigated.[20] Although much remains to be done I think it is fair to say that the basic theory for hydrogen and Rydberg states in constant intensity fields is well founded. Recent research has concerned itself with the effects of pulsed and fluctuating fields along with other relaxation mechanisms.[23]

For situations where many channels are coupled the computational problems are enormous. The problem of above threshold ionization in intense fields is particulary fierce.[24] The recent work of Eberly and Deng[25] has shown a simple model can give many of the important features observed in the experiments.

In weak fields one can use perturbation theory[26] to calculate multiphoton ionization rates. A great deal of detailed work has been done using perturbation theory for hydrogen and effectively one electron systems (or many electron systems that can be described using MCQDT).

There is as yet very little detailed work on laser induced autoionization in intense fields. The experiments are in general very difficult to perform and to interpret.[20,27]

Complex-atoms in Intense Laser Fields

The theory of ionization of complex atoms in the presence of intense laser fields is a matter of intense debate.[28] The field has been fed by several experimental studies, in particular those of Luk et al.[29] In these experiments highly charged fragments have been observed. It seems most unlikely that any sequential absorption of photons can hope to explain these phenomena. This means that perturbation theory or even all-order single electron theory is unlikely to be useful in explaining the production of highly charged fragments.

Crance[30] has developed an intriguing model in which she deals with photon absorption in terms of effective cross-sections along with statistical sharing of absorbed energy. This theory seems to give an adequate explanation of the dependence of the charge states produced on power. Its theoretical basis is, however, somewhat obscure. It must contain more than a germ of the truth, especially with regard to the rapid statistical sharing of energy amongst electrons in a given shell.

Geltman[31] has given a discussion which assumes each electron in a shell ionizes independently following a Hartree model, where each electron has its own private wavefunction and energy. The statistics then enters when one calculates the probabilities for calculating different charge states. Rhodes[32] has emphasised that the marked Z dependence of the effective cross-sections and fragment spectra imply that shell structure plays an important role. He has suggested that collective motion of a whole shell is important in the process, as observed in the production of giant dipole resonances.[33]

Rhodes[32] further suggests that it may be essentially non-linear coupling to giant-dipole resonances that produces the highly charged fragments. It is obviously important that more systematics are needed in these experiments. It would for example be most useful to have studies of the wavelength dependence of the processes.

It will be very difficult to make rigorous calculations of these processes. Random Phase Approximation and the time dependant local density approximation are able to give good results for giant dipole resonances, in weak fields.[34] It is obviously, however, a gargantuan task to convert these linear response theories to the non-linear regime.

One could think of this as interaction of an intense laser-field with a dense intromogeneous degenerate plasma

The possibilities for research that quite justifiably can be termed new and exciting are manifold. Future experiments should also be able to reach field strengths where $E \sim \frac{e^2}{4\pi\varepsilon_0 a_0^2}$ (10^{19}-10^{20} W/cm^2). This will be a marvelous challenge for experimentalists and theoreticians alike!

Influence of Intense Fields on Atomic&Molecular Collision Processes

Intense infra-red radiation can produce multiphoton dissociation of molecules.[35] For a while it was thought that this could be used to dissociate specific bonds in a molecule. For most molecules randomization of vibrational excitation is too rapid to enable one to break specific bonds. Miller has considered the effect of intense infra-red radiation on a reaction complex. It has been shown that infra-red radiation can indeed influence chemical reactions. It is however difficult to find a case where these modifications can be clearly observed in the presence of the heating and fragmentation produced by the field.

Intense-field Photo-dissociation

We can discuss photo-dissociation in the same manner as we did for intense-field photo-ionization.[37] A dressed bound electronic state potential curve will cross a dressed excited dissociative curve. Photo-dissociation is then described by curve crossing between potential curves. That is by predissociation of dressed states. This parallels closely the description of ionization in terms of autoionization of dressed atomic states.[20]

In sufficiently intense fields, when the crossing becomes strongly avoided, the positions of the predissociation resonances will be changed completely.

Photon-Catalysed Reactions

We discussed above how the presence of a laser field could imbed a bound state into a continuum. This produces a new resonance in the continuum and dramatically modifies the cross-section in the vicinity of the resonance. One could ask whether this mechanism could be used to modify chemical reaction dynamics.

It has been shown in the work of T F George's group how this can be done.[38] In the chemical reaction case it is often true that photon absorption can couple a repulsive curve to an attractive excimer curve. An intense field will then imbed resonances in a reactive channel. It has yet to be seen whether this can be exploited usefully. Work on absorption by photons during reactive encounters is also being developed as a probe of reaction dynamics.[39]

Acknowledgements

I should like to thank J P Connerade, M H R Hutchinson and P L Knight for helpful discussions on the topics discussed above.

References

1. Cohen-Tannoudji, C, 1976, "Frontiers in Laser Spectroscopy", Les Houches Summer School 1975, Edited by R Balian, S Haroche and S Liberman, North Holland, Amsterdam.

2. Stark J, 1913, Sitzungsber Akad. Wiss. Berlin, 47 : 932.

3. Oppenheimer, J R, 1928, Phys. Rev., 31 : 66

4. Lanczos, C, 1931, Z Physik, 68 : 204.

5. Junker, B R, 1982, Adv. in At. & Mol. Phys. Vol 18, Academic Press, New York. H G Muller & A Tip, 1984, Phys. Rev. A 30:3039

6. Reinhardt, W P, 1982, Annu. Rev. Phys. Chem 33 : 223

7. Silverstone, H J and Koch, P M, 1979, J. Phys B, 12 : L537.

8. Harmin, D A, 1982, Phys. Rev. A, 26 : 2656

9. Freeman, R R, 1981 in "Atomic Physics 7", page 209, Plenum, New York.

10. Janev R K, 1974, J. Phys. B, 7 : 1506

11. Brako, R and Newns, D M, 1982, Vacuum vol 32, 1 : 39

12. See the Lectures by R More in this volume.

13. Ibbertson, V, 1984, New Scientist, 26 July : 19.

14. Rafelski, J, Fulcher, L and Klein, A, 1978, Physics Report 38C : 228.

15. Chu, S I and Reinhardt, 1977 W P, Phys. Rev. Lett. 39 : 1195.

16. Shirley, J H, 1965, Phys. Rev. B, 138 : 979 and Ben-Reuven. A, 1980, Phys. Rev. A, 22 : 6

17. Reynaud, S and Cohen-Tannoudji, C, 1982, J Physique, 43 : 1021.

18. Haan, S L and Cooper, J, 1984, J. Phys. B. 3481-3492. Javanainen, J, 1983 Optics Communications 46 : 175.

19. Fano, U, 1961, Phys. Rev. A 124 : 1866.

20. Knight, P L, 1984, Comments At. Mol. Phys. 15 : 193.

21. Seaton, M J, 1983, Rep. Prog. Phys. 46:167.

22. Zoller, P, 1985, Abstract 2nd ECAMP Amsterdam and unpublished work.

23. Dalton, B J, Knight, P L and Lauder, M, 1985 to be published.

24. Agostini, P, Fabre F, Mainfray, G, Petite, G and Rahman, N K , Phys. Rev. Lett 42 : 1127.

25. Deng, Z and Eberly, J H, 1985, J. Phys. B 18 : L287.

26. Lambropoulos, P and Chin, S L (Eds.), 1984, "Multiphoton Ionization of Atoms", Academic Press, New York.

27. Pavlov, L I, Dimov, S S, Metchkov, D I, Milera, G M, Stamenov, K V and Altshuller, G B, 1982 Phys. Lett 89A : 441.

28. Lambropoulos, P and Smith, S J (Eds.), 1984, "Multiphoton Processes", Springer, Berlin.

29. Luk, T S, Pummer, H, Boyer, K, Shahidi, M, Egger, H and Rhodes, C K, Phys. Rev. Lett 51 :110.

30. Crance, M, 1984, J. Phys. B L635 and L355.

31. Geltman, S, 1985, Phys. Rev. Lett 54 : 1909.

32. Rhodes, C K, in 'Multiphoton Processes', 1984, Edited by Lambropoulos, P and Smith, S J, Springer, Berlin.

33. Connérade, J P, 1984 in Les Houches Session XXXVIII 1982, Edited by G Grynberg and R Stora, Elsevier, Amsterdam.

34. Lundguist, S, 1983 in "Theory of the Inhomogeneous Elec-

tron Gas" Edited by Lundquist, S and March, N H, Plenum, New York.

35. Evans, D K, McAlpine, R D, 1984, ref: 28.

36. Orel, A E and Miller, W H, 1980, J. Chem. Phys, 73 : 241.

37. Bandrauk, A D and Turcotte, G, 1982, J. Chem. Phys, 77 : 3867.

38. Lam, K S and George, T F, 1984, Phys. Rev. A, 29 : 242.

39. Maquire, T C, Brooks, P R and Curl, R F, 1983, Phys. Rev. Lett 50 : 1918.

ATOMIC PROCESSES IN HIGH-INTENSITY, HIGH-FREQUENCY LASER FIELDS

M. Gavrila

FOM-Institute for Atomic and Molecular Physics
Kruislaan 407, 1098 SJ Amsterdam, The Netherlands

ABSTRACT

A nonperturbative theory is described for electron-atom interactions in intense, high-frequency laser fields. It is illustrated on the case of free-free transitions of electrons colliding with a potential, and on the multiphoton ionization of one-electron atoms. Transition amplitudes are obtained, and their validity is discussed. The theory applies at already existing excimer laser frequencies (and intensities), but extends beyond, into the XUV range. It is shown that in this regime collisions are dominated by the elastic channel, and that the atom, although possibly strongly distorted, has a small decay rate. Numerical results are presented for the laser-modified elastic scattering from a Coulomb potential.

I. INTRODUCTION

The development of high-intensity lasers, ranging from the IR to the VUV, has stimulated the study of atomic processes occurring in such fields. Intensities of more than 10^{16} W/cm^2 are now available, and still higher values are expected. At these high intensities atomic transitions abundantly involve multiphoton absorption and emission (for a review of these processes see ref. 1). The description by perturbation theory is no longer valid, and new methods of solution of the Schrödinger equation are needed. A nonperturbative theory was developed earlier by Kroll and Watson for the *low-frequency regime* [2,1], well suited for the range of the intense IR lasers. We have recently developed a nonperturbative approach to deal with the opposite case, of the *high-frequency regime* [3,4,1]. It applies to the intense excimer lasers already in operation in the VUV (e.g. see refs. 5,6), but extends beyond, into the XUV range.

In the following we present our theory for the high-frequency regime. We begin with the case of electron-atom (ion) collisions in the radiative field, also termed *free-free transitions* (FFT). The formalism is illustrated in Sec.II on the case of an atom represented by a potential model. It is shown that the elastic scattering from a static "dressed" potential (which is a modification of the original one) represents the dominant channel of the collision. Expressions for the FFT amplitudes are derived, and an analysis is presented of the conditions of validity of the theory. In Sec.III we apply the formalism to elastic scattering from the dressed Coulomb potential. However, we consider here only the case of a direction-averaged dressed potential. Numerical results are presented. Sec.IV contains an extension of the formalism to describe the behavior of a one-electron atom in an intense, high-frequency field. It is shown that in the limit of extremely high frequencies the atom becomes stable (its decay by multiphoton ionization is quenched). The corresponding eigenvalue equation contains the dressed potential mentioned before,and has real eigenvalues. At high, but finite frequencies, the eigenvalue equation contains a complex, optical potential, and its eigenvalues are also complex (decaying states). Multiphoton ionization amplitudes are derived. Sec.V contains conclusions and outlines some of the current work.

II. FORMALISM FOR FREE-FREE TRANSITIONS

For the purpose of describing a FFT process, the physical space may be divided conveniently into three regions. The first one is the interior of the atom, where the electron-atom interaction takes place. The second, overlapping the previous one but orders of magnitude larger, is that in which the laser pulse is focused. This may be of a few μm diameter, or more. Finally, the third one is that of the surrounding macroscopic world containing the electron-source and -detector.

An electron travelling from the source towards a target-atom, can be considered as free in the outside region then under the action of the laser field only, when in the laser focus, and subsequently undergoing the FFT process when in the atomic region. On its way out it will proceed through the three regions in reversed order.

Appropriate quantum mechanical descriptions should be given to the electron as it passes through them. We assume that the electron has a given momentum $\vec{p}'$ in the region outside, and hence can be represented by the plane wave (in atomic units):

$$\phi_{\vec{p}'}(\vec{r},t) = \exp\, i(\vec{p}'\vec{r} - \frac{p'^2}{2}\, t) \quad . \tag{1}$$

When in the laser focus and on the verge of entering the collision process, the electron will experience the action of the electromagnetic field. This is considered monochromatic, and can be assimilated to a plane wave over the distances of interest. For simplicity we shall assume linear polarization, and adopt the dipole approximation, which neglects the space-dependence of the electromagnetic phase over the volume where the photons are absorbed or emitted. This is quite justified in the frequency range from the visible to the XUV we are interested in. Consequently, we take the electrodynamic potentials of the laser field in this region of the form $\vec{A} = \vec{a}\cos\omega t$ (with $\vec{a}$ real), and the scalar potential equal to zero.

An electron of given (canonical) momentum $\vec{p}$ in this electromagnetic wave can be represented by the modulated electron-plane wave

$$\chi_{\vec{p}}(\vec{r},t) = \exp\, i\left\{ \vec{p}\vec{r} - \tfrac{1}{2} \int_0^t [\vec{p} + \frac{1}{c} A(t')]^2\, dt' \right\} \quad . \tag{2}$$

When the amplitude $\vec{A}$ varies adiabatically over the transition region between the free space outside and the laser-focus, the space and time continuity of the wave function requires that the magnitude of $\vec{p}'$ and $\vec{p}$ appearing in Eqs.(1),(2) be related by $p'^2 = p^2 + \frac{1}{2}(a/c)^2$. Thus, upon entering the laser field the free particle undergoes a change of momentum [7,19].

The considerations above hold for an isolated atom in the laser focus. Since the target is made of many atoms and the laser pulse is generally inhomogeneous in space and time over the target, results (e.g. a cross section) derived on the basis of the fixed amplitude plane wave assumption Eq.(2) should be averaged in space and time and, if need may be, also over the statistics of the photon field (case of multimode operation). For the single-mode case, see Ref. 8.

Thus, Eq.(2) describes the incoming electron before colliding with the atom. A fully realistic description of the ensuing electron-atom interaction is quite difficult. We shall represent it here by a potential, which will be taken of the self-consistent type: Coulomb-like at the nucleus $(V(r) \simeq -Z/r)$, short-range or ionic $(V(r) \simeq -Z'/r)$ at large distances, but unspecified elsewhere. With this in mind, Eq.(2) can be taken to represent the asymptotic incoming state of the collision only for a short range potential $V(r)$. For a Coulomb-tail potential the asymptotic state Eq.(2) should be modified according to the method used in radiationless scattering.

The scattered particle, on its way out to the detector will be represented by a spherical outgoing wave, of a form corresponding to Eq.(2), and, further out, of a form corresponding to Eq.(1). Its energy E_n and momentum p_n may differ from the original ones E,p due to the absorption $(n > 0)$ or emission $(n < 0)$ of n photons:

$$E_n = E + n\omega, \qquad E_n = p_n^2/2, \qquad n = 0, \pm 1, \pm 2, \ldots \tag{3}$$

The collision in the laser field is described by the time-dependent Schrödinger equation:

$$\{\tfrac{1}{2}[\vec{P} + \frac{1}{c}\vec{A}(t)]^2 + V(r)\}\Psi = i\,\frac{\partial\Psi}{\partial t}\ . \tag{4}$$

According to the discussion above, the appropriate solution should satisfy the boundary condition (case of a short range potential)

$$\Psi_{\vec{p}}(\vec{r},t) \underset{r\to\infty}{\longrightarrow} \chi_{\vec{p}}(\vec{r},t) + $$

$$+ \sum_n \frac{f_n(\hat{r})}{r} \exp i\left\{p_n r - \tfrac{1}{2}\int_0^t [p_n\hat{r} + \frac{1}{c}\vec{A}(t')]^2\, dt'\right\}\ . \tag{5}$$

with $\chi_{\vec{p}}$ defined by Eq.(2).

Rather than use Eq.(4) with the boundary condition Eq.(5), it is more convenient for the following to apply the "space-translation transformation", discussed by Kramers [9], and Henneberger [10]. This is a time dependent translation of the coordinate $\vec{r}$ to $\vec{r} + \vec{\alpha}(t)$, where

$$\vec{\alpha}(t) = \frac{1}{c}\int_0^t \vec{A}(t')dt' = \vec{\alpha}_0 \sin\omega t\ , \tag{6}$$

$$\vec{\alpha}_0 = \alpha_0\vec{e}, \qquad \alpha_0 = a/\omega c = I^{\frac{1}{2}}\ \omega^{-2}\ ,$$

and $\vec{e}$ and I are the real polarization vector and time averaged intensity of the plane wave. All our formulas are written in atomic units; the a.u. of time averaged intensity is $I_0 = 3.51 \times 10^{16}$ W/cm^2.

By also isolating a time-dependent phase factor, we introduce the new wave function $\psi(r,t)$:

$$\psi(\vec{r},t) = \Psi(\vec{r} + \vec{\alpha}(t),t)\ \exp\left(-\frac{i}{2c^2}\int_0^t \vec{A}^2(t')dt'\right)\ . \tag{7}$$

This satisfies the new Schrödinger equation:

$$[\tfrac{1}{2}\vec{P}^2 + V(\vec{r} + \vec{\alpha}(t))]\psi = \frac{\partial\psi}{\partial t}\ . \tag{8}$$

The boundary condition corresponding to Eq.(5) now takes the simpler form:

$$\psi_{\vec{p}}(\vec{r},t) \underset{r\to\infty}{\longrightarrow} \phi_{\vec{p}}(\vec{r},t) + \sum_n \frac{f_n(\hat{r})}{r} \exp i(p_n r - \frac{p_n^2}{2}\ t)\ , \tag{9}$$

with $\phi_{\vec{p}}$ of the form given in Eq.(1).

Equation (8) has periodic time-dependent coefficients. As usual, we seek a quasiperiodic solution of the Floquet form

$$\psi(\vec{r},t) = e^{-iEt}\ \Sigma_{n=-\infty}^{+\infty}\ \psi_n(\vec{r})e^{-in\omega t}\ . \tag{10}$$

Then, we Fourier analyze the potential:

$$V(\vec{r} + \vec{\alpha}(t)) = \Sigma_{n=-\infty}^{+\infty}\ V_n(\vec{\alpha}_0;\vec{r})e^{-in\omega t}\ . \tag{11}$$

By some algebraic manipulations the coefficients can be written as

$$V_n(\vec{\alpha}_0;\vec{r}) = (i^n/\pi)\int_{-1}^{+1} V(\vec{r} + \vec{\alpha}_0 u)T_n(u)(1-u^2)^{-\frac{1}{2}}\ du\ , \tag{12}$$

where $T_n(u)$ are Chebyshev polynomials.

Insertion of Eqs.(10) and (11) into Eq.(8) leads to a system of coupled differential equations for the components $\psi_n(\vec{r})$, which we write

$$[\tfrac{1}{2}P^2 + V_o - (E+n\omega)]\psi_n = - \sum_{\substack{m=-\infty \\ (m \neq n)}}^{+\infty} V_{n-m}\psi_m \ . \tag{13}$$

The boundary condition Eq.(9) requires that our solutions $\psi_n(\alpha_o,\omega;\vec{r})$ behave asymptotically as follows:

$$\psi_o(\vec{\alpha}_o,\omega;\vec{r}) \rightarrow \exp\left\{i[\vec{p}\vec{r} + \gamma_o \ln(pr-\vec{p}\vec{r})]\right\} +$$

$$+ \frac{1}{r} f_o(\alpha_o,\omega;\hat{r}) \exp[i(pr - \gamma_o \ln 2pr)] \ , \tag{14}$$

$$\psi_n(\vec{\alpha}_o,\omega;\vec{r}) \rightarrow \frac{1}{r} f_n(\vec{\alpha}_o,\omega;\hat{r}) \exp[i(p_n r - \gamma_n \ln 2p_n r)]$$

$$(n \neq 0) \ . \tag{15}$$

We have considered here the more general case of an ionic potential V(r) with asymptotic charge Z' (for a short range potential Z' = 0), and have defined $\gamma_n = -Z'/p_n$. Equation (14) contains the elastic scattering amplitude $f_o(\vec{\alpha}_o,\omega;\hat{r})$, and Eq.(15) that for absorption/emission $f_n(\vec{\alpha}_o,\omega;\hat{r})$. The associated scattering cross sections are [2]

$$d\sigma_n/d\Omega = (p_n/p)\,|f_n(\vec{\alpha}_o,\omega;\hat{r})|^2 \qquad (n = 0,\pm1,\pm2,\dots) \ . \tag{16}$$

For a single-mode laser pulse of adiabatically varying intensity, Eq.(16) should be time-averaged appropriately [8].

We shall now describe a method for handling the system Eq.(13) [3]. The left-hand side contains the Hamiltonian

$$H = \tfrac{1}{2}P^2 + V_o(\vec{\alpha}_o;\vec{r}) \ . \tag{17}$$

By use of the Green's operator $G(\Omega)$ associated to it, where Ω is the energy parameter, Eq.(13) may be formally solved as

$$\psi_n = \psi_{\vec{p}}^{(+)}\delta_{no} - G^{(+)}(E_n) \sum_{\substack{m \\ (m \neq n)}} V_{n-m}\psi_m \ . \tag{18}$$

Here $\psi_{\vec{p}}^{(+)}$ is the ($\vec{\alpha}_o$-dependent) solution of the equation

$$H\psi_{\vec{p}} = E\psi_{\vec{p}} \ , \tag{19}$$

satisfying the boundary condition Eq.(14) with an amplitude $f_o^{(o)}(\vec{\alpha}_o;\vec{r})$. It then follows from Eq.(18) that the ψ_n satisfy the boundary condition required by Eqs.(14) and (15), with the following expressions for the scattering amplitudes:

$$f_n(\vec{\alpha}_o,\omega;\hat{r}) = f_o^{(o)}(\vec{\alpha}_o;\hat{r})\,\delta_{no} - (1-\delta_{no})\,\frac{1}{2\pi}\,\langle\psi^{(-)}_{\vec{p}_n}|V_n|\psi^{(+)}_{\vec{p}}\rangle +$$

$$+\frac{1}{2\pi}\sum_{\substack{m\\(m\neq n)}}\;\sum_{\substack{m'\\(m'\neq m)}}\langle\psi^{(-)}_{\vec{p}_n}|V_{n-m}\,G^+(E_m)\,V_{m-m'}|\psi_{m'}\rangle\;. \tag{20}$$

Besides $\psi^{(+)}_{\vec{p}}$, Eq.(20) also contains $\psi^{(-)}_{\vec{p}_n}$, which is an incoming-wave solution of Eq.(19), as well as the unknown set of components $\psi_{m'}(r)$ satisfying Eq.(18).

By repeated insertion of Eq.(18) into Eq.(20) an expansion can be derived for f_n. Obviously, the iteration will have practical significance only if the successive terms decrease sufficiently rapidly. Since this will not be true in general, it is important to establish the conditions under which the first nonvanishing term of Eq.(20) represents a good approximation.

It was possible to extract the *exact* form of the dominant contribution to the last term in Eq.(20) for a potential of the self consistent type described before, under the three simultaneous conditions: (a) $\omega \gg |E_o(\alpha_o)|$, where $E_o(\alpha_o)$ is the ground state energy of the modified Hamiltonian, Eq.(17); (b) $\alpha_o^2\omega \gg 1$; (c) $\omega \gg E$.[3] It was found that essentially the same conditions insure the dominance of the first term of Eq.(20) over the corrective terms. However, this conclusion may hold under wider conditions than we were able to prove so far. The significance of these conditions is the following:

(a) This is a *high-frequency requirement*. The fact that the photon energy ω should be large with respect to the ground state of the *modified* Hamiltonian Eq.(17), rather than the unperturbed one, represents a relatively much weaker condition. This is because, from Eqs.(12),(21) it follows that by *increasing* α_o the dressed potential becomes shallower, and therefore the binding energy $|E_o(\alpha_o)|$ *decreases* from its unperturbed value at $\alpha_o = 0$. Numerical calculations[11,12,13] (see also Sec.IV) indicate that for $\alpha_o = 10$ (a value attained in experiments[5,6]) $E_o(10)/E_o(0) \simeq 0.1$, and thus the condition can be approximately satisfied for an excimer laser. (When the laser of Rhodes and collaborators[5,6] is operated at $I = 10^{15}$ W/cm^2, we have $\alpha_o = 3.1$, and at $I = 10^{16}$ W/cm^2 $\alpha_o = 9.8$).

(b) This is a *high-intensity requirement*. Indeed, it excludes the case of vanishing α_o. Consequently, α_o should be finite or large, and from its definition Eq.(6) it follows that

I should be large, since ω is such. For an excimer laser ($\omega \approx 0.2$ a.u.) this condition can be easily satisfied (e.g. $\alpha_o > 4$) at the present intensities.

(c) This is a *low initial electron energy requirement*. With an excimer laser it can be satisfied with an electron of a few eV energy. It excludes the possibility of free-free emission ($n < 0$) in our case.

From Eq.(20) the *elastic amplitude* f_o reduces, to lowest order (in the sense discussed above), to $f_o^{(o)}$, which is that calculated from the *time-dependent* Schrödinger equation Eq.(19). This shows that in the high-frequency, high-intensity regime the incoming electron feels only the static distorted potential $V_o(\vec{\alpha}_o;\vec{r})$, the "dressed" potential associated to V(r):

$$V_o(\vec{\alpha}_o,\vec{r}) = \frac{1}{\pi} \int_{-1}^{+1} V(\vec{r} + \vec{\alpha}_o u) \frac{du}{\sqrt{1-u^2}} . \tag{21}$$

The dressed potential, Eq.(21), can be looked upon as created by a linear distribution of "charges" extending from $-\alpha_o$ to $+\alpha_o$ along $\vec{e}$, with density

$$\sigma(\xi) = \frac{1}{\pi\alpha_o} [1 - (\frac{\xi}{\alpha_o})^2]^{-\frac{1}{2}} , \tag{22}$$

the unit of "charge" generating the potential V(r). This behavior appears natural due to the rapid oscillations of the center of force in Eq.(8). In the regime we are considering, ω and I enter the scattering problem only through α_o.

For the *absorption amplitude* ($n > o$) we get to lowest order from Eq.(20):

$$f_n(\vec{\alpha}_o,\omega;\hat{r}) = -\frac{1}{2\pi} \langle\psi^{(-)}_{\vec{p}_n} | V_n | \psi^{(+)}_{\vec{p}} \rangle . \tag{23}$$

Thus, the Fourier component V_n acts in our regime as a transition operator between scattering states $\psi^{(\pm)}_{\vec{q}}$ of the dressed potential V_o. The amplitude f_n depends on ω via the final momentum $\vec{p}_n$ (see Eq.(3)). Because ω was assumed to be large, all f_n will be small with respect to f_o. (This contrasts with the low-frequency case where many f_n may be larger than f_o.[2])

III. ELASTIC SCATTERING FROM THE DRESSED COULOMB POTENTIAL

Since the original potential V(r) is spherically symmetric, $V_0(\vec{\alpha}_0;\vec{r})$ has axial symmetry around $\vec{\alpha}_0$ (we have assumed linear polarization). The axial symmetry of V_0 complicates the elastic scattering problem, as the azimuthal quantum number ℓ is no longer conserved. Computations taking this into account have been recently performed [14]. In this section we shall describe a simplified approach [4] in which $V_0(\vec{\alpha}_0,\vec{r})$ is modelled by its spherical average $\bar{V}_0(\alpha_0,r)$. The problem is thus reduced to a phase shift calculation. This gives the right order of magnitude for the elastic scattering, particularly at low energies, when the electron wavelength is larger than the extension of the linear charges, i.e. $p\alpha_0 \lesssim 1$ (p is the electron momentum). On the other hand, interesting features related to the dependence of the cross section on the polarization vector $\vec{e}$ (or $\vec{\alpha}_0$) will thus be lost [14].

We shall consider the case of a Coulomb potential V(r) = -Z/r. By averaging Eq.(21) over all directions of $\vec{\alpha}_0$, one finds:

$$\bar{V}_0(\alpha_0,r) = \frac{-Z}{\pi\alpha_0\rho}\left[2\arcsin\rho - \rho\ln\frac{1-(1-\rho^2)^{\frac{1}{2}}}{1+(1-\rho^2)^{\frac{1}{2}}}\right], \qquad \rho \leqq 1$$

$$= -\frac{Z}{\alpha_0\rho}, \qquad \rho \geqq 1, \tag{24}$$

where $\rho = r/\alpha_0$. For $r \geqq \alpha_0$, $\bar{V}_0$ coincides with the original Coulomb potential. At the origin $\bar{V}_0$ has a *logarithmic singularity*. This is much weaker than that of the Coulomb potential.

As is easily seen, $\bar{V}_0(\alpha_0,r)$ is the potential energy due to a spherically symmetric distribution of electric charges extending up to $r = \alpha_0$, of radial density:

$$\tau(r) = 2Z\sigma(r), \tag{25}$$

with the function σ defined by Eq.(22). Our scattering problem resembles thus that of an electron probing an extended charge nucleus [15] of density $\tau(r)$ and radius α_0; in contrast to this, however, we are dealing here with a nonrelativistic case.

Because of the long range behaviour of the potential, Eq. (24), the scattering amplitude is given by the theory for *modified Coulomb scattering*:

$$f(\theta) = f_c(\theta) + f'(\theta), \tag{26}$$

$$f_c(\theta) = \frac{(-\gamma)}{2p\sin^2\frac{\theta}{2}} \exp(-i\gamma \ln \sin^2\frac{\theta}{2} + 2i\sigma_o) \;, \tag{27}$$

$$f'(\theta) = \frac{1}{2ip} \sum_\ell (2\ell+1)e^{2i\sigma_\ell}(e^{2i\delta_\ell} - 1)P_\ell(\cos\theta) \;. \tag{28}$$

Here θ is the scattering angle, f_c is the Coulomb amplitude, f' is the extra contribution due to $\bar{V}_o - V$, δ_ℓ is the total phase and σ_ℓ is the Coulomb phase. The modified cross section is

$$\frac{d\sigma}{d\Omega} = \frac{d\sigma_c}{d\Omega} + 2\,\mathrm{Re}\, f_c^* f' + |f'(\theta)|^2 \;, \tag{29}$$

where the first term at the right is the Rutherford cross section and the second represents the interference of the Coulomb and short-range amplitudes. Eqs.(26)-(29) are similar to those used in the analysis of the classical proton-proton collision experiments at low (nuclear) energies with the short range nuclear force taken into account; see Ref. 15, chapter 10, §§ 5 and 9 (exchange effects are absent here).

In Fig. 1 we present the angular dependence of the ratio $R = (d\sigma/d\Omega)/(d\sigma_c/d\Omega)$ of the modified cross section, Eq.(29), to the Rutherford cross section, at $Z = 1$ for a number of electron energies E and values of α_o. The energies E chosen satisfy the condition of validity (b) of our theory given in Sec.II, for ω attainable with some existing high-frequency lasers [5,6], and the values considered for α_o have also been achieved experimentally[5,6].

In all cases $R \to 1$ as $\theta \to 0$. This is due to the fact that $f_c(\theta)$ and $d\sigma_c/d\Omega$ become infinite as $\theta \to 0$, whereas $f'(\theta)$ stays finite. Moreover, for $\theta \to 0$, R has infinitely many oscillations of increasing frequency and decreasing amplitude. These are due to the presence of the Coulomb phase factor, see Eq.(27), in the interference term of Eq.(29) (but not in $d\sigma_c/d\Omega$). The amplitude of these *Coulomb-interference oscillations* decreases for $\theta \to 0$ because the relative importance of the interference term in Eq.(29) diminishes compared to $d\sigma_c/d\Omega$. Note that, as seen in Fig. 1, the Coulomb-interference oscillations occur in an accessible range of experimental parameters. (They were not detected in the proton-proton collision experiments, since there the energy had to be high enough, $\gamma \simeq 0$, to overcome the Coulomb repulsion so that the particles could approach within the range of nuclear forces).

For large scattering angles Fig. 1 shows that the modified

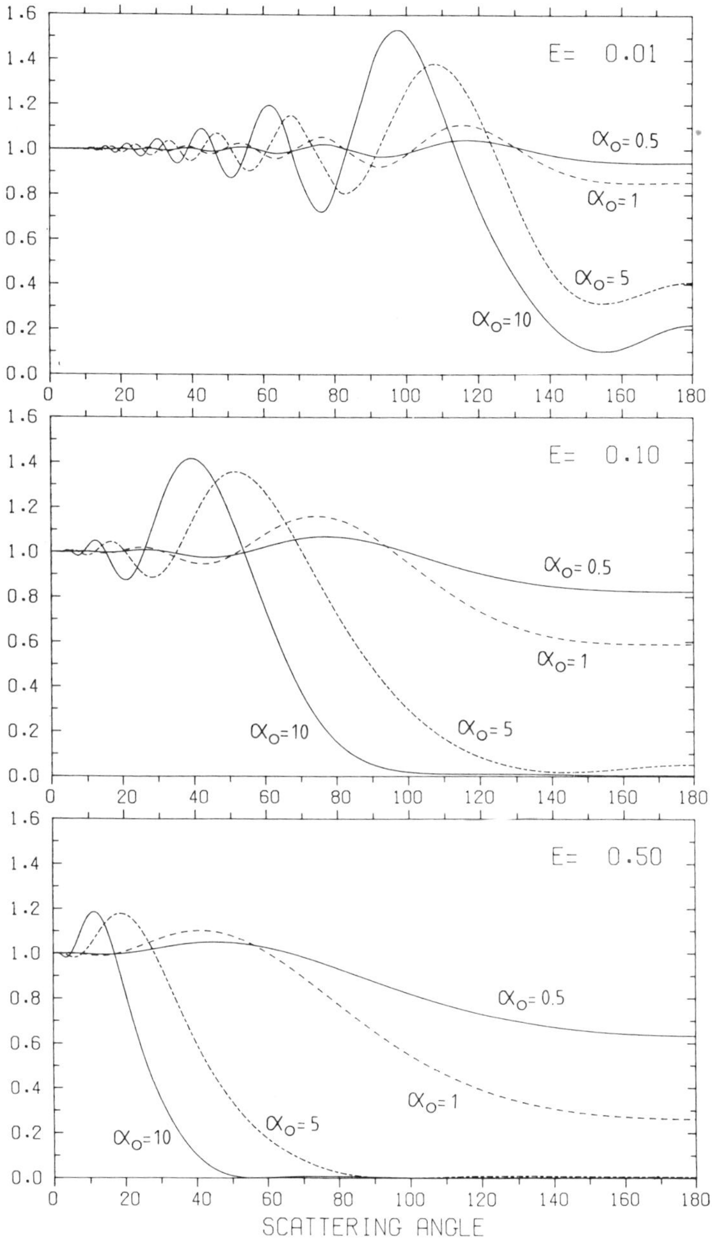

Fig. 1. Ratio R of the elastic scattering cross section for the averaged dressed Coulomb potential, Eq.(24), to the Rutherford cross section. Nuclear charge Z = 1. Values of the electron energy E (in Ry) and of α_0, as indicated (for the definition of α_0 see Eq. (6)).

cross section is considerably reduced with respect to the original value, because of destructive interference in Eq.(29).

IV. FORMALISM FOR ATOMIC STRUCTURE AND MULTIPHOTON IONIZATION

An atom driven by an intense laser field undergoes distortion, while decaying by multiphoton ionization. We are interested in analyzing these phenomena in the high-frequency regime. We shall consider here the simplest case of a one-electron atom, with a binding potential of the general form described in Sec. II. The assumptions concerning the radiation field will be the same as there.

We start again from the space-translated Schrödinger equation Eq.(8), and seek a *quasienergy solution*, Eq.(10), where the quasienergy E remains to be determined (for the background of this type of approach see Refs. 17,18). As before one obtains the system of coupled equations Eq.(13). However, in the present case we want to solve it with asymptotic conditions describing the decay of a bound state:

$$\psi_n(\vec{\alpha}_0,\omega;\vec{r}) \to \frac{1}{r} f_n(\vec{\alpha}_0,\omega;\hat{r})\exp[i(p_n r - \gamma_n \ln 2p_n r)] , \quad n > 0,$$
$$\psi_n(\vec{\alpha}_0,\omega;\vec{r}) \to 0 \text{ (exponentially)} , \quad n \leq 0 , \tag{30}$$

where $f_n(\vec{\alpha}_0,\omega;\hat{r})$ is the n-photon ionization amplitude. For $n > 0$ we have the open channels (we assume that ω is larger than the ionization potential of the ground state), and for $n \leq 0$ the closed ones (note that the incident channel $n = 0$ is closed). The energies E_n and the corresponding momenta p_n are given by Eq.(3) in terms of the real part of E. The existence of the open channels forces the quasienergy E to be complex. The theory of decaying states shows that the quasienergies are poles of the S matrix (resonances)[17].

Using Eq.(30) one finds for the angular decay rate by n-photon ionization the expression:

$$\frac{d\Gamma^{(n)}}{d\Omega} = p_n |f_n(\vec{\alpha}_0,\omega;\hat{r})|^2 , \tag{31}$$

if the solution Eq.(10) is properly normalized.

The system of coupled equations Eq.(13) can be solved similarly to Eq.(18):

$$\psi_n = u_\lambda \delta_{no} - G^{(+)}(E_n) \sum_{\substack{m \\ (m \neq n)}} V_{n-m}\psi_m \, . \tag{32}$$

Here u_λ is an eigenfunction of the (Hermitian) Hamiltonian Eq. (17) with the (real) eigenvalue W :

$$Hu_\lambda = W_\lambda u_\lambda . \tag{33}$$

ψ_n satisfies the boundary conditions Eq.(30). By iteration of Eq.(32), ψ_n can be expressed as a series in terms of u_λ and E. Upon inserting it into the equation for ψ_o of the system Eq. (13), one can write:

$$[\tfrac{1}{2}\vec{P}^2 + V_o - \sum_{\substack{m \\ (m \neq o)}} V_{-m} \, G^{(+)}(W_\lambda + m\omega) V_m + \cdots]\psi_o = E\psi_o \, . \tag{34}$$

This equation determines the complex quasienergy E.

Assuming ω to be sufficiently large, Eq.(34) yields, *to lowest order* in $1/\omega$, Eq.(33). This shows that *in the high-frequency limit* ($\omega \to \infty$) *the atom is stable* no matter the intensity of the radiation, but may be strongly distorted as a consequence of the "dressing" of its potential by the field. Moreover, due to the axial symmetry of the dressed potential V_o, the labeling of the levels changes (ℓ is no longer a good quantum number).

To *next order* in $1/\omega$, Eq.(34) can be solved by perturbation theory. By denoting

$$E = W_\lambda + E_\lambda^{(1)} + \cdots \, , \tag{35}$$

one finds for the complex $E_\lambda^{(1)}$:

$$E_\lambda^{(1)} = \Delta_\lambda - i \, \frac{\Gamma_\lambda}{2} \; ; \tag{36}$$

$$\Delta_\lambda = - \sum_{\substack{m \\ (m \neq o)}} \langle u_\lambda | V_{-m} \, \mathcal{P}(H - W_\lambda - m\omega)^{-1} \, V_m | u_\lambda \rangle \, , \tag{37}$$

$$\Gamma_\lambda = 2\pi \sum_{\substack{m \\ (m \neq o)}} \langle u_\lambda | V_{-m} \, \delta(H - W_\lambda - m\omega) V_m | u_\lambda \rangle \, . \tag{38}$$

The eigenvalue equation Eq.(34) for the high frequency limit, together with Eqs.(36)-(38), were obtained earlier by Gersten and Mittleman [12] by a different method (see also Ref. 19, § 3.6).

From the asymptotic behavior of the *corrective* term to the $\psi_o = u_\lambda$ approximation, Eq.(34), we obtain for the n-photon ionization amplitude

$$f_{\vec{p}_n \lambda} = - \frac{1}{2\pi} \langle \psi^{(-)}_{\vec{p}_n} | V_n | u_\lambda \rangle \quad . \tag{39}$$

This is similar in form to the free-free transition amplitude Eq.(23). Both decrease rapidly with ω.

We are now solving eq.(33) numerically for model potentials (Coulomb, Yukawa), to study the α_o dependence of the levels [13].

V. CONCLUSIONS, OUTLOOK

The theory presented here for the high-intensity, high-frequency regime reveals new features, different from those known at lower frequencies. The atomic structure is strongly distorted (because of the high intensity), while the decay by multiphoton ionization is strongly quenched (because of the high frequency). The atom is thus quasi "frozen" in a distorted state. Due to the axial symmetry of the field (linear polarization) the classification scheme of the levels is modified. In the case of free-free transitions the elastic scattering dominates, while again multiphoton absorption is strongly suppressed. These features appear in ranges of frequency and intensity now opening up to experiment.

We are now extending the formalism to treat the case of the decay of a several electron atom in the radiation field. Again, in the high-frequency limit a modified, time independent Schrödinger equation emerges, describing the distortion of the atom. However, this presents new features as compared to Eq. (33). In the FFT case, we are considering the collision of an electron with a realistic atom. At very high frequencies, we are dealing with the distorted, but stable atom considered before, and the general methods of scattering theory can be applied. At high, but finite frequencies, the target atom is decaying, and this raises interesting conceptual problems.

REFERENCES

1. M. Gavrila, in *Atomic Physics 9*, Eds. N. Fortson and R. van Dyck (World Scientific Publishing Co, 1985) p.523.
2. N.M. Kroll and K.M. Watson, Phys.Rev. A *8*, 804 (1973).
3. M. Gavrila and J.Z. Kaminski, Phys.Rev.Lett. *52*, 613 (1984) and to be published.
4. M.J. Offerhaus, J.Z. Kaminski and M. Gavrila, Phys. Lett. 112A, 151(1985).
5. T.S. Luk, H. Pummer, K. Boyer, M. Shahidi, H. Egger and C.K. Rhodes, Phys.Rev.Lett. *51*, 110 (1983).
6. K. Boyer, H. Egger, T.S. Luk, H. Pummer and C.K. Rhodes, J.Opt.Soc.Am. B *1*, 3 (1984).
7. H.G. Muller and A. Tip, Phys.Rev. A *30*, 3039 (1984), and unpublished.
8. H. Krüger and Ch. Jung, Phys.Rev. A *17*, 1706 (1978).
9. H.A. Kramers, *Collected Scientific Papers* (North-Holland, Amsterdam, 1956), p.866
10. W.C. Henneberger, Phys.Rev.Lett. *21*, 838 (1968).
11. Chan K. Choi, W. Henneberger and F.C. Sanders, Phys.Rev. A *9*, 1895 (1974).
12. J.I. Gersten and M.H. Mittleman, J.Phys. B *9*, 2561 (1976).
13. M. Pont and M. Gavrila, to be published.
14. J. van de Ree, J. Kaminski and M. Gavrila, to be published.
15. L.R. Elton, *Nuclear Sizes* (Oxford University Press, 1961).
16. R. Evans, *The Atomic Nucleus* (McGraw-Hill, 1955).
17. A.I. Baz, Ya.B. Zeldovich and A.M. Perelomov, *Scattering, Reactions and Decay in Non-relativistic Quantum Mechanics* (translated from the Russian by the Israel Program for Scientific Translations, Jerusalem 1969), Chap. 5.
18. Ya.B. Zeldovich, Sov.Phys.Usp. *16*, 427 (1974).
19. M.H. Mittleman, *Introduction to the Theory of Laser-Atom Interactions* (Plenum, 1982).

MULTIPLE IONISATION OF ATOMS IN INTENSE LASER FIELDS

P. Agostini, A. L'Huillier, and G. Petite

Service de Physique des Atomes et des Surfaces
CEN Saclay
91191 Gif sur Yvette Cedex

INTRODUCTION

Interaction of strong laser beams with atoms results in multiphoton excitation and ionisation (MPI), i.e bound-bound and bound-free transitions involving the simultaneous absorption of several photons. Such processes have been quite well described within the framework of time-dependent perturbation theory and single-electron approximation [1]. The validity of the latter is expected to be very good for alkali atoms, but is clearly questionable for alkaline-earth atoms, in which the two outermost electrons can be strongly correlated so that both are expected to participate to the interaction. Of course, it is even more questionable for rare gases, in spite of several MPI experimental results which do not contradict the prediction of the one-electron approximation. Multiple ionisation in MPI, recently discovered for both alkaline-earth atoms [2] and rare gases [3] is an obvious manifestation of the multi-electron nature of these atoms. This collision-free process has now been reported in wide ranges of intensity (10^{11} W.cm^{-2}-10^{17} W.cm^{-2}) and of wavelengths (193 nm-10000 nm). It can require the absorption of hundreds of photons (triple ionisation of Xe at 10000 nm[4]) and, at very high intensity and short wavelength, it can go along with inner-shell excitations (Auger lines have been reported in photoelectron spectra [5]) . Such processses are obviously far from being completely understood and the theoretical approaches are still unpolished . The aim of this paper is to briefly review the most significant experiments and the available theoretical methods in this field. Experiments can be classified into two main groups : those performed at "moderate" intensity (up to 10^{13} W.cm^{-2}) with tunable lasers and electron spectroscopy on alkaline-earths (Sec. I) and those performed at different fixed laser wavelengths and very high intensity chiefly on rare gases (Sec.II). Theories range from statistical models (when large numbers of photons are absorbed) to many-body, multiphoton perturbative approaches (two-photon double-ionisation of Helium), tunneling, and more speculative mechanims (Sec.III).

I. MULTIPLE IONISATION OF ALKALINE-EARTH ATOMS

The first experimental evidence of double ionisation in MPI was reported in 1975 for Barium [2]. It was shortly followed by similar observations on Strontium, Magnesium, Europium and Lead [6]. These experiments were performed with a Nd laser (10^{11} W.cm^{-2}) and a time-of-flight charge spectrometer. Ion yields dependences on the laser intensity, wavelength and polarisation were measured . The double ionisation mechanism emerging from these results was

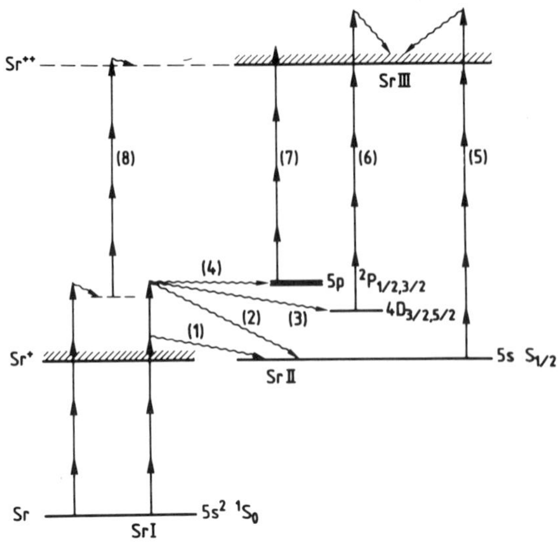

Fig. 1. Schematics of energy levels of Strontium and of some multiphoton processes leading to double ionisation.

the following [6] : the two electrons are driven from the ground state up to the low-lying autoionising states by multiphoton absorption with possible resonances on bound two-electron states . Then, they "diffuse" very rapidly through the dense spectrum of two-electron states up to the double-ionisation limit [7]. The cascade process ($A \rightarrow A^+$, $A^+ \rightarrow A^{++}$) was apparently ruled out by two experimental facts : the ratio A^{++}/A^+ was several orders of magnitude larger than expected from known MPI cross-sections and the intensity dependence was not that predicted by perturbation theory.

Subsequently, experiments performed on Strontium with a pulsed Dye laser led to a different conclusion [8]. Although the power dependences had the same characteristics as those mentioned above (low index power law and high value of the ratio) the authors concluded to a stepwise mechanism for the production of Sr^{++}. This opposite conclusion was drawn from a tentative assignment of an observed resonance to a transition between excited states of Sr^+. Analysis of the photoelectron energy spectrum with a retarding potential analyser seemed to support this conclusion : in a first step an ion is created, possibly in an excited state, the ion being doubly ionised in a second step by a MPI process which can be resonnant.

Recent experiments on Calcium and Strontium with picosecond pulses from a Nd:Yag system or a Nd:Yag-pumped Dye laser strongly support this model. Several channels leading to double ionisation have been identified with help of electron spectroscopy and resonance patterns, as shown on Fig.1. Both doubly excited states and ion excited states can play an important role in these processes. Two-photon resonances on bound two-electron states in three-photon ionisation have been identified in the Sr^+ yield as well as three-photon resonances on autoionising states [8,9] (these states are not shown on Fig.1). They demonstrate that the low-lying two-electron states can enhance the MPI (process labelled (1) on Fig.1) either as intermediate states or as final states in the continuum . Photoelectron spectra have revealed the processes labelled (2),(3) and (4) [9,10]. All are four-photon MPI leaving the ion in its ground state or in excited states. The process (2) in which only one electron is excited (Above-Threshold Ionisation [1]) is much less probable than processes (3) and (4) in which both electrons participate. The latter shows a characteristic threshold behaviour when the energy of four photons becomes smal-

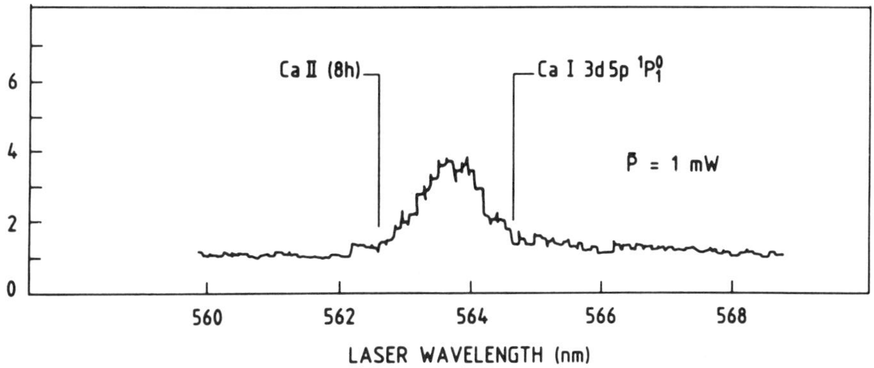

Fig. 2. Ca^{++} ion yield versus laser wavelength : perhaps the first evidence of a "direct resonant" double ionisation.

ler than the energy of the 5P states [10]. Three-photon resonances on intermediate two-electron states have also been observed in four-photon MPI (3) and (4). These processes are characteristic of core excitation and obviously cannot be observed in alkali at such laser wavelengths and intensity. Excited states of Sr^+ lead to the double ionisation through processes (6) and (7) which can be resonant [9]. Of course the same is observed in the process of ionisation of the ground state (5). All processes have been identified on electron spectra and by assigning the ionic resonances. The resonance shown on the Ca^{++} yield (Fig.2) has been assigned to the three-photon excitation of the 3d5p ^1F state (also a low-lying autoionisig state) [12]. This could be the first evidence of a "direct resonant" double ionisation although ionic resonances cannot be totally ruled out. In summary, these experiments stress the role of low-lying two-electron states and of ionic states in double ionisation. The "non-resonant" process (8), which is in principle possible, has not been observed yet. Furthermore, it is very doubtful that high-lying two-electron states (so-called Wannier states [13]) play a significant role if no special care is taken to excite such states. This can be inferred from recent experiments in Barium[14] in which Ba^{++} was produced through resonant stepwise excitation of doubly excited Rydberg states : the corresponding resonances are much weaker than the neighbouring ionic resonances.

In conclusion, in the competition between upward transitions driving the two electrons to the double escape, and autoionisation followed by ion ionisation, the latter usually prevails. This is due to the specific structure of alkaline-earths spectrum and multiple ionisation of rare gases, analysed in the next section, can be expected to proceed differently.

II. MULTIPLE IONISATION OF RARE GASES

Besides the alkaline earths, multiphoton multiple ionisation has been observed in a large variety of atoms : alkali [15] and Actinides [16] are among them. But the type of atoms for which the most extensive ensemble of data is available is that of rare gases . Because of their low ionisation rate and some particular features of their spectrum , they are well fitted to MPI studies under very high laser intensity, and they are also easy to handle experimentally . They were studied for wavelengths ranging from 193 nm to 9000 nm and intensities between 10^{11} and 10^{17} W.cm^{-2}. Fig.3 shows a typical result obtained with Xenon at 532 nm [17]. It displays the dependences of the different ion yields on the laser intensity. Up to Xe^{5+} is detected for an intensity of 10^{13} W.cm^{-2}. On such a log-log plot, the straight lines superposed to the Xe^+ and Xe^{++} experimental data represent a I^N power law dependence, where N is the number of photons necessary to ionise the specie (atom

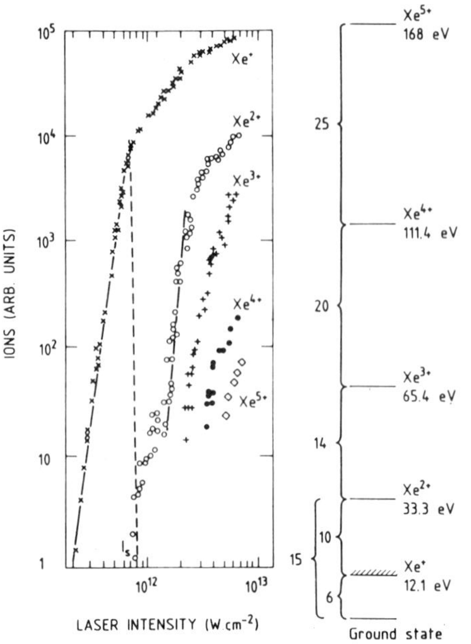

Fig. 3. Xe^{n+} ions yields versus laser intensity at 532 nm[17].

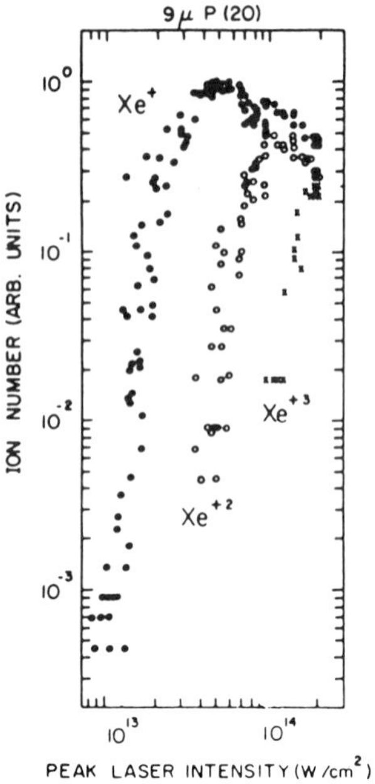

Fig. 4. Xe^{n+} ions yields versus laser intensity at 9000 nm[4].

or ion). In this case, (upper part of the Xe^{++} curve), Xe^{++} was considered to originate from MPI of singly charged ions already widely present in the interaction volume (stepwise ionisation). At lower intensities, Xe^{++} ions are also observed. They are believed to be created directly from the ground state neutral atom (direct ionisation). The saturation of this process at the same intensity than the single ionisation (occuring in this case before the onset of sequential ionisation) indicates that both processes pump on the same population. The number of doubly charged ions created at the saturation intensity, through direct ionisation, can be increased by decreasing the pulse duration. This fact together with the reasonable fit between the Nth-power law dependences and the experimental data was considered as a proof that multiphoton multiple ionisation processes could be thought of in terms of perturbation theory (i.e subsequent photon absorptions) even if it is not sure in which case high order corrections to the lowest non vanishing order have to be considered. The same authors have observed up to the 6th ionisation of Xe at 1064 nm which corresponds to a total absorption of 218 photons. Triple ionisation of Xe observed at 9000 nm (Fig. 4) amounts to a total of 475 photons absorbed ! Simple ionisation requires also more than a hundred photons and the shape of the ion yield curves does not support the multiphoton interpretation. The authors' interpretation is that ionisation is due to optical tunneling. The decrease, at high intensity, of the simple (and even double) ionisation signals is a proof that multiple ionisation is essentially a stepwise process.

For near I.R. (1064 nm) and U.V. light (193 nm), the different rare gases have been studied and some general remarks can be made. For both wavelengths, the strength of the coupling between the atom and the radiation increases with the atomic number. Infrared light seems to be more efficient for light atoms (He,Ne,Ar) and less for heavy atoms (Kr,Xe) than U.V. light. Up to a complete shell (8 electrons) can be removed from the Xenon atom in the case of U.V. light, but it seems so far impossible to go beyond this point. This shell effect is not observed in the case of other atoms since double ionisation of alkali and triple ionisation of alkaline earths were observed.

There is also a marked effect of the atomic structure: Iodine, the closest atom to Xenon concerning the atomic number and the ionisation potentials, seems very hard to multiply ionise (fifth ionisation only is observed, in very small quantity [18]).

Not only ion yield measurements but also electron spectroscopy has been used in these experiments. Electron spectra have been reported in the case of Xe at 193 nm and for intensities between 5 10^{13} and 1.5 10^{15}. At low intensities, these spectra essentially show single two-photon ionisation of Xe with a small contribution of high energy electron (around 5 eV), which seems to include both Above Threshold Ionisation [19] of Xe and sequential double ionisation. This component grows with increasing intensity, while the low energy one is considerably broadened. At the maximum intensity a new component appears, for electron energies between 10 and 20 eV, which, when analysed in details shows several lines which can be attributed to the Auger decay of 4d vacancies of Xe, indicating that some new excitation mechanisms have to be considered in these experiments. It should be remembered, concerning this point, that the 4d shell has an ionisation threshold of 67 eV, followed by a broad maximum at 100 eV, an energy well in the range of the ones considered in these experiments.

III. THEORY

In this last section, we want to point out the theoretical difficulties of this topic, and survey the few attempts which have been made, essentially to describe the rare gases experiments

The theoretical description of the interaction of a many-electron atom with an intense laser field brings up two difficulties . As long as a few photons are absorbed by a one-electron atom (alkali), multiphoton processes are well described by perturbation theory [20], either to lowest order or to an arbitrary high order, by the use of summation techniques[21]. The validity of these methods for very strong fields, of the order of the intra-atomic field, and when many quanta are involved, can be questioned. Optical tunneling has been proposed as an alternate mechanism overtaking multiphoton ionisation at high intensities and low frequencies[22], but this point is still very controversial and no doubtless experimental signature of such an effect has ever been observed. More recently some non-pertubative treatments of MPI have been proposed [23] for one electron atoms.

The description of an atom with several elctrons on the outermost shell (rare gases, alkaline earths,...) submitted to an electromagnetic field is by far not as easy as that of a one electron atom. Independent particle approximation (like average configuration Hartree-Fock) are sometimes unable to explain some experimental data. A famous example is that of the photoabsorption cross-section of the 5p,4d shells of Ba and Xe [24], which exhibit a strong collective behaviour called "giant dipole resonances". However, many electron effects in weak fields,i.e. in the case of single photon ionisation are now well understood and correctly described by theories going beyond independent-electron approximations, an example of such theories being the Random Phase Approximation [24,25]. Some work has been done to describe, in a similar way, non linear optical properties [26] and 2-3 photon ionisation [27] of rare gases but the understanding of these problems is still quite limited. However, many electron effects could play an important role in MPI of heavy atoms and this problem presently receives much attention [28].

Taking advantage of the complexity of the process itself (large number of quanta, even larger number of interfering quantum paths) some simple statistical models have been recently proposed. We shall briefly describe their main ideas and conclusions, refering the reader to the original papers for more details . Aberg & al. [29] present a statistical analysis of the rare gases experimental data[17], based on the maximum-entropy formalism. The conclusion of this analysis is that the multiply charged ions distribution has a very statistical behaviour: it follows a binomial law and is only governed by the Pauli principle and the energy available in the interaction (and thus the laser intensity).

Statistical models also provide a very simple way of calculating the multiple ionisation probabilities[30]. At a given laser intensity (I) an atom absorbs a mean number of photons (p) estimated by:

$$p = I \pi r^2 \tau$$

r is the mean radius of an outer electron orbit and τ the time during which a coherent absorption can take place, i.e. $\hbar/E$ (E: ionisation energy). These photons are statistically shared among the outer shell electrons. If one electron absorbs enough photons, it can be removed from the atom. The ionisation probabilities are then simply the ratio of the number of distributions leading to ionisation to the total number of possible distributions. This model yields reasonable agreement with experimental data for the saturation intensities and multiply charged ion distributions at high intensities.

Lastly, it was shown [31] that an acceptable interpretation of some experimental results on the relative production of charged ionic states may be obtained on the basis of Hartree's independent-electron model of the atom. A surprising consequence of this model is that the multiple ionisation potentials are multiples of the first ionisation energy.

Statistical models are based on the assumption that the interaction takes place in the outermost shell only and they do not take into account the energy structure of this shell. If they yield reasonable results, this must be attributed to the use of very high laser fields. It is well known that resonance effects, which dramatically enhance the ionisation probability in weak fields, are damped and even wiped out in strong fields (and at short interaction times) due to enormous Stark shifts and broadenings. Details of the atomic spectra are not important any more, which result in a statistical behaviour of complex atoms in intense laser fields.

For completeness, one should mention the following model based on very simple physical concepts [32] : an atom submitted to a VERY strong laser field (i.e. much larger than e/a_0^2, the intra-atomic field) is modelled as a two-parts system : (A) an external shell of "free" electrons which coherently oscillate in the external field ; (B) the remaining core (nucleus and deep shells) which is not directly coupled to the external field. Nevertheless, some of the energy gained by the outer electrons can be communicated to the core by various processes like inverse Auger transitions. These processes are modelled as inelastic collisions at high energies between the two parts of the atom ((A) and (B)).

IV. CONCLUSION

Multiphoton multiple ionisation is a process which dominates the interaction of a complex atom with a very intense laser field. Its understanding is still far from being complete and theory is developing along two lines : i) perturbative calculations including many-body effects (for small numbers of absorbed photons) ; ii) statistical descriptions when large numbers of photons are involved so that quantum calculations are untractable. From the experimental point of view, multiphoton multiple ionisation establishes a new approach to the fundamental physics of complex atoms : from two-electron states to shell collective oscillation and inner-shell excitation. On the other hand, emission of high energy photons by excited deep levels offers interesting perspectives for X-ray laser schemes.

REFERENCES

1. J. Morellec, D. Normand and G. Petite, Non Resonant Multiphoton Ionization of atoms, in : "Adv. At. Mol. Phys.", B. Bederson, ed., Academic Press, New York (1982).
2. V. V. Suran and I. P. Zapesochnyi, Sov. Tech. Phys. Lett., 1: 420 (1975)
3. A. L'Huillier, L. A. Lompre, G. Mainfray and C. Manus, Phys. Rev. Lett., 48:1814 (1982)
4. S. L. Chin, F. Yergeau and P. Lavigne, J. Phys. B. At. Mol. Phys., 18:L213 (1985)
5. C. K. Rhodes, Studies of Multiquantum Processes in Atoms, in : "Fundamentals of Laser Interactions, Proceedings, Obergurgl, Austria" F. Ehlotzky, ed., Springer Verlag, Berlin (1985)
6. N. B. Delone, V. V. Suran and B. A. Zon, Many-Electron Processes in Non Linear Ionization of Atoms, in : "Multiphoton Ionization of Atoms", S. L. Chin and P. Lambropoulos, eds., Academic Press, Toronto (1984)
7. N. B. Delone, B. A. Zon, V. P. Krainov and M. A. Preobrazhensky, JETP Lett., 30:260 (1979)
8. D. Feldmann, H. J. Krautwald and K. H. Welge, in : "Multiphoton Ionization of Atoms" , op. cit.
9. P. Agostini and G. Petite, J. Phys. B. At. Mol. Phys., 18:L281 (1985)
10. P. Agostini and G. Petite, To Be Published
11. P. Agostini and G. Petite, J. Phys. B. At. Mol. Phys., 17:L811 (1984)
12. C. L. Cromer and C. W. Clark, J. Phys. B. AT. Mol. Phys., Letter to the Editor,
13. A. R. P. Rau, This Volume
14. P. Camus, P. Pillet and J. Boulmer J. Phys. B At. Mol. Phys. (1985)
15. S. L. Chin, K. X. He and F. Yergeau, J. Opt. Soc. Am. B., 1:505 (1984)
16. T. S. Luk, H. Pummer, K. Boyer, M. Shadidi, M. Egger and C. K. Rhodes, Phys. Rev. Lett., 51:110 (1983)
17. A. L'Huillier, L. A. Lompre, G. Mainfray and C. Manus, Phys. Rev. A., 27:2503 (1983);J. Phys. B At. Mol. Phys 16:1363 (1983)
18. C. K. Rhodes, Studies of collision-free nonlinear processes in the Ultraviolet range, in: "Multiphoton processes", P.Lambropoulos and S. J. Smith, Springer-Verlag, Berlin (1984)
19. P. Agostini and G. Petite, Multiphoton Transitions in the Ionization Continuum of Atoms, in :"Multiphoton Processes", op. cit.
20. Y. Gontier and M. Trahin, Theory of Multiphoton Ionization of Atoms, in: "Multiphoton Ionization of Atoms", op. cit.
21. Y. Gontier, N. K. Rahman and M. Trahin, Phys. Rev. A, 14:2109 (1976)
22. E. Arnous, J. Bastian and A. Maquet, Phys. Rev. A, 27:2946 (1983)
23. L. V. Keldysh, Sov. Phys. JETP, 20:1307 (1965)
24. G. Wendin, Application of Many-Body problems to Atomic Physics in: "New Trends in Atomic Physics", G. Grinberg and R. Sora ed., Elsevier Science Publishers (1984)
25. M. Ya. Amusia, Comm. At. Mol. Phys., 10:155 (1981)
26. A. Zangwill, J. Chem. Phys., 78:5926 (1983)
27. M. S. Pindzola and H. P. Kelly, Phys. Rev. A 1:1543 (1975)
28. M. Crance, Multiphoton ionization of complex atoms ; A. F. Starace and P. Zoller, Transition matrix methods for multiphoton ionization processes in: "Fundamental of Laser Interactions, Proceedings, Obergurgl, Austria", op. cit.
29. J. Aberg, A. Bloemberg, J. Tulki and O. Goscinski, Phys. Rev. Lett. 52: 1207 (1984)
30. M. Crance, J. Phys. B At. Mol. Phys., 17:3503 (1984)
31. S. Geltman, Phys. Rev. Lett. 54:1909 (1985)
32. K. Boyer and C. K. Rhodes, Phys. Rev. Lett. 54:1490 (1985)

PART III

Quantum Electrodynamics and Relativity

in elementary systems or highly stripped very heavy ions

QED AND RELATIVITY IN ATOMIC PHYSICS

Joseph Sucher

Department of Physics and Astronomy
University of Maryland
College Park, MD 20742

I. INTRODUCTION; REVIEW OF NONRELATIVISTIC QED

My task in these lectures, set by the organizers, is to review basic concepts of quantum electrodynamics (QED) and to describe some applications of QED to problems of current interest in atomic physics in which relativistic effects are important. Since the subject had its beginnings almost sixty years ago, with Dirac's quantization of the radiation field and his discovery of the relativistic wave equation for the electron, five hours is too small by at least an order of magnitude to do it even partial justice. In the review part, I hope to avoid displeasing some of the audience all of the time or all of the audience some of the time - not to mention all of the audience all of the time - by assuming only an average memory for things learned long ago by some of you but not much used, by taking shortcuts wherever possible and by emphasizing aspects which are either not well known or not discussed clearly in textbooks.

By way of a rough outline, in this lecture I will first review nonrelativistic QED, in a nonstandard way designed to lead to a definition of a "genuine relativistic effect" in atomic physics and to set the stage for the second lecture, which will introduce the fundamentals of a relativistic description of electrons. The third lecture will deal with full-fledged QED and with a general QED-based framework for handling relativistic effects in atomic bound states. The last two lectures will be devoted to a survey of applications of this formalism to a variety of physical problems, including the theory of the α^3 Ryd corrections to the low-lying levels of He or He-like ions, the theory of the fine structure of Rydberg levels, which is of interest for the quantum theory of long-range forces, and the theory of forbidden or relativistic M1 transitions, such as $2^3S_1 \to 1^1S_0 + \gamma$, for systems of this kind.

A. Nonrelativistic Quantum Theory of Many-Electron Atoms

1. Preliminaries

The extreme nonrelativistic theory of the discrete energy levels of an N-electron or ion (or molecule) is a beautiful but, from a physical point of view, a more or less closed chapter in quantum theory. One looks for normalizable solutions $\varphi(\vec{r}_1, \ldots, \vec{r}_N)$ of the eigenvalue equation

$$H_{nr}\varphi = W\varphi \tag{1.1a}$$

where

$$H_{nr} \equiv \sum_{i=1}^{N} [\vec{p}_i^{\,2}/2m + V_{ext}(r_i)] + \sum_{i<j} V^C_{e-e}(r_{ij}), \tag{1.1b}$$

with $V_{ext}(r_i) = -Z\alpha/r_i$ and $V^C_{e-e}(r_{ij}) = \alpha/r_{ij}$ the potentials describing the electron-nucleus and electron-electron Coulomb interactions, respectively $[\hbar = c = 1, \alpha = e^2/4\pi\hbar c \simeq 1/137]$. Of course the mathematical difficulties of finding accurate solutions to (1.1) when $N \geqslant 2$ and $Z \sim N$ are truly formidable and have been keeping theorists busy for over half a century. For the sake of discussion let us imagine that the eigenfunction and eigenvalues of (1.1) are known exactly.

Three vital elements are missing so far. The first of these is electron spin. The second is the antisymmetry principle, the generalized form of the Pauli exclusion principle. In the present case these two combine to yield the following picture for the physically relevant wave functions associated with an eigenvalue W_n: The Hilbert space in which H_{nr} acts is enlarged to a space of functions of space and spin variables, denoted by $\mathcal{H}_{nr}$. The degenerate spatial wave functions $\varphi_{n;\gamma}$ associated with W_n are multiplied by Pauli-spin wave functions χ_δ to form linear combinations $\phi = \sum C_{\gamma\delta}\,\varphi_{n;\gamma}\chi_\delta$ which are eigenstates of $\vec{J}^2 = (\vec{L} + \vec{S})^2$ and $J_z = L_z + S_z$ and which are totally antisymmetric under interchange of any pair of electrons. This will normally lead to a set of $2J + 1$ Schroedinger-Pauli (SP) wave functions $\{\phi_n^{J,m_J}\}$, with J uniquely determined by n. (In practice, things are more complicated because one constructs approximate ϕ's as products of spatial orbitals, thereby creating spurious degeneracies, but we are after the big picture here.) Henceforth the symbol ϕ or ϕ_n will refer to such an SP wavefunction; the J,m_J label will be understood.

What is the third ingredient? So far the physical instability of the excited states of the atom and in fact any interaction of the atom with the environment is totally unaccounted for. It is not enough that Bohr say "Jump!" for the atom to jump and to emit a photon. It was Dirac who first showed how to describe photon emission and absorption within a coherent quantum-theoretic framework. Once again one enlarges the Hilbert space of states, to include not only a (fixed) number of electrons but also one or more photons. At the same time one introduces a specific interaction between electrons and photons, by using the principle of minimal electromagnetic interaction (MEI). This interaction conserves the number of electrons but not the number of photons and can therefore mediate radiative decay, photon absorption and scattering, etc.

2. Quantization of the electromagnetic field[1]

Formally one proceeds as follows: One first considers the free transverse radiation field $\vec{A}_T(\vec{x},t)$ in a box of volume V, satisfying $(\partial_t^2-\vec{\nabla}^2)\,\vec{A}_T = 0$ and $\vec{\nabla}\cdot\vec{A}_T = 0$, with associated electric and magnetic fields $\vec{E}_T(\vec{x},t) = -\partial\vec{A}_T/\partial t$ and $\vec{H}_T(\vec{x},t) = \vec{\nabla}\times\vec{A}_T$. The field $\vec{A}_T$ is reinterpreted as an operator (strictly speaking, as an operator-valued distribution) acting on a Hilbert space, $\mathcal{H}_{rad}$, spanned by states containing one, two, ... photons and a state containing <u>no</u> photons, the "photon vacuum". Mathematically this is done by expanding $\vec{A}_T$ in terms of a complete set of classical solutions of the above equations for A_T, i.e. by writing

$$\vec{A}_T(\vec{x},t) = \frac{1}{\sqrt{V}}\sum_{\vec{k}}\sum_{\lambda=1}^{2}\left[A_\lambda(\vec{k})(2\omega)^{-1/2}\,\vec{\varepsilon}_\lambda\, e^{i(\vec{k}\cdot\vec{r}-\omega t)} + \text{h.c}\right] \tag{1.2}$$

with $\vec{k}\cdot\vec{\varepsilon}_\lambda = 0$ $(\lambda=1,2)$, $\omega = |\vec{k}|$, and imposing on the (operator) coefficients $A_\lambda(\vec{k})$ and $A_\lambda^\dagger(\vec{\lambda})$ in (1.2) the commutation relations (CR)

$$[A_\lambda(\vec{k}),A^\dagger_{\lambda'}(\vec{k}')] = \delta_{\lambda\lambda'}\,\delta_{\vec{k}\vec{k}'}\ . \tag{1.3}$$

There is a standard procedure called canonical quantization for getting to these CR but a lot of time is saved by just imposing them. The operators $A^\dagger$ and A are interpreted as photon creation and destruction operators, for reasons which the following discussion will make clear. The energy operator for the field is defined, in close analogy with the classical case, by

$$H_{rad} = \frac{1}{2}\int d\vec{x}\ :\vec{E}_T^2 + \vec{H}_T^2: \tag{1.4}$$

Here the colons mean "normal order" - i.e. move $A^\dagger$'s to the left of A's, after substituting the expansions for $\vec{E}_T$ and $\vec{H}_T$ obtained from (1.2) into (1.4). This is just a fancy way of removing an additive (but infinite) constant from H_{rad}, associated with zero-point energy. The result is

$$H_{rad} = \sum_{\vec{k}} \sum_{\lambda} N_\lambda(\vec{k})\, \omega \tag{1.4'}$$

where $N_\lambda(\vec{k}) = A_\lambda^\dagger(\vec{k})\, A_\lambda(\vec{k})$. The manifest positivity of $N_\lambda(\vec{k})$ together with the CR

$$[N_\lambda(\vec{k}), A_{\lambda'}(\vec{k}')] = -\,\delta_{\lambda\lambda'}\, \delta_{\vec{k}\vec{k}'}\, A_\lambda(\vec{k}) \tag{1.5}$$

allows one to verify that the eigenvalues of $N_\lambda\,(\vec{k})$ are just the non-negative integers; thus $N_\lambda\,(\vec{k})$ is interpreted as a <u>number operator</u> for photons of momentum $\vec{k}$ and polarization $\vec{\varepsilon}_\lambda$ and

$$N_\gamma^{op} = \sum_{\vec{k},\lambda} N_\lambda(\vec{k}) \tag{1.6}$$

as the (total) photon-number operator. This interpretation is supported by calculation of the analogue of the Poynting vector

$$\vec{P}_{rad} = \int d\vec{x}\, \vec{E}_T \times \vec{H}_T = \sum_{\vec{k},\lambda} N_\lambda(\vec{k})\, \vec{k} \ ,$$

which is then manifestly the operator associated with the total momentum carried by the photons. One can now also verify that

$$\vec{A}_T(\vec{x},t) = e^{iH_{rad}t}\, \vec{A}_T(\vec{x},o)\, e^{-iH_{rad}t}$$

$$\vec{A}_T(\vec{x},t) = e^{-i\vec{P}_{rad}\cdot\vec{x}}\, \vec{A}_T(o,t)\, e^{i\vec{P}_{rad}\cdot\vec{x}}$$

as befits the interpretation of H_{rad} and $\vec{P}_{rad}$ as time and space displacement operators.

Finally, it follows from the CR (1.5) that for each $(\vec{k},\lambda)$ there is at least one state, denote it by $|0\rangle_{\vec{k},\lambda}$, such that

$$N_\lambda(\vec{k})|0\rangle_{\vec{k},\vec{\lambda}} = 0 \ .$$

Since the $N_\lambda(\vec{k})$ commute with each other one expects that there will be at least one state in $\mathcal{H}_{rad}$ which will be a simultaneous eigenvector with eigenvalues zero for all of the $N_\lambda(\vec{k})$. Let us assume that up to a phase, there is <u>only</u> one such (normalizable) state; call it the "photon vacuum" and denote it by $|vac\rangle_\gamma$. From the relations

$$N_\lambda(\vec{k})|vac\rangle_\gamma = 0$$

it follows, on the one hand, that

$$A_\lambda(\vec{k})|vac\rangle_\gamma = 0,$$

i.e. $|vac\rangle_\gamma$ is annihilated by each of the A's and, on the other, that states of the form $\Psi = |\vec{k}_1,\lambda_1; \vec{k}_2,k_2; \ldots; \vec{k}_N,\lambda_N\rangle$, containing photons with quantum members $\vec{k}_1,\vec{\lambda}_1 \ldots \vec{k}_N,\lambda_N$, can be constructed from $|vac\rangle_\gamma$ via

$$\Psi = A^\dagger_{\lambda_1}(\vec{k}_1) \ldots A^\dagger_{\lambda_N}(\vec{k}_N)|vac\rangle_\gamma ,$$

having eigenvalues N, $\Sigma\omega_i$ and $\Sigma\vec{k}_i$ for N^{op}_γ , H_{rad} and $\vec{P}_{rad}$, respectively.

Before continuing it is convenient to go from the Heisenberg picture, in which we have been working, to the Schroedinger picture, in which operators are time independent, by dealing just with $\vec{A}_T(\vec{x}) \equiv \vec{A}_T(\vec{x},0)$. The time development of the state of the photon system is then governed by the equation

$$i \frac{\partial}{\partial t} \Psi_\gamma = H_{rad} \Psi_\gamma \tag{1.7}$$

where Ψ_γ is a vector belonging to $\mathcal{H}_{rad}$. Finally it is convenient to pass to the $V = \infty$ limit in which case one writes, instead of (1.2),

$$\vec{A}_T(\vec{x}) = \frac{1}{(2\pi)^{3/2}} \int \frac{d\vec{k}}{\sqrt{2\omega}} \sum_{\lambda=1}^{2} (A_\lambda(\vec{k})\, e^{i\vec{k}\cdot\vec{x}} + \text{h.c.}) \tag{1.8}$$

and, instead of (1.3)

$$[\vec{A}_\lambda(\vec{k}),\, A^\dagger_{\lambda'}(\vec{k}\ )] = \delta_{\lambda\lambda'}\, \delta(\vec{k} - \vec{k}') . \tag{1.9}$$

3. Interaction

We now consider the system of N (atomic) electrons interacting with the radiation field. The Hilbert space is the tensor product $\mathcal{H}_{nr} \times \mathcal{H}_{rad}$, the space of linear combinations of product states of $\mathcal{H}_{nr}$ and $\mathcal{H}_{rad}$, and the Hamiltonian of the combined but noninteracting systems is the sum

$$H^{(o)}_{qed} = H_{nr} + H_{rad} \tag{1.10}$$

Interaction is introduced by using the MEI principle mentioned above, which ensures that a gauge transformation of $\vec{A}_T$, $\vec{A}_T \rightarrow \vec{A}_T + \vec{\nabla}\Lambda(\vec{x})$ can be compensated by a phase transformation of the atomic wave function: $\phi \to e^{-ie\Sigma\Lambda(\vec{x}_i)}\phi$. The prescription which does the job is to make the replacement

$$\vec{p}_i \to \vec{\pi}_i \equiv \vec{p}_i + e\,\vec{A}_T(\vec{x}_i) \tag{1.11}$$

in $H_{nr} = H_{nr}(\vec{x}_i, \vec{p}_i)$. However, a too-naive application of (1.11) yields a theory which is also too naive because it misses the interaction of the electron magnetic moment with the quantized magnetic field $\vec{H}_T$, and one would get wrong answers even for ordinary magnetic dipole transitions. This can be remedied by an old trick which allows one, in this context, to maintain the logical coherence of the presentation - but also shows that the application of the MEI principle is not without ambiguity. One writes $\vec{p}_i^{\,2}$, which is now really $\vec{p}_i^{\,2}\,\underline{1}$ where $\underline{1}$ is the 2x2 unit matrix, in the form $\vec{p}_i^{\,2} = (\vec{\sigma}_i \cdot \vec{p}_i)^2$ in (1.10) and <u>then</u> makes the replacement (1.11). When this is done and the relation

$$\vec{\sigma} \cdot \vec{A}\,\vec{\sigma} \cdot \vec{B} = \vec{A} \cdot \vec{B} + i\vec{\sigma} \cdot \vec{A} \times \vec{B}$$

is used one gets

$$H_{qed} = H^{(o)}_{qed} + H' \tag{1.12}$$

where

$$H' = \sum_i \frac{e}{2m}(\vec{p}_i \cdot \vec{A}_T(\vec{x}_i) + \vec{A}_T(\vec{x}_i) \cdot \vec{p}_i) + \sum_i -\vec{\mu}_i \cdot \vec{H}_T(\vec{x}_i)$$
$$+ \sum_i \frac{e^2}{2m} : \vec{A}_T^{\,2}(\vec{x}_i) : \tag{1.13}$$

with $\vec{\mu}_i = -(e/2m)\,\vec{\sigma}_i$ the electron spin-magnetic moment operator; a (divergent) constant which arises from normal ordering of the $\vec{A}_T^{\,2}$ term has been dropped. Thus electron spin now achieves a <u>dynamical</u> significance. Alternatively one could simply add the $\vec{\mu} \cdot \vec{H}_T$ terms by hand, since they are gauge independent; a merit of the approach sketched is that the value of μ is correctly "predicted", without any appeal to the Dirac equation.

The introduction of H' allows the basic processes of emission or absorption of a virtual transverse photon by an electron to be described mathematically, because H' has nonvanishing matrix elements between eigenstates of $H^{(o)}_{qed}$ containing different numbers of photons. This leads among other things to (a) the decays of excited atomic states and (b) to a modification of the electron-electron interaction via the exchange of transverse photons.

(a) Decays

As an example, the matrix element of H' between an atomic state $|\phi_n\rangle$, with no photons present and an atomic state

$$|\phi_{n'};\vec{k},\vec{\varepsilon}\rangle \equiv \vec{A}^{\dagger}_{\lambda}(\vec{k})\ |vac\rangle_{\gamma} \times \ |\phi_{n'}\rangle$$

with one photon present is given by $M/(2\pi)^{3/2}(2\omega)^{1/2}$ with

$$M = \langle\phi_{n'}|\sum_{i}\vec{\Gamma}_i^{\,op}(\vec{k})\cdot\vec{\varepsilon}^{\,*}|\phi_n\rangle \tag{1.14}$$

where

$$\vec{\Gamma}_i^{\,op}(\vec{k}) = \frac{1}{2}\{e\frac{\vec{p}_i^{\,op}}{m} + i\,\vec{\mu}_i \times \vec{k}\ ,\ e^{-i\vec{k}\cdot\vec{r}_i}\}$$

is a vector radiation operator associated with emission of a photon of momentum $\vec{k}$ (or absorption of a photon of momentum $-\vec{k}$) by electron "i". The excited states of the atom can now decay, the rate for the process $|\phi_n\rangle \to |\phi_{n'}\rangle + \gamma$ being given in this approximation by

$$R = \int \frac{d\vec{k}}{(2\pi)^3 2\omega}\sum_{\lambda}\delta(W_{n'} + \omega - W_n)\ |M|^2,$$

which is a standard starting point for rate calculations, discussion of selection rules, etc.

Let us note, en passant, that the matrix element of $\vec{\Gamma}_i^{\,op}\cdot\vec{\varepsilon}^{\,*}$ between electron Pauli-plane waves $e^{i\vec{p}\cdot\vec{r}}\chi$ and $e^{i\vec{p}'\cdot\vec{r}}\chi'$ is, apart from a factor $(2\pi)^3\,\delta(\vec{p}' + \vec{k} - \vec{p})$, just

$$\langle\chi'|\Gamma(\vec{p}',\vec{p})\cdot\vec{\varepsilon}^{\,*}|\chi\rangle$$

where

$$\vec{\Gamma}(\vec{p}',\vec{p}) = \frac{e}{2m}[\vec{p}' + \vec{p} + i(\vec{p}' - \vec{p}) \times \vec{\sigma}] \tag{1.15}$$

and the constraint $\vec{\varepsilon}\cdot\vec{k} = 0$ has been used. The vector $\vec{\Gamma}$ is just the nonrelativistic limit of the plane-wave matrix element of the Dirac matrix vector $e\vec{\alpha}$. In this language the matrix element M defined by (1.14) has the form, in $\vec{p}$-space,

$$M = \int d\vec{p}_1\cdots d\vec{p}_N[\tilde{\phi}^{\dagger}_{n'}(\vec{p}_1',\vec{p}_2,\cdots\vec{p}_n)\vec{\varepsilon}^{\,*}\cdot\vec{\Gamma}(\vec{p}_1',\vec{p}_1)\tilde{\phi}_n(\vec{p}_1,\vec{p}_2\cdots\vec{p}_N) + \cdots] \tag{1.16}$$

where $\vec{p}_1' = \vec{p}_1 - \vec{k}$, $\tilde{\phi}_n$ is the Fourier transform of ϕ_n and the dots represent contributions from electrons 2,3, ··· N. The matrix element M may

Fig. 1 Nonrelativistic time-ordered diagrams representing contributions to the amplitude $|n\rangle \to |n'\rangle + \gamma$ for a two-electron atom.

be represented graphically, say for N=2, as shown in Fig. 1. Of course the two contributions are equal because of the antisymmetry under exchange of electron coordinates. The basic ingredients in these pictures are the photon-emission vertex of Fig. 1, which represents in an obvious way the factor $\vec{\Gamma}(\vec{p}',\vec{p}) \cdot \vec{\varepsilon}^*$ and the oval bubbles at the bottom and top, which represent the initial atomic wave function ϕ_n and the conjugate of the final wave function $\phi_{n'}$, respectively.

b) Transverse-photon exchange

The introduction of the interaction H' not only permits decay but also gives rise to level shifts, which in turn affect the frequency of the observed photons. In fact it affects the description of the atomic system itself, because each of the atomic states, including the ground state, is now accompanied by a cloud of virtual photons. As an example, let us study the level shift ΔW_n induced by H'. To lowest order in H' this is given formally by second order perturbation theory. Since H' has terms both of order e and order e^2, ΔW_n contains terms of order e^2 and order e^4. Let us restrict our attention to the term of order e^2, coming from the first line in (1.13). On inserting a complete set of eigenstate of $H^{(o)}_{qed}$ of the type $|\phi_m;\vec{k},\vec{\varepsilon}\rangle$ (the only type which can contribute) and using (1.14) we get (again take N=2)

$$\Delta W_n^{(2)} = \int \frac{d\vec{k}}{(2\pi)^3\, 2\omega} \sum_{\lambda=1}^{2} I_n(\vec{k},\lambda) \tag{1.17a}$$

with

$$I_n(\vec{k},\lambda) =$$

$$\sum_m \frac{\langle\phi_n|(\vec{\Gamma}_1^{op} + \vec{\Gamma}_2^{op})^{\dagger} \cdot \vec{\varepsilon}_\lambda|\phi_m\rangle\langle\phi_m|(\vec{\Gamma}_1^{op} + \vec{\Gamma}_2^{op}) \cdot \vec{\varepsilon}_\lambda^*|\phi_n\rangle}{W_n - W_m - \omega} . \tag{1.17b}$$

Now comes a crucial and tricky point, which is usually not discussed

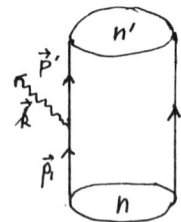

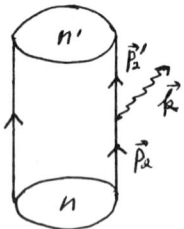

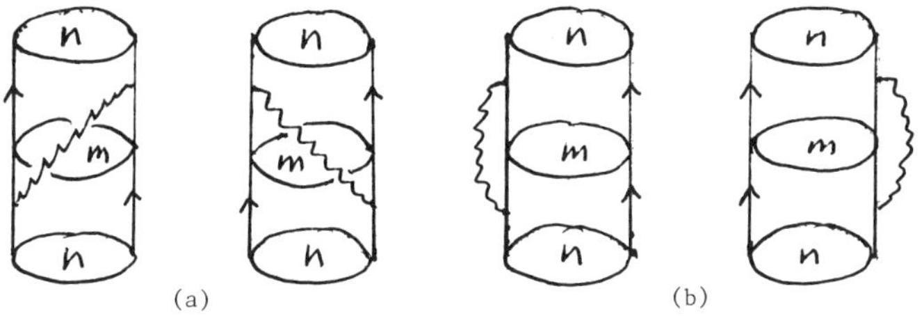

Fig. 2 Nonrelativistic time-ordered diagrams for the second-order contribution of H_T, corresponding to (a) transverse photon exchange and (b) self-energy corrections. A summation over a complete set of eigenstates $|m\rangle$ of $H_{nr}(1,2)$ is understood.

in textbooks. It may appear obvious (but it really isn't!) that $\Delta W_n^{(2)}$ contains two kinds of contributions, one corresponding to interaction between the electrons and another corresponding to electron self-energy. These can be separated by summing in (1.17b) not only over the eigenstates ϕ_m of H_{nr} which satisfy the antisymmetry principle (AP) but also - for the case N=2 - over the symmetric ones. Note that this extension of the summation does not change the value of I_n. This is because the sum $\vec{\Gamma}_1 + \vec{\Gamma}_2$ is symmetric so that $(\vec{\Gamma}_1 + \vec{\Gamma}_2)|\phi_n\rangle$ is still antisymmetric and $\langle\phi|(\vec{\Gamma}_1 + \vec{\Gamma}_2)|\phi_n\rangle$ is zero if $|\phi\rangle$ is symmetric. Thus we can write

$$I_n = (I_n)_{int} + (I_n)_{s.e.}$$

and, correspondingly,

$$\Delta W_n^{(2)} = (\Delta W_n^{(2)})_{int} + (\Delta W_n^{(2)})_{s.e.} \qquad (1.18)$$

where

$$(I_n)_{int} = \sum_{\text{all } \phi_m} \frac{\langle\phi_n|\vec{\Gamma}_1^{\,op\dagger}\cdot\vec{\varepsilon}_\lambda|\phi_m\rangle\langle\phi_m|\vec{\Gamma}_2^{\,op}\cdot\vec{\varepsilon}_\lambda^{\,*}|\phi_n\rangle}{W_o - W_m - \omega} + (1 \leftrightarrow 2) \qquad (1.19a)$$

$$(I_n)_{s.e.} = \sum_{\text{all } \phi_m} \frac{\langle\phi_n|\vec{\Gamma}_1^{\,op\dagger}\cdot\vec{\varepsilon}_\lambda|\phi_m\rangle\langle\phi_m|\vec{\Gamma}_1^{\,op}\cdot\vec{\varepsilon}^{\,*}|\phi_n\rangle}{W_o - W_m - \omega} + (1 \leftrightarrow 2). \qquad (1.19b)$$

In terms of diagrams the corresponding contributions can be described as shown in Figs. 2a and 2b, for the interaction part and the self-energy part of $\Delta W_n^{(2)}$, respectively.

We see here the rules for time-ordered Feynman-like diagrams for level shifts emerging: i) factors $\vec{\Gamma}\cdot\vec{\varepsilon}$ or $(\vec{\Gamma}\cdot\vec{\varepsilon})^\dagger$ for transverse-photon absorption or emission; ii) a factor $(E_o - E_{int})^{-1}$ for an intermediate state of total energy E_{int} ($E_{int} = W_m + \omega$ for our case) and (iii) summation over intermediate states, unrestricted by the AP. For each virtual photon there is an integration with a factor $d\vec{k}/(2\pi)^3\, 2\omega$ and a sum on λ.

The evaluation of $(\Delta W_n^{(2)})_{s.e.}$ requires the addition of mass counter terms to H'. It would take us too far afield to discuss this here. The main point is that a clean separation between interaction and self-interaction effects can be made if (and only if!) the antisymmetry principle is ignored in intermediate states. Note that this does not change the total answer (if a cutoff is introduced to make everything finite) so it is not an <u>ad hoc</u> description. Note also that the AP is not to be ignored for the initial or final state.

The separation of self-energy from interaction effects can be carried out in a similar way for more than two electrons, by using the orthogonality of subspaces carrying different irreps of the symmetric group S_n.

4. Effective potential from transverse-photon exchange

The contribution $\Delta W_{int}^{(2)}$ can be written as the expectation value of a potential, provided we make an approximation in the denominator of (1.19a): <u>Neglect $W_o - W_n$ relative to ω.</u> I will refer to this as "neglect of recoil". This approximation is justified on the following grounds, for low-lying levels ϕ_n of the atom. The mean distance of an electron from the nucleus is of order $Z^{-1/3}a_1$ with $a_1 = (\alpha m)^{-1}$ the Bohr radius. Because of the factors $e^{i\vec{k}\cdot\vec{r}_i}$, the important values of $|\vec{k}| = \omega$ in the integration over $\vec{k}$ satisfy $k \lesssim \alpha m/Z^{-1/3}$. But the mean excitation energy $(W_o - W_n)_{av}$ is of order $\alpha^2 m$, so that its ratio to ω_{av} is $\alpha Z^{1/3}$, which is much less than unity even if Z is relatively large. If recoil is neglected, completeness can be used for the sum on <u>m</u> and

$$(\Delta W_n^{(2)})_{int} \rightarrow \langle\phi_n|H'_{fs}|\phi_n\rangle \tag{1.20}$$

where

$$H'_{fs} = -\int \frac{d\vec{k}}{(2\pi)^3\, 2\omega^2}\, [(\vec{\Gamma}_1^{op\dagger}\cdot\vec{\Gamma}_2^{op} - \vec{\Gamma}_1^{op\dagger}\cdot\hat{k}\,\vec{\Gamma}_2^{op}\cdot\hat{k} + (1\leftrightarrow 2)] \quad . \tag{1.21}$$

The integration over $\vec{k}$ can be carried out with the result that

$$H'_{fs}(1,2) = H_{o-o}(1,2) + H_{s-o-o}(1,2) + H_{s-s}(1,2) \tag{1.22}$$

where the H(i,j) operators are the familiar orbit-orbit, spin-other-orbit and spin-spin interactions, respectively.

These operators are usually described as coming from the reduction to Schroedinger-Pauli (SP) of the Breit operator involving the Dirac matrices $\vec{\alpha}_i$ which I discuss later, but we see here that they can be obtained from the nonrelativistic theory without knowing anything about relativistic quantum field theory or even the Dirac equation. Moreover, the spin-spin contact term -- which was missed by Breit when he carried out the reduction -- is impossible to overlook in this approach. I should add that all these terms were guessed before the invention of QED on the basis of semiclassical argument, except for the contact term which in fact can also be obtained from such arguments.[2]

$\langle H'_{fs}\rangle$ gives rise to a level shift of order α^2 Ryd and so is part of the fine structure. What about the rest? This is <u>not</u> contained in $H^{(o)}_{qed}$ and must be grafted on, by adding to H'_{fs} two kinds of terms: First there are the one-body operators

$$H''(i) \equiv H_{s-o}(i) + \frac{"-\vec{p}_i{}^4"}{8m^3} + H^{cont}_{s.i.}(i) \ , \tag{1.23}$$

corresponding respectively to the spin-(self)-orbit interaction, the relativistic correction to the kinetic-energy operator, obtained by the expansion of $(m^2 + \vec{p}_i^{\,2})^{1/2}$, and a spin-independent contact term given by

$$H^{cont}_{s.i.}(i) = \frac{\pi\alpha}{2m^2} \cdot Z\delta(\vec{r}_i) \ . \tag{1.24}$$

(1.24) is obtainable from the reduction of the <u>one-electron</u> Dirac equation (as are the other two terms -- but for these the Dirac equation is not needed!) Given the operators $H_{s-o}(i)$ and $H^{cont.}_{s.i.}(i)$, both of which arise from the presence of a charge Ze at the origin, it is clear on physical grounds that there must be counterparts of this in which the source is an electron j at $\vec{r}_j$, obtained from (1.24) by replacing Ze by $-e$ and $\vec{r}_i$ by $\vec{r}_{ij}$:

$$H_{s-o}(i,j) = \frac{-\alpha}{4m^2 r_{ij}^3} (\vec{\sigma}_i \cdot \vec{r}_{ij} \times \vec{p}_j + i \leftrightarrow j) \tag{1.25}$$

and

$$H^{cont.}_{s-i}(i,j) = \frac{-\pi\alpha}{2m^2} \cdot 2\delta(\vec{r}_{ij}) \ . \tag{1.26}$$

The total for an N-electron atom is therefore

$$H''_{fs} = \sum_i H''(i) + \sum_{i<j} \left(H_{s-o}(i,j) + H^{cont.}_{s.i.}(i,j)\right) \tag{1.27}$$

and the total fine-structure level shifts to order α^2 Ryd is given by

$$\Delta W = \langle \phi_n | H_{fs} | \phi_n \rangle \tag{1.28a}$$

with

$$H_{fs} = H'_{fs} + H''_{fs} \ . \tag{1.28b}$$

In a way, the most interesting part of H_{fs} is $H^{cont.}_{s.i.}$, because it appears to have no semiclassical origin. Thus in principle one could have detected the failure of this "hybrid theory" by making measurements of level splittings to an accuracy of only α^2Ryd.

C. Limitations

How good is the theory for decays, say one-photon radiative transitions? As we will see (1.16) yields the correct values in leading order for E1 transition and for ordinary M1 transitions but not for the so-called hindered or relativistic M1 transitions. Although some further patch work can be done to fix this, the theory essentially ends at this stage. With regard to energy levels, note that

$$H_{nr} + H_{fs} \tag{1.29}$$

is not a good starting point from which corrections of order higher than α^2 Ryd could in principle be calculated, because the operator H_{fs} is too singular to be used in higher-order perturbation theory. Although the main part of the Lamb shift, of order $\alpha(Z\alpha)^4 \log Z\alpha \ m$, can be obtained from the hybrid theory, there are already "genuine relativistic effects," of order α^3 Ryd and higher which this theory knows nothing about, because they involve the two major qualitative aspects missing so far: The first is the relativistic description of an electron, purely from a kinematic point of view. The second is the existence of the positron and in this connection the contributions of intermediate states involving virtual electron-positron pairs. We now turn to these topics.

II. RELATIVISTIC THEORY: A NEW LOOK AT THE OLD DIRAC EQUATION

Apart from kinematical aspects, the feature which most distinguishes the relativistic theory from the nonrelativistic theory is the existence of positrons. In the latter theory one can say that, in analogy with a rose, "an electron is an electron is an electron", at least if the interaction with $\vec{A}_T$, which leads to a photon cloud accompanying the physical electron, is turned off. This is true even if external fields are present. In contrast, in the relativistic theory it is more accurate to say that "an electron is an electron, except when it is an electron plus an electron-positron pair" (or an electron plus two pairs, etc.)

By way of an easy introduction to $e^- - e^+$ pairs I will consider Dirac's one-electron equation from a non-standard point of view which makes the existence of pairs explicit and motivates the introduction of time-ordered diagrams involving such pairs. This will pave the way for study of the full QED in the next lecture.

A. "Free" Electrons

The natural relativistic analogue of the nonrelativistic Schroedinger equation for a free particle, $i\partial\phi/\partial t = (\vec{p}_{op}^{\,2}/2m)\phi$, is

$$i \frac{\partial}{\partial t} \psi = E_{op}\psi \tag{2.1}$$

where

$$E_{op} = E(\vec{p}_{op}) = \left(\vec{p}_{op}^{\,2} + m^2\right)^{1/2} . \tag{2.2}$$

Although this equation is perfectly covariant, with ψ transforming as a scalar, Dirac showed that an equation for a particle of spin-1/2 arises naturally from it if one reinterprets ψ as a four-component column vector ("Dirac spinor") and the symbol $\vec{p}_{op}^{\,2} + m^2$ as $(\vec{p}_{op}^{\,2} + m^2)\, \underline{1}_4$ where $\underline{1}_4$ is the four by four unit matrix.

One can then find a square root of this matrix, call it h_D, which involves the components p_x, p_y and p_z of $\vec{p}$ linearly, like $\partial/\partial t$ in (2.1), and restore a symmetry between space and time which is absent in (2.1). h_D has the form

$$h_D = \vec{\alpha}\cdot\vec{p}_{op} + \beta m \tag{2.3}$$

where $\alpha_x, \alpha_y, \alpha_z$ and β are any hermitian four-by-four matrices which square to $\underline{1}_4$ and anticommute with each other. The <u>standard choice</u> for these is

$$\vec{\alpha} = \begin{pmatrix} 0 & \vec{\sigma} \\ \vec{\sigma} & 0 \end{pmatrix} , \quad \beta = \begin{pmatrix} \underline{1}_2 & 0 \\ 0 & -\underline{1}_2 \end{pmatrix} \tag{2.4}$$

with $\vec{\sigma}$ the Pauli-matrix vector, but physical quantities turn out to be independent of any such choice. The Dirac equation for a free electron is then given by

$$i \frac{\partial}{\partial t} \psi = h_D \psi \quad . \tag{2.5}$$

For each three-momentum $\vec{p}$ there are two linearly independent (ℓ.i.) plane wave solutions, of the form

$$u_r(\vec{p}) e^{-iE(\vec{p})t} e^{i\vec{p}\cdot\vec{x}} \qquad (r = 1,2) \ , \tag{2.6a}$$

where

$$h_D(\vec{p}) u_r(\vec{p}) = E(\vec{p}) u_r(\vec{p}) \ , \tag{2.6b}$$

corresponding to the two spin degrees of freedom of a spin-1/2 particle with momentum $\vec{p}$ and positive energy $E(\vec{p})$. However, there are also two ℓ.i. negative-energy solutions

$$v_r(\vec{p}) e^{iE(\vec{p})t} e^{i\vec{p}\cdot\vec{x}} \qquad (r = 1,2) \tag{2.7a}$$

with

$$h_D(\vec{p}) v_r(\vec{p}) = -E(\vec{p}) v_r(\vec{p}) \ . \tag{2.7b}$$

The existence of these eventually makes trouble for the one-electron theory point of view. However, for a free electron there is no real problem.

A general wave packet solution of (2.1) will consist of a positive-energy part ψ_+ and a negative-energy part ψ_- and these will propagate <u>independently</u>. To make this explicit, we use the operator form of the Casimir projection operators:

$$\Lambda_\pm^{op} = \frac{E_{op} \pm h_D}{2E_{op}} \ . \tag{2.8}$$

Acting on a plane wave of momentum $\vec{p}$ or, equivalently, in $\vec{p}$ space, these are just the $\vec{p}$-dependent 4×4 matrices introduced by Casimir:

$$\Lambda_\pm(\vec{p}) = \frac{E(\vec{p}) \pm (\vec{\alpha}\cdot\vec{p}+\beta m)}{2E(\vec{p})} \ . \tag{2.9}$$

Since

$$[h_D, \Lambda_\pm^{op}] = 0$$

the Dirac equation (2.5) separates into two uncoupled equations for the parts ψ_+ and ψ_- , formally defined by

$$\psi_\pm = \Lambda_\pm^{op}\psi \ , \tag{2.10}$$

viz.,

$$i\frac{\partial}{\partial t}\psi_+ = E_{op}\psi_+ \ , \quad \Lambda_+^{op}\psi_+ = \psi_+ \tag{2.11a}$$

and

$$i\frac{\partial}{\partial t}\psi_- = -E_{op}\psi_- \ , \quad \Lambda_-^{op}\psi_- = \psi_- \ , \tag{2.11b}$$

with $\psi = \psi_+ + \psi_-$. Thus if at $t = 0$ we impose the boundary condition $\psi_- = 0$, then $\psi = \psi_+$ for all $t > 0$ and the negative-energy states play no role. The modified equation (2.11a) is by itself Lorentz invariant.

Although ψ_+ still has four components, only two of them are independent. To make this explicit, define projection-matrices $\beta^{(\pm)}$ and, for any ψ, associated spinors $\psi^{(\pm)}$ via

$$\beta^{(\pm)} = \frac{1 \pm \beta}{2} \ , \qquad \psi^{(\pm)} = \beta^{(\pm)}\psi \ . \tag{2.12}$$

Note that if we use the standard rep for β then $\psi^{(+)}$ has only non-zero "upper" components, ψ_1, ψ_2 and $\psi^{(-)}$ has only "lower" components ψ_3, ψ_4. If ψ is a ψ_+, , i.e. if

$$\Lambda_+^{op}\psi = \psi , \tag{2.13}$$

then $(E_{op}-\beta m)\psi = \vec{\alpha}\cdot\vec{p}_{op}\psi$ and multiplication of this equation on the left by $\beta^{(+)}$ yields,after some rearrangement,

$$\psi^{(-)} = R_{op}\psi^{(+)} , \qquad R_{op} \equiv \frac{\vec{\alpha}\cdot\vec{p}_{op}}{E_{op}+m} \ . \tag{2.14}$$

Thus if ψ satisfies (2.13) it is completely determined by $\psi^{(+)}$:

$$\psi = \psi^{(-)} + \psi^{(+)} = (1 + R_{op})\psi^{(+)} \ . \tag{2.15}$$

From the relations $\beta\psi^{(+)} = \psi^{(+)}$ and $\{\beta , R_{op}\} = 0$ it follows that

$$\langle\psi^{(+)}|R_{op}|\psi^{(+)}\rangle = 0 \ , \tag{2.16}$$

and hence the norm of ψ is given by

$$\langle\psi|\psi\rangle = \langle\psi^{(+)}|1 + R_{op}^\dagger R_{op}|\psi^{(+)}\rangle = \langle\psi^{(+)}|A_{op}^{-2}|\psi^{(+)}\rangle \ . \tag{2.17}$$

where

$$A_{op} = \left(\frac{E_{op}+m}{2E_{op}}\right)^{1/2} \ . \tag{2.18}$$

This may be rewritten in the form

$$\langle\psi|\psi\rangle = \langle\phi|\phi\rangle \tag{2.19}$$

where ϕ is a new wave function defined by

$$\phi = A_{op}^{-1}\,\psi^{(+)} , \tag{2.20}$$

which also satisfies

$$\beta^{(+)}\phi = \phi \tag{2.21}$$

but has the same norm as ψ. Note that in $\vec{p}$-space ϕ differs from $\psi^{(+)}$ only by the slowly varying factor $A(\vec{p})$ which is unity for $\vec{p} = 0$ and $1/\sqrt{2} \approx 0.7$ for $\vec{p} = \infty$.

The relation between ψ and ϕ can be expressed in terms of an operator U,

$$\psi = U\phi \tag{2.22}$$

with

$$U \equiv (1 + R_{op})A_{op}\beta^{(+)} . \tag{2.23}$$

The extra factor of $\beta^{(+)}$ in (2.23) ensures that U satisfies

$$U^{\dagger}U = \beta^{(+)} , \tag{2.24a}$$

which makes manifest the norm-preserving nature of the mapping (2.22), if (2.21) is satisfied. Calculation shows that in contrast to (2.24a) one has

$$UU^{\dagger} = \Lambda_{+}^{op} . \tag{2.24b}$$

In $\vec{p}$-space this coincides with (2.24a) only for $\vec{p} = 0$. We will come back to the operator U later.

B. External-field Dirac Equation

For an electron moving in an external electromagnetic field described by the four-potential $A^{\mu}(x)$ one replaces $i\partial/\partial t$ by $i\partial/\partial t + eA^{0}$ and $\vec{p}_{op}$ by $\vec{\pi}_{op} = \vec{p}_{op} + e\vec{A}$ in (2.5) to get

$$i\frac{\partial}{\partial t}\psi = h_{D;ext}\psi \tag{2.25}$$

with

$$h_{D;ext} = h_{D} + V_{ext} \tag{2.26}$$

where

$$V_{ext} = -eA^o + e\vec{\alpha}\cdot\vec{A} \ . \tag{2.27}$$

Let us restrict our attention to static fields and stationary states $\psi(x) = \psi(\vec{x})e^{-iEt}$. Then (2.25) reduces to

$$E\psi = h_{D;ext}\psi \ . \tag{2.28}$$

As you all know, for $V_{ext} = -Z\alpha/|\vec{x}|$, $h_{D;ext}$ has normalizable eigenfunctions and associated eigenvalues which describe the energy levels of hydrogen or H-like ions with great accuracy.

The nonrelativistic limit of (2.28) is usually discussed via a "reduction to large components". In the present language this is equivalent to the following procedure. Multiply (2.28) by $\beta^{(+)}$ and $\beta^{(-)}$ in turn to get coupled equations for $\psi^{(\pm)}$ and solve for $\psi^{(-)}$ in terms of $\psi^{(+)}$. This yields

$$E\psi^{(+)} = h^{red}_{D;ext}\psi^{(+)} \tag{2.29}$$

where

$$h^{red}_{D;ext} = \vec{\alpha}\cdot\vec{\pi}_{op} \frac{1}{E+m+eA^0} \vec{\alpha}\cdot\vec{\pi}_{op} + -eA^0 + m \ . \tag{2.30}$$

For $E \approx m$ and $|eA^o| << m$, (2.30) then reduces to $\vec{\pi}^2/2m + V_{ext} + m$. By writing $E = m + W + \Delta W$ and expanding in powers of $\Delta W/m$ one can, for $\vec{A}=0$, obtain the usual operators whose expectation value yields the α^2Ryd correction to the hydrogen fine structure, provided one makes suitable use of the nonrelativistic Schroedinger equation to eliminate W. The result for ΔW is, of course, in agreement with the expansion in powers of α of the exact E for a Coulomb potential. Note that no choice of a representation of the Dirac matrices was made in arriving at (2.30).

1. Positive-energy form of the external field equation and time- ordered diagrams.

To learn more let us proceed somewhat differently. For simplicity take $\vec{A} = 0$. Multiply (2.28) by Λ_+ and Λ_- in turn to get

$$E\psi_+ = E_{op}\psi_+ + \Lambda_+V_{ext}(\psi_+ + \psi_-), \tag{2.31+}$$

$$E\psi_- = -E_{op}\psi_- + \Lambda_-V_{ext}(\psi_+ + \psi_-) \ , \tag{2.31-}$$

where, for ease of writing, I have dropped the superscript "op" on $\Lambda_\pm$. On solving (2.31+) for ψ_- one gets

$$\psi_- = \frac{1}{E+E_{op}-\Lambda_-V_{ext}\Lambda_-} \Lambda_-V_{ext}\psi_+ \ , \tag{2.32}$$

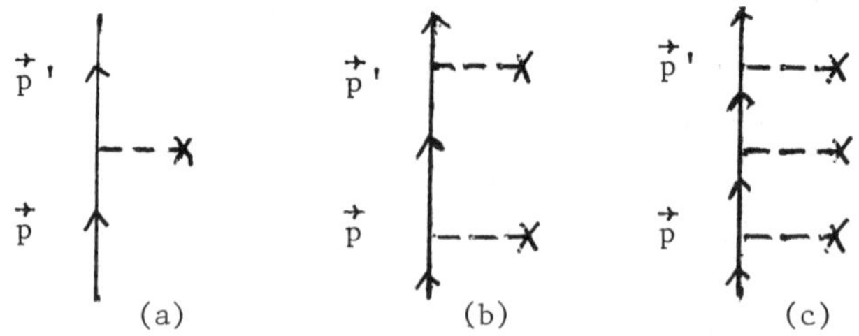

Fig. 3 Relativistic time-ordered diagrams corresponding to an electron scattering in an external field, through third order. Intermediate states are positive energy plane waves; spin labels are suppressed

and substitution into (2.31+) gives

$$E\psi_+ = (h_+ + \delta h_+)\psi_+ \tag{2.33}$$

where

$$h_+ = E_{op} + \Lambda_+ V_{ext} \Lambda_+ \tag{2.34}$$

and

$$\delta h_+ = \Lambda_+ V_{ext} \frac{\Lambda_-}{E + E_{op} - \Lambda_- V_{ext} \Lambda_-} V_{ext} \ . \tag{2.35}$$

Although this Eq. (2.33) looks (and is) more complicated than (2.28) it has a simpler physical interpretation. To analyze this, let us imagine that instead of bound states we look at the scattering of an electron by V_{ext}, from the viewpoint of perturbation theory (PT). In first order we get a contribution to the transition amplitude given by

$$T_1 = \langle \vec{p}',\sigma' | V_{ext} | \vec{p},\vec{\sigma}\rangle \equiv$$

$$= \int d\vec{x} (u_\sigma(\vec{p}')e^{i\vec{p}'\cdot\vec{x}})^\dagger \, V_{ext} (u_\sigma(\vec{p})e^{i\vec{p}\cdot\vec{x}}) \tag{2.36}$$

which is represented graphically in Fig. 3a.

In second order we get a contribution

$$T_2 = \langle \vec{p}',\sigma' | V_{ext} \frac{1}{E - h_D + i\varepsilon} V_{ext} | \vec{p},\sigma\rangle \ . \tag{2.37}$$

Both positive- and negative-energy eigenfunctions of h_D contribute in the sum over intermediate states implicit in (2.37). By writing $1 = \Lambda_+ + \Lambda_-$ we can separate T_2 into a "no-pair part",

$$T_2^{np} \equiv \langle\vec{p}',\sigma'|V_{ext} \frac{\Lambda_+}{E-E_{op}+i\varepsilon} V_{ext}|\vec{p},\sigma\rangle$$

and a "pair part",

$$T_2^{pair} \equiv \langle\vec{p}',\sigma'|V_{ext} \frac{\Lambda_-}{E+E_{op}+i\varepsilon} V_{ext}|\vec{p},\sigma\rangle . \tag{2.38b}$$

The notation anticipates the interpretation of the terms involving virtual transitions to negative-energy states as processes in which a virtual electron-positron pair is involved. The graph associated with T_2^{np} is given by Fig. 3b. If we go to third and higher orders and for the moment keep only the no-pair parts, as above, we get (see Fig. 3c),

$$T_3^{np} = \langle\vec{p}',\sigma'|V_{ext} \frac{\Lambda_+}{E-E_{op}+i\varepsilon} V_{ext} \frac{\Lambda_+}{E-E_{op}+i\varepsilon} V_{ext}|\vec{p},\sigma\rangle \tag{2.39}$$

and similar expressions for the higher-order terms. The sum

$$T^{np} = T_1 + T_2^{np} + T_3^{np} + \dots \tag{2.40a}$$

therefore represents all contributions in which the electron "remains an electron". Because of its geometric character the series (2.40a) may be formally summed:

$$T^{np} = \langle\vec{p}',\sigma'|V_{ext}|\chi_{\vec{p}\sigma}\rangle \tag{2.40b}$$

where

$$\chi_{\vec{p}\sigma} = (1 - \frac{\Lambda_+}{E-E_{op}+i\varepsilon} V_{ext}\Lambda_+)^{-1} |\vec{p},\sigma\rangle . \tag{2.41}$$

Since $(E-E_{op})|\vec{p},\sigma\rangle = 0$ we get

$$E\chi_{\vec{p}\sigma} = h_+\chi_{\vec{p}\sigma} , \tag{2.42}$$

where h_+ is defined by (2.39) and $\chi_{\vec{p}\sigma}$ satisfies outgoing wave boundary conditions. It follows that h_+ takes into account precisely all virtual processes which involve only positive-energy states.

2. Virtual pairs and pair diagrams.

What about δh_+ ? Consider first the lowest-order approximation $\delta h_+^{(2)}$, obtained by neglecting V_{ext} in the denominator:

$$\delta h_+^{(2)} = \Lambda_+ V_{ext} \frac{\Lambda_-}{E+E_{op}+i\varepsilon} V_{ext}\Lambda_+ \tag{2.43}$$

By comparison with (2.38b) we see that $\delta h_+^{(2)}$ generates, in lowest order,

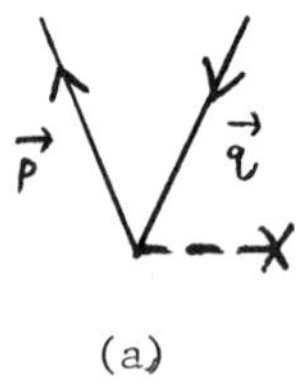

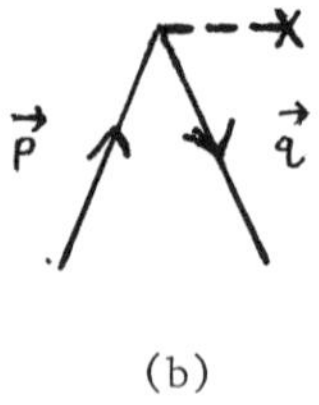

Fig. 4 Diagrams representing (a) pair creation and (b) pair annihilation by an external field.

the leading pair contribution:

$$\langle\vec{p}',\vec{\sigma}'|\delta h_{+}^{(2)}|\vec{p}\sigma\rangle = T_2^{\text{pair}} \quad . \tag{2.44}$$

Suppose we introduce the pictures shown in Figs. 4a and 4b to denote a matrix element of V_{ext} between plane waves $|\vec{q},s;-\rangle = v_s(\vec{q})e^{i\vec{q}\cdot\vec{x}}$ and $|\vec{p},r;+\rangle = u_r(\vec{p})e^{i\vec{p}\cdot\vec{x}}$ and the reverse, respectively. This suggests that we associate the graph in Fig. 5 with T_2^{pair} ; for brevity spin labels are omitted. If, following the ideas of Dirac's hole theory, we interpret a "hole" in the filled negative-energy sea as a positron and consequently a transition of an electron into a negative-energy state as e^--e^+ annihilation, we can reinterpret Fig.5 as follows, reading from bottom to top: The initial electron comes into the neighborhood of V_{ext}, which creates the final electron and a positron; the latter then annihilates with the initial electron. The energy of the intermediate state is just $E + E(\vec{q}) + E$, since both initial and final electron have energy E. Thus the expected energy denominator is just

$$\frac{1}{E-E_{int}} = \frac{-1}{E+E(\vec{q})} \quad .$$

On combining this factor with the numerator factors from the vertices and noting that

$$-\int \frac{d\vec{q}}{(2\pi)^3}\sum_r \frac{|v_r(\vec{q})\rangle\langle v_r(\vec{q})|}{E+E(\vec{q})} = -\frac{\Lambda_-^{op}}{E+E_{op}} = \frac{-\Lambda_-^{op}}{E-h_D}$$

we get T_2^{pair}, except for a minus sign. Let us add a <u>rule</u>: Include a factor −1 for every "ziz-zag" of the type shown in Fig. 5. With this understanding not only does Fig. 5 give back T_2^{pair}, but <u>all higher order contributions to T^{pair} are obtained correctly from more complicated diagrams of this type</u>. Here T^{pair} is defined by

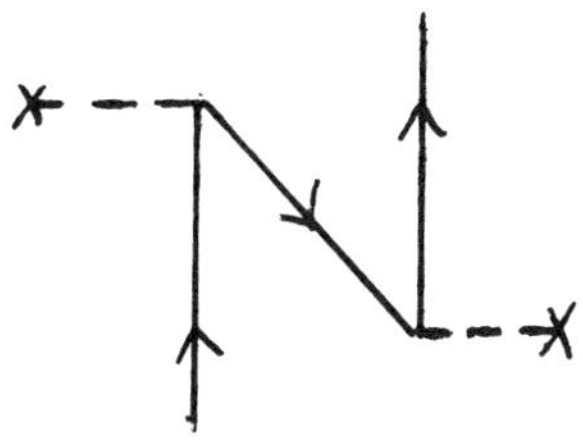

Fig. 5 Diagram representing the second-order contribution T_2^{pair} to an electron scattering in a external field, which arises from pair creation followed by pair annihilation.

$$T^{pair} = T - T^{np} \ . \qquad (2.45)$$

In particular, it is fairly straightforward to show with this rule, for example, that the sum of the graphs shown in Fig. 6 yields T_3^{pair}, the contribution to T^{pair} of third order in V_{ext}. Note that there are five terms in the sum, whereas use of the identity $1 = \Lambda_+^{op} + \Lambda_-^{op}$ in T_3 yields only three terms which involve at least one factor Λ_-^{op} in the numerator, so a little algebra is required to verify the identity in question. A lot more algebra is needed to give a direct proof to all orders that the combination of $\Lambda_+ V_{ext} \Lambda_+$ and δh_+ (See (2.33) and (2.34)) generates all the diagrams corresponding to intermediate states with one or more virtual pairs, e.g. such as the ones shown in Fig. 7.[3]

It follows that δh_+ can be interpreted as an effective interaction operator which takes into account all (irreducible) virtual processes involving virtual electron-positron pairs. The form (2.33) thus reveals the many-body aspect of the "one-electron" Dirac equation, an aspect which is hidden in the original form (2.25) or (2.28).

C. The "No-Pair Approximation" and the Magnitude of Pair Effects

Having identified δh_+ in Eq. (2.33) as representing the effects of

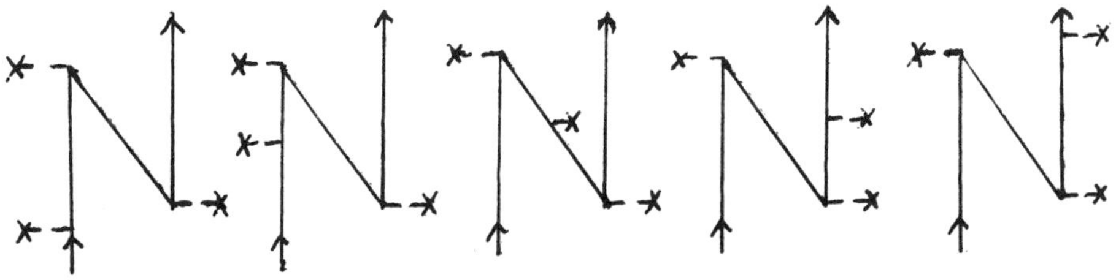

Fig. 6 Diagrams representing contributions to T^{pair} which are of third order in V_{ext}.

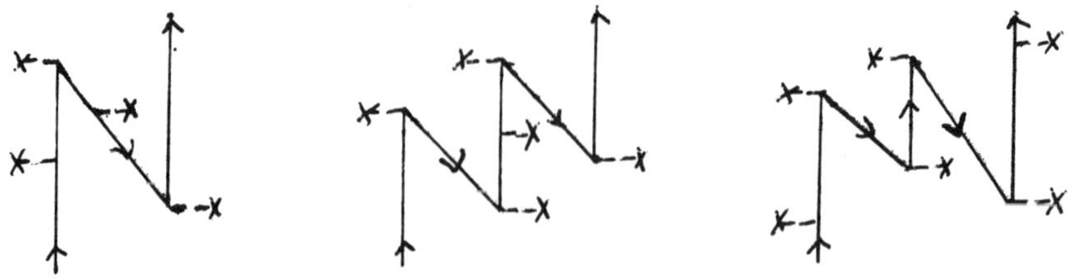

Fig. 7 Examples of fourth and higher-order contributions to T^{pair}.

virtual pairs, it is natural to study the equation obtained by dropping this term, viz. the "no-pair approximation" defined by

$$E\psi = (h_D + \Lambda_+ V \Lambda_+)\psi \ , \tag{2.46a}$$

with the subscript "ext" on V dropped, and

$$\Lambda_+\psi = \psi \tag{2.46b}$$

for the case of interest. Because of (2.46b), Eq. (2.46a) can be reduced to a relativistic SP form without approximation; as we will see, it contains all fine structure effects of order α^2Ryd and these emerge in a transparent manner. This exercise will also enable us to estimate the magnitude of virtual-pair effects in one-electron atoms.

1. Reduction to SP form and the fine-structure operators

Because of (2.46b) we may write ψ in the form

$$\psi = U\phi \tag{2.47a}$$

where U is defined by (2.23) and ϕ satisfies

$$\beta^{(+)}\phi = \phi \ . \tag{2.47b}$$

Multiplication of (2.46) on the left by $U^\dagger$ gives, on noting that (2.24a) and (2.47a,b) yield $U^\dagger\psi = \phi$, the equation

$$E_+\phi = h_+^{red}\phi \tag{2.48}$$

where

$$h_+^{red} = "U^\dagger h_+ U" \ . \tag{2.49}$$

The quotes on the right hand side mean that when there is no $\vec{\alpha}$ matrix to the right of a β matrix, β is to be replaced by unity, in view of (2.47b). On carrying out the multiplications involved, one gets

$$h_+^{red} = E_{op} + V_{eff} \tag{2.50}$$

where, if V is even in Dirac matrices, e.g. $\vec{A} = 0$ and $V = -eA^o$ for the case at hand,

$$V_{eff} = A_{op}[V + \frac{\vec{\sigma}\cdot\vec{p}}{E_p+m} V \frac{\vec{\sigma}\cdot\vec{p}}{E_p+m}] A_{op}. \tag{2.51}$$

$\vec{\sigma}$ is the doubled Pauli matrix, in the first instance, but in the standard rep the lower components of ϕ vanish and (2.48) may be regarded as a Pauli-type equation. To order v^2/c^2, the spin-independent term arising from (2.51) is given by

$$\frac{-1}{8m^2}(\vec{p}^2V + V\vec{p}^2 - 2\vec{p}V\cdot\vec{p}) = \frac{-1}{8m^2}[\vec{p},[\vec{p},V] = \frac{1}{8m^2}(\nabla^2 V) , \tag{2.52}$$

which represents an interaction of the electron with the charge density of the source of the electrostatic potential and reduces to, for V = $-Z\alpha/r$,

$$H_{s.i.}^{cont.} = \frac{Z\alpha\pi}{2m^2}\delta(\vec{r}) , \tag{2.53}$$

in agreement with (1.24). Note that in this formalism it emerges without any reference the nonrelativistic equation satisfied by ϕ_{nr} , which in the standard approach via (2.29) must be used to convert some terms involving the n.r. energy W to state-independent operators.

A feature of physical interest emerges from this analysis: The usual fine-structure operators do not involve (and hence cannot reveal) the existence of intermediate pair states. This is obvious for the $\vec{p}^4$ and spin-orbit operators, which have an immediate classical origin, but we have now shown it to be true also for the spin-independent contact term. [However, the fact that a certain term emerges naturally in a classical setting does not mean that it may not have a deeper quantum-theoretic origin. An example of this is provided by the $\vec{A}^2$ term in H_{qed} which, from the viewpoint of relativistic QED, arises from an approximation to processes involving virtual pairs].

2. Magnitude of pair effects

Finally, let us consider the order of magnitude of δh_+ in the context of a hydrogenic bound state. Use of the leading term (2.43) gives

$$\delta E = \langle\phi_m|\delta h_+^{(2)}|\phi\rangle . \tag{2.54}$$

Replacing E_{op} by m in the denominator of (2.43), we get

$$\delta E \approx \frac{\langle\phi|V\Lambda_- V|\phi\rangle}{2m} . \qquad (2.55a)$$

Since $\Lambda_-\phi = 0$, this can be rewritten in the form

$$\delta E \approx \frac{-\langle\phi|[[\Lambda_-,V],V]|\phi\rangle}{4m} . \qquad (2.55b)$$

If we use $\phi \to \phi_{nr}$, and $\langle\phi_{nr}|\vec{\alpha}|\phi_{nr}\rangle = 0$ we get, for $\ell \neq 0$ states,

$$\delta E \approx \langle\phi_{nr}| \frac{(\vec{\nabla}V)^2}{8m^3} |\phi_{nr}\rangle . \qquad (\ell \neq 0) \qquad (2.55c)$$

Since $\vec{\nabla}V = e\vec{E}$ where $\vec{E}$ is the electric field produced by the source of A^0 this has the form

$$\delta E \approx \frac{-1}{2} \langle\phi_{nr}| \alpha_E^{pair} \vec{E}^2 | \phi_{nr}\rangle \qquad (2.56)$$

where

$$\alpha_E^{pair} = \frac{-\alpha}{4m^3} = \frac{-\alpha}{4} \chi_e^3 = \frac{-\alpha^4}{4} a_o^3 . \qquad (2.57)$$

The quantity α_E^{pair} may be thought of as a (negative!) electric polarizability associated with virtual-pair creation by the external field. It is easy to see that, with $V = -Z\alpha/r$,

$$\delta E \sim (Z\alpha)^6 \, m = 2(Z\alpha)^4 \, Z^2 \text{ Ryd}. \qquad (2.58)$$

For $\ell = 0$, (2.56) diverges -- not all of the approximations made are justified. A more careful evaluation yields

$$\delta E \sim (Z\alpha)^5 \, m . \qquad (2.59)$$

However, this term is canceled by an opposite term arising from the expansion of E_+ in powers of αZ , consistent with the fact that the αZ expansion of the Dirac eigenvalue (2.28) has only even powers. There are of course also $\alpha^4 Z^6$ Ryd terms in E_+ but these do <u>not</u> cancel against (2.58), for $\ell \neq 0$.

With this we end our study of the c-number Dirac equation and turn to QED, the field theoretic formulation of hole theory, which incorporates the positron concept from the outset.

III. QED AND MANY-ELECTRON BOUND STATES

The main differences between the hybrid q.e.d. of Lecture #1 and QED itself lie in (i) the description of free electrons by means of the Dirac equation (really only the positive-energy part), i.e. in the incorporation of relativistic kinematics for the electron and (ii) the a priori recognition of the positron degree of freedom and its association with the negative-energy solutions. Both of these features are included in QED at the same time by the introduction of the Dirac matter field - denoted by $\psi_D(\vec{x})$ in the S-picture which we shall use - also called the "electron-positron field." We will consider $\psi_D(\vec{x})$ in IIIA, the construction of the Hamiltonian H_{QED} of QED in IIIB, and the derivation of relativistic many-electron configuration space-equations in IIIC. In IVD we will compare these with the usual equations and in IIIE we consider DHF-type equations.

A. The Electron-Positron Field $\psi_D(\vec{x})$.

The general solution $\psi(x)$ of the c-number Dirac equation, (2.5) can be written as a superposition of plane-wave solutions (2.6) and (2.7) so that, for a box of volume V,

$$\psi(x) = \psi(\vec{x},t) = \frac{1}{\sqrt{V}} \sum_{\vec{p},r} \Big[C_r(\vec{p};+)\, u_r(\vec{p})\, e^{-iE(\vec{p})t} e^{i\vec{p}\cdot\vec{x}} + C_r(\vec{p};-) v_r(\vec{p}) e^{iE(\vec{p})t}\, e^{i\vec{p}\cdot\vec{x}} \Big] . \tag{3.1}$$

We turn $\psi(x)$ into an operator (-valued distribution) by replacing the coefficient $C_r(\vec{p};\pm)$ by operators, analogous to the photon operators $A_\lambda(\vec{p})$ and $A^\dagger_\lambda(\vec{p})$, which act in a Hilbert space $\mathcal{H}_e$ analogous to $\mathcal{H}_\gamma$. Specifically we make the replacements

$$C_r(\vec{p};+) \to a_r(\vec{p}) , \qquad C_r(-\vec{p};-) \to b^\dagger_r(\vec{p}) \tag{3.2}$$

with $a_r(\vec{p})$ to be interpreted as an electron destruction operator and $b^\dagger_r(\vec{p})$ as a positron creation operator - vice versa for their hermitian conjugates. The a's and b's are taken to satisfy anti-commutation relations, which in the limit of infinite volume are

$$\{a_r(\vec{p}),\, a^\dagger_{r'}(\vec{p}\,')\} = \delta_{rr'}\delta(\vec{p}-\vec{p}\,')$$

$$\{b_r(\vec{p}),\, b^\dagger_r(\vec{p}\,')\} = \delta_{rr'}\delta(\vec{p}-\vec{p}\,') , \tag{3.3a}$$

to satisfy the antisymmetry principle separately for electrons and positrons, and also to anticommute with each other:

$$\{a_r(\vec{p}),\, b_{r'}(\vec{p}\,')\} = 0 \tag{3.3b}$$

etc. With the normalization $u^\dagger_r u_s = \delta_{rs}$, $v^\dagger_r v_s = \delta_{rs}$, the field operator

$\psi_D(x)$ is defined by

$$\psi_D(x) = \frac{1}{(2\pi)^{3/2}} \int d\vec{p} \sum_r \left(a_r(\vec{p}) u_r(\vec{p}) e^{-ip\cdot x} + b_r^\dagger(\vec{p}) v_r(-\vec{p}) e^{ip\cdot x} \right) . \quad (3.4)$$

$\psi_D(x)$ and the hermitian conjugate $\psi_D^\dagger(x)$ then satisfy <u>local</u> equal-time anti-CR:

$$\{\psi_{D\alpha}(\vec{x},t), \psi_{D\beta}^\dagger(\vec{x}',t)\} = \delta_{\alpha\beta}\delta(\vec{x}-\vec{x}') . \quad (3.5)$$

This in turn leads to the desired commutativity at space-like separations of currents like $\psi_D^\dagger \vec{\alpha} \psi_D$, which enter the theory when interactions are introduced.

The algebraic description is completed by assuming that H_e contains a state $|vac\rangle_e$ with the property

$$a_r(\vec{p})|vac\rangle_e = 0, \quad b_r(\vec{p})|vac\rangle_e = 0 , \quad (3.6)$$

which is identified as the fermionic vacuum, a state containing no electrons or positrons. Number operators N_-^{op} and N_+^{op} for electrons and positrons are defined by

$$N_-^{op} = \int d\vec{p} \sum_r a_r^\dagger(p) a_r(p), \quad N_+^{op} = \int d\vec{p} \sum_r b_r^\dagger(\vec{p}) b_r(\vec{p})$$

A state of the form

$$\Psi_{1e} = \int d\vec{p} \sum_r f_r(\vec{p})\, a_r^\dagger(\vec{p})|vac\rangle_e , \quad (3.8a)$$

having eigenvalue one and zero for N_-^{op} and N_+^{op} respectively is identified as a "one-electron state," with momentum-space wave function $f_r(\vec{p})$. The more familiar Dirac-spinor wave function can be defined, in $\vec{p}$-space, by

$$\tilde{\psi}(\vec{p}) = \sum_r f_r(\vec{p})\, u_r(\vec{p}) . \quad (3.8b)$$

It has the same norm as Ψ_{1e} and it satisfies

$$(\vec{\alpha}\cdot\vec{p}+\beta m)\tilde{\psi}(\vec{p}) = E(\vec{p})\tilde{\psi}(\vec{p}) \quad (3.8c)$$

or equivalently

$$\Lambda_+(\vec{p})\psi(\vec{p}) = \tilde{\psi}(\vec{p}) . \quad (3.8d)$$

Hence its $\vec{x}$-space transform

$$\psi(\vec{x}) = \int \frac{d\vec{p}}{(2\pi)^{3/2}} e^{i\vec{p}\cdot\vec{x}}\, \tilde{\psi}(\vec{p}) \quad (3.9)$$

satisfies

$$\Lambda_+^{op}\psi(\vec{x}) = \psi(\vec{x}). \tag{3.10}$$

States involving several electrons are defined by an obvious extension of (3.8a,b).

The total energy operator, naturally defined by

$$H_D = \int d\vec{p} \sum_r E(\vec{p})[a_r^\dagger(1)a_r(\vec{p}) + b_r^\dagger(\vec{p})b_r(\vec{p})] , \tag{3.11a}$$

can be written in the compact form in terms of $\psi_D(\vec{x}) \equiv \psi_D(\vec{x},o)$

$$H_D = \int d\vec{x} \; :\psi_D^\dagger(\vec{x})(\vec{\alpha}\cdot\vec{p}_{op} + \beta m)\psi_D(\vec{x}): \tag{3.11b}$$

where the colons denote normal ordering : Move creation operators to the left of destruction operators and include a factor -1 on for each commutation of fermion operators.

B. QED in Coulomb Gauge[1]

The total free hamiltonian of the (non-interacting) radiation field $\vec{A}_T(\vec{x})$ and electron-positron field $\psi_D(\vec{x})$ is given by

$$H_{QED}^{(o)} = H_{rad} + H_D \tag{3.12}$$

and acts in the tensor product space $\mathcal{H}_{\gamma e} = \mathcal{H}_\gamma \times \mathcal{H}_e$, i.e. on linear combinations of products of photon and electron-positron states. The number operators N_γ^{op} , N_-^{op} and N_+^{op} all commute with $H_{QED}^{(0)}$. Interaction is introduced by analogy with classical electrodynamics. Recall that in the c-number theory of the Dirac equation the quantity $-e\bar{\psi}\gamma^\mu\psi$ is identified as the electromagnetic current density associated with the electron. This suggests that (in the S-picture) we identify the quantities

$$j^o(\vec{x}) = \rho(\vec{x}) = -e :\psi_D^\dagger(\vec{x})\psi_D(\vec{x}): \tag{3.13a}$$

and

$$\vec{j}(\vec{x}) = -e :\psi_D^\dagger(\vec{x})\vec{\alpha}\,\psi_D(\vec{x}): \tag{3.13b}$$

as operator analogs of the classical charge and current densities, respectively. One then expects that there is an operator analog of the Coulomb interactions of the charge density with itself, viz.

$$H_C = \frac{1}{2}\iint d\vec{x}d\vec{y} : \frac{j^o(\vec{x})j^o(\vec{y})}{|\vec{x}-\vec{y}|} : \tag{3.14}$$

and of the current $\vec{j}(\vec{x})$ with the radiation field:

$$H_T = - \int d\vec{x}\, \vec{j}(\vec{x})\cdot\vec{A}_T(\vec{x}). \tag{3.15}$$

Thus one takes for the hamiltonian of "pure" QED the operator

$$H_{QED}^{pure} = H_{QED}^{(o)} + H_C + H_T + c.t. \tag{3.16}$$

where "c.t." again denotes counter terms needed to preserve the meaning of the parameters $-\underline{e}$ and $\underline{m}$ as the physical charge and mass of the electron and to cancel divergences which appear in perturbative calculations.

By "pure QED", I refer to the theory of the electromagnetic interactions of photons, electrons, and positrons with each other in the absence of any other matter. It suffices for the study of scattering processes such as $\gamma + e^- \to \gamma + e^-$, $e^- + e^- \to e^- + e^-$, $e^- + e^+ \to e^- + e^+$ (Compton, Möller and Bhabha scattering, respectively), pair annihilation ($e^- + e^+ \to \gamma + \gamma$) , pair creation, e.g. $\gamma + e^- \to e^- + e^- + e^+$, and processes related to these by additional photons in the final state. To deal with these it is most convenient to use a manifestly covariant quantization procedure for the e.m. field by describing it via a four-potential $A^\mu(x)$, all of whose components are kept and quantized, and introducing an interaction of the form $\int j_\mu(x)\, A^\mu(x)\, d\vec{x}$. This leads to the manifestly covariant Feyman rules and associated diagrams, described in many texts,for the calculation of scattering and reaction amplitudes.

In pure Q.E.D. there are apparently no true bound states, i.e. the spectrum of the Hamiltonian consists of scattering states only. There are metastable hydrogenlike (e^-,e^+) states referred to as states of positronium. However even the lowest-lying of these are unstable, since the processes $(e^-,e^+) \to \gamma + \gamma$ and $(e^-,e^+) \to \gamma + \gamma + \gamma$ are energetically allowed.

We shall be interested in the modification of pure QED provided by the presence of an external (static) field, described by a four-potential $A^\mu_{ext}(\vec{x})$, which corresponds to the addition of an interaction term

$$H_{ext} = \int j_\mu(\vec{x})\, A^\mu_{ext}(\vec{x})\, d\vec{x} \tag{3.17}$$

or, equivalently, to the replacement

$$H_D \to H_{D;ext} \equiv \int :\psi_D^\dagger(\vec{x})(\vec{\alpha}\cdot\vec{\pi}_{op} + \beta m - e\, A^o_{ext}(\vec{x}))\psi_D(\vec{x}): d\vec{x}\ . \tag{3.18}$$

With $\vec{A}_{ext} = 0$ and $A^o_{ext}(\vec{x}) = -Z\alpha/|\vec{x}|$ this yields an extremely good starting point for the study of the properties of atoms. The effects arising

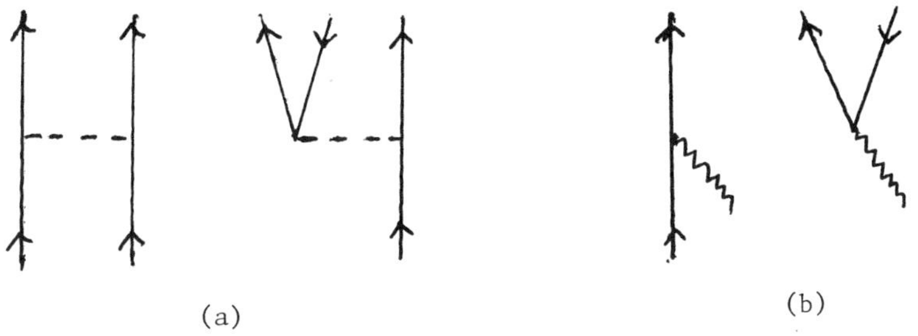

Fig. 8 Examples of diagrams representing some of the plane wave matrix elements of (a) the Coulomb interaction Operator H_C and (b) the transverse photon interaction operator H_T.

from nuclear recoil, structure, and magnetic field can be treated perturbatively and the effects of nuclear charge distribution can be included from the outset by modifying the form of $A^o_{ext}(\vec{x})$. The total Hamiltonian for "impure QED" is therefore taken to be

$$H_{QED} = H^{(o)}_{QED} + H_{ext} + H_C + H_T + c.t. \tag{3.19a}$$

$$= H_{rad} + H_{D;ext} + H_C + H_T + c.t. \tag{3.19b}$$

The major new ingredient in (3.19) relative to H_{qed} is not so much the relativistic description of the electrons or even the introduction of positrons but the fact that now not only is the number of photons not conserved but neither is the number of electrons nor the number of positrons. So a kind of democracy has been restored. Although this complicates the theory considerably, there is one simplifying feature: <u>charge conservation</u>. Consider the operator

$$Q_{op} = -e\int d\vec{x}\ \rho(\vec{x}) = -e\int d\vec{p}\ \sum_r\ [a^\dagger_r(\vec{p})a_r(\vec{p})-b^\dagger_r(\vec{p})b(\vec{p})] \ . \tag{3.20}$$

It is readily verified that Q_{op} commutes not only with $H^{(o)}_{QED}$ but also with H_C and H_T. This is associated with the circumstance that H_C and H_T create or destroy fermions only in <u>pairs</u>: for every e^- created or destroyed, an e^+ is also created or destroyed.

This and other features of H_C and H_T are incorporated in the graphical description of their matrix elements between the simplest plane wave eigenstates of $H^{(o)}_{QED}$, as illustrated in Fig. 8. These graphs can moreover be used to formulate rules for computation in configuration space, which make life easier by leaving Fock space behind.

C. Return to Configuration Space: No-pair Equations[4]

For atomic systems, it is well known that the effects of H_T, involving the emission and/or absorption of transverse photons, can usually be treated perturbatively. Unfortunately the eigenvalue problem

$$H_{QED}\Psi = E\Psi \qquad (3.21)$$

involves an infinite number of coupled equations even if H_T is dropped, because the fermion number operators $N_{\pm}^{op}$ are not conserved by H_{QED}. However, it is also true, but not so well recognized, that the emission or absorption of virtual pairs by an electron can be treated by PT. Moreover, this is true even for virtual-pair effects arising from the external field, unless Z is very large indeed, as we already saw in Sec. II.

One can take advantage of this circumstance by separating H_{QED} into a no-pair part H_{QED}^{np} and a remainder,

$$H_{QED} = H_{QED}^{np} + H_{QED}^{rem} \qquad (3.22)$$

where

$$H_{QED}^{np} = H_{rad} + H_{D;ext}^{np} + H_C^{np} \ . \qquad (3.23)$$

Here the superscript "np" signifies that one is to retain only that part of the operator in question which does not involve creation or destruction of (e^-,e^+) pairs. To make this notion precise it is necessary to specify the expansion of $\psi_D(\vec{x})$ being used. Until further notice, we shall take it to be the plane wave expansion (3.1). Thus the operators $a_r^\dagger(\vec{p})$, $b_r^\dagger(\vec{p}')$ are creation operators for "free" electrons and positrons. The operator H_{QED}^{np} then has the property

$$[N_-^{op}, H_{QED}^{np}] = [N_+^{op}, H_{QED}^{np}] = 0 \qquad (3.24a)$$

and, trivially, because the photons are now decoupled.

$$[N_\gamma^{op}, H_{QED}^{np}] = 0 \ . \qquad (3.24b)$$

The eigenvalue problem

$$H_{QED}^{np}\,\Psi = E\Psi \qquad (3.25)$$

breaks up into "sectors," specified by definite values for the number operators:

$$N_\gamma^{op}\,\Psi = N_\gamma \psi \ , \qquad N_\pm^{op}\,\Psi = N_\pm \Psi \ . \tag{3.26}$$

A zeroth-order description of atomic bound states is therefore obtained by studying the sector $N_- = N$, $N_+ = N_\gamma = 0$. The general form of a state in this sector is given by

$$\Psi = \sum_{\vec{n}} f(\vec{n})\, a^\dagger(n_1)\ldots a^\dagger(n_N)\,|\mathrm{vac}\rangle_{\gamma,e^+e^-} \tag{3.27}$$

where $\vec{n} = (n_1,\ldots n_N)$ and n_i denotes the labels specifying the functions used in the expansion of $\psi_D(\vec{x})$. In our case, these are just the momentum $\vec{p}$ and spin label r; in the $V = \infty$ limit, the symbol $\sum$ is to be understood as an integration with weight $d\vec{p}/(2\pi)^{3/2}$ and summation over $r = 1,2$ for each electron. For the sector at hand, the eigenvalue problem (3.25) then yields without approximation an equation which determines the Fock-space amplitude $f(\vec{n})$. From this equation it is straightforward to obtain an equation for the configuration space wavefunction $\psi(\vec{r}_1\ldots\vec{r}_N)$ defined by

$$\psi = \sum_{\vec{n}} f(\vec{n})\, u_{n_1}(\vec{r}_1)\ldots u_{n_N}(\vec{r}_N) \ . \tag{3.28}$$

Because the function ψ satisfies the condition

$$\Lambda_+(i)\,\psi = \psi \qquad (i = 1,2,\ldots,N) \tag{3.29}$$

regardless of the nature of $f(\vec{n})$, this equation can be written in a variety of ways, one of which is

$$H_+\psi = E\psi \tag{3.30a}$$

where

$$H_+ = \sum h_D(i) + \Lambda_+^{tot}\left(V_{ext}^{tot} + V_{ee}^{tot}\right)\Lambda_+^{tot} \tag{3.30b}$$

with

$$V_{ext}^{tot} = \sum_i V_{ext}(i) \ , \quad V_{ee}^{tot} = \frac{1}{2}\sum_{i\neq j}\alpha/r_{ij} \ , \quad \Lambda_+^{tot} = \Lambda_+(1)\ldots\Lambda_+(N) \ . \tag{3.31}$$

Because of (3.29), we may also write (3.30a) in the equivalent form

$$\hat{H}_+\psi = E\psi \tag{3.32a}$$

where

$$\hat{H}_+ = \Lambda_+^{tot}\, H_{DC}\, \Lambda_+^{tot} \tag{3.32b}$$

where H_{DC} is defined by (3.41) below, or by

$$\tilde{H}_+ \psi = E\psi \tag{3.33a}$$

where

$$\tilde{H}_+ = \sum_i E(\vec{p}_i^{\,op}) + \Lambda_+^{tot} \left(V_{exp}^{tot} + V_{ee}^{tot} \right) \Lambda_+^{tot} \ . \tag{3.33b}$$

Note also that in <u>any</u> of the equations (3.30), (3.32) or (3.33), the right-hand factor Λ_+^{tot} could be omitted but it is convenient to retain them to make manifest the hermiticity of the operators involved.

If we denote by S the linear space of <u>all</u> multi-Dirac spinors ψ , and by S_+ the subspace of S of ψ's which satisfy (3.29) the operators H_+, $\hat{H}_+$ and $\tilde{H}_+$, as formally defined by the above expression differ on the orthogonal complement S_+' of S_+ --- for example, $\hat{H}_+$ annihilates S_+' --- but they coincide on S_+ , which is the only part of S which is of physical relevance at this stage.

Although eq. (3.30) appears to be both conceptually and technically the simplest no-pair many-electron wave equation one can obtain from QED, which reduces to the usual Schrodinger equation in the n.r. limit, it is only one of a whole class of such equations. These equations are obtained by expanding the field operator $\psi_D(\vec{x})$ into a complete orthogonal set of spinor wave functions, each of which is a (proper or improper) eigenfunction of a one-body operator h_D^U of the form

$$h_D^U = h_D + U \tag{3.34}$$

where U is a one-body "potential". The choice $U(1) = V_{ext}(1)$ is equivalent to using the time-independent version of the so-called "Furry picture" or "bound state interaction picture." For $V_{ext} = -Z\alpha/r$ and $Z\alpha < 1$, the spectrum of $h_{D;ext} = h_D + V_{ext}$ is then continuous in the interval $(-\infty,-m)$, discrete in $(0,m)$ and continuous in (m,∞) . The eigenfunctions $u_n(\vec{x})$ associated with the positive-energy part now have operator coefficients a_n in the expansion

$$\psi_D(\vec{x}) = \sum_n a_n u_n(\vec{x}) + \Sigma b_m^\dagger v_m(\vec{x}) \tag{3.35}$$

which are interpreted as destruction operators for "bound-state electrons" in the $(0,m)$ interval and "scattering-state electrons" in the (m,∞) interval, respectively. The charge conjugates of the eigenfunctions $v_m(\vec{x})$ associated with the $(-\infty,-m)$ interval are associated with positron scattering states, and b_m is interpreted as a destruction operator for scattering-state positrons. A vacuum state $|0\rangle_{ext}$ is defined by

$$a_n|vac\rangle_{ext} = 0 \quad , \quad b_m|vac\rangle_{ext} = 0 \, . \tag{3.36}$$

Note that $|vac\rangle_{ext} \neq |vac\rangle$, because V_{ext} can create (free) virtual electron-positron pairs. The advantage of such an expansion is that it diagonalizes all of the operator

$$H_{D;ext} = \int d\vec{x} : \psi_D^\dagger(\vec{x}) \, (\vec{\alpha}\cdot\vec{p} + \beta m + V_{ext}) \, \psi_D(\vec{x}) : \tag{3.37}$$

not just the part in which V_{ext} is replaced by

$$\Lambda_+ V_{ext} \Lambda_+ + \Lambda_- V_{ext} \Lambda_- \, .$$

This means that only virtual-pair effects coming from electron-electron interaction are included in the pair-part of H_{QED}. The disadvantage is that the projection operators $\Lambda_{+;ext}(i)$ which now appear in the corresponding C.S. hamiltonian, defined by

$$H_+^{ext} = \sum_{i=1}^{N} h_{D;ext}(i) + \Lambda_{+;ext}^{tot} \, V_{ee} \, \Lambda_{+;ext}^{tot} \tag{3.38}$$

are considerably more complicated than before. Nevertheless the equations,

$$H_+^{ext}\psi = E\psi \quad , \quad \Lambda_+^{ext}(i) \, \psi = \psi \tag{3.39}$$

has been successfully applied to several physical problems, as we shall see.

More generally, for any reasonable U, be it 0 or V_{ext} or something in between, the corresponding C.S. hamiltonian is given by

$$H_+^U = \sum_{i=1}^{N} h_D^U(i) + \Lambda_{+;U}^{tot} \left(V_{ext}^{tot} - U^{tot} + V_{ee} \right) \Lambda_{+;U}^{tot} \tag{3.40a}$$

or equivalently by

$$\hat{H}_+^U = \Lambda_{+;U}^{tot} \, H_{DC} \, \Lambda_{+;U}^{tot} \, , \tag{3.40b}$$

where H_{DC} is the Dirac-Coulomb Hamiltonian

$$H_{DC} = \sum_i h_{D;ext}(i) + V_{ee} \, . \tag{3.41}$$

Note that if $U \neq V_{ext}$ virtual-pair effects arising from the difference potential $\delta U = V_{ext} - U$ need to be considered separately. For any such choice the equation to be solved is

$$H_+^U\psi = E\psi \tag{3.42a}$$

with the constraint

$$\Lambda_+^U(i)\psi = \psi \ . \qquad (3.42b)$$

The question of the "best choice" of U has been discussed elsewhere[5] and I will not expand on it here. Note that U need not be local.

D. Comparison with the Traditional Equation: $H_{DC}\,\psi = E\psi$.

The traditional starting point for the joint treatment of relativistic and correlation effects in atomic physics has been the equation

$$H_{DC}\,\psi = E\psi \qquad (3.43)$$

where H_{DC} is the operator defined by (3.41). However, it has been known since the 1951 paper of Brown and Ravenhall that (3.43) has no normalizable solutions.[6] To see this most simply, take N=2 and turn off the electron-electron Coulomb interaction V_1. The spectrum of the resulting operator $H_{DC}^{(o)} = h_{D;ext}(1) + h_{D;ext}(2)$ is then the whole real line. Although this operator has normalizable eigenfunctions their energies are continuously degenerate with non-normalizable functions in which one electron is in the positive-energy continuum and the other is in the negative-energy continuum. The turning on of V_{ee} will then lead to a mixing of these states and cause the bound state to "dissolve" into the continuum. A solvable model which provides an explicit illustration of this has been constructed recently.[7]

The phenomenon of "continuum dissolution" (CD) is analogous in its mathematical aspects to the process of auto-ionization but, unlike the latter, it has no physical counterpart. The fatal flaw of H_{DC} is related to the fact that it corresponds to an extension of Dirac's "one-electron theory" rather than Dirac's "hole-theory" to many electrons. We have already seen that QED, which is the mathematical formulation of many-electron hole-theory, leads to CS Hamiltonians in which V_{ee} is surrounded by positive-energy projection operators. For example, in H_+^{ext} , the electron-electron interaction is described by the operator $\Lambda_{+;ext}^{tot}\, V_{ee}\, \Lambda_{+;ext}^{tot}$ which has no matrix elements between mixed positive- and negative-energy product states. As a consequence, H_+^{ext} does not suffer from CD. We now also see that the projection operators appearing in the various no-pair Hamiltonians considered have a simple origin: Within a framework that describes electrons in terms of four-component Dirac spinors, they express the constraints imposed by the exclusion principle as applied to the filled negative-energy sea.

As discussed elsewhere [4], the CD phenomenon raises serious methodological questions about the interpretation of calculations of atomic properties which appear to be based on H_{DC}. To some extent these have been resolved, as we will see in the next section.

E. Variational Approximation Schemes

The traditional relativistic central-field-approximation (CFA) schemes are based on equations obtained (or obtainable) by applying the variational principle to the functional

$$F[\tilde{\Phi}] = \langle\tilde{\Phi}|H_{DC}|\tilde{\Phi}\rangle \tag{3.44a}$$

where

$$\tilde{\Phi} = \Sigma\ C_\Gamma \tilde{\Psi}_\Gamma \tag{3.44b}$$

and $\tilde{\Phi}_\Gamma$ is a Slater determinant formed from in Dirac-type orbitals $\tilde{\phi}_a(i)$.

In one version of the so-called relativistic configuration interactions (RCI) method the orbitals are fixed, taken (say, as a result of a preliminary calculation) as positive-energy eigenfunctions of the same hermitian one-body operator $h_D' = h_D + U'$. This kind of calculation has a simple field-theoretic interpretation, because one can identify U' with the operator U of the preceding section, and therefore, a posteriori, regard the calculation as based on H_+^U and $\tilde{\Phi}$ as an approximation to an eigenfunction of H_+^U : The equations determining the C_Γ are the same in either case, since

$$\Lambda_+^U(i)\tilde{\phi}_a(i) = \tilde{\phi}_a(i)\ . \tag{3.45}$$

However, the interpretation of RCI calculations involving orbitals $\tilde{\phi}_a$ which satisfy coupled one-body equations, with right-hand sides of the form $\sum_b \lambda_{ab}\tilde{\phi}_b$ (λ_{ab} = Lagrange multiplier) is not so clear cut and requires additional study.[8] The multi-configuration Dirac-Fock (MCDF) type of calculation[9] also requires further elucidation. In this approach the radial wave functions $P_a(r)$ and $Q_n(r)$ in a Dirac orbital are both left free to vary, along with the coefficients C_Γ . This is in contrast to equations one gets by applying the variational principle to H_+^U , with an ansatz (3.44b) and orbitals satisfying (3.45). If we write such an orbital in the form

$$\tilde{\phi}_a = \begin{pmatrix} P_a(r)\ Y^m_{j;\ell} \\ -iQ_a(r)\ \vec{\sigma}\cdot\hat{r}\ Y^m_{j;\ell} \end{pmatrix} \tag{3.46}$$

the constraint (3.45) fixes the relation between Q and P. For example, for the choice U = 0, we get

$$Q_a(r) = X_{op}^{j,\ell} P_a(r) \tag{3.47}$$

where

$$X_{op}^{j,\ell} = \frac{1}{E_{op}+m}\left(\frac{d}{dr} - \frac{\eta^{j,\ell}}{r}\right) \tag{3.48}$$

with $\eta^{j,\ell} = \ell$ if $\ell = j-1/2$ and $-(\ell+1)$ if $\ell = j + 1/2$.

Very recently, a variational calculation based on the free projection-operator hamiltonian has been carried out for bromine by B. Hess, with encouraging results.[10]

IV. Applications

A. Low-lying States of He and He-like Ions

An early application of eq. (3.39) was made for N = 2, to calculate the α^3Ryd corrections to the fine-structure of low-lying states in helium.[11] Let us make a brief survey of this calculation. We imagine that ψ and E are known --- in practice approximations will of course have to be made --- and begin to compute the left-over effects from H_T and H_C^{pair}, using ordinary perturbation theory.

For example, use of H_T in second order yields

$$\Delta E_T = \langle\Psi| H_T \frac{1}{E-H_{QED}^{np}} H_T|\Psi\rangle \tag{4.1}$$

where Ψ is the Fock-space counterpart of ψ . Insertion of a complete set of intermediate states and retention of only the "interaction part" of ΔE_T , i.e. dropping of self-energy terms, yields

$$(\Delta E_T)_{int} =$$

$$e^2\int\frac{d\vec{k}}{(2\pi)^3 2\omega}\sum_\lambda \langle\psi|\vec{\varepsilon}_\lambda\cdot\vec{\alpha}_2 e^{i\vec{k}\cdot\vec{r}_2}\frac{\Lambda_{+;ext}^{tot}}{E-H_+-\omega}\vec{\varepsilon}_\lambda^{\,*}\cdot\vec{\alpha}_1 e^{-i\vec{k}\cdot\vec{r}_1} + (1\leftrightarrow 2)|\psi\rangle . \tag{4.2}$$

Note that if we use the expansion

$$\frac{1}{E-H_+-\omega} = -\frac{1}{\omega} + \frac{E-H_+}{\omega^2} + \cdots , \tag{4.3}$$

the first term yields the Breit operator B(1,2) and the higher order terms correspond to recoil corrections:

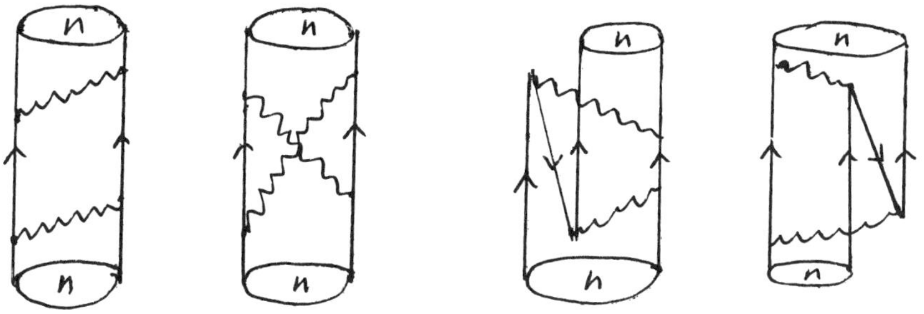

Fig. 9 Examples of two-photon exchange diagrams, which contribute to the α^3 Ryd level shift in He-like ions, arising from use of H_T in fourth order.

$$(\Delta E_T)_{int} = \langle\phi|B(1,2)|\phi\rangle + \text{recoil terms} \tag{4.4}$$

where

$$B(1,2) = -\frac{\alpha}{2r_{12}}(\vec{\alpha}_1\cdot\vec{\alpha}_2 + \vec{\alpha}_1\cdot\hat{r}\,\vec{\alpha}_2\cdot\hat{r}) \ . \tag{4.5}$$

The first term yields, in the nonrelativistic limit, the expectation value of the reduced Breit operator (1.22) with the wavefunction ϕ_{nr}, which is of order α^2Ryd , plus an α^3Ryd correction. The recoil terms yield an α^3Ryd correction only.

As a more complicated example consider the effect of H_T in fourth order (again self-energy corrections are dropped):

$$(\Delta E_{TT})_{int} = \langle\Psi|H_T \frac{1}{E-H^{np}_{QED}} H_T \frac{1}{E-H^{np}_{QED}} H_T \frac{1}{E-H^{np}_{QED}} H_T|\Psi\rangle_{int} \tag{4.6}$$

This gives rise to many terms some of which are shown graphically in Fig. 9; for clarity, the fact that the electrons interact via α/r_{12} while the photons are "in the air" is not shown explicitly. An example involving H_C^{pair} is given by

$$\Delta E^{pair}_{CC} = \langle\Psi|H^{pair}_C \frac{1}{E-H^{np}_{QCD}} H^{pair}_C|\Psi\rangle_{int} , \tag{4.7}$$

which corresponds to the graphs shown in Fig. 10, and an example involving both H_T and H_C^{pair} is shown in Fig. 11.

All these contribute to the α^3Ry level shift. The sum of the leading contributions from each of these diagrams may be expressed as the expectation value of an operator of the form

$$H_{eff} = a\,\delta(\vec{r}_{12}) + b\,\vec{\sigma}_1\cdot\vec{\sigma}_2\,Y(\vec{r}_{12}) , \tag{4.8}$$

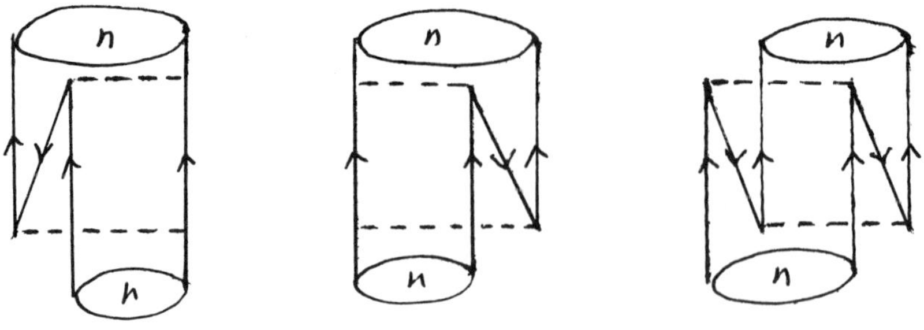

Fig. 10 The diagrams which arise from using H_C^{pair} in second order.

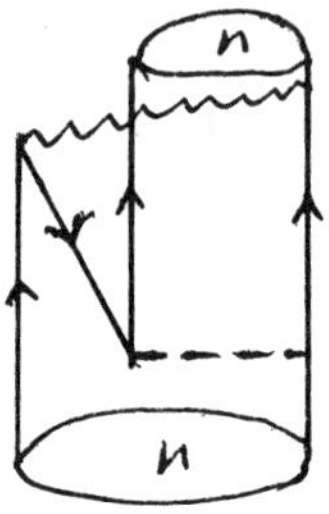

Fig. 11 A diagram representing a contribution arising from the combined use of H_C^{pair} and H_T.

where $Y(\vec{r})$ is a singular function defined by a limiting procedure. However for ^1S states $\phi_{nr}(\vec{r}_1,\vec{r}_2)$ vanishes at $\vec{r}_{12} = 0$ and Y may be replaced by r_{12}^{-3}. To all this the contributions from diagrams corresponding to radiative corrections (Lamb shift) must of course be added. Reviews and updated comparison with experiment have been given recently by several authors; I refer you to these for further details.[12]

B. Rydberg states of He or He-like Ions and Long-Range Forces.

Rydberg states in atomic systems have become the object of intense experimental investigation in recent years and this topic has been amply discussed by other lecturers here. There is however one aspect which I believe has not been addressed: the opportunity that some Rydberg systems afford for testing --- albeit indirectly --- some predictions of QED which are off the beaten path but which have a beauty and life of their own. I refer to the quantum theory of long-range forces (LRF) between electromagnetic systems. In particular, it turns out that recent measurements[13] of the fine-structure of the n = 10 Rydberg states of He may be capable of checking predictions of QED for the forces arising from two-photon exchange between electrons, at distances which are large compared to ordinary atomic dimensions. For the case of two (neutral) atoms such forces go by the familiar name of van der Waals forces. I will first review briefly the ideas and some results of the dispersion theoretic approach to the calculation of the van der Waals potential for neutral systems, following ancient work of Gary Feinberg and myself.[14] Then I will sketch some of the modifications needed to apply this approach to charged systems and to experiment.

1. Two-photon exchange between neutral systems

Consider the scattering amplitude M for spinless particles A and B, with initial and final four-momenta p_A, p_B and p'_A, p'_B respectively. Define

$$s = (p_A+p_B)^2 \ , \quad t = (p_A-p'_A)^2 \ , \quad u = (p_A-p'_B)^2 \ , \tag{4.9}$$

as invariant squared energies and momentum transfers. Note that $s + t + u = p_A^2 + p'^2_A + p_B^2 + p'^2_B \to 2m_A^2 + 2m_B^2$ when the particles are on their mass shells. We can regard the on-shell amplitude M as a function of s and t only: $M = M(s,t)$.

Let Γ^μ_A denote the four-vector amplitude for emission of a virtual photon by A. Current conservation requires that

$$q_\mu \Gamma^\mu_A = 0 \tag{4.10}$$

for A on shell. The general form of Γ^μ_A is, from Lorentz covariance

$$\Gamma^\mu_A = f_A P^\mu_A + g_A q^\mu \tag{4.11}$$

where $P_A = p_A + p'_A$, $q = p_A - p'_A$ and f and g are functions of say q^2, p_A^2 and p'^2_A only. Since $q \cdot P_A = p_A^2 - p'^2_A$ vanishes on-shell, we get from (4.10), $f \cdot 0 + g_A q^2 = 0$ or $g_A = 0$ on shell, so that

$$\Gamma^\mu_A = f_A(q^2)(p^\mu_A + p'^\mu_A) \tag{4.12}$$

on the mass shell. The quantity $f_A(q^2)$ is called the charge form factor and $f_A(0) = e_A$ is the charge of A. [To see this note that $\Gamma_\mu = \langle p'|j_\mu(0)|p\rangle$ where $j_\mu(x)$ is the electromagnetic current density operator, then integrate $\Gamma_0 e^{iq\cdot x}$ over all $\vec{x}$.]

The contribution of <u>one-photon exchange</u> to M is given by

$$M_{(1\gamma)} = - \frac{\Gamma^\mu_A \, g_{\mu\nu} \, \Gamma^\nu_B}{q^2} \tag{4.13}$$

which on use of (4.12) reduces to

$$M_{(1\gamma)} = -f_A(t) f_B(t) \frac{(s-u)}{t} \ . \tag{4.14}$$

For small t and low energy this behaves as $e_A e_B / \vec{q}^{\,2}$ in the c.m. system and is proportional to the Fourier transform (F.T.) of the Coulomb potential $e_A e_B/r$. However, if either e_A and/or $e_B = 0$, then f_A and/or f_B vanishes for $t \to 0$ and the pole at $t = 0$ is canceled; the F.T. of M is then a short-range interaction. This is physically plausible, since a spinless neutral particle does not have a long-range static field associated with it in classical physics.

The situation changes drastically if we consider two-photon exchange. Then even if both particles are neutral there can be an LRF. To see this, we shall need the analogue of (4.12), i.e. in tensor amplitude $T_A^{\mu\nu}$ for two-photon emission, or equivalently, the Compton amplitude tensor $M_A^{\mu\nu}$ for absorption of a photon of momentum k and emission of a photon of momentum k' by particle A. The contribution to M from two-photon exchange can be written in the form of a Feynman integral involving two photon propagators

$$M_{2\gamma} = \text{const.}\int d^4k\ d^4k'\ \delta(q-k-k')\left(T_A^{\mu\mu'} \frac{g_{\mu\nu}}{k^2} \frac{g_{\mu'\nu'}}{k'^2} T_B^{\nu\nu'}\right) . \tag{4.15}$$

It turns out that in order to evaluate $M_{2\gamma} = M_{2\gamma}(s,t)$ it is only necessary to know the values of the T's on the photon mass shell $k^2 = k'^2 = 0$. Moreover in the case of a neutral particle A, $T_A^{\mu\nu}(k,k')$ is determined by use of current conservation and Bose symmetry for the photons in terms of just two invariant amplitudes. In terms of a suitably chosen "electric tensor" $T_E^{\mu\nu}$ and a "magnetic tensor" $T_M^{\mu\nu}$ one may write the on-shell Compton amplitude in the form

$$M^{\mu\nu} = F_E T_E^{\mu\nu} + F_M T_M^{\mu\nu} \tag{4.16}$$

where $F_x = F_x(\sigma,t)$ $(x = E,M)$ with $\sigma = (p+k)^2$ and $t = (p-p')^2$. It can be shown that the threshold values of the F_x are just polarizabilities:

$$F_E(m^2,0) = 4\pi\alpha_E\ ,\quad F_M(m;0) = 4\pi\alpha_M\ . \tag{4.17}$$

Moreover the $F_x(\sigma,t)$ satisfy dispersion relations, considered as functions of σ , regarded as a complex variable, and can be written in the form

$$F_x(\sigma,t) = \frac{1}{\pi}\int_m^\infty d\sigma'\ \left(\frac{1}{\sigma'-\sigma} + \frac{1}{\sigma'-\bar{\sigma}}\right)\rho_x(\sigma',t) \tag{4.18a}$$

where $\bar{\sigma} + \sigma + t = 2m^2$, and the "spectral function" $\rho_x(\sigma,t)$ is proportional to the discontinuity of F_x across the branch cut from m to $+\infty$:

$$\rho_x(\sigma,t) = \lim_{\varepsilon\to 0} \frac{1}{2i}\left[F_x\ (\sigma + i\varepsilon,t) - F_x(\sigma - i\varepsilon,t)\right] . \tag{4.18b}$$

The spectral functions ρ_x^A and ρ_x^B are determined by the electromagnetic structure of A and B, respectively. Furthermore $M_{2\gamma}$ itself admits a spectral representation of the form

$$M_{2\gamma}(s,t) = \frac{1}{\pi} \int \frac{\rho_{2\gamma}(s,t')}{t'-t} dt' . \qquad (4.19)$$

The use of (generalized) unitarity allows one to express $\rho_{2\gamma}$ in terms of the on-shell amplitudes T^A and T^B as an integral obtained from (4.15) by replacing each photon propagator $(k^2-i\varepsilon)^{-1}$ by its positive-energy on-shell part $\delta(k^2)\theta(k^o)$. It follows that $\rho_{2\gamma}$ may be expressed in terms of ρ^A and ρ^B, and hence so can an effective low-energy potential $V_{2\gamma}(r)$, defined as the Fourier-transform of $M_{2\gamma}(s,t)$ via

$$V_{2\gamma}(r) = \text{const.} \int e^{i\vec{q}\cdot\vec{r}} M_{2\gamma}(s_o,-\vec{q}^2) . \qquad (4.20)$$

Here $s_o = (m_A + m_B)^2$ is the threshold value of s. In particular, one finds that

$$V_{2\gamma}(r) = \frac{1}{(4m_A m_B)(4\pi^2 r)} \int_{t_o}^{\infty} dt\, \rho_{2\gamma}(t) e^{-\sqrt{t}\, r} \qquad (4.21a)$$

where

$$\rho_{2\gamma}(t) = -\frac{1}{16\pi} \int \frac{d\hat{k}}{4} (M_A : M_B) . \qquad (4.21b)$$

Here $d\hat{k}$ is the element of integration over the direction $\hat{k}$ of one of the two photons in their c.m. system and the colon denotes a contraction over Lorentz indices. It is easy to see from (4.21a) that if $\rho(t) \sim t^N$ for $t \to 0$, then $V_{2\gamma}(r) \sim r^{-(2N+3)}$ for $r \to \infty$. In particular, one finds for the case at hand (two neutral spinless particles), that for large r

$$V_{2\gamma}(r) \sim -\frac{D}{r^7} \qquad (4.22a)$$

where

$$D = \frac{23}{4\pi}\left(\alpha_E^A \alpha_E^B + \alpha_M^A \alpha_M^B\right) - \frac{7}{4\pi}\left(\alpha_E^A \alpha_M^B + \alpha_M^A \alpha_E^B\right) , \qquad (4.22b)$$

which is a model-independent generalization of the original result of Casimir and Polder.[15] Similar results may be derived for particles with spin.

2. Extension to charged systems

The techniques described above are readily extended to the calculation of $V_{2\gamma}(r)$ for a neutral particle A and a charged particle B.[16] In particular, if B is a point particle, with spin-1/2, it suffices to calculate $M_B^{\mu\nu}$ in Born approximation and to compute from this the discontinuity of the spin-independent part of M_B. The spectral functions associated with the corresponding form factors F_E^B and F_M^B then turn out to be

$$\rho_E^B(\sigma,t) = \pi\ f_E^B(t)\ \delta(\sigma-m_B^2)\ ,\quad \rho_M^B = 0 \tag{4.23}$$

where $f_E^B = 16e^2m_B^3/(4m_B^2-t)t$. For the case where B is an electron and A a heavy neutral spinless particle ("atom") one finds, that

$$V_{2\gamma}(r) = -\frac{1}{2}\frac{\alpha\alpha_E^A}{r^4} + \frac{11}{4\pi}\frac{\alpha\alpha_E^A}{r^4}\frac{\chi_e}{r} + \frac{5}{4\pi}\frac{\alpha\alpha_M^A}{r^4}\frac{\chi_e}{r} + Y_{2\gamma}(r) \tag{4.24}$$

where

$$Y_{2\gamma}(r) = \frac{\alpha}{16\pi^2}\frac{\chi_e}{r^5}\int_o^\infty \frac{dk}{\pi}\left(\frac{\rho_E^A(k)}{k}J_E(kr) + \frac{\rho_M^A(k)}{k}J_M(kr)\right) . \tag{4.25}$$

The $J_x(y)$ are functions defined in Ref. 16, which we need not record here. All the effects of the structure of A are taken into account by the spectral functions ρ_E^A and ρ_M^A entering (4.25). Note that the leading term in (4.24) is just the classical interaction $-\vec{d}\cdot\vec{E}/2$ of an induced electric dipole $\vec{d} = \alpha_E^A\vec{E}$ with the electrostatic field $\vec{E} = \alpha\hat{r}/r^2$ produced by the electron. The next two terms are retardation corrections found earlier by Kelsey and Spruch[17] and Bernabeu and Tarrach[18].

3. Application to He Rydberg states.

The calculation of the level shift arising from two-photon exchange in Rydberg states of He or He-like ions is relatively intricate, for two reasons. Strictly speaking, eq. (4.24) is not immediately applicable to this case because the role of system A is then played by the He^+ core, which is not electrically neutral. Second, the quantity $V_{2\gamma}$ was computed covariantly, which means that it includes, from the point of view of a Coulomb gauge calculation, not only the effects of transverse photon exchange but also direct Coulomb interactions between the outer and inner electrons. These latter have already been largely taken into account in work of R. Drachman,[19] who computed to great accuracy the energy levels $W_{n,\ell}$ of the nonrelativistic two-electron Hamiltonian $H_{nr}(1,2)$ for $n \sim 10$. It is therefore necessary to make some subtractions from the quantity $\langle n,\ell|V_{2\gamma}|n,\ell\rangle$, in order to avoid double counting. Both these difficulties have been overcome, but I shall spare you the nasty details.[20] The important points are the following: (i) The calculated values of the n = 10 level splittings in He are consistent with experiment, but the experimental accuracy needs improvement by a factor of ten or so before the theory faces a stringent test. (ii) The experiments are already sufficiently accurate to show clearly that the full formula (4.24), including the structure-dependent term (4.25) must be used. This is a consequence of the fact that if n = 10, the distance of the outer electron from the core is not large enough for the r^{-5} term to be a good approximation to the retardation correction.

A comprehensive discussion of the quantum field-theoretic approach to the energy levels of Rydberg states in He-like ions, justifying and extending the approach described in Ref. 20 will appear soon.[21]

V. FURTHER APPLICATIONS; CONCLUDING REMARKS

A. Forbidden M1 Transitions in He-like Ions.

Another example of the formalism described in III involves the decay

$$2^3S_1 \rightarrow 1^1S_o + \gamma \tag{5.1}$$

in He or He-like ions. This decay is highly forbidden for small Z because in a nonrelativistic description it involves orthogonal spatial wave functions as well as a change in the total spin quantum number. Thus with initial and final wavefunctions of the form

$$\psi_i = \phi_i\chi_i \ , \quad \psi_f = \phi_f\chi_f \tag{5.2}$$

the matrix element

$$M_{nr} = i\langle\psi_f|(\vec{\sigma}_1 + \vec{\sigma}_2)|\psi_i\rangle\cdot(\vec{k} \times \vec{\varepsilon}^*/m) \tag{5.3a}$$

vanishes, both the factors involved being zero:

$$\langle\phi_f|\phi_i\rangle = \langle\chi_f|\vec{\sigma}_1 + \vec{\sigma}_2|\chi_i\rangle = 0 \ . \tag{5.3b}$$

If retardation is taken into account by restoring factors $\exp(-i\vec{k}\cdot\vec{r}_i)$, M_{nr} no longer vanishes but is smaller than naive expectation by a factor of order $(Z\alpha)^2$.

However, it turns out that the amplitude which one gets this way, while having the right order of magnitude, is quantitatively incorrect. From the point of view of the formalism described in Sec. III this is because contributions of comparable magnitude come from intermediate states involving virtual pairs. If we use, as suffices for the case at hand, the free no-pair hamiltonian (3.30b), the time-ordered diagrams which contribute are those shown in Fig. 12 and 13. To the accuracy needed these may all be evaluated in terms of the nonrelativistic wave

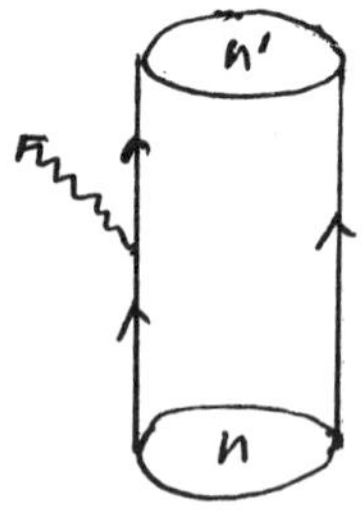

Fig. 12 Lowest order diagram for the decay (5.1) not involving virtual pairs.

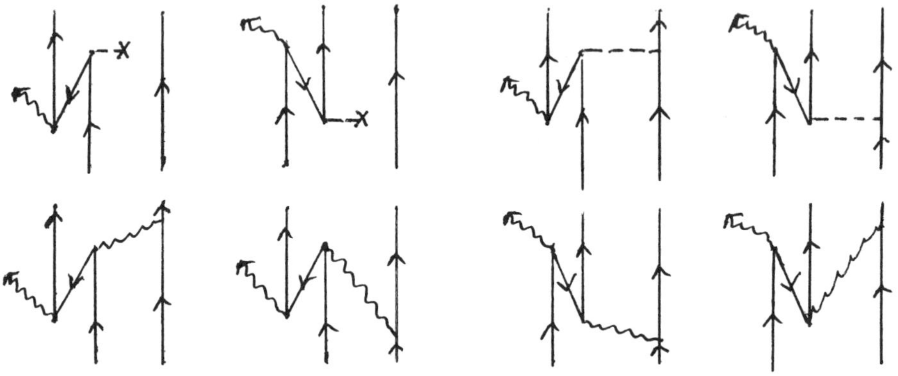

Fig. 13 Lowest order diagrams for the decay (5.1) involving virtual pairs. Bubbles indicating initial and final states are omitted.

functions (5.2). In leading order the contributions of the pair-effects involving e-e Coulomb and transverse-photon interactions cancel, and the final result takes the form

$$M = 2i\langle\chi_f|\vec{\sigma}_1\cdot\frac{\vec{k}\times\vec{\varepsilon}^{\,*}}{m}|\chi_i\rangle\ I \tag{5.4}$$

where

$$I = \langle\phi_f|\vec{p}_1^{\,2}/3m^2 + k^2r_1^2/12 - r_1V'_{ext}(r_1)/6m|\phi_i\rangle\ . \tag{5.5}$$

The term involving $V'(r_1)$ arises from virtual-pair creation by the external field.

The decay rate computed from (5.4) is in excellent agreement with experimental values over a range of lifetimes which vary from 10^4 sec for He to 0.17×10^{-9} sec for Kr, corresponding to a range of fourteen orders of magnitude.[22] Some improvement in experimental accuracy will be necessary before the theory, including radiative corrections to the leading term, can be put to a more stringent test.

B. Remarks on strong-field QED

As you know, the subject of strong-field electrodynamics has become of great interest in recent years, particularly in connection with the experimental search for anomalous positron production in heavy ion collisions, as will be discussed in detail by other lecturers.

Here I want to just touch on some aspects of this problem which seem to have been neglected in the literature.

Recall first that for an electron moving in a pure Coulomb field, $V = -Z\alpha/r$, the eigenvalue associated with the lowest lying S-wave bound state is

$$E(1s) = m(1-Z^2\alpha^2)^{1/2} ,$$

according to the Dirac equation. For $Z > 137$ this becomes imaginary and the physical interpretation of the Dirac equation becomes obscure. However,if the nuclear charge is spread out in a physically reasonable way then, according to the results of numerical calculation, the 1s-state eigenvalue is still positive at $Z = 137$, decreases to zero at $Z \sim 150$ and for still larger Z remains real, decreasing to $-m$ as Z approaches 170 or so. For Z exceeding this critical value, there appear states in the negative-energy continuum which, while not normalizable, have an appreciable localized part. If one thinks of these states as quasi-bound states with binding energies exceeding 2m, rather than as resonances in positron scattering, one is led to imagine that in a collision of say, two bare uranium nuclei, such a state is some of the time temporarily formed, after the creation of an e^--e^+ pair, with the e^+ escaping to infinity. If so, one has available an exotic mechanism for positron production.[23]

In connection with these issues a number of questions come to mind which are at least of mathematical interest:

i) In the pure Coulomb case is the existence of a critical value for the coupling constant $\gamma = Z\alpha$ connected with the fact that the spectrum of $h_{D;ext}$ is unbounded below?

ii) What is the effect of recoil, i.e. of the finite mass of the physical system to which the electron binds, on critical values?

iii) For the finite mass case, what is the effect of including transverse photon exchange?

iv) More generally, how does one define critical values for the case of two interacting finite mass systems.[2]

These questions have been studied recently by Gene Hardekopf and myself within the context of the two-body no-pair Hamiltonian defined by

$$h_{++} = h_D(1) + h_D(2) + \Lambda_{++} V \Lambda_{++} \tag{5.6a}$$

and the eigenvalue equation

$$h_{++}\phi = E\phi \tag{5.6b}$$

with $V = V_C$ or $V = V_C + V_B$, i.e. a Coulomb potential or a Coulomb plus Breit potential.[24]

Let us begin with the last question posed. Suppose that for small coupling the lowest lying bound state of (5.6) has mass

$$E = E(m_1, m_2; \gamma) \; . \tag{5.7}$$

With $m_1 \leq m_2$ it may be that for sufficiently large γ , say $\gamma > \gamma_{dec}$, we have

$$E < m_2 - m_1 \; , \tag{5.8}$$

so that the binding energy exceeds $2m_1$. Then the inclusion of any interaction which permits virtual production of $1,\bar{1}$ pairs will lead to instability of particle 2 via the process

$$2 \rightarrow (1,2) + 1 + \bar{1} \; . \tag{5.9}$$

There may also be a value of γ , call it γ_{max} , beyond which E becomes complex or beyond which h_{++} ceases to be self adjoint on a dense domain.

To answer the first question we first take the limit $m_2 = \infty$ in (5.6). With

$$E' = \lim_{m_2 \to \infty} \left[E(m_1, m_2; \gamma) - m_2 \right] \; . \tag{5.10}$$

we get

$$[E_{op}(1) + \Lambda_+(1) V_C \Lambda_+(1)]\phi(1) = E'\phi(1) \tag{5.11}$$

which is equivalent to the eq. (2.46) studied in Sec. II, and involves an operator whose spectrum is bounded below. Nevertheless, there is a maximum value for γ given by

$$\gamma_{max} = 2/(\tfrac{\pi}{2} + \tfrac{2}{\pi}) \approx .91 \; . \tag{5.12}$$

The fact that γ_{max} is less than unity is not surprising, because the pair effects omitted in (5.11) correspond to a repulsive rather than an attractive potential.

If we keep m_2 finite and look for the maximum value of γ there is a surprise: The result is, for $V = V_c$,

$$\gamma_{max} = 4/(\frac{\pi}{2} + \frac{2}{\pi}) \simeq 1.81 \qquad (5.13)$$

which is larger than (5.12) by a factor of two and independent of the value of m_2. This shows that the limits involved don't commute or, to put it another way, that γ_{max} is a discontinuous function of m_2/m_1 at $(m_2/m_1)^{-1} = 0$.

Turning to γ_{dec}, one finds that in the equal mass case it coincides with γ_{max} and that as m_2 increases it decreases, but very slowly. For example for $m_2/m_1 = 100$, γ_{dec} is about 1.25. Presumably it remains larger than .91, the $m_2 = \infty$ value of γ_{max}; it would be interesting to study this question in more detail.

When the Breit interaction is included, there is drastic change in γ_{max}: It is still independent of the value of m_2 but smaller than the pure Coulomb value by more than a factor of two; our numerical work indicates that γ_{dec} exists for this case also, but our accuracy is not sufficient to be certain of this.

It must be stressed that all these results have been obtained for the case of point-like interactions, with 1/r singularities. It may well be that the replacement of point charges by distributed charges will have a drastic effect on the critical values involved and the results described are only of academic interest. Further study of these questions would, I believe, be very worthwhile.

C. Concluding remarks

I hope that with these examples I have convinced you that the no-pair formalism is a convenient tool for handling a wide variety of problems in atomic physics which require the simultaneous consideration of relativistic and correlation effects. With regard to heavy atoms the philosophy of this approach is to delay the making of any central field approximation as long as possible, by providing a many-body configuration space Hamiltonian which does not suffer from continuum dissolution and which has bona-fide eigenfunctions with a clear-cut field theoretic interpretation.[25] Decay amplitudes and further level shifts may at least in principle be calculated from these, as matrix elements of operators which can be constructed by returning to field theory, without fear of double counting. For example, for radiative transitions which satisfy the selection rules for E1 decay the leading contribution to the decay amplitude is contained in the expression

$$M = \langle\psi_f|R(\vec{k})|\psi_i\rangle \cdot \vec{\varepsilon}^* \tag{5.14}$$

where

$$\vec{R}(\vec{k}) = \sum_i \vec{\alpha}_i e^{-i\vec{k}\cdot\vec{r}_i} . \tag{5.15}$$

This is also true for ordinary M1 decays, but for forbidden M1 decays one must add contributions from virtual-pair effects, as discussed above.

Note that the identity

$$\vec{\alpha}_i = i[h_D(i), \vec{r}_i] \tag{5.16}$$

would permit one to rewrite the dipole approximation to M, defined by

$$M_{dip} = \langle\psi_f|\vec{R}(0)|\psi_i\rangle\cdot\vec{\varepsilon}^* , \tag{5.17}$$

in the form

$$M'_{dip} = i(E_f-E_i) \langle\psi_f|\vec{S}(0)|\psi_i\rangle\cdot\vec{\varepsilon}^* \tag{5.18}$$

where

$$\vec{S}(0) = \sum_i \vec{r}_i , \tag{5.19}$$

<u>if</u> the wavefunctions were eigenfunctions of a local Hamiltonian. However, H_+ is not local even for the choice $U = V_{ext}$ because of the V_{e-e} term. Thus the commutator between, say, H_+^{ext} and $\vec{S}(0)$ is given by

$$i[H_+^{ext}, \vec{S}(0)] = \vec{R}(0) + i[\Lambda_+^{ext} V_{ee} \Lambda_+^{ext}, \vec{S}(0)] \tag{5.20}$$

and the relation between M_{dip} and M'_{dip} is given by

$$M_{dip} = M''_{dip} \tag{5.21}$$

where

$$M''_{dip} = M'_{dip} - i\langle\psi_+|V_{ee}[\Lambda_+^{ext}, \vec{S}(0)] + [\Lambda_+^{ext}, \vec{S}(0)]V_{ee}|\psi_i\rangle\cdot\vec{\varepsilon}^* . \tag{5.22}$$

The fact that the so-called velocity and length forms, M_{dip} and M'_{dip}, are not equal is not a "failure of gauge invariance". We have chosen the transverse or radiation gauge in the beginning and the question of gauge invariance simply no longer arises. The electromagnetic interaction is $\vec{\alpha}\cdot\vec{A}_T$ and not $\vec{r}\cdot\vec{E}_T$. The fact that under simplified circumstances these give equivalent results should not be allowed to obscure this fundamental fact. The difference between M_{dip} and M''_{dip} when only approximate wavefunctions are available can be used as a measure of the uncertainty in the values of dipole decay matrix elements, in a way analogous to the situation in the nonrelativistic case. Note that the commutator term in (5.22) becomes small compared to M'_{dip} in the nonrelativistic regime.

Finally, let us briefly consider the question of the calculation of higher-order corrections to energy levels. The calculation of the effects of transverse-photon exchange between electrons offers no problems of principle. As discussed elsewhere[4] a large part of these effects can be taken into account by adding the Breit operator to the e-e Coulomb interaction in H_+. The inclusion of the leading parts of contributions coming from virtual-pair creation is also straightforward. The only delicate aspect is the calculation of true radiative corrections, i.e. self-energy, vertex and vacuum polarization effects for atoms with more than one electron. It was shown long ago how these could be handled, with the help of a symmetrized form of the Gell-Mann Low level shift formula,[26] at least for the case of two-electron atoms. This formula allows the use of formally covariant methods of calculation; in particular, it can be used to show that one can mix gauges in the calculation of level shifts, i.e. one can use Feynman propagators for photons which are emitted and absorbed by the same electron and non-covariant propagators for exchanged photons. It seems likely that similar methods will work in the many-electron case.

Acknowledgement

This work was supported in part by the National Science Foundation, by the General Research Board of the University of Maryland and by the Ernest Kempton Adams Fund of Columbia University.

References

1. For an excellent discussion, see J.D. Bjorken and S.D. Drell, Relativistic Quantum Fields (McGraw-Hill Inc., New York, 1956), pp. 68-93.
2. A.M. Sessler and H.M. Foley, Phys. Rev. 92, 1321 (1953). See also J. Sucher and H.M. Foley, Phys. Rev. 95, 966 (1954).
3. S. Weiskop and J. Sucher, U. of MD report, 1983 (unpublished).
4. This section is based on the paper J. Sucher, Phys. Rev. D22, 348 (1980) and on later reviews and extensions described in (a) Proceedings of the Agronne Workshop on the Relativistic Theory of Atomic Structure, H.G. Berry, K.T. Cheng, W.K. Johnson and Y.-K. Kim, Eds. ANL-80-116 (Argonne National Laboratory, Argonne, IL, 1980); (b) Proceedings of the NATO Advanced Study Institute on

Relativistic Effects in Atoms, Molecules and Solids, G. Malli, Ed. (Plenum, New York, (1982) and (c) Int. Jour. Quant. Chem. XXV, 3 (1984). For closely related work see M. Mittleman, Phys. Rev. A24, 1167 (1981).

5. J. Sucher, UMPP/86-29, to appear in Proceedings of the Atomic Theory Workshop on Relativity and QED Effects in Heavy Atoms, (National Bureau of Standards, Gaithersburg, MD, May 1985); referred to as NBS Workshop 1985 hereafter.
6. G.E. Brown and D.G. Ravenhall, Proc. Roy. Soc. London, Sec. A208, 552 (1951).
7. J. Sucher, Phys. Rev. Letters 55, 1033 (1985).
8. I thank Y.-K. Kim for a discussion of this point.
9. For reviews, see e.g. J.-P. Desclaux in Ref. 4a and I.P. Grant, ref 4b.
10. B. Hess, to appear in Phys. Rev. A.
11. J. Sucher, Columbia University Ph.D. Dissertation (1957, unpublished) and Phys. Rev. 109, 1010 (1958); H. Araki, Prog. Theoret. Phys. 17, 619 (1957). These papers use the Bethe-Salpeter equation rather than the simpler approach sketched below.
12. G.W.F. Drake and A.J. Makowski, J. Phys. B, 18, L103 (1985); A. Ermolaev, Durham University preprint.
13. S.L. Palfrey and S. Lundeen, Phys. Rev. Lett. 53, 1141 (1924).
14. G. Feinberg and J. Sucher, Phys. Rev. A2, 2395 (1970); J. Sucher, in Cargese Lectures in Physics, edited by M. Levy (Gordon and Breach, New York, 1977), Vol. VII, pp. 43-110.
15. H.B.G. Casimir and D. Polder, Phys. Rev. 73, 360 (1948).
16. G. Feinberg and J. Sucher, Phys. Rev. 27, 1958 (1983).
17. E.J. Kelsey and L. Spruch, Phys. Rev. A18, 1055 (1978).
18. J. Bernabeu and R. Tarrach, Ann. Phys. (N.Y.) 102, 323 (1976).
19. R. Drachman, Phys. Rev. A26, 1228 (1982).
20. C.-K. Au, G. Feinberg and J. Sucher, Phys. Rev. Lett. 53, 1145 (1984).
21. C.-K. Au, G. Feinberg and J. Sucher, paper in preparation.
22. G. Feinberg and J. Sucher, Phys. Rev. Lett. 26, 681 (1971); G.W.F. Drake, Phys. Rev. A5, 1979 (1972); for a review, with extensive references see J. Sucher, Rep. Prog. Phys. 41, 1781 (1978).
23. For a comprehensive review, see G. Soff, NBS Workshop 1985.
24. G. Hardekopf and J. Sucher, Phys. Rev. A31, 2020 (1985).
25. For other points of view, see K. Dietz, in NBS Workshop 1985, and G. Feldman and T. Fulton, Johns Hopkins University preprint, August 1985.
26. J. Sucher, Phys. Rev. 107, 1448 (1957).

RELATIVISTIC AND QED CALCULATIONS FOR MANY-ELECTRON AND FEW-ELECTRON ATOMS

Peter J. Mohr

Physics Department
Yale University
New Haven, CT 06511

I. PRESENT STATUS

Over the past two decades, calculations of atomic energy levels have reached a high degree of technical refinement. For example, for many-electron atoms, relativistic Dirac-Fock calculations together with approximate estimates of quantum electrodynamic corrections give precise predictions for inner-shell transitions that agree with experiment to within parts in 10^4. With this approach, the numerical precision of the calculations is not the factor that limits the agreement with experiment, but rather it is the approximate nature of the formulation that produces the current uncertainty in theory. To make further progress it will be necessary to refine the formulation to include corrections to the approximate equations, or perhaps to make a more radical departure from traditional Dirac-Fock methods. In any case, it will be necessary to include radiative (QED) corrections in the formulation, since the level of precision already achieved is well past the magnitude of these effects.

MANY-ELECTRON ATOMS

In many-electron atoms, relativistic and quantum electrodynamic effects are most pronounced in inner shells. Sophisticated computer codes that calculate these energy levels based on the multiconfiguration relativistic Dirac-Fock formalism have been written both by Desclaux[1] and by Grant et al.,[2] and an extensive tabulation of energy levels based on a Dirac-Fock-Slater approximation has been made by Chen et al.[3,4] These methods take into account the finite size of the nucleus exactly in the calculation of the wavefunctions. In addition to the Dirac-Fock energy level, smaller corrections consisting of the Breit interaction, the self energy, and the vacuum polarization are included in the calculations. The Breit interaction, including retardation, is added as a perturbation, and the self energy is calculated on the basis of a hydrogenic approximation with a phenomenological correction for screening by the other electrons. The vacuum polarization is treated in perturbation theory with the main effect arising from the Uehling potential.

Despite the various approximations that are built into this formulation, the theoretical values for inner-shell electron transition frequencies predicted by these calculations are in excellent agreement with experiment.

A comparison of theory[4] and experiment[5] is given in Fig. 1, where the radiative correction to each level is approximated by the appropriate Coulomb field value of the self energy of the corresponding one-electron state with or without a phenomenological screening correction as indicated in the figure. It is clear from the figure that the precision of the theory is at the level where details of the electron interaction corrections to the radiative corrections are important. The screening correction improves the agreement between theory and experiment in this particular case, but a sound theoretical treatment needs to be developed.

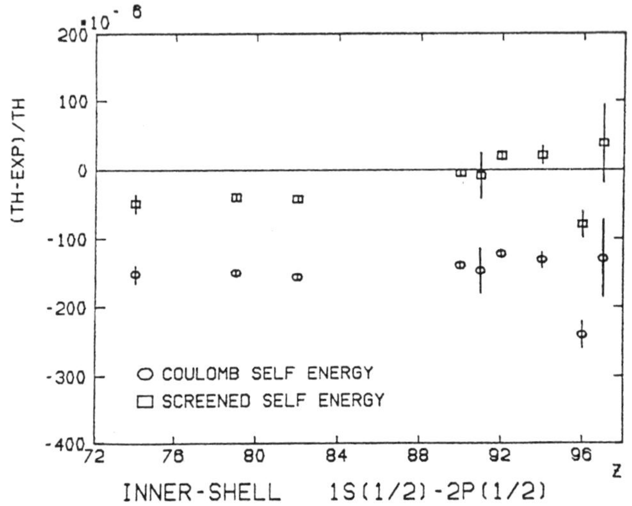

Figure 1. Comparison of theory and experiment for inner-shell transitions in heavy atoms. The vertical scale is the fractional difference between theory and experiment in parts per million.

FEW-ELECTRON ATOMS

High-Z few-electron atoms provide a system on which methods of calculation can be tested without the complication of dealing with many electrons. Because Z is large it is necessary to include relativistic effects and radiative corrections in the calculations. The development of sources of highly-ionized atoms has made experiments on such systems feasible, so that meaningful checks of the calculations are possible.

In high-Z few-electron atoms, there is a small parameter not available in many-electron atoms. In the few-electron case, the ratio of the strength of interactions among the electrons is small compared to the strength of the interaction of each electron with the nucleus. This ratio, which is proportional to 1/Z, is a useful expansion parameter for calculating energy levels, and evaluations based on such an expansion have been made both nonrelativistically and relativistically.[6] In addition, multiconfiguration

Dirac-Fock calculations analogous to those made for heavy neutral atoms have been done for few-electron atoms,[7] and the Relativistic Random Phase Approximation has been applied to few-electron atoms.[8] Breit interaction, self energy, and vacuum polarization effects are included as corrections in the evaluation of energy levels. There is good agreement between theory and experiment for Z near 20 at the level of precision where relativistic electron interaction effects on the self energy, the largest corrections not included in the calculations, play a role.[6]

II. OPEN QUESTIONS

Despite the excellent agreement between theory and experiment for the calculations described above, there are important basic questions that need to be answered in order to improve the accuracy of the theory. Three examples are listed here.

i. What is the correction for pair effects at high Z? It can be shown that the starting Hamiltonian on which the Dirac-Fock calculations are based can be derived from QED by making a no-pair approximation in which the production of virtual electron-positron pairs is neglected, and that the pair corrections to this approximation are higher order in $Z\alpha$. At high Z, these corrections may be important, and an estimate of their size is needed.

ii. How can corrections from the combined effects of relativity, electron interactions, and self energy be taken into account? Such corrections appear to be the largest known theoretical uncertainty in the atomic levels described above.

iii. What is the relation between the projection operator formalism as discussed by Sucher[9] and the imposition of boundary conditions as implemented in the actual calculations? The boundary conditions are imposed in calculations of Desclaux and Grant by fixing the ratio of small to large radial components of the Dirac wavefunction to the appropriate hydrogenic values in the initial trial functions. At short distances, the ratio is fixed to the hydrogenic value for charge number Z, while at large distances from the nucleus, the ratio is fixed to the Coulomb value for a unit charge source, which takes the screening effect of the other electrons into account. This procedure leads to stable solutions of the Dirac-Fock equations that are closly related, if not equal, to the no-pair solutions.

III. PERTURBATION THEORY

The theory of relativistic and QED effects in high-Z few-electron atoms can be formulated in terms of bound-state quantum electrodynamic perturbation theory. In this framework, pair effects are included exactly, corrections corresponding to the combined effects of relativity, electron interactions, and self energy can be calculated; and effects due to negative energy states are treated properly. The formulation is based on the following physical idea. In a highly-ionized atom, the interaction of each electron with the nucleus of charge number Z dominates over the interactions between the electrons and the radiative corrections. Hence, the latter corrections are treated as perturbations to zero-order Dirac hydrogenic electrons. In this approach, the perturbation series for the energy takes the form

$$E = [f_0(Z\alpha) + 1/Z\ f_2(Z\alpha) + 1/Z^2\ f_4(Z\alpha) + \ldots](Z\alpha)^2 mc^2 \qquad (1)$$

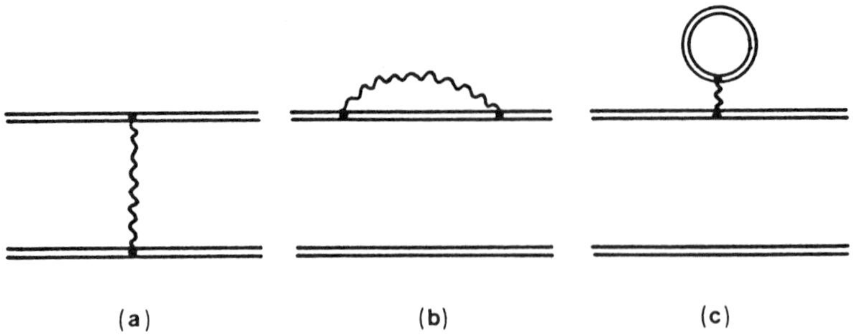

Figure 2. Feynman diagrams for the one-photon corrections in high-Z two-electron atoms.

where the functions $f_n(Z\alpha)$ approach constants as $Z\alpha$ approaches zero, and the series converges rapidly when Z is large. In this series, the first term is the zero-order Dirac hydrogenic approximation, and the second term is the second-order correction corresponding to Feynman diagrams with one photon. The third term corresponds to two photons, etc. The one-photon Feynman diagrams appear in Fig. 2. In that figure, the double lines represent bound hydrogenic electrons and the wavy lines represent Feynman gauge photons, i.e., complete photons that in the Coulomb gauge correspond to the sum of the instantaneous Coulomb interaction and retarded transverse photon exchange.

The one-photon corrections have been evaluated in detail.[10,11] In the remainder of this section, we examine the leading two-photon correction in order to illustrate the connection between the perturbation theory formulation described here and the more familiar no-pair Hamiltonian approach.[9]

Figure 3 shows the two-exchanged-photon diagrams that correspond to the dominant fourth-order corrections. In that figure, the double lines correspond to a contraction of the electron-positron field which gives the propagation function (in the notation of Ref. 10)

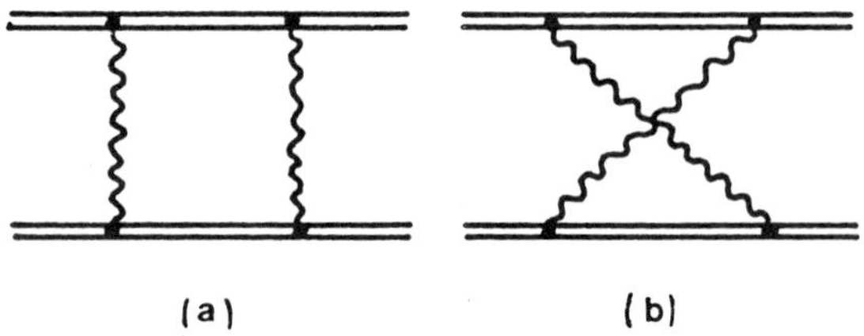

Figure 3. Feynman diagrams for the two-exchanged-photon corrections in high-Z two-electron atoms.

$$\underbrace{\psi(x_2)\bar{\psi}(x_1)} = S_F(x_2,x_1) = \frac{1}{2\pi i}\int_{-\infty}^{\infty} dz \sum_n \frac{\phi_n(\vec{x}_2)\bar{\phi}_n(\vec{x}_1)}{(1-i\delta)E_n - z} e^{-iz(t_2-t_1)}$$

$$= \begin{cases} \sum_{n+} \phi_n(\vec{x}_2)\bar{\phi}_n(\vec{x}_1)e^{-iE_n(t_2-t_1)}; & t_2 > t_1 \\ -\sum_{n-} \phi_n(\vec{x}_2)\bar{\phi}_n(\vec{x}_1)e^{-iE_n(t_2-t_1)}; & t_2 < t_1 \end{cases} \tag{2}$$

The last equality in (2) follows by contour integration where the location of the poles of the integrand is determined by the parameter δ, in the energy denominator of the integrand, which fixes the boundary conditions. The expression for S_F on the first line is amenable to calculation since the integrand is the solution of a differential equation ($\hbar$=c=m=1)

$$[-i\vec{\alpha}\cdot\vec{\nabla}_2 + \beta + V(x_2) - z]\sum_n \frac{\phi_n(\vec{x}_2)\bar{\phi}_n(\vec{x}_1)}{E_n - z} = \sum_n \phi_n(\vec{x}_2)\bar{\phi}_n(\vec{x}_1) = \delta(\vec{x}_2-\vec{x}_1)\gamma^\circ \tag{3}$$

The fourth-order level shift corresponding to the Feynman diagram in Fig. 3(a) is

$$\Delta E^{(4)} \sim \frac{i\alpha^2}{4\pi}\int d\vec{x}_4\int d\vec{x}_3\int d\vec{x}_2\int d\vec{x}_1 \sum_{\substack{km\ell n\\ ij}}{}' \int_{-\infty}^{\infty} dz\, \bar{\phi}_k(\vec{x}_4)\gamma_\mu \frac{\phi_i(\vec{x}_4)\bar{\phi}_i(\vec{x}_3)}{z-(1-i\delta)E_i}\gamma_\nu\phi_\ell(\vec{x}_3)$$

$$\times \bar{\phi}_m(\vec{x}_2)\gamma^\mu \frac{\phi_j(\vec{x}_2)\bar{\phi}_j(\vec{x}_1)}{E_n+E_\ell-(1-i\delta)E_j-z}\gamma^\nu\phi_n(\vec{x}_1)\frac{e^{-b_2x_{42}}}{x_{42}}\frac{e^{-b_1x_{31}}}{x_{31}} \tag{4}$$

$$\times\, \delta(E_m + E_k,\, E_n + E_\ell)\langle a_k^\dagger a_m^\dagger a_n a_\ell\rangle$$

where

$$b_2 = -i\sqrt{(E_k-z)^2+i\varepsilon}\ ;\ \mathrm{Re}(b_2)>0$$

$$b_1 = -i\sqrt{(z-E_\ell)^2+i\varepsilon}\ ;\ \mathrm{Re}(b_1)>0 \tag{5}$$

$$\delta(a,b) = \begin{cases} 1 \text{ if } a=b \\ 0 \text{ if } a\neq b\end{cases}$$

and $a_n^\dagger$ is the creation operator for an electron in state n. The prime on the summation in (4) indicates that terms for which $E_i+E_j = E_n+E_\ell$ are not

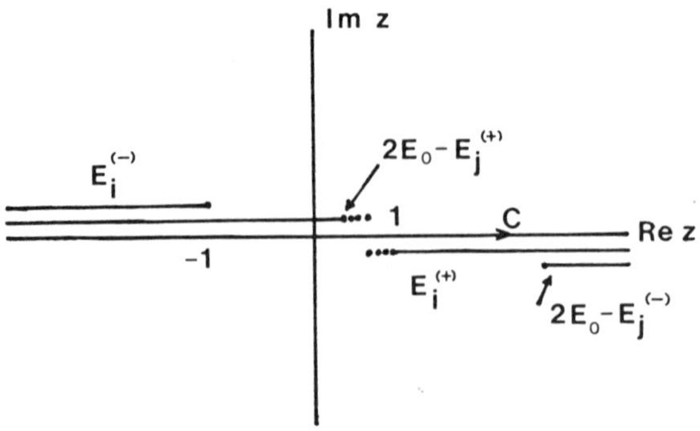

Figure 4. Complex z plane for the integrand of (6). The lines and dots indicate possible locations of the poles for all values of i and j.

included. For the two-electron ground state, to lowest order ($b_2 = b_1 = 0$), the integral over z in (4) is

$$I = \frac{1}{2\pi i}\int_{-\infty}^{\infty} dz \, \frac{1}{z-(1-i\delta)E_i} \, \frac{1}{2E_0-(1-i\delta)E_j-z} \tag{6}$$

Figure 4 shows the complex z plane and the possible locations of the poles of the integrand for positive $E^{(+)}$ and negative $E^{(-)}$ energy eigenvalues. The values of the integral for the four possible combinations of positive or negative values of E_i and E_j are

$$\begin{aligned}
&E_i>0,\ E_j>0: \quad I = -\frac{1}{2E_o-E_i-E_j} \\
&E_i>0,\ E_j<0: \quad I = 0 \\
&E_i<0,\ E_j>0: \quad I = 0 \\
&E_i<0,\ E_j<0: \quad I = \frac{1}{2E_o-E_i-E_j} \quad \rightarrow \text{pair correction}
\end{aligned} \tag{7}$$

The lowest order level shift is thus

$$\Delta E^{(4)} \sim \int d\vec{x}_4 \int d\vec{x}_3 \int d\vec{x}_2 \int d\vec{x}_1 \, \Phi^\dagger(\vec{x}_4,\vec{x}_2)\, \frac{\alpha}{x_{42}} \times \sum_{\substack{E_i>0 \\ E_j>0}}' \frac{\phi_i(\vec{x}_4)\phi_j(\vec{x}_2)\phi_i^\dagger(\vec{x}_3)\phi_j^\dagger(\vec{x}_1)}{2E_o-E_i-E_j} \, \frac{\alpha}{x_{31}} \, \Phi(\vec{x}_3,\vec{x}_1) \tag{8}$$

where Φ is the two-electron hydrogenic product wave function. This correction is the same as the result of ordinary second-order perturbation theory applied to the Hamiltonian

$$H = H_1^D + H_2^D + L^{++} \frac{\alpha}{x_{21}} L^{++} \tag{9}$$

where H_i^D is the one-particle Dirac Hamiltonian, and the projection operators L^{++} exclude the negative-energy intermediate states.

From a comparison of Eqs. (8) and (4), it is evident that the boundary conditions imposed by the parameter δ are closely related to the appearance of projection operators. For an appropriate deformation of the contour of integration over z in (4), the parameter δ can be taken to zero. Thus the boundary conditions can be implemented by solving Eq. (3) for a set of complex values of z and carrying out the integration over the corresponding contour. The feasibility of this general approach has been demonstrated in a related non-relativistic calculation.[12]

This research was supported by the National Science Foundation, Grant No. PHY-8403322.

REFERENCES

1. J. P. Desclaux, Compt. Phys. Commun. 9, 31 (1975).
2. I. P. Grant, B. J. McKenzie, P. H. Norrington, D. F. Mayers, and N. C. Pyper, Compt. Phys. Commun. 21, 207 (1980).
3. K.-N. Huang, M. Aoyagi, M. H. Chen, B. Crasemann, and H. Mark, At. Data Nucl. Data Tables 18, 243 (1976).
4. M. H. Chen, B. Crasemann, M. Aoyagi, K.-N. Huang, and H. Mark, At. Data Nucl. Data Tables 26, 561 (1981).
5. E. G. Kessler, Jr., R. D. Deslattes, D. Girard, W. Schwitz, L. Jacobs, and O. Renner, Phys. Rev. A 26, 2696 (1982).
6. For a review, see P. J. Mohr, Nucl. Inst. and Meth. in Phys. Res. B9, 459 (1985).
7. This approach is reviewed by I. P. Grant, Nucl. Inst. and Meth. in Phys. Res. B9, 471 (1985).
8. W. R. Johnson and C. D. Lin, Phys. Rev. A 14, 565 (1976).
9. J. Sucher, in these proceedings.
10. P. J. Mohr, Phys. Rev. A 32, 1949 (1985).
11. P. J. Mohr, At. Data and Nucl. Data Tables 29, 453 (1983).
12. J. D. Lakdawala and P. J. Mohr, Phys. Rev. A 29, 1047 (1984).

POSTER SESSION ON RELATIVISTIC ATOMIC STRUCTURE

I. RELATIVISTIC METHODS FOR FEW-ELECTRON IONS

J.-P. Desclaux and P. Indelicato

Centre D'Etudes Nucléaires de Grenoble
B. P. 85X, 38041 Grenoble Cedex, and
Institut Curie
11, rue Pierre et Marie Curie, 75231 Paris Cedex 05
France

Various methods of calculating energy levels in few-electron ions are compared. Methods considered are expansions in powers of 1/Z and Zα, and approaches that take into account nonrelativistic correlation effects and relativistic and quantum electrodynamic effects separately. These approaches are contrasted to methods that deal with correlation and relativistic effects simultaneously, such as multiconfiguration Dirac-Fock and Relativistic Random Phase Approximation calculations.

II. MEANINGFUL COMPARISON OF THEORY AND EXPERIMENT IN HELIUMLIKE IONS

P. Indelicato and J.-P. Desclaux

Institut Curie
11, rue Pierre et Marie Curie, 75231 Paris Cedex 05, and
Centre D'Etudes Nucléaires de Grenoble
B. P. 85X, 38041 Grenoble Cedex
France

A comparison of theory and experiment is made for the transitions $1s^2\ ^1S_0$ - $1s2p\ ^3P$ in heliumlike argon (Z = 18), based on a multiconfiguration Dirac-Fock calculation of the levels.

III. EFFECT OF THE BREIT INTERACTION ON SUPERHEAVY ATOMS

P. Indelicato

Institut Curie
11, rue Pierre et Marie Curie, 75231 Paris Cedex 05,
France

The effect of the Breit interaction on superheavy atoms in the range $100 \leq Z \leq 170$ is examined. Particular attention is given to the magnitude of the corrections to energy levels due to the magnetic and retardation parts of the Breit interaction. It is found that there is a substantial magnetic correction that is of the order of 20% of the electrostatic interaction of the electrons.

IV. INTERPLAY BETWEEN CORRELATION AND RELATIVISTIC EFFECTS: $2s^2$ - 2s3p TRANSITIONS IN Be-LIKE IONS

Y.-K. Kim and A. W. Weiss

National Bureau of Standards
Gaithersburg, MD 20899

Striking effects of relativity and correlation in the beryllium isoelectronic series are illustrated. For the transitions $2s^2$ - 2s2p $^3P_1$ and $^1P_1$, the singlet is higher in energy than the triplet and the intercombination line is relatively weak, which is the expected result. On the other hand, for the transitions $2s^2$ - 2s3p $^3P_1$ and $^1P_1$, there are values of Z in the isoelectronic series for which the transition strengths are of comparable magnitude, and values for which the triplet level is higher in energy than the singlet level. In order to make qualitatively correct predictions of these effects, a calculation must include correlation and relativity, as is the case for multiconfiguration Dirac-Fock methods.

V. IMPLICIT PROJECTION OPERATORS IN BASIS-SET EXPANSIONS OF THE MOLECULAR DIRAC-FOCK-SLATER PROBLEM

W. Sepp

Universität Kassel, Institut für
Theoretische Physik, Kassel
Federal Republic of Germany

For relativistic molecular structure calculations, basis sets that are linear combinations of atomic orbitals are useful. An implicit positive energy projection operator is defined by summing over only positive energy atomic orbitals. Variational collapse is thereby avoided. The relation between such an implicit projection operator and a complete molecular projection operator is examined.

VI. NON-RELATIVISTIC MANY-BODY PERTURBATION THEORY CALCULATION: TOWARD RELATIVISTIC MANY-BODY PERTURBATION THEORY

I. Lindgren and E. Lindroth

Department of Physics
Chalmers University of Technology and University of Gothenburg
S-41296 Göteborg, Sweden

A. Non-relativistic many-body perturbation theory is discussed. Methods and results in the solution of inhomogeneous 1- and 2-particle equations are presented. B. Similar programs for the Dirac equation are considered. The 1-particle equation is equivalent to the relativistic random phase approximation (work by A.-M. Martensson-Pendrill), and the 2-particle equation is under study. C. Matrix diagonalization of the Dirac equation is being explored as a method of isolating positive energy solutions. For a weak external field, the upper components of the diagonal equation correspond to positive energy solutions.

VII. QUANTUM ELECTRODYNAMICS OF HIGH-Z FEW-ELECTRON ATOMS

P. J. Mohr

Physics Department
Yale University
New Haven, CT 06511

Calculation of energy levels of high-Z few-electron atoms based on relativistic quantum electrodynamic perturbation theory is discussed. A more detailed account is given elsewhere in these proceedings.

VIII. g-HARTREE METHOD

J. P. Connerade

Physikalisches Institut der Universität Bonn
Nussallee 12, 5300 Bonn 1, West Germany
Blackett Laboratory, Imperial College
London SW7 2AZ, England

A method of calculating atomic properties is discussed, and a comparison to experiment is made. A more detailed account is given elsewhere in these proceedings.

SUPERHEAVY ATOMS - ELECTRONS IN STRONG FIELDS

Fritz Bosch

Gesellschaft für Schwerionenforschung mbH
P.O. Box 11 05 41
D-6100 Darmstadt, Fed. Rep. of Germany

1. INTRODUCTION AND OVERLOOK

The production and investigation of stable superheavy atoms ($\alpha Z \gtrsim 1$) has to remain still for a long time a "Gedankenexperiment" only. This restriction of scientific development seems to be an act of deepest philanthropy on the part of Nature, however, when considering the "gifts" that followed to date from the man-made extension of the periodic system of elements. We may therefore speculate peacefully on the properties of superheavy atoms being quite sure not to meet the day after tomorrow with superheavy bombs or rockets. On the other hand, Nature allows a short (and harmless) glance onto the shadows of superheavy atoms which are formed transiently in adiabatic heavy ion-atom collisions for tiny 10^{-21}s, i.e. the time where the innermost electrons are exposed to the strong field of the combined nuclear charges ($Z_1 + Z_2 = Z_u \gtrsim 137$). It is one of the scopes of this lecture to show that some quantities measured in heavy ion collisions may render a direct 'snapshot' of the transient superheavy atom and, therewith, of the properties of electrons bound in very strong electromagnetic fields.

The lecture is organized as follows: First (Section 2) I will shortly present the predictions concerning electrons that are bound with a coupling strength $\alpha Z > 1$, with emphasis on the points listed below:

(1) The dependence of the binding energy of the innermost electron orbitals on charge number Z;

(2) The 'diving' of the 1s-orbital as a localized resonance into the negative energy continuum ("Dirac sea") at $Z_u \approx 173$;

(3) The decay of the neutral vacuum ground state by spontaneous positron creation.

Then (Section 3) I will describe the way we obtain information about strong binding from an analysis of heavy ion-atom collisions, especially from:

(1) The impact parameter dependence of inner shell vacancy production;

(2) The spectroscopy of high energy δ-electrons;

(3) The 'atomic' positron creation in close collisions.

The central part of the lecture will be the presentation and interpretation of positron spectra from supercritical ($Z_u > 173$) collision systems where narrow structures have been found quite unexpectedly (Section 4). The origin of these lines will be discussed, especially in the context of a recently proposed hypothesis (Section 5). According to it the narrow positron lines indicate both the formation of long-lived ($\gtrsim 10^{-19}$ s) giant nuclear molecules **and** the creation of spontaneous positrons giving, therefore, direct evidence for the predicted 'diving' at supercritical field strengths. Finally, the pro's and con's of this hypothesis (and also other hypotheses) will be critically analyzed.

2. PROPERTIES OF ATOMS WITH Z > 137

From the famous Dirac-Sommerfeld formula for the 1s-binding energy ($K = -1$)

$$E_{1s} = mc^2 (1-(\alpha Z)^2)^{1/2} \tag{2.1}$$

follows that $\alpha Z = 1$ is the ultimate coupling strength for a point-like nuclear charge center, caused by the singularity at the origin. State-of-the-art-calculations (Fri77) which take into account the finite nuclear size predict that the 1s-state acquires a binding energy of mc^2 at $Z \gtrsim 150$ (i.e. the total energy of the electron becomes zero) and of twice the electron rest mass at $Z \gtrsim 173$ (see Fig. 2.1).

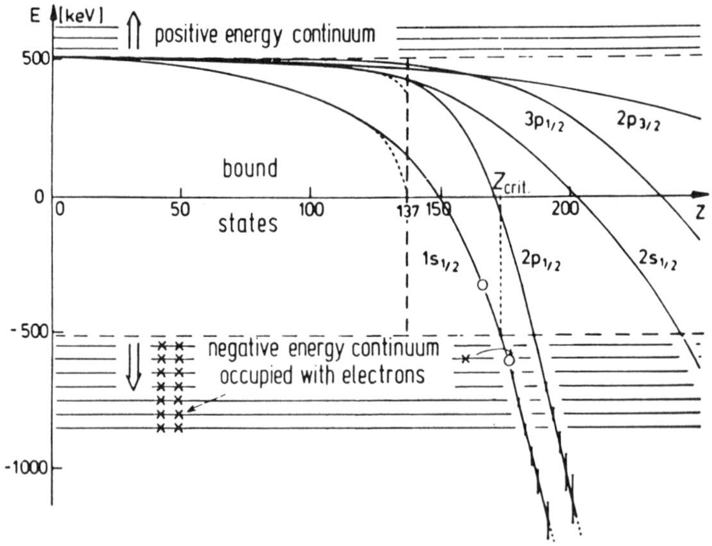

Fig. 2.1 Binding energies of electronic states in atoms as function of nuclear charge Z. At $Z_{crit} = 173$ the 1s-state dives into the negative energy continuum.

It has been shown that QED-corrections to the binding energy of the 1s-state, i.e. vacuum polarization and self-energy, remain very small at

Z = 173. The total energy shift produced by effects to first order in $\alpha = e^2/\hbar c$ is even less than 1 keV, because the contributions of vacuum polarisation (- 10.7 keV, Guy75) and of electron self-energy (+ 11 keV, Sof82) are cancelling almost completely one another. Hence, beyond this 'supercritical' charge Z = 173 the 1s-state becomes immersed in the negative energy continuum ('diving') as a narrow, localized resonance. The width of this resonance which is determined by the coupling between the localized 1s-state and the negative continuum is found (Rei81) to be of the order of a few keV for cases of interest (Z ∿ 180-190). Therefore, the empty 1s-state of any atom with Z > 173 would be 'spontaneously' filled by two electrons of the negative continuum in a time of ∿ 10^{-19} s, leaving behind two holes which escape as positrons with a kinetic energy of $E_T = |E_{1s}| - 2\,mc^2$ (see Fig. 2.2). This decay-time of a 1s-hole by spontaneous positron creation is much shorter than its expected radiative decay-time in the order of ∿ 10^{-18} s.

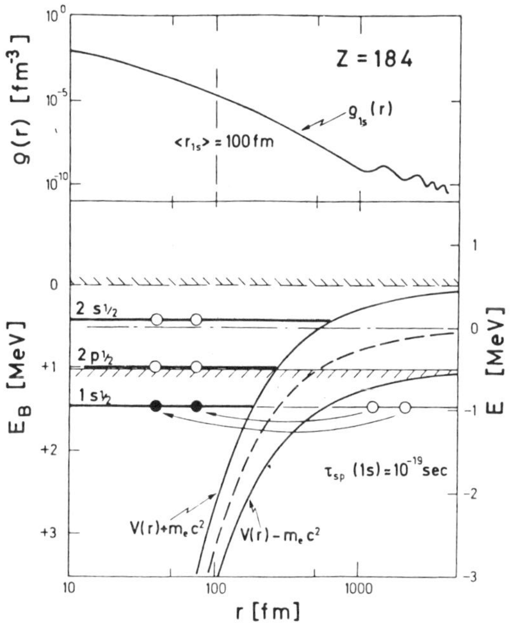

Fig. 2.2 Coulomb potential V(r) together with the gap borders $V(r)\pm mc^2$ for a fictive, supercritical nucleus Z = 184. An empty 1s-state decays spontaneously by emitting a monochromatic positron. Also shown (top) is the radial density of the 1s-resonance state embedded in the negative continuum.

The observation of such monochromatic positrons with a kinetic energy that corresponds exactly to the resonance energy in the negative continuum would give, therefore, a direct and clear evidence for the predicted 'diving'. In the language of field-theory, spontaneous positron creation can be understood as phase transition of the vacuum, from a neutral ground state to a new, charged ground state in the presence of supercritical fields. It is this context which gives the search for spontaneous positron creation its general and princicpal importance.

Stable nuclei with charge $Z > 137$ or even $Z > 173$ do not exist, but electric fields of the required strength can be produced in close collisions of two very heavy ions with $Z_1 + Z_2 = Z_u > 137$ or 173, respectively (Pb+Pb, U+U, etc.). Here, the binding energies of the quasimolecular electron orbitals depend on the time-varying nuclear separation R, and spontaneous positron emission is possible as long as R is smaller than a critical distance R_c, and if there is at least one hole in the deepest orbital. For the example of elastic U+U collisions nearby the Coulomb barrier, R_c is about 30 fm (for the 1sσ-orbital) and the time of supercritical binding is in the order of 10^{-21} s (see Fig. 3.1). The energy required to bring two uranium nuclei as close together as 20 fm is about 5 MeV per nucleon in the laboratory system. A comparison of the velocities of inner shell electrons and of the scattering nuclei then indicates that the motion of the electrons may be successfully described in an adiabatic picture. Assuming the electron (coordinate r) to adjust instantaneously to the nuclear motion (coordinate R) one obtains a complete basis of states ϕ_i, the quasimolecular states, which are eigenstates of the instantaneous relativistic two-center-Dirac Hamiltonian H_{TCD} (Mue76).

$$H_{TCD}(r,R)\ \phi_i(r,R) = E_i(R)\ \phi_i(r,R) \qquad (3.1)$$

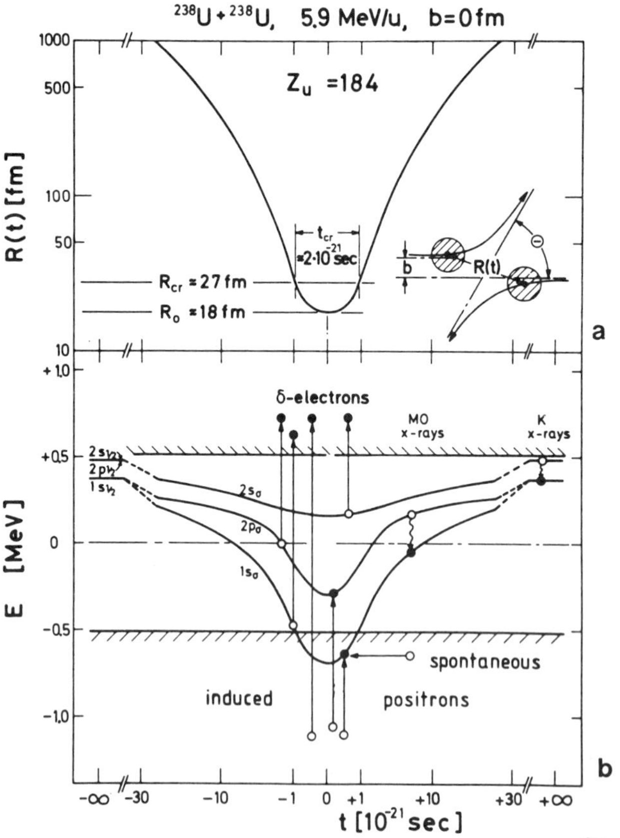

Fig. 3.1 Internuclear distance (a) and binding energies of quasimolecular orbitals as a function of time in the collision for elastic scattering processes.

with

$$H_{TCD}(r,R) = c \cdot \vec{\alpha} \cdot \vec{p} + \beta\, mc^2 + V(r,R) \tag{3.2}$$

Assuming now that the nuclei move on classical trajectories R(t) we get the time-dependent two-centre-Dirac equation

$$i\, \partial/\partial t\; \phi_i\; (\vec{R}(t)) = H_{TCD}\; (\vec{R}(t))\; \phi_i\; (\vec{R}(t)) \tag{3.3}$$

where we expand the wave function ϕ_i using the molecular eigenstates ϕ_j of H_{TCD}:

$$\phi_i\; (\vec{R}(t)) = \sum_j a_{ij}(t)\; \phi_j\; (\vec{R}(t)) \exp\{i/\hbar \int^t dt' E_j\} \tag{3.4}$$

The sum runs over all bound and continuum states of positive and negative energy.

Inserting (3.4) into (3.3) we obtain a set of coupled differential equations for the transition amplitudes a_{ij} (t):

$$a_{ij}(t) = -\sum_{k \neq j} a_{ik}(t)\; \langle \Phi_j\; | \partial/\partial t |\; \Phi_k \rangle\; \exp\{i/\hbar \int^t dt'\; (E_j - E_k)\} \tag{3.5}$$

Equation (3.5) contains in the brittle style of mathematics all these processes occuring in the collision which I tried to illustrate more simply in Fig. 3.2:

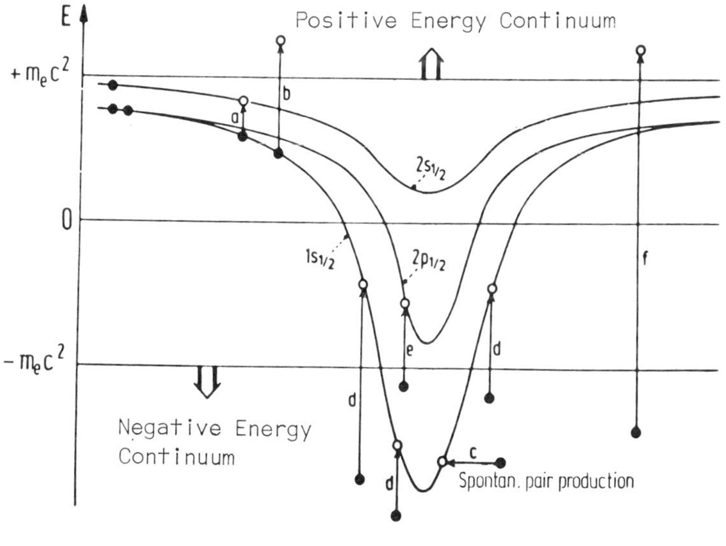

Fig. 3.2 The various processes for electron excitation in quasimolecular states. (a,b) Ejection of electrons out of the 1s-state.
(c) Spontaneous positron emission
(d,e) Dynamical positron production involving vacant bound states
(f) Direct excitation from the negative into the positive continuum.

(a,b) Ejection of electrons out of the 1s-state into bound states (a) or positive continuum states.

(d,e,f) Dynamical positron production via transitions of negative continuum states into vacant bound states (d,e) or into positive continuum states (f).

(c) Spontaneous positron production via transitions of negative continuum states into a vacant state of a 'dived' orbital without energy transfer.

As spontaneous positron creation represents only a small part of the dynamics coming up during the collision, we may justly ask how it could be separated experimentally from the other processes. Before coming back to this question, we solve Eq. (3.5) in first order perturbation theory for discussing the physics on a mathematical but still comprehensible basis.

3.1 Electron Excitation in First Order Perturbation Theory

In first order perturbation theory we get from Eq. (3.5) for the transition amplitude form an initial electron state i to a final state f:

$$a_{fi} = - \int_{\infty}^{\infty} < \phi_f | \partial/\partial_t | \phi_i > \exp\{i/\hbar \int^t (E_f - E(t')dt'\}dt \qquad (3.6)$$

The following assumptions seem to be fairly well justified (Sof79) at small internuclear distances and for very heavy systems ($\alpha Z \gtrsim 1$):

(i) Restriction onto the monopole part of the two-center internuclear potential.

(ii) Restriction onto radial coupling only, i.e.

$$\partial/\partial t = \dot{R}\ \partial/\partial R$$

and finally a scaling of the transition matrix element as

(iii) $<\phi_f |\partial/\partial R| \phi_i> \propto 1/R$

Therewith we obtain a simple, analytical expression for the transition probability:

$$|a_{fi}|^2 = D(Z_u) \exp\{-2 R_o/\hbar v \cdot \Delta E(R_o)\} \qquad (3.7)$$

where $D(Z_u)$ is a coupling strength that depends only on the combined nuclear charge Z_u of the collision system, and $\Delta E(R_o)$ is the energy transfer onto the bound electron at the distance of closest approach R_o, and finally v is the projectile velocity at infinity.

Equation (3.7) is - at closer inspection - a master-key for understanding the process of electron excitation at close encounters in very heavy collision systems:

(i) The electron excitation is determined in this regime by the combined charge Z_u only,

(ii) The excitation depends only on the energy transfer ΔE at R_o, irrespective of whether the electron was initially in a bound state (1sσ, 2pσ, etc.) or in a negative continuum state.

Therewith three different quantities which may be observed in such collisions, namely inner shell vacancy production, δ-electrons, and, finally, dynamically induced holes in the negative continuum (i.e. positrons), are brought into a natural connection.

How are the experimental quantities connected with the parameters of Eq. (3.7) ? The scattering angle of the nuclei defines the nearest internuclear distance R_o. Since $\Delta E(R_o)$ is the difference between final and initial electron energy, $E_f - E_i(R_o)$, we have to measure E_f **and** to know from which initial state the electron was ejected. Now, an excitation of the electron from the deepest-lying $1s\sigma$- and $2p\sigma$-quasimolecular orbitals leads to an 1s-vacancy in the heavier or lighter atom, respectively. Hence, by measuring simultaneously the characteristic K-X-radiation from the scattered nuclei, it can be decided from which quasimolecular orbital the electron was kicked out. In the case a positron of energy E_{e^+} is created in the collision, its initial energy is simply $E_{e^+} + 2mc^2$. There are, thus, two kinds of 'complete' experiments where **all** parameters of Eq. (3.7) become determined:

(i) a triple coincidence between scattered nuclei (R_o), δ-electron (E_f), and characteristic K-x radiation (rendering the initial electron state),

and

(ii) a triple coincidence between the scattered nuclei, the created positron (E_i), and the emitted electron (E_f).

In practice, only a few "complete" experiments of type (i) have been performed (see Fig. 3.5), whereas in all positron experiments and in most of inner shell vacancy measurements the final electron state E_f was **not** observed.

Hence, in all these cases we have to make an assumption concerning the unobserved final electron energy E_f. Setting somewhat roughly the final energy to zero, because E_f should be in the average much smaller then the binding energy, and because the electron might go as well into empty bound states ($E_f < 0$) as into continuum states ($E_f > 0$), we obtain for the excitation probability of the $1s\sigma$-shell, $P_{1s\sigma}$, and the creation probability of positrons, respectively, the expression:

$$|a_{fi}|^2_{1s\sigma} = P_{1s\sigma} \propto \exp\{-2\, R_o/\hbar v\, |E_{1s\sigma}(R_o)|\}$$

$$|a_{fi}|^2_{e^+} \propto \exp\{-2\, R_o/\hbar v\, (E_{e^+} + 2\, mc^2)\} \tag{3.8}$$

From Eq. (3.8) we draw the conclusion that, for small R_o and very heavy collision systems, the $1s\sigma$-excitation reflects directly the binding energy of the (almost) **combined** system. Since the $1s\sigma$-binding energy should reach several hundred keV in very heavy systems, and since the typical projectile velocity is in the order of ~ 0.1 c, we easily show that noticeable excitation is restricted onto internuclear distances R_o of a few ten fm ! In other words, the heavier the collision systems are, the smaller the range of excitation should be.

It is time now to compare these predictions with experiments. In Fig. (3.3) we show $1s\sigma$-excitation data for the medium-heavy systems Xe+Au ($Z_u = 133$) and for the very heavy Pb+Cm-system ($Z_u = 178$) as a function of the impact parameter $b \gtrsim R_o - a$ (a is half the minimum distance in head-on collision). The Xe+Au-data fit perfectly to straight lines in a logarithmic scale which indicates that $E_{1s\sigma}(R_o)$ is nearly constant for this system within a R_o-range from 30 to 120 fm. We deduce for this range immediately an average $1s\sigma$-binding energy of 250 ± 35 keV, as compared to the theoretical 1s-binding energy of 290 keV (Fri77) for a Z = 133 atom. The Pb+Cm-data, however, do not fit as well to a straight line, because here the $1s\sigma$-binding energy strongly depends on R_o.

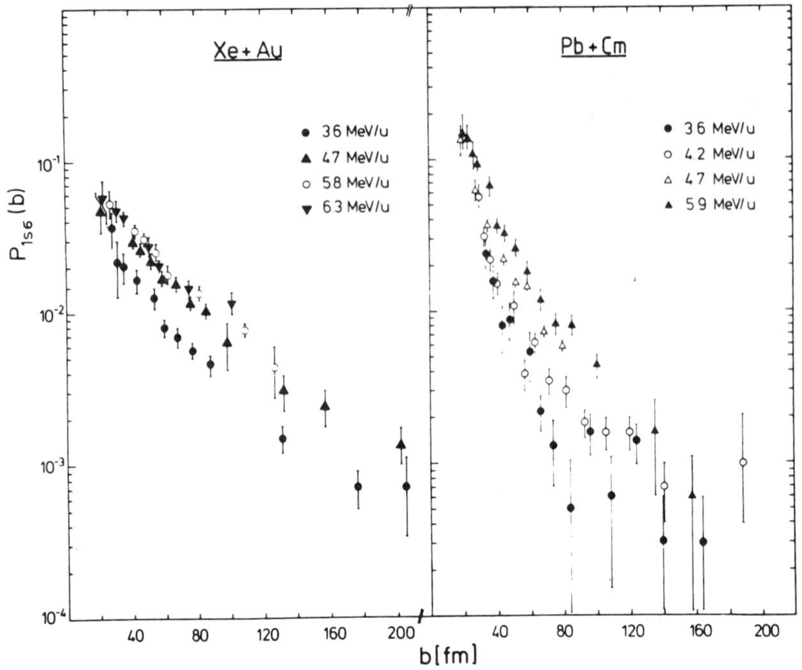

Fig. 3.3 Vacancy production probability P(b) in the 1sσ-shell of Xe+Au (Z = 133) and Pb+Cm (Z = 178) as a function of the impact parameter b at several projectile energies.

On the other hand, if Eq. (3.8) describes correctly the 1sσ-excitation of superheavy systems, then in reverse, **all** experimental data (normalized in absolute height) should fit, drawn versus $R_o \cdot E_{1s\sigma}(R_o)/\hbar v$ on **one** universal straight line when the 1sσ-binding energies are inserted from theory. In Fig. 3.4 this plot is shown for almost all data obtained up to now.

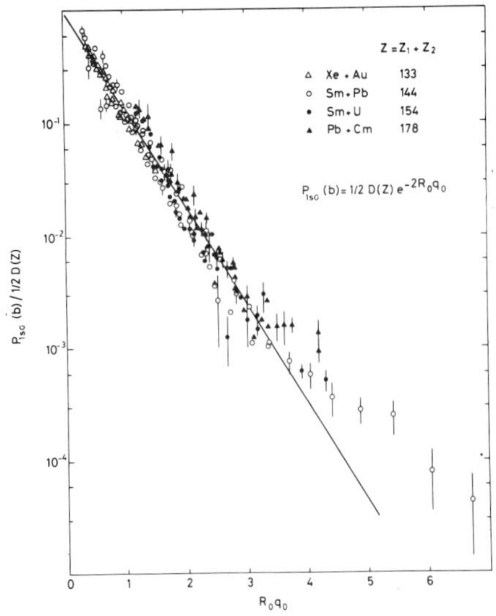

Fig. 3.4 Universal plot of normalized 1sσ-excitation data $P_{1s\sigma}$ / 1/2 D(Z) versus $R_o q_o = R_o \cdot E_{1s\sigma}(R_o) / \hbar v$ in the region 133 ≤ Z ≤ 178 using theoretical values (Fri77) for the quasimolecular 1sσ-binding energies $E_{1s\sigma}(R_o)$.

In the same manner the impact parameter dependence of atomic positron production may be tested by a comparison of the data with Eq. (3.8b). In Fig. 3.5 we show the probability for atomic positron creation within 423 $\leq E_{e^+} \leq$ 532 keV versus R_ρ for U+Pb collisions. The dashed curve represents the normalized Eq. (3.8b) with 2 $mc^2 + E_{e^+}$ = 1.5 MeV. We show again an excellent agreement, and conclude, therefore, that Eqs. (3.8) describe correctly the impact parameter dependence of excitation in heavy ion collisions, as well for strongly bound electrons as for the negative continuum states.

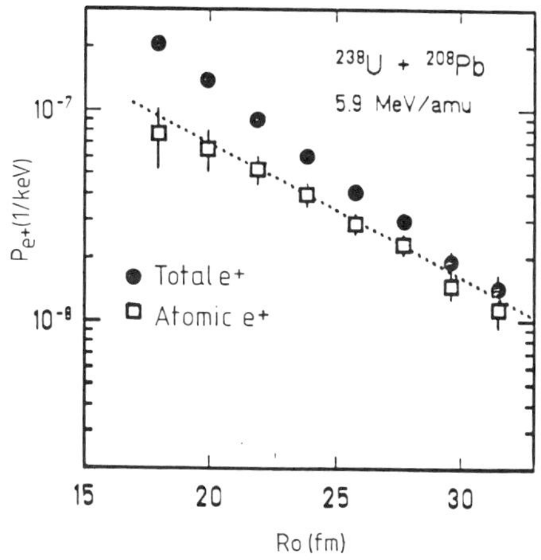

Fig. 3.5 Energy-averaged positron production probabilities P_{e^+} versus the distance of closest approach R_o for the U+Pb (Z_u = 174)-system at 5.9 MeV/u bombarding energy. Full symbols: Measured P_{e^+}-values. Open symbols: Nuclear background subtracted. Dashed Line: Atomic theory for elastic collisions.

In Fig. 3.6 we show one of the few "complete" experiments, a triple coincide measurement of particles, δ-electrons and X rays in the I+Pb (Z_u = 135) system. Here the excitation probability is shown for electrons from the 1sσ-shell versus the final kinetic electron energy, for fixed internuclear distances R_ρ. Since in this case the final electron state E_f is known, we can extract from the slope constant at any R_o **directly** the binding energies $E_{1s\sigma}(R_o)$, i.e. we can really perform a spectroscopy of superheavy bound states. The values extracted from Fig. 3.7 (e.g. 270 keV for R_o = 49 fm) are in excellent agreement with theoretical calculations.

We should now try to understand why the physics in these superheavy collision systems becomes as simple, and why the measured quantities reflect the properties of the **combined** system rather than these of the separated atoms. For that we look onto Fig. 3.7 where δ-electron spectra are shown for Pb+Pb and Pb+Sn-collisions without any additional coincidence requirement. We note

(i) The large difference in the intensities of the Pb+Sn and Pb+Pb spectra

Obviously, the spectra cannot be understood in terms of an excitation from orbitals of the Pb-like or Sn-like atoms.

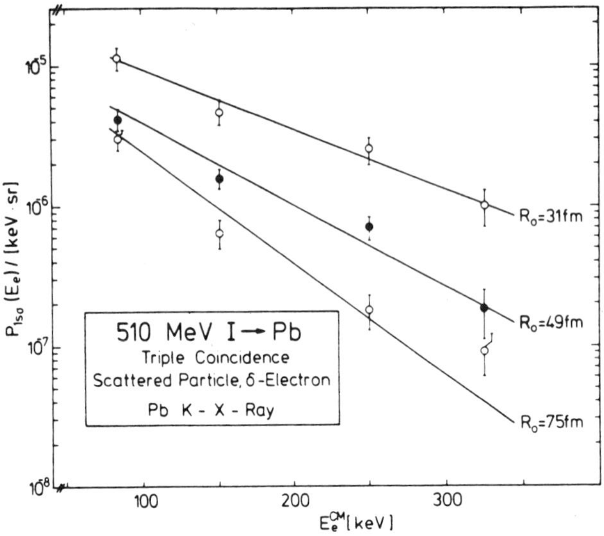

Fig. 3.6 Triple coincide measurement (Gue82) for the exctiation of δ-electrons from the 1sσ-shell at fixed internuclear distance R_o versus the final kinetic energy of the electrons.

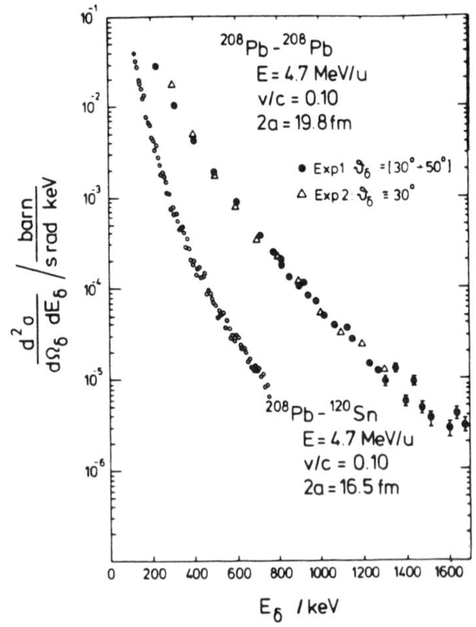

Fig. 3.7 δ-electron production probabilities from Pb+Pb (Z_u = 164) and Pb+Sn (Z_u = 132) collision systems measured at the same projectile velocity v/c = 0.1, versus the final δ-electron energy. Indicated are the distances of closest approach in a head-on collision 2a.

We further note

(ii) The very high kinetic energy of the δ-electrons observed: Up to 750 keV for the Pb+Sn (Z_u = 132) and over 1.7 MeV for the Pb+Pb (Z_u = 164)-system.

and finally

(iii) The much steeper fall-off of the Pb+Sn spectrum as compared with the Pb+Pb distribution.

Why do we observe these high energy electrons, whereas the maximum energy transfer onto a **free** electron is $\Delta E_{max} = 4\ m_e/m_p \cdot E_p$, which corresponds for projectile energies of $E_p \approx 5$ MeV/u to $E_{max}^p = 10$ keV ? The answer comes from simple kinematical considerations: Since the momentum transfer q_o onto a bound electron is connected with the energy transfer by $q_o = \Delta E/\hbar v$, and since momentum conservation holds, i.e. $\vec{k}_i + \vec{q}_o = \vec{k}_f$ (with k_i, k_f the initial and final electron momenta) it follows immediately that for large $\Delta E (\Delta E \gtrsim 0.1$ MeV) the initial momentum has to be very high, $|\vec{k}_i| \approx |\vec{q}_o|$ ($|\vec{k}_f|$ is very small as compared with $|\vec{q}_o|$ for all realistic cases). Such high momentum components are not available but next to the charge center of the combined system due to the strong relativistic shrinking of the electron wave functions. Hence, 1sσ-excitation and, a fortiori, high energy δ-electron creation, can happen in highly relativistic systems $\alpha Z \gtrsim 1$ only at very small internuclear distances R. In other words, the higher the combined charge Z_u is, the closer the collisions should be in order to can 'find' bound electrons with high momentum components. This 'unusual situation', however, is the real reason that allows us some kind of spectroscopy of transient superheavy atoms.

3.2 Positron Spectroscopy

We are addressing now to the central part of this lecture, the analysis of positrons from heavy ion-atom collisions.

For in-beam positron spectroscopy an experimental set-up is desirable that allows to measure highly resolved positron energy spectra with high efficiency while suppressing background radiation of electrons, neutrons, and photons. In addition, for each positron event the actual trajectory of the colliding nuclei, i.e. the distance of closest approach R_o at elastic scattering, and, for inelastic processes, mass transfers and energy losses, should be determined as accurately as possible. Finally, also the γ- and X-radiation associated with the scattering process has to be monitored. For that purpose a set-up has to be installed which combines a highly efficient, broad energy band positron detection system, fast position sensitive particle counters, and large γ-detectors. Because not all of these - to some extent contradictory - requirements can be fulfilled at once and because compromises have to be made, three different set-ups have been developed and put into operation at GSI.

The EPOS-instrument (Fig. 3.8) utilizes the broad band transport property of a solenoidal magnetic field to guide a positron away from the target to an almost background-free region where its energy is analyzed by a high resolution detector (Bok83, Epo85). The positrons are detected in a cooled pencil-like Si(Li)-diode of 1 cm diameter, the axis of which coincides with the symmetry axis of the solenoid. This configuration combines maximum detection efficiency for positrons created in the target with strong suppression for positrons produced elsewhere.

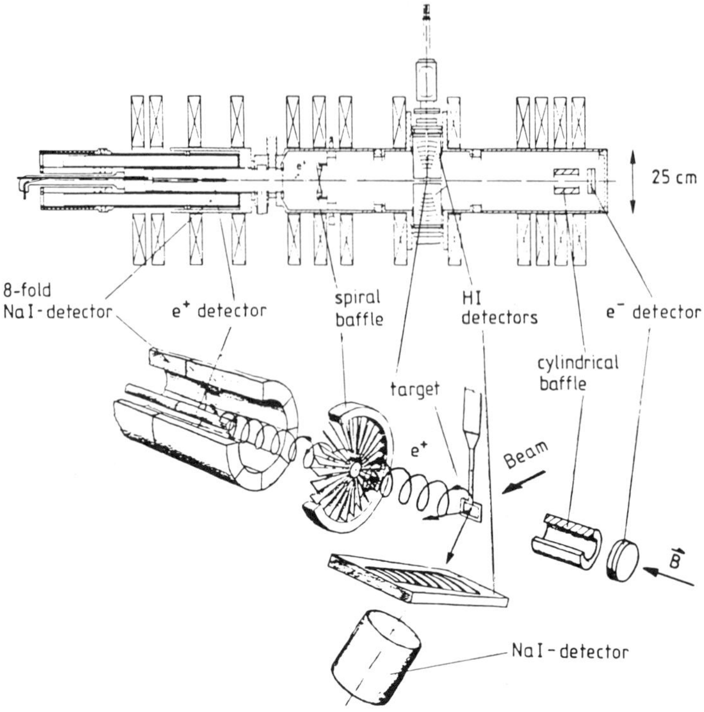

Fig. 3.8 Schematic view of the EPOS set-up (upper part) together with a detail drawing of the main components (only one out of two heavy-ion detectors is shown). Positrons are identified by a signal of the cooled, pencil-like Si(Li)-diode, associated with an annihilation γ-quantum of the NaI-crystal.

Electrons have to be prevented from reaching the detector, as they are generated by many orders of magnitude more frequently than positrons of comparable energy. For this purpose a spiral baffle has been installed between target and detector. Because the suppression cannot be complete, however, for a proper positron identification in addition to the Si(Li)-signal, at least one coincident 511 keV annihilation γ-quantum is demanded from an eightfold NaI-counter which surrounds the Si(Li)-diode.

In coincidence with positrons, both the scattered projectile and the recoiling target nucleus are being detected by two position sensitive parallel plate avalance counters (PPAC) with delay-line read out covering scattering angles ϑ from 15° to 75° in the laboratory system at a resolution $\Delta\vartheta$ better than 1° (at ϑ_{lab} = 45°).

In the TORI-apparatus (Fig. 3.9) the high detection efficiency of a solenoid field was combined with a very elegant discrimination between positrons and electrons (Bac83, Kan85, Tor85). An S-shaped field-configuration made up by pancake coils surrounding two quarters of a toroidal tube is used. In the inhomogeneous field electrons and positrons experience a drift perpendicularly to the field lines and to the field gradient but **opposite** in direction, which leads to their spatial separation at the end of the first quarter. There, the electrons (or positrons at a reversed field) can be absorbed in a detector while the positrons (electrons) are allowed to pass to the second quarter torus and finally reach the planar Si(Li)-diode at its end. Besides its very efficient electron suppression, this instrument allows a clean simultaneous measurement of both electrons and positrons emitted in the same collision process.

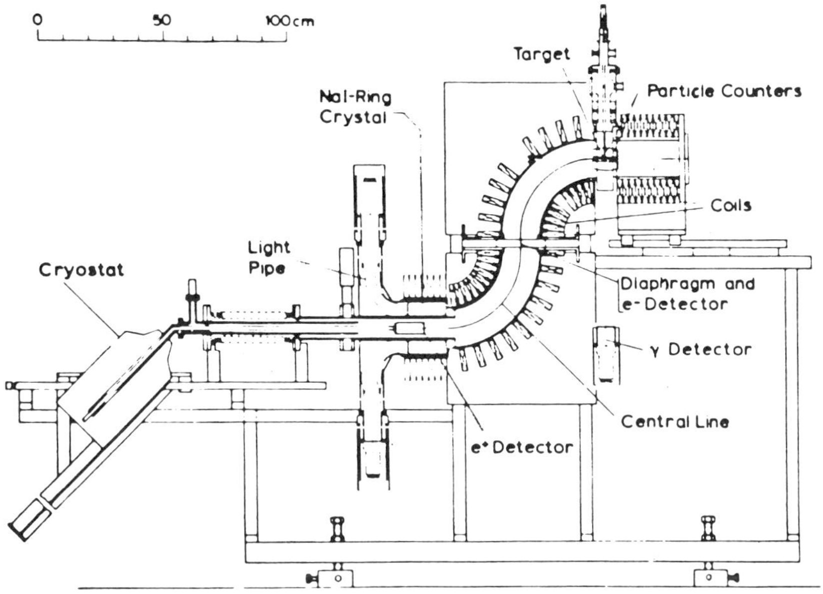

Fig. 3.9 Side-view of the TORI transport system showing its main components. The beam runs perpendicularly to the drawing plane. Positrons are detected by a cooled Si(Li)-diode at the end of the second quarter torus. As in the EPOS-system a coincidence with an annihilation quantum is required for positron identification.

Similarly to the EPOS-system, the proper signature of a positron requires a coincident 511 keV γ-event from the NaI-crystal which surrounds the positron detector. The scattered particles are monitored by two position sensitive PPAC accepting scattering angles ϑ_{lab} from 18° to 70° at an angular resolution $\Delta\vartheta$ of $\approx$ 2°.

A third apparatus (Koz79, Cle84, Ora85) that, finally, turned out to work successfully for in-beam positron spectroscopy, is an iron-free Orange-type β-spectrometer (Fig. 3.10). This spectrometer profits by the focussing power of a toroidal magnetic field produced by 60 properly shaped current coils (Cu) which are arranged regularly around the spectrometer axis (which coincides in the actual geometry with the beam axis).

This instrument works - in contrast e.g. to conventional solenoids - as a charge and momentum filter by focussing at a given field setting only positrons (defocussing at once electrons) onto a ring focus of about 4 cm around the axis (Mol65). Along the axis of the focal cylinder a conically shaped plastic scintillator of 12 cm length was installed surrounded by a position sensitive annular proportional counter. Therewith, the spectrometer could be used as a spectrograph with a momentum resolution $\Delta P/P$ of $\approx$ 3 % (corresponding to an energy resolution of ~ 15 keV at E_{e^+} = 300 keV) at an overall momentum acceptance $\Delta P/P$ of $\approx$ 13 %. Positrons are discriminated from γ-background by demanding a signal from both the proportional and the plastic detector. An annular, position sensitive PPAC placed 7 cm behind the target allows a determination of the particle scattering angle to better than 2° at an acceptance from 15° to 52° for polar and of 2π for azimuthal scattering angles. A unique feature of this β-spectrometer is its symmetry around the beam axis which minimizes Doppler broadening for positrons emitted from a source moving in beam direction (< 15 keV for a positron energy of 300 keV). On the other hand, the restricted momentum acceptance makes it necessary to measure at several magnetic field settings in order to scan the full positron spectrum.

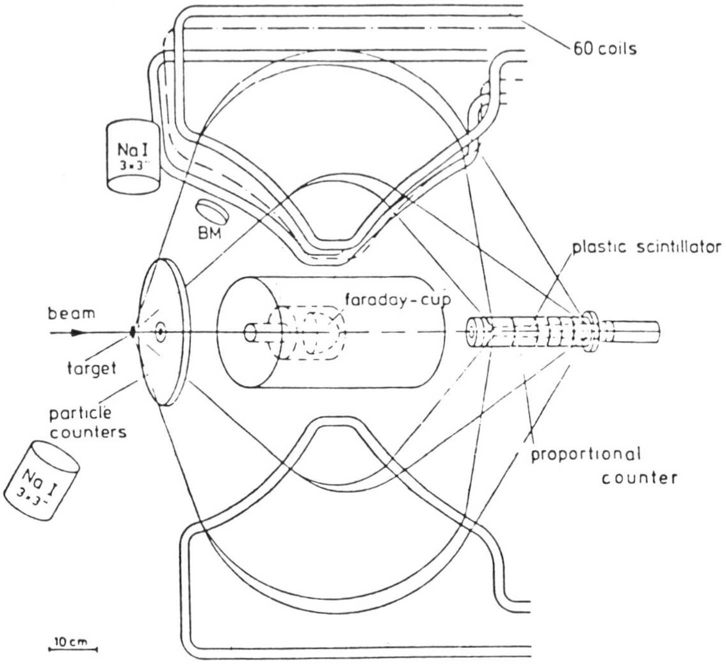

Fig. 3.10 Schematic view of the 'Orange'-type β-spectrometer. Positrons emitted at 30° to 70° with respect to the beam axis are focussed onto a cylindrical plastic detector surrounded by a position sensitive proportional counter. The spectrometer works as a charge- and momentum filter. For positron identification coincident signals of the two counters are required.

In all three set-ups also the γ-spectrum associated with the scattered particles was recorded by large NaI-crystals. As well, in all experiments the effective beam energy was controlled by monitoring the energy distribution of the scattered particles with surface barrier detectors. This steady control turned out to be very important for the proper interpretation of the observed positron spectra.

In-beam positron spectroscopy is confronted with the fact that inevitably a "background" of those positrons emerges which have to be attributed to pair converted nuclear transitions following Coulomb excitation. These 'nuclear' positrons cannot be discriminated easily from 'atomic' positrons by standard timing or recoil techniques, because they are emitted within 10^{-15} to 10^{-12} after collision (whereas atomic positrons are created during the formation of the quasiatomic collision system itself, i.e. within 10^{-21} s to 10^{-20} s). In principle, however, by applying blocking techniques, a separation of the two kinds of positrons could be possible, as Kaun (Kau78) has suggested first. Considerable problems, e.g. the production of the thin heavy-atom crystals or the very small solid angle available for positron detection, up to now prevented an experimental realisation of this very interesting idea. Therefore, following Meyerhof (Mey77), a less direct method has been adopted by all groups for evaluating the amount of nuclear positrons in the measured spectra. By this method, the unfolded γ-spectra above $E = 2\ mc^2$, recorded at exactly the same kinematical conditions as the positrons, are transformed into positron spectra using theoretical conversion coefficients for internal pair production (Slu79).

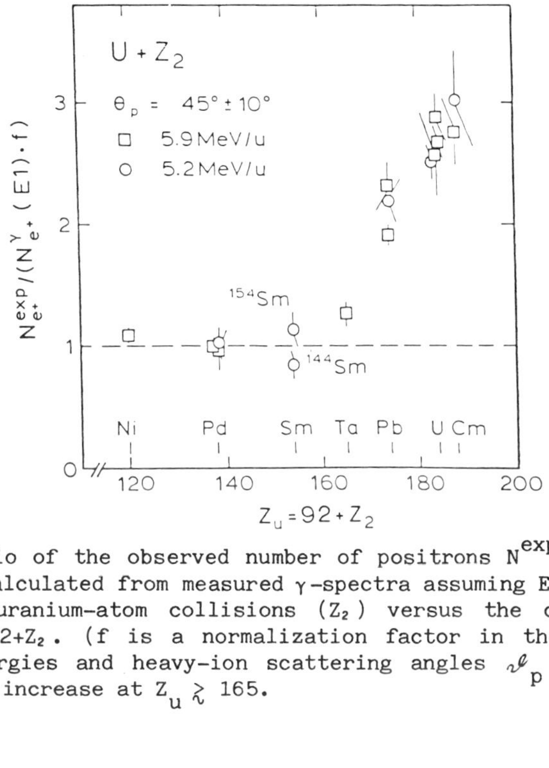

Fig. 3.11 Ratio of the observed number of positrons N^{exp} to the number of positrons calculated from measured γ-spectra assuming E1-multipolarity $N_{e^+}(E1)$ for uranium-atom collisions (Z_2) versus the combined charge number $Z_u = 92+Z_2$. (f is a normalization factor in the order of 1.) Bombarding energies and heavy-ion scattering angles ϑ_p are indicated. Note the steep increase at $Z_u \gtrsim 165$.

As shown in Fig. (3.11) (Bac81), beyond $Z_u \gtrsim 165$ the total positron yield surpasses the nuclear 'background', rising steeply as $\sim Z_u^{18}$ in fair agreement to the expected (Rei81) Z_u^{20}-dependence for dynamic positron production. Hence, the experiments give strong evidence for the appearance of a second positron source at $Z_u \gtrsim 165$, namely dynamic positron creation in the strong di-nuclear Coulomb field of transient superheavy quasiatoms.

The evaluation of nuclear positrons in a clean and reproduceable manner marked the decisive step toward a systematic investigation of atomic positron creation. Meanwhile, many experiments have been performed investigating especially the dependence of positron energy spectra on combined nuclear charge Z_u, bombarding energy and impact parameter(Bac78, Koz79, Bac83a, Swe83, Cle84). In the focus, however, stood the search for a signature of spontaneous positron creation, i.e. the question whether the critical charge $Z_c = 173$ and the critical distance R_c would be reflected in the data or not.

First we show in Fig. (3.12) energy differential positron production probabilities per collision dP_{e^+}/dE_{e^+}, measured by the EPOS-group for collision systems from $Z_u = 154$ to $Z_u = 188$ at bombarding energies around the respective Coulomb barrier (Swa84). These spectra present the impact parameter averaged yields as no specific kinematical cuts have been selected. No obvious change of the smooth spectral form is found when comparing the 'light' U+Sm ($Z_u = 154$)-system with the still subcritical U+Pb ($Z_u = 174$) and the supercritical (at small impact parameters U+U ($Z_u = 184$)- and U+Cm ($Z_u = 188$)-systems. The only significant difference is the strong increase of nonnuclear positrons for the higher Z_u. Moreover, the shapes as well as the absolute yields perfectly correspond to theoretical expectations for atomic positron creation at **elastic** scattering (Mue83).

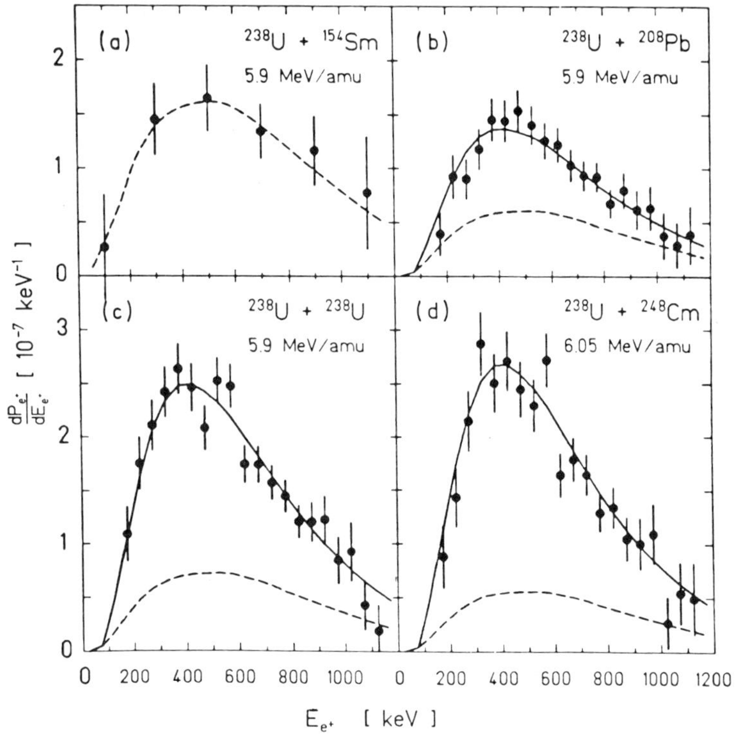

Fig. 3.12: Energy differential positron production probability per collision dP_{e^+}/dE_{e^+} versus positron kinetic energy. E_{e^+} in the laboratory system, averaged over heavy-ion scattering angles $25° < \vartheta_{lab} < 65°$ for four different collision systems and at bombarding energies around the repsective Coulomb barrier. Dashed lines: Nuclear background, Solid lines: Atomic theory (Mue84) assuming Rutherford trajectories and including nculear background.

These results are corroborated by data obtained with the Orange-spectrometer. As examples in Fig. (3.13) (Cle84, Tse85) positron spectra are shown from the subcritical Pb+Th ($Z_u = 172$)-system. Again a smooth positron energy distribution can be observed in quantitative agreement with theory. It should be emphasized, however, that in contrast to Fig. (3.12) these spectra show positrons associated with rather restricted angular ranges of scattered particles. Even for narrow kinematical cuts, **all subcritical** systems investigated up to now do not show any kind of structure in addition to the expected smooth distribution of nuclear and dynamic positrons.

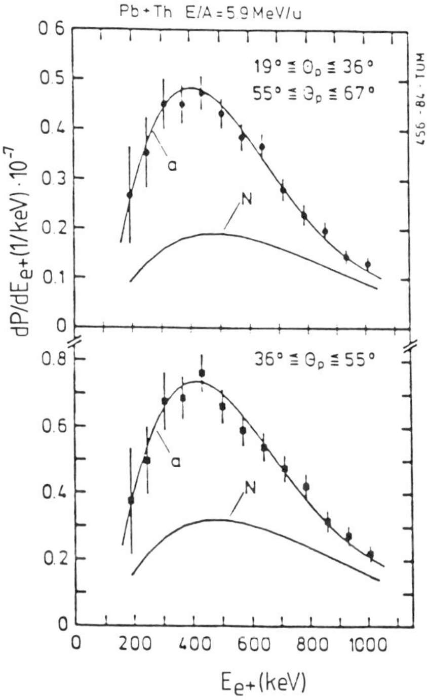

Fig. 3.13 Energy differential positron production probability per collision dP_{e^+}/dE_{e^+} versus positron kinetic energy in the c.m.-system for Pb+Th ($Z_u = 172$)-collisions at E/A = 5.9 MeV/u, associated with heavy ions scattered into the angular ranges as indicated. The nuclear background (N) is already subtracted. Curve a: Atomic theory for Rutherford trajectories (Mue83) without fitting any scale.

Altogether, the **global** results obtained with different experimental set-ups at bombarding energies around the Coulomb barrier can be summarized in the following conclusions:

(i) The predicted process of dynamic positron creation in strong fields has been confirmed.

(ii) The observed yields and energy-differential shapes of dynamic positrons correspond quantitatively to theoretical expectations.

(iii) Especially the dependence of dynamic positron creation on the combined nuclear charge Z_u and on the distance of closest approach R_o agrees well with the prediction for **elastic** collisions.

(iv) Neither the scattering angle-averaged positron spectra nor the energy-averaged impact parameter dependent yields show significant differences when going from subcritical (Z_u 173) to supercritical collision systems, i.e. in the **global** spectra a clear signature of spontaneous positron creation has **not** been found.

Now, the most important motivation for positron spectroscopy was the hope to find a clear signature for **spontaneous** positron creation and, hence, for the long-searched diving process. We have to ask whether there should be at all a qualitative change of positron spectra in going to supercritical systems. Looking once again to Fig. 3.1, we obtain immediately an answer: As long as **elastic** collisions are considered, the time where supercritical binding exists is in the order of 10^{-21} s. On the other hand, the decay-time of an empty 1s- resonance state is about 10^{-19} s. Therefore, the 'diving'-time is simply too short to allow for a significant contribution from spontaneous decay of the neutral vacuum. Moreover, a diving-time of $\sim 10^{-21}$ s corresponds to an energy-width of $\sim$600 keV. Hence, in **elastic** collisions we could never observe monochromatic positrons as a clear signature of 'diving'.

4. NARROW POSITRON LINES

4.1 The Observation of Narrow Structures in Supercritical Collision Systems

Quite surprisingly, in 5.9 MeV/u U+U (Z_u = 184)-collisions, superimposed on the bell-shaped spectrum predicted for elastic scattering, a narrow positron line has been found (Kie83) - under restricted kinematical conditions for the scattered particles - at an energy of about 300 keV with an experimental width of $\sim$ 70 keV (Fig. 4.1). Also in U on Th-encounters, at a bombarding energy close to the Coulomb barrier and at restricted angular ranges of scattered ions, a narrow line emerged (Fig. 4.1) at about the energy found for the U+U-system and with comprable width, differing only by a somewhat smaller production yield (Cle84).

Independently, narrow structures were observed by the EPOS-group (Fig. 4.2, Swe83, Cow85). Here, associated with selected, but from case to case different particle angular regions, prominent positron peaks were found in five supercritical collision systems ($180 \leq Z_u \leq 188$). These lines all show similar energies ($316 \leq E_e \leq 357$ keV), similar widths (experimental FWHM $\sim$ 75 keV) and comparable intensities. It is worth noting that the positron lines detected up to now are associated (i) with supercritical collision systems, (ii) with scattered particles which have to be attributed to elastic or nearly elastic collisions (experimentally deduced energy loss $\Delta E < 40$ MeV), and (iii) with bombarding energies very close to the respective Coulomb barrier.

Since these lines were observed with two set-ups, differing considerably with respect to their positron detection method, instrumental mistakes can be excluded almost with certainty. The existence of narrow structures must be regarded as an experimentally established fact. Already from the experimental line-widths (i.e. without having corrected for detector resolution and Doppler broadening) a lifetime of the emitting source of more than 10^{-20} s can be deduced, solely on the basis of the uncertainty relation.

As such lifetimes readily occur in the final excited nuclei produced in such collisions, we have to discuss first whether or not these lines should be attributed to a **nuclear** origin.

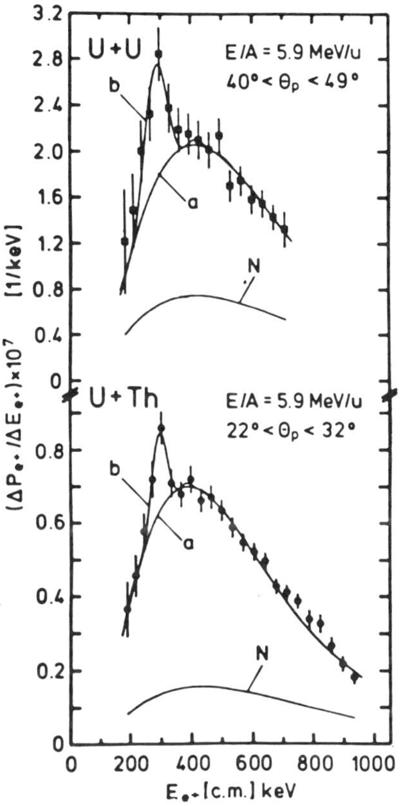

Fig. 4.1 Positron creation probabilities per collision for U+U- and U+Th-collisions at 5.9 MeV bombarding energy versus positron kinetic energy in the c.m. system, for selected angular ranges of scattered particles. The solid line (a) represents the theoretical expectation for elastic collisions including nuclear background (N) (Mue84). In addition narrow lines are fitted (b) for spontaneous positron creation during long nuclear contact (with a sharp contact time T), according to the semiclassical theory of Section 5.2. Orange- data (Cle84, Tse85).

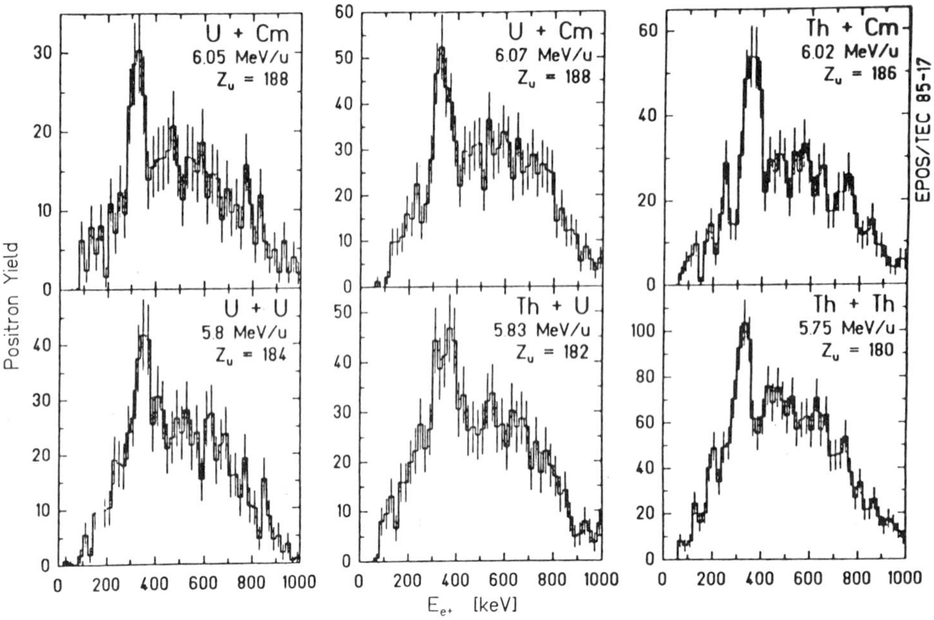

Fig. 4.2 Positron spectra for the five supercritical collision systems investigated by the EPOS-collaboration, taken at different kinematical constraints and at bombarding energies as indicated. The U+Cm data at 6.05 MeV/u are from (Swe83), all other data from (Cow85).

4.2 Discussion of a Possible Nuclear Origin of the Positron Lines

For a given nuclear transition energy E of multipolarity Mλ, mainly four processes may occur:

(i) γ-emission with $E_\gamma = E$ ($M\lambda \neq E\emptyset$);

(ii) electron-emission due to internal conversion with $E_{e^-} = E - |E_B|$ (with E_B the binding energy of the electron to be ejected);

(iii) **continuous** internal e^+-e^- pair conversion with $0 \leq E_{e^-} \leq E - 2\,mc^2$ showing a broad, triangularly shaped e^+ spectrum, and

(iv) **monoenergetic** internal e^+-e^- pair conversion, with $E_{e^+} = E - 2\,mc^2 + |E_B|$ (i.e. pair conversion into a bound final electron state).

For M λ fixed, all these processes are related by known branching ratios which are independent of the specific structure of initial and final nuclear states. Hence, if the narrow positron lines are attributed (via processes (iii) and (iv)) to a strongly excited nuclear state in one of the reaction products, then the strengths of the corresponding γ- (for $M\lambda \neq E\emptyset$) and electron-lines can be calculated from the known branching ratios (Slu79). That obviously is the mirror- image of the method pursued in evaluating nuclear positrons from the measured γ -yields (cf. Section 3.2).

For the two systems U+U (Orange-group) and U+Cm (EPOS-collaboration) on the basis of the assumption that the entire observed positron line is of nuclear origin, the corresponding γ - and electron-lines have been calculated. This was possible as for both systems γ - and electron-spectra were simultaneously recorded (Cle84, Bok84).

Continuous e^+-e^- pair creation can be excluded rightaway as possible origin of the lines, since it leads to a very broad positron energy distribution with a width (FWHM) in the order of ∿ 150 keV (in both set-ups) which does not fit to the observed narrow width of 60 to 80 keV. For completeness we show in Fig. (4.3), for the U+U system (Orange), the hypothetical γ-lines to be expected if the observed positron line originated from continuous pair conversion of E1 or E2 γ-transitions (Cle84). Even for the most unfavourable case of E1, the expected γ-line intensity (drawn with detector resolution and Doppler spread taken into account) is of similar order as the continuous background and should have been detected (note that the additional γ-lines are not added to the measured spectrum but drawn in an absolute scale). Similarily, for an assumed pair converted EØ transition, Fig. (4.3, top) shows the corresponding K-conversion line which has to be expected in the electron-spectrum (in this case the hypothetical e^--Line is **superimposed** onto the smooth background) (Cle84). Obviously, the data exclude this hypothesis.

There is, however, a nuclear process which would, in principle, lead to such narrow positron lines as they were observed: monochromatic pair creation. In this case the nuclear transition energy E is not split continuously onto both electron and positron, but rather the electron will be captured into a vacant inner-shell state emitting a **monoenergetic** positron of energy $E_{e^+} = E - 2\,mc^2 + |E_B|$ (Sli49). As an inner-shell (mainly K,L)-vacancy has to be available at the time the nuclear transition occurs, monochromatic pair creation normally has a very small probability to take place. Even, if in a heavy ion collision several inner-shell vacancies are generated, they would be refilled after 10^{-17} s (for heavy atoms) to be compared with typical nuclear lifetimes in the order of 10^{-12} s to 10^{-15} s. Thus, for usual K-hole life-times, the process of monoenergetic pair conversion is suppressed by two to five orders of magnitude.

Inspecting now the γ- and electron spectra for possible indications of a monoenergetic pair conversion, an assumption has to be made about the mean number of inner-shell vacancies which are available during the nuclear transitions. In Fig. (4.4, top) the γ-lines are shown (Bok84) to be expected if the observed yield of the positron line would be due to monoenergetic pair conversion of E1 or E2 transitions (taking into account experimental resolution and Doppler spread). Even for **one** K-vacancy permanently availabe, still a very strong γ-line should emerge at the expected energy of $E = E_{e^+} + 2\,m_e c^2 - |E_K|$ (with E_{e^+} the energy of the positron line as found in the U+Cm system) in complete disagreement to the observation. Similarily, assuming EØ transition, the hypothetical electron K-conversion line (Sof83) is inserted into the measured electron spectrum (Bok84). Only for the mean number of K-holes exceeding 1.85, the corresponding electron line would drop below the detection limit (for empty K-shells, K-conversion disappears completely). A comparable limit for the mean number of K-holes was extracted from the electron spectrum of the U+U-system, obtained with the Orange spectrometer.

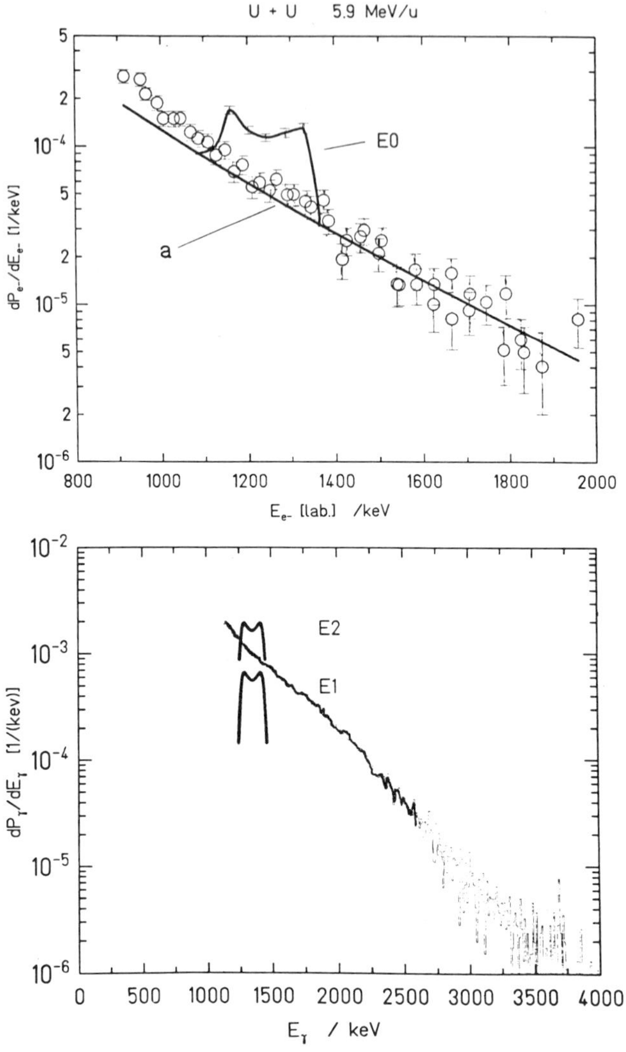

Fig. 4.3 Electron- (top) and γ-emission (bottom) probabilities per collision for U+U encounters at 5.9 MeV/u and scattering angles between 40.5° and 49°. The double humped, Doppler broadened lines would be expected, if the observed U+U-positron line in Fig. 4.1 would originate from nuclear pair-conversion-decays in the separated reaction products for multipolarities as indicated. The expected electron K-conversion line (corresponding to a hypothetical nuclear EØ-transition) is superimposed on a calculated e-spectrum scaled by 1.25 (solid line (Sof83)), whereas the lines in the γ-spectrum are drawn in an absolute scale. Orange-data (Cle84).

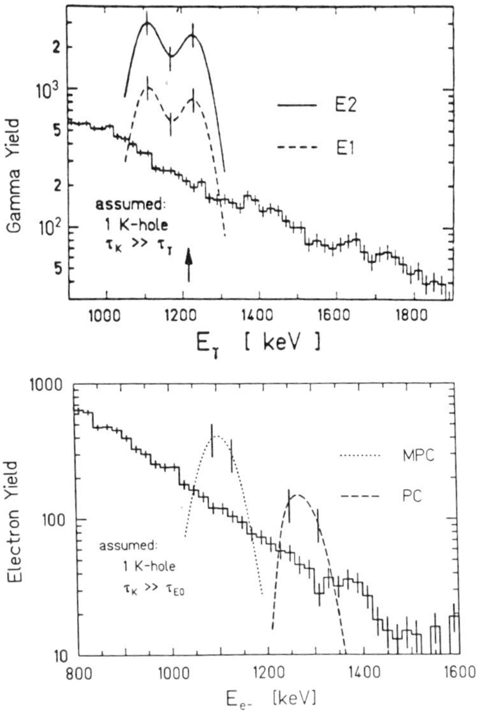

Fig. 4.4 γ-(top) and electron-spectrum of the U+Cm collision system recorded for the same ion scattering angles as the positron spectrum of Fig. 4.2. The intensity of the superimposed γ-lines are calculated assuming monoenergetic e^+-e^- pair conversion (of E1- or E2-nuclear transitions) to be responsible for the positron peak. For a hypothetical EØ-transition the expected electron K-conversion lines are shown in an absolute scale (**not** added). Dashed line (PC): Continuous e^+-e^- pair conversion assumed. Dotted line (MPC): Monenergetic pair conversion assumed with initially 1 K hole of infinite life-time. EPOS-data (Bok84).

Thus, we have to conclude that all nuclear processes - insofar they are in the scope of our present understanding - are very unlikely candidates for the origin of the narrow positron structures.

The observed positron lines necessarily are affected by Doppler-shift and broadening which signals the velocity of the emitting source. This provides an additional method to check whether the lines are caused by a nuclear process or whether they are associated, e.g. with the formation of a superheavy collision system: In the former case, one out of the two scattered particles is the emitting source. In the latter case, where the positrons are emitted from a system moving into the beam direction with the c.m. velocity, the line profile has to be almost independent on the emission angle of the final reaction products.

Figure 4.5 shows at the example of the U+Cm and Th+Cm-collision systems the normalized line widths as a function of the particle scattering angle (Cow85). The observed widths correspond well to the Doppler broadening expected for a positron source moving with the velocity of the center of mass v_{CM}. Moreover, the independence of the width on the scattering angle obviously rules out both the scattered projectile and ejectile as the emitting source (cf. first of all the data near ϑ_{Lab} = 45°). One might object, however, that this method of discovering the source is - to some extent - indirect, since it supposes a precise knowledge of experiments energy resolution and of the differential detection efficiency for all positron emission angles.

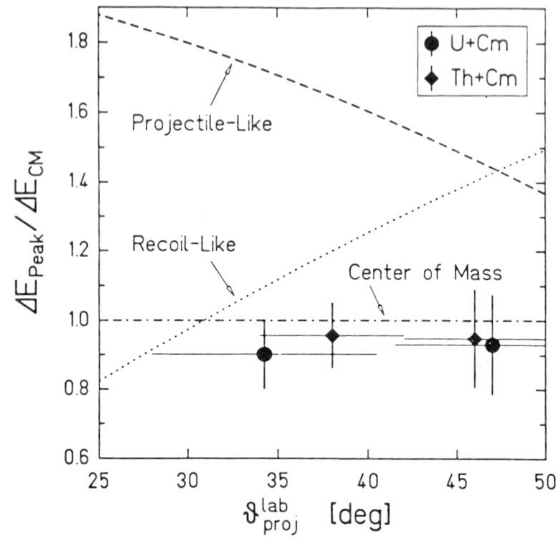

Fig. 4.5 Ratio of measured positron line width (ΔEpeak) to the line width expected for positron emission from a source moving with c.m.-velocity (ΔE_{CM}), versus the laboratory projectile scattering angle for the U+Cm- and Th+Cm-collision systems. The dotted and dashed lines indicate the values expected for positron emission from the recoil- or projectile-like scattered nucleus, respectively. EPOS-data (Cow85).

The main points of this section can be summarized as follows:

(i) In addition to the smooth positron energy distributions, expected for dynamic positron creation, narrow lines were found in several supercritical collision systems with experimental FWHM-widths of about 60 to 80 keV.

(ii) The lines were found using two different positron detection systems giving them a high degree of confidence;

(iii) continuous pair creation of a nuclear transition giving broad positron distributions, can be excluded as origin of the lines.

(iv) monenergetic pair conversion of an EØ-transition also can be excluded, except the nuclear transition occurs almost exclusively in completely stripped ions - which has to be considered as rather unlikely at our present understanding;

(v) line shape analysis also is **not** in favour of a nuclear origin of the positron structure.

Since known nuclear processes do not give a reasonable mechanism for the posible origin of the positron lines we have to search for other explanations.

5. The Hypothesis of Long Lived Giant Nuclear Molecules

Whereas in elastic collisions no clear signature for spontaneous positron creation is expected, the picture changes drastically if projectile and target nucleus stick together for a time T comparable to the typical decay-time $\sim 10^{-19}$ s of the neutral vacuum. Then for a supercritical collision system, supposed only that in the dived orbital a vacancy is available, a narrow positron line is expected to occur as the binding energy will stay constant during the nuclear contact time T. Following this idea, first Reinhardt et al. (Rei81) suggested that the positron lines might indeed originate from spontaneous positron creation in the supercritical field of long-lived giant nuclear molecules formed in central collisions close to the Coulomb barrier.

The authors show that for any reasonable distribution of the contact time T a peak-like structure will still be present if only the expectation value $\langle T \rangle$ is large enough ($\langle T \rangle \gtrsim 10^{-20}$ s). The width of the line should be, depending on the distribution function of T, in the order of $(2 \ldots 5)\ \hbar/T$. Thus, the observed line widths of about 60 keV would correspond to astonishing contact times between 2×10^{-20} s and 10^{-19} s. The position of the line should lie at the diving depth of the K-hole during the nuclear contact.

An inherent problem of this giant molecule model is the fact that all observed position lines are associated with scattered nuclei that bear all the properties specific for elastic scattering, i.e. no or at least very small energy loss ($\Delta E \lesssim 40$ MeV) and/or mass transfer. On the other hand, when extracting from the measured yield of the positron lines a reaction cross section for giant molecule formation (10 mb), one might argue that only 1 out of 10^3 events within the selected particle range would proceed via a long-lived composite system. In other words, spontaneous positron creation could be the only process filtering out such very rare nuclear events never seen before in other reaction channels. There are first attempts (Rho83) to understand the long contact times using potential models which generate pockets for special angular momentum windows in the entrance channel of supercritical collision systems. In the context of giant molecule formation, the energy of the line should depend very strongly on the combined nuclear charge Z_u and on the shape of the molecule.

For the system U+Cm the measured peak energy of 316 keV correspond to two deformed nuclei which are just in touch in an elongated orientation (internuclear distance $R \gtrsim 16.5$ fm). For the systems U+U and Th+Cm one has to assume distances of 11-12 fm which correspond to strongly deformed giant compound nuclei. The U+Th-system must be almost spherical, while the line position in the Th+Th system cannot be explained, if electron screening is taken into account (see Fig. 5.1). For fully ionized systems the binding energy of the 1s-state is higher by about 100 keV (as compared with an ionization state of q = 50). The position of the Th+Th-line is then compatible with the assumption of a bare, spherical nucleus $Z_u = 180$.

However, the fact that all measured positron lines have nearly the same energy is certainly the most severe problem within the framework of the hypothesis of spontaneous positron creation in the field of giant nuclear molecules. A simple average of all EPOS-data gives 336±10 keV, the corresponding mean value of the Orange-data lies at 288±10 keV. The

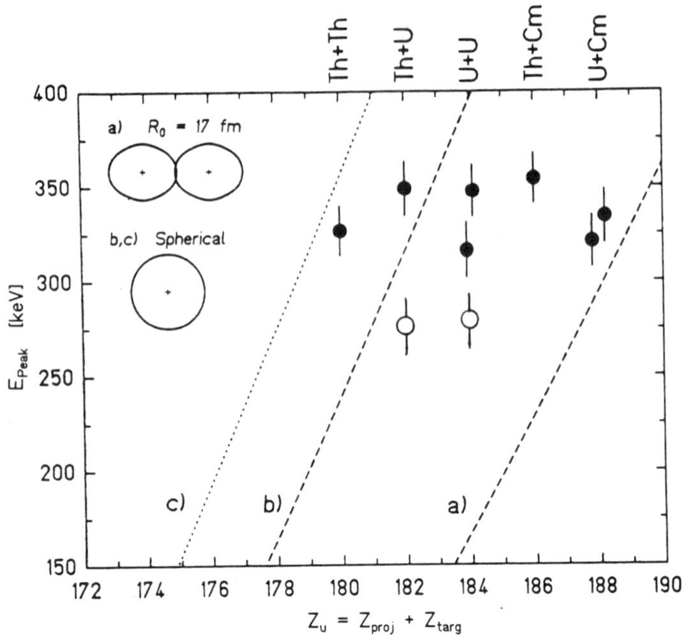

Fig. 5.1 Mean energies of all positron peaks measured up to now as a function of Z_u (black points: EPOS-data, open points: Orange-data). The lines indicate peak energies calculated for spontaneous positron emission from nuclear complexes with either (a) deformed shapes (nuclear centers separated by 17 fm at touching) or with spherical shapes (b), both with 50-fold ionization. In (c) bare nuclei with spherical shape are shown.

distinct and systematic difference of Orange- and EPOS-values is not understood and has to be cleared up. Putting aside for the time being this serious problem, it cannot be strictly excluded from the whole of existing data that the peak-energy is essentially the **same** for all collision systems. In this - at present still speculative - case of a **common** source for the positron lines at all Z_u the critical charge Z_c = 173 would no longer remain in all likelihood a 'magic' threshold - underlining the urgency to investigate more systematically the subcritical systems. A common line-source could be e.g. a hitherto unknown particle, produced in the collision and decaying into an e^+-e^--pair. An unambiguous signature for it would be correlation of the line-positrons with electrons of the same energy, reflecting the two body break-up of a particle with well defined mass. At the TORI- and EPOS-spectrometers the search for such correlated events will be prepared, since these devices are especially well suited for e^+-e^- coincidence measurements.

6. CONCLUSIONS CONCERNING THE NARROW POSITRON LINES

The existence of narrow line structures in the positron spectra emitted from collisions of U, Th, and Cm nuclei is by now firmly established. The analysis based on simultaneously measured electron and photon spectra has shown that pair conversion processes of excited states in the heavy nuclear fragments can be almost completely ruled out as possible origin of the lines. The exceptional process that cannot be excluded at present, is monoenergetic pair-conversion in the field of always completely stripped nuclei. This possibility must be, however, considered as very unlikely.

Spontaneous positron emission from long-lived di-nuclear states, on the other hand, provide a reasonable explanation of many features of the experimental data, both qualitatively and quantitatively. While the length of the contact time required to explain the narrow line-width - almost 10^{-19} s - is surprising, there are no data from other experiments that rule out the existence of such long times for quasielastic reactions at a cross-section level of several millibarn. Model calculations based on elastic scattering of an attractive pocket in the nuclear potential even support this assumption. The main unexplained feature is the lack of variation in the line position with total nuclear charge, which requires rather special assumptions about the shape and degree of ionization of the di-nuclear systems. Indeed, the shapes have to be substantially different since, for similar shapes of the giant nuclear complexes, the line-energy should scale roughly as Z_u^{20}. On the other hand, as long as (non-nuclear) positron lines are not confirmed without doubt in subcritical systems, adn/or as long as no clear signature of an invariant mass is found in the e^+-e^- correlations, spontaneous psoitron creation from long-lived di-nuclear systems remains, in our opinion, a probable and comprehensive interpretation of the narrow positron structures.

The essential test of this far-reaching hypothesis, however, will be effected by the experiments just starting: In particlular, by the systematic investigation of subcritical systems, and by the search for e^+ -e^- correlations. Extrapolating previous experiences in positron spectroscopy to the future we have to expect, almost certainly, unexpected new results.

ACKNOWLEDGEMENT

I wish to thank many colleagues for fruitful discussions. I am very grateful to W. Greiner, U. Müller, J. Rafelski, J. Reinhardt, T. de Reus, G. Soff, to my colleagues of the Orange-group E. Berdermann, M. Clemente, M. Franosch, S. Huchler, P. Kienle, W. Koenig, C. Kozhuharov, H. Tsertos, and W. Wagner. I am especially indebted to P. Armbruster, D. Liesen, and P.H. Mokler for their invaluable help in putting this lecture together.

REFERENCES

Bac78 H. Backe, L. Handschug, F. Hessberger, E. Kankeleit, L. Richter, F. Weik, R. Willwater, H. Bokemeyer, P. Vincent, Y. Nakayama, and J.S. Greenberg, Phys. Rev. Lett. 40, 1443 (1978)

Bac81 H. Backe in: Present Status and Aims of Quantum Electrodynamics. Lecture Notes in Physics, ed. by E. Gräff, E. Klempf, G. Werth, Vol. 143, p. 277., Springer Verlag, Berlin, Heidelberg, New York (1981)

Bac83a H. Backe, W. Bonin, E. Kankeleit, M. Krämer, R. Krieg, V. Metag, P. Senger, N. Trautmann, F. Weik, and J.B. Wilhelmy in: Gre83, p. 107.

Bac83b H. Backe, P. Senger, W. Bonin, E. Kankeleit, M. Krämer, R. Krieg, V. Metag, N. Trautmann, and J.B. Wilhelmy, Phys. Rev. Lett. 50, 1838 (1983).

Bok83 H. Bokemeyer, K. Bethge, F. Folger, J.S. Greenberg, H. Grein, A. Gruppe, S. Ito, R. Schule, D. Schwalm, J. Schweppe, N. Trautmann, P. Vincent, and M. Waldschmidt, in: Gre83, p. 273.

Bok84 H. Bokemeyer, Invited Lecture given at XIX Winter School on Physics, Zakopane, Poland, 1984; GSI-preprint 84-43.

Cle84 M. Clemente, E. Berdermann, P. Kienle, H. Tsertos, W. Wagner, C. Kozhuharov, F. Bosch, and W. Koenig, Phys. Lett. 137B, 41 (1984)

Cow85 T. Cowan, H. Backe, M. Begemann, K. Bethge, H. Bokemeyer, H. Folger, J.S. Greenberg, H. Grein, A. Gruppe, Y. Kido, M. Klüver, D. Schwalm, J. Schweppe, K.E. Stiebing, N. Trautmann, and P. Vincent, Phys. Rev. Lett., 54, 1761 (1985)

Epo85 The members of the EPOS group, a GSI - Yale - Frankfurt - Heidelberg - Mainz collaboration, are: H. Backe, M. Begemann, K. Bethge, H. Bokemeyer, T. Cowan, H. Folger, J.S. Greenberg, H. Grein, A. Gruppe, M. Kluever, D. Schwalm, J. Schweppe, K.E. Stiebing, N. Trautmann, and P. Vincent.

Fri77 B. Fricke and G. Soff, Atom. Nucl. Data Tables 19, 83 (1977)

Gre83 W. Greiner (ed.): 'Quantum Electrodynamics of Strong Fields' Lahnstein/Rhein 1981, Proc. NATO Advanced Study Institute Programme, Plenum New York 1983.

Gue82 F. Güttner, W. Koenig, B. Martin, B. Povh, H. Skapa, J. Soltani, Th. Walcher, F. Bosch, and C. Kozhuharov, Z. Phys. A304, 207 (1982)

Gyu75 M. Gyulassi, Nucl. Phys. A244, 497 (1975).

Kan85 E. Kankeleit, U. Gollerthan, G. Klotz, M. Kollatz, M. Krämer, R. Krieg, U. Meyer, H. Oeschler, and P. Senger, Nucl. Instr. Meth. A234, 81 (1985)

Kau78 K.H. Kaun and S.A. Karamyan, Dubna Preprint JINR-P7-11420 (1978).

Kie83 P. Kienle, in: Gre83, p. 293.

Koz79 C. Kozhuharov, P. Kienle, E. Berdermann, H. Bokemeyer, J.S. Greenberg, Y. Nakayama, P. Vincent, H. Backe, L. Handschug, and E. Kankeleit, Phys. Rev. Lett. 42, 376 (1979)

Mey77 W.E. Meyerhof, R. Anholt, Y. El Masri, D. Cline, F.S. Stephens, and R. Diamond, Phys. Lett. 69B, 41 (1977).

Mol65 E. Moll and E. Kankeleit, Nukleonik 7, 180 (1965).

Mue76 B. Müller and W. Greiner, Z. Naturforsch. 35a, 1 (1976).

Mue83 U. Müller, G. Soff, T. de Reus, J. Reinhardt, B. Müller, and W. Greiner, Z. Phys. A313, 263 (1983).

Mue84 U. Müller, private communication.

Ora85 Members of the Orange-group, a TU München - GSI - Heidelberg collaboration, are: E. Berdermann, F. Bosch, M. Clemente, M. Franosch, S. Huchler, P. Kienle, C. Kozhuharov, W. Koenig, H. Tsertos, and W. Wagner.

Rei81 J. Reinhardt, U. Müller, B. Müller, and W. Greiner, Z. Phys. A303, 173 (1981).

Rho83 M.J. Rhoades-Brown, V.E. Oberacker, M. Seiwert, and W. Greiner, Z. Phys. A310, 287 (1983).

Sli49 L.A. Sliv, Dokl. Akad. SSSR, 64, 521 (1949).

Slu79 P. Schlüter and G. Soff, Atom. Nucl. Data Tables 24, 509 (1979).

Sof79 G. Soff, J. Reinhardt, B. Müller, and W. Greiner, Phys. Rev. Lett. 43, 1981 (1979).

Sof82 G. Soff, P. Schlüter, B. Müller, and W. Greiner, Phys. Rev. Lett. 48, 1465 (1982).

Sof83 G. Soff and W. Greiner, J. Phys. B15, 1681 (1983).

Swa84 D. Schwalm, in: Electronic and Atomic Collisions (XIII ICPEAC 1983), p. 295;
J. Eichler, V. Hertel, N. Stolterfoht (Eds.) North-Holland, Amsterdam 1984.

Swe83 J. Schweppe, A. Gruppe, K. Bethge, H. Bokemeyer, T. Cowan, F. Folger, J.S. Greenberg, H. Grein, S. Ito, R. Schule, D. Schwalm, K.E. Stiebing, N. Trautmann, P. Vincent, and M. Waldschmidt, Phys. Rev. Lett. 51, 2261 (1983).

Tor85 Present members of the TORI group are: U. Gollerthan, E. Kankeleit, G. Klotz, M. Kollatz, M. Krämer, R. Krieg, U. Meyer, H. Oeschler, and P. Senger (TH Darmstadt).

Tse85 H. Tsertos, private communication.

EXOTIC PHENOMENA IN COLLISIONS OF VERY HEAVY IONS

G. Soff*, U. Müller**, S. Schramm, T. de Reus, G. Mehler, J. Reinhardt, B. Müller, and W. Greiner

Institut für Theoretische Physik der Johann Wolfgang Goethe-Universität, Robert-Mayer-Straße 8-10, Postfach 111 932 D-6000 Frankfurt am Main, West Germany

*Institut für Theoretische Physik, Justus-Liebig-Universität Heinrich-Buff-Ring 16, D-6300 Gießen, West Germany

**Gesellschaft für Schwerionenforschung (GSI), Planckstraße 1 Postfach 110 541, D-6100 Darmstadt, West Germany

INTRODUCTION

Over the last decade our knowledge on atomic structure of superheavy quasimolecules in the range $110 \leq Z_{tot} \leq 188$ has increased considerably[1-26]. Heavy ion collisions, in which superheavy quasimolecules are formed for a short period of time, offer us a unique tool to investigate the electronic structure of ultra-high Z-systems, which are not otherwise accessible to experiment. Comparison of K-vacancy formation, δ-electron and positron emission with available experimental data suggests the validity of the quasimolecular picture, which will be taken as the theoretical framework of our calculations.

One of the pivotal points still under discussion in this context is the question, whether in systems with $Z_{tot} \geq 174$ a vacant 1sσ-state dives into the lower continuum and subsequently decays into adjacent positron states. In case of nuclear time delay this process could cause a pronounced structure in the positron spectra[13,14]. Such a structure has been observed in different independent experiments[22,23,25,26]. However, the Z-dependence of this structure is not yet[25,26] understood. The observed independence of the narrow positron lines of the combined nuclear charge number led to the speculation whether a new, previously undetected particle could be created in collisions of very heavy ions. Corresponding theoretical investigations, however without stringent conclusions were reported in refs. 5, 19 and 21. As a more conservative explanation conversion processes in superheavy atoms were considered[5,10]. But also this presumably more realistic alternative displays several deficiencies in relation to the measured data. In summary one has to state that the detected constancy of the e^+-peak structure at E_{e^+} ≃ 320 keV has no rigorous theoretical basis. The scenario of spontaneous positron emission including the underlying nuclear models for time delay[27-29] and a survey of other possible origins of the positron line structure are discussed elsewhere[30].

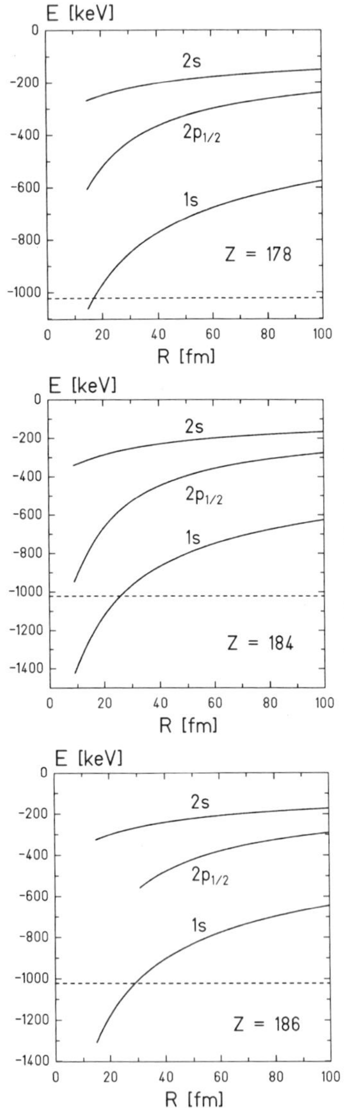

Fig. 1: Binding energies of the 1s-, $2p_{\frac{1}{2}}$- and 2s-electrons versus internuclear separation R in the giant quasiatomic systems Z = 178, 184 and 186. The dashed line indicates the border line to the negative energy continuum.

To demonstrate the enormous binding energies of inner-shell electron states in giant quasiatomic systems and their strong variation as function of the internuclear separation R we plot in figure 1 this quantity for the 1s-, $2p_{\frac{1}{2}}$- and 2s-states in the colliding systems Pb + Cm (Z = 178), U + U (Z = 184) and Th + Cm (Z = 186), respectively. Electron screening corrections being described within the relativistic Hartree-Fock-Slater formalism are taken into account. In the range 20 fm $\leq$ R $\leq$ 100 fm the change in the binding energy is comparable with the electron rest mass.

To exemplify current theoretical investigations we will discuss three different topics. After a presentation of the underlying theoretical framework for ionization processes we will sketch the possibility to employ δ-electron emission as a clock to measure nuclear reaction times in intermediate energy collisions[9,11] of very heavy ions. Besides the phenomenon of vacuum decay into a new twofold negatively charged stable vacuum ground state[1,2], electron excitation in heavy ion collisions may be employed for the determination of delay and deceleration times on the nuclear time scale, i.e. offering an atomic clock, operating in the range 10^{-21} - 10^{-24} s. In deep-inelastic heavy ion collisions this provides a test for classical nuclear reaction models. In collisions at intermediate energies an independent measurement of the deceleration time is of interest for comparison, e.g., with the results of the pion bremsstrahlung model[31]. After that we investigate the influence of one or more pockets in the ion scattering potential on the energy distribution of emitted positrons within a quantum mechanical framework. Finally we very briefly consider some phenomenological corrections to the Dirac equation and its consequences on electron binding energies in heavy and superheavy atoms.

DELTA-ELECTRON PRODUCTION IN COLLISIONS OF VERY HEAVY IONS

In the framework of the Hartree-Fock approximation, the dynamical evolution of the electrons in the time varying Coulomb field of two colliding nuclei is described by the solutions Φ_j of the time-dependent two-centre Dirac equation:

$$i(\partial/\partial t)\, \Phi_j(R(t)) = H_{TCD}(R(t))\, \Phi_j(R(t)) \tag{1}$$

Since the relativistic two-centre Hamiltonian $H_{TCD}(R(t))$ depends sensitively on the internuclear separation R(t), we expand the total wavefunction Φ_j into Born-Oppenheimer states, represented by the stationary molecular eigenstates of the Hamiltonian

$$\Phi_j(t) = \Sigma_k\, a_{jk}(t)\, \phi_k(R(t)) \exp\{-i\chi_k(t)\}, \tag{2}$$

with

$$\chi_k(t) = \int^t E_k(t')\, dt'. \tag{3}$$

The sum in equation (2) includes an integration over continuum states with respect to positive and negative frequencies. Inserting the expansion of equation (2) into equation (1) and projecting with stationary eigenfunctions ϕ_j, a set of first order coupled differential equations for the occupation amplitudes $a_{jk}(t)$ is obtained

$$\dot{a}_{ij}(t) = - \Sigma_{k\neq j}\, a_{ik}(t)\, \langle\phi_j|\partial/\partial t|\phi_k\rangle\, \exp\{i(\chi_j - \chi_k)\} \tag{4}$$

The initial condition is $a_{ij}(-\infty) = \delta_{ij}$. Dividing the time derivative operator into a radial part and a rotational coupling and neglecting the latter, the coupled equations (4) are solved by numerical integration. In the independent-particle approximation excitations of many-electron systems are described by incoherent summation over one-electron transition probabilities. After the collision the number of particles p occupying a state above the Fermi level, up to which the quasimolecular levels were initially filled, is given by

$$N_p = \Sigma_{r<F}\, |a_{rp}(\infty)|^2 \qquad (p > F) \tag{5}$$

while the number of holes in a state q below the Fermi level is given by

$$N_q = \Sigma_{s>F}\, |a_{sq}(\infty)|^2 \qquad (q < F) \tag{6}$$

A positron is represented as a hole in the negative energy continuum. For the number of correlated particle-hole pairs N_{pq} one finds

$$N_{pq} = N_p N_q + |\Sigma_{r<F}\, a^*_{rp} a_{rq}|^2. \tag{7}$$

In the formulae (5) and (6) the indices r and s indicate particle and hole states with the same angular momentum quantum numbers κ and the same spin. In formula (7) particle and hole have corresponding spins, in the sense that an emitted particle with spin up results in a hole state with spin down. Spin, however, is not measured in this kind of experiments we want to consider. Additionally it is impossible to distinguish electrons emitted from the $1s_{\frac{1}{2}}\sigma$ or the $2p_{\frac{1}{2}}\sigma$ state. This has also to be taken into account in an expression which should describe the experimentally observed coincidence rate. Furthermore we have to consider the fact that holes created in a 1sσ molecular level do not exclusively lead to holes in a 1s atomic level of the heavier collision partner, but may be shared[32] by the 1s level of the lighter partner. In view of these arguments we want to derive an expression for the probability to measure a δ - electron in coincidence with a hole, e.g. a K - vacancy.

Let us for the sake of simplicity take a symmetric system, that is $Z_1 = Z_2$. Consider $N_{\alpha\beta\gamma\delta,\kappa\lambda\mu\nu}$ as a probability for particle - hole coincidences. The greek letters have possible values 1 and 0 for detecting a particle (hole), or for detecting no particle (no hole), respectively, in the given state. $\alpha\beta\gamma\delta$ are symbols related to particles, $\kappa\lambda\mu\nu$ are those for holes. The indices are associated with the following combinations: $s_{\frac{1}{2}}$ - states with spin up, s↑, $s_{\frac{1}{2}}$ - states with spin down, s↓, and the same for $p_{\frac{1}{2}}$ - states, i.e. p↑ and p↓. In this convention $\alpha = 1$ means that one can detect a particle in the s↑ - state, $\mu = 0$ has the meaning that there is no hole in the p↑ - state recognized. $N_{0000,0000}$ indicates that there are no particles and no holes in the investigated states.

Since in our calculations these four properties are independent, $N_{\alpha\beta\gamma\delta,\kappa\lambda\mu\nu}$ factorizes in $N_{\alpha\kappa} \cdot N_{\beta\lambda} \cdot N_{\gamma\mu} \cdot N_{\delta\nu}$. In the following, summation has to be understood over all possible combinations of greek indices. Thus a sum $\Sigma\, N_{\alpha\beta\gamma\delta,\kappa\lambda\mu\nu}$ can be written as a product of sums

$$\Sigma\, N_{\alpha\beta\gamma\delta,\kappa\lambda\mu\nu} = (\Sigma\, N_{\alpha\kappa})\,(\Sigma\, N_{\beta\lambda})\,(\Sigma\, N_{\gamma\mu})\,(\Sigma\, N_{\delta\nu})\ . \tag{8}$$

Each term can take on four possible values: N_{00}, N_{01}, N_{10} and N_{11}. For these the following expressions have been derived[33]

$$N_{11} = N_{pq} = N_p \cdot N_q + |\sum_{r<F} a^*_{rp}(\infty)\, a_{rq}(\infty)|^2\ , \tag{9a}$$

$$N_{10} = N_p - N_{pq}\ , \tag{9b}$$

$$N_{01} = N_q - N_{pq}\ , \tag{9c}$$

$$N_{00} = 1 - N_p - N_q + N_{pq} \tag{9d}$$

Thus

$$\Sigma\, N_{\alpha\kappa} = \Sigma\, N_{\beta\lambda} = \Sigma\, N_{\gamma\mu} = \Sigma\, N_{\delta\nu} = 1, \tag{10}$$

and

$$\Sigma N_{\alpha\beta\gamma\delta,\kappa\lambda\mu\nu} = 1 .$$

We are interested in those terms $N_{\alpha\beta\gamma\delta,\kappa\lambda\mu\nu}$, where at least one of the first and at least one of the second sequence of indices is 1, indicated as $N_{(1),(1)}$. We assume that holes and particles with distinguishable properties in a multi - coincidence event can be measured simultaneously and can be separated.

$$N_{(1),(1)} = 1 - \tilde{\Sigma} N_{\alpha\beta\gamma\delta,\kappa\lambda\mu\nu} , \tag{11}$$

where $\tilde{\Sigma}$ represents the terms where either no particles or no holes can be detected

$$\tilde{\Sigma} N_{\alpha\beta\gamma\delta,\kappa\lambda\mu\nu} = \Sigma N_{\alpha\beta\gamma\delta,0000} + \Sigma N_{0000,\kappa\lambda\mu\nu} - N_{0000,0000} . \tag{12}$$

The last term has to be subtracted to compensate double counting in the two sums. Using eqs. (8) and (9) we derive

$$\Sigma N_{\alpha\beta\gamma\delta,0000} = (\Sigma N_{\alpha 0}) \bullet (\Sigma N_{\beta 0}) \bullet (\Sigma N_{\gamma 0}) \bullet (\Sigma N_{\delta 0}) \tag{13}$$

and

$$\Sigma N_{\alpha 0} = N_{00} + N_{10} = 1 - N_q = 1 - N_\kappa , \tag{14a}$$

and equivalently the term $\Sigma N_{0000,\kappa\lambda\mu\nu}$

$$\Sigma N_{0\kappa} = N_{00} + N_{01} = 1 - N_p = 1 - N_\alpha . \tag{14b}$$

If we consider that in our calculations there is no difference between the excitation rates of states with spin up and those with spin down, we arrive at

$$\Sigma N_{\alpha\beta\gamma\delta,0000} = (\Sigma N_{\alpha 0})^2 (\Sigma N_{\gamma 0})^2 = (1 - N_\kappa)^2 (1 - N_\mu)^2 . \tag{15}$$

Thus

$$\begin{aligned} N_{(1),(1)} = 1 &- (1 - N_\kappa)^2 (1 - N_\mu)^2 - (1 - N_\alpha)^2 (1 - N_\gamma)^2 \\ &+ (1 - N_\alpha - N_\kappa + N_{\alpha\kappa})^2 \bullet (1 - N_\gamma - N_\mu + N_{\gamma\mu})^2 . \end{aligned} \tag{16}$$

All terms in eq. (16) have to be understood as measurable quantities within a solid angular window $\Delta\Omega$ and (for continuum states) a certain energy range ΔE. Thus one can rewrite eq. (16) with the abbreviations

$$\Delta p = \Delta E_p \, \Delta\Omega_p \ , \quad \Delta q = \Delta\Omega_q$$

$$N_{(1),(1)} \, \Delta p \Delta q = 1 - (1 - N_\kappa \Delta q)^2 \, (1 - N_\mu \Delta q)^2 - (1 - N_\alpha \Delta p)^2 \, (1 - N_\gamma \Delta p)^2$$
$$+ (1 - N_\alpha \Delta p - N_\kappa \Delta q + N_{\alpha\kappa} \Delta p \Delta q)^2 \bullet (1 - N_\gamma \Delta p - N_\mu \Delta q + N_{\gamma\mu} \Delta p \Delta q)^2 \ . \quad (17)$$

The terms Δp, Δq can easily be extended to include the detector efficiency ε_p, ε_q as well as other experimental quantities. For the theoretically interesting limit Δp, $\Delta q \to 0$ we derive

$$N_{(1),(1)} = 2\,(N_{\alpha\kappa} + N_\alpha N_\kappa) + 4\, N_\alpha N_\mu + 2\,(N_{\gamma\mu} + N_\gamma N_\mu) + 4\, N_\gamma N_\kappa \quad (18)$$

This expression was also derived in ref. 33 and given there as

$$N_{p=E,q=1s} = N_{E,1s\sigma} + N_{E,2p_{\frac{1}{2}}\sigma} \ , \quad (19)$$

with

$$N_{E,1s\sigma} = 2\,(N_{Es\sigma,1s\sigma} + N_{Es\sigma} \bullet N_{1s\sigma}) + 4\, N_{Ep_{\frac{1}{2}}\sigma} \bullet N_{1s\sigma} \ , \quad (20a)$$

$$N_{E,2p_{\frac{1}{2}}\sigma} = 2\,(N_{Ep_{\frac{1}{2}}\sigma,2p_{\frac{1}{2}}\sigma} + N_{Ep_{\frac{1}{2}}\sigma} \bullet N_{2p_{\frac{1}{2}}\sigma}) + 4\, N_{Es\sigma} \bullet N_{2p_{\frac{1}{2}}\sigma} \ . \quad (20b)$$

Note that $N_{Es\sigma,1s\sigma}$ contains also a term $N_p N_q$ for accidental coincidences for an electron and a K - hole with corresponding spins. So the probability for the accidental coincidence of δ-electrons in a $s_{\frac{1}{2}}\sigma$ state with a K - hole is weighted by a factor 4, due to the four possible spin combinations. The same argument holds for the third term in the sum of eq. (20a) which gives the number of accidental coincidences of a δ-electron in a $p_{\frac{1}{2}}\sigma$ state and a K - hole. Eq. (20b) describes coincidences of electrons with holes in the $2p_{\frac{1}{2}}\sigma$ state. Its terms have a similar meaning.

Until now we have considered symmetric systems, where created vacancies in the molecular levels are shared equally between the collision partners to create K - holes in the separate atoms. In case of asymmetric systems we have coincidences with the K-x-rays of the collision partner with higher nuclear charge and those with the collision partner having lower nuclear charge. In this paper we consider only coincidences with K-x-rays originating from the collision partner with higher nuclear charge. In this case we have to weight the contributions from the $1s\sigma$ and the $2p_{\frac{1}{2}}\sigma$ molecular levels according to Meyerhof[32]. This yields

$$N_{p=E,q=1s}(H) = (1-w)\cdot N_{E,1s\sigma} + w\cdot N_{E,2p_{\frac{1}{2}}\sigma} \quad . \tag{21}$$

$N_{p=E,q=1s}(H)$ indicates the number of created particles measured in coincidence with the K-x-rays from the collision partner with higher nuclear charge. w is defined by the equations

$$w = \frac{e^{-2x}}{1 + e^{-2x}} \quad , \tag{22a}$$

$$x \equiv \frac{\sqrt{2}\,\pi\,(\sqrt{E_H} - \sqrt{E_L})}{\sqrt{m_e}\; v} \quad . \tag{22b}$$

E_H and E_L denote the energies of the 1s atomic levels of the collision partners with higher and lower nuclear charge, respectively. m_e is the electron mass and v the projectile velocity. For very asymmetric systems with a large difference $\sqrt{E_H} - \sqrt{E_L}$, the contribution from the $2p_{\frac{1}{2}}\sigma$ hole disappears.

To allow for a comparison with the experimental data of Herath-Banda et al[34] we still have to perform an integration over the impact parameter b. Thus the double differential cross section for the production of a δ-electron in coincidence with a K-x-ray of the collision partner with the higher nuclear charge reads

$$\frac{d^2\sigma}{dE\,d\Omega} = \int_0^\infty [(1-w)\cdot N_{E,1s\sigma} + w\cdot N_{E,2p_{\frac{1}{2}}\sigma}]\; 2\pi b\; db \tag{23}$$

Our calculations were performed using a basis of 8 $s\sigma$-bound and 18 $s\sigma$-continuum states, 6 $p_{\frac{1}{2}}\sigma$-bound and also 18 $p_{\frac{1}{2}}\sigma$-continuum states. All calculations have been performed twice using a Fermi surface of F = 3 and F = 4 (i.e. all levels up to $4p_{\frac{1}{2}}\sigma$ and $5p_{\frac{1}{2}}\sigma$, respectively, are filled) to get an estimate of the influence of the Fermi level. The differences in the coincident spectra were of the order of 1%. In the total spectra the contribution of the higher Fermi level varies from 35% in the electron energy region of 200 keV down to 15% at 600 keV.

In figure 2 we compare our results for the double differential cross section with the experimental data of Herath-Banda et al[34] for the system Au-U.

The double differential cross section is shown versus the kinetic electron energy in the center of mass system. The lower curve represents coincidences between electrons and K-holes. The upper curve indicates our result for the total δ-electron emission rate. Remarkable agreement is achieved for the total and the coincident spectra for the slope as well as for the absolute numbers. Note, that in the calculations no scaling or fitting has been applied.

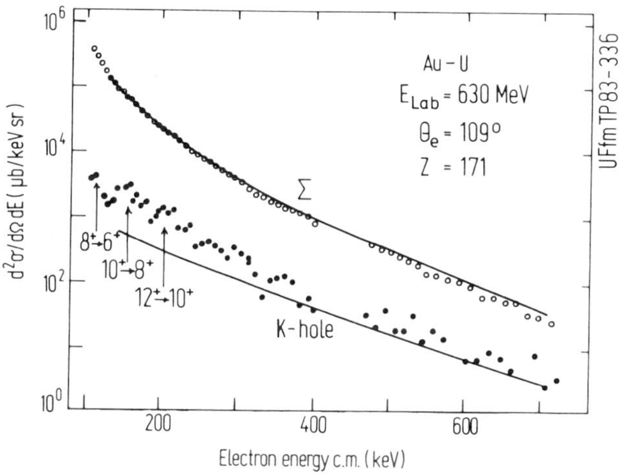

Fig. 2: Experimental double differential cross sections for the total δ-electron yield (Σ, open circles) and for the 1s atomic coincident yield (K-hole, full dots) versus kinetic electron energy. The system Au-U at 630 MeV bombarding energy is considered. The full lines are the results of our coupled channel calculations. Experimental data by M.A. Herath-Banda and collaborators[34].

For the system Au-U the vacancy sharing raises the coincident rate 50% for the low energy part (≃150 keV) and 20% for the high energy part (≃ 600 keV) of the spectrum. The bumps in the coincidence spectrum arise from nuclear Coulomb excitation and subsequent internal conversion in the U-nucleus. Corresponding nuclear transitions are indicated. The electrons were observed at certain angles with respect to the beam axis. No asymmetries in the coincident spectra were detected for these systems.

In conclusion we may state that our theoretical considerations correctly describe inner-shell excitation and δ-electron production in superheavy systems. The strong relativistic effects are reflected in the rather high production rate of high energy electrons.

The recently observed subthreshold production of neutral pions motivates a determination of the reaction time scale. Since neutral pion production has been measured at bombarding energies as low as 20 MeV/u it is believed to be of cooperative resp. collective origin because the independent nucleon model with Fermi motion fails in predicting pions below 50 MeV/u. Hence the pions should depend strongly on the evolution of the nuclear collision zone respectively on the deceleration time.

In our treatment of electron emission we will assume that the nuclear charge and current density during the reaction can be described classically. This is probably a reasonable approximation as long as the energy carried away by the particle is a small fraction of the total centre-of-mass energy, and if the measurement averages over many nuclear final states. As complementary channel to photon and pion emission we propose the investigation of electron spectra in intermediate energy collisions (E_{Lab} = 20- 100 MeV/u). While in atomic collisions with relativistic heavy ions (E_{Lab} = 82- 670 MeV/u) K-vacancy production has been successfully explained using atomic models such as plane-wave Born approximation, we retain the molecular model described in the introduction which should still be valid in the considered energy range.

In first order perturbation theory the differential emission probability of a quasimolecular bound electron with energy E_i into a continuum state with energy E becomes ($A_o \equiv V_o$)

$$dN^{e^-}/dE = \Sigma_i |\int dt \langle\phi_E|(E_i - E)^{-1} e\partial V_o/\partial t - ie/\hbar\vec{\alpha}\cdot\vec{A}|\phi_i\rangle \exp\{i(E - E_i)t/\hbar\}|^2 \qquad (24)$$

Equation (24) can be compared[11] with the classical expression for the distribution of the bremsstrahlung photons. In contrast to the bremsstrahlung, photons which are directly given by the Fourier transform of the nuclear currents $\vec{j}$, δ-electrons depend on $j_\mu(\vec{r},t)$ indirectly via the four-potential of the transition currents. Thus high energetic δ- electrons emitted in intermediate energy heavy ion collisions may be envisaged as supplementary messengers carrying information on the nuclear current besides the photons.

To be more specific we may ask whether the mean nuclear deceleration time τ in central intermediate energy heavy ion collisions is reflected in the spectra of emitted δ-electrons. In order to deal with this question we use a simplified semiclassical model assuming that the relevant information on the reaction can be described by a nuclear trajectory R(t). As a first step we consider only head-on (b = 0) collisions and assume that the relative motion comes to a complete stop. Such a trajectory may be parametrized by the ansatz

$$R(t) = v_\infty \tau/2 \ \ln[\exp\{2t/\tau\}/(1 + \exp\{2t/\tau\})]. \tag{25}$$

In first order perturbation theory the amplitude for the transition of, e.g., a quasimolecular 1sσ- electron to the continuum state E reads

$$a(E) = -\int_{-\infty}^{\infty} \dot{R}(t)/(E_{1s\sigma} - E) \langle\phi_E | e\partial V_o/\partial R | \phi_{1s\sigma}\rangle \exp\{i(E - E_{1s\sigma})t/\hbar\} \tag{26}$$

Since for large continuum energies $|E_{1s\sigma}(R(t)| \ll E$, we set $E - E_{1s\sigma}(R(t)) = E - E_{1s\sigma}(R_{min}) = \Delta E = \text{const.}$, furthermore we neglected magnetic effects due to the vector potential $\vec{A}$. The R-dependence of the radial coupling matrix element can be approximated by

$$\langle\phi_E | e\partial V_o/\partial R | \phi_{1s\sigma}\rangle \ / \ \Delta E \sim R \ / \ (R_m{}^2 + R^2). \tag{27}$$

The parameter R_m indicates the position of the maximum in the matrix element. This is dependent on the Coulomb potential and thus on the chosen nuclear charge distribution in the overlap region. Considering, e.g., the system Pb + Pb a value of $R_m \sim 11$ fm is found when the nuclear densities are assumed to add up (sudden approximation). In the opposite limit of nuclear volume conservation (adiabatic potential) the maximum is reached earlier at $R_m \sim 14$ fm.

One may attempt to integrate equation (26) in the complex t- plane. From such considerations, i.e. inspecting the location of the poles of the integrand in the complex time plane, we conclude that δ- electron emission from the 1sσ- state may be approximately described by

$$|a(E)|^2 \sim \pi^2 \exp\{-\Delta E(R_m/v_\infty + 3\pi\tau/2)/\hbar\} \tag{28}$$

Equation (28) is no exact solution of equation (26), because equation (26) carries two types of singularities in the complex t- plane: one originating from the matrix element, yielding the discussed expression and another originating from the trajectory which allows no elementary solution. The latter expression which might interfere with the evaluated residuum was assumed to be small and neglected. Although we have no mathematical justification for this approximation, a numerical evaluation of equation (26), using the trajectory in equation (29) suggests the validity of equation (28) concerning the dependence on the deceleration time τ.

In fig. 3 we show an example for the numerical integration of equation (26) using the full numerical wavefunctions and energies.

In contrast to our approximation yielding equation (28), for the numerical calculations we used a Rutherford trajectory with exponential velocity loss for $R(t) < R_c = 20$ fm according to

$$R(t) = R(t)_{Rutherford}/(1 + \exp\{(t - t_o)/\tau\}), \tag{29}$$

where t_o was fixed by the condition

$$\dot{R}(t_o(R_o)) = (1/2)\, \dot{R}_{Rutherford}(t(R_c)). \tag{30}$$

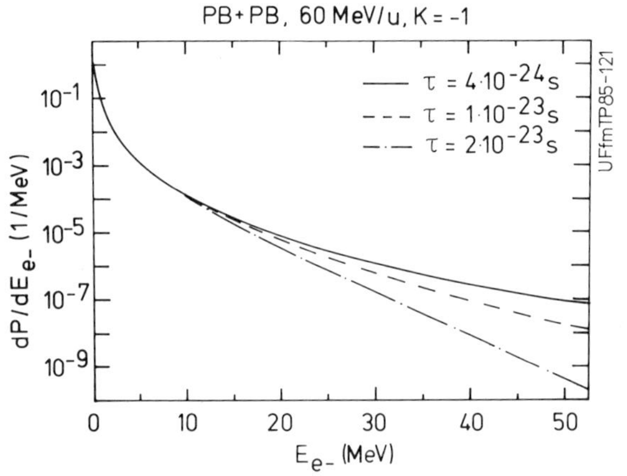

Fig. 3: Differential probability of δ-electron emission from the 1sσ-state in a 60 MeV/u Pb + Pb collision versus kinetic electron energy. Different curves belong to different deceleration times τ: full curve $\tau = 4 \cdot 10^{-24}$ s, dashed curve $\tau = 10^{-23}$ s and dashed-dotted curve $\tau = 2 \cdot 10^{-23}$ s.

The nuclei are assumed to stop at $R_o = 7$ fm. The systematics in the slope of the high energy tail of the δ- electron spectrum is surprisingly well described by equation (28). It is steepest for large values of τ and decreases if the deceleration becomes more abrupt (shorter times τ) with a tendency towards a limiting slope for $\tau \ll 10^{-23}$ s. For very small deceleration times the curves deviate from the exponential given by equation (28).

In the limit $\tau \to 0$ equation (26) can be integrated analytically, yielding the result

$$|a(E)|^2 \sim \pi^2 \exp\{-2\Delta E\ R_m/(\hbar v_\infty)\} + 2E_i(\Delta E\ R_m/(\hbar v_\infty))\ E_i(-\Delta E\ R_m/(\hbar v_\infty))$$
$$+ \exp\{-2\Delta E\ R_m/(\hbar v_\infty)\}\ \ E^2_i(\Delta E\ R_m/(\hbar v_\infty))$$
$$+ \exp\{2\Delta E\ R_m/(\hbar v_\infty)\}\ E^2_i(-\Delta E\ R_m/(\hbar v_\infty)) \qquad (31)$$

being a function similar to the curve for $\tau = 4 \cdot 10^{-24}$ s in fig. 3.

At intermediate energy collisions of heavy ions, where subthreshold pion production is observed we stressed how electron emission depends on the space-time evolution of the internuclear collision zone. Restricting ourselves to first-order perturbation theory and a simple Fermi-type model for the trajectory, valid only for head-on (b = 0) collisions, we predict the emission of electrons having kinetic energies up to 50 MeV in 60 MeV/u Pb + Pb collisions. The slope of the δ-electron spectra in the high energy wing provides a tool to determine the underlying deceleration time scale: for increasing deceleration times the slope becomes steeper while it decreases in case of more abrupt decelerations, i.e. shorter deceleration times.

In order to obtain quantitative predictions which can be compared with measurements, a full coupled channel calculation also including $p_{\frac{1}{2}}$ states is required which could increase the probabilities by about an order of magnitude. Furthermore an ansatz for the impact parameter dependence of close collisions and the outgoing trajectory are needed. Electron screening effects are negligible. In future calculations a more refined model for the time- evolution of the nuclear charge and current distribution will be studied.

The experimental feasibility depends on the intensity of background effects producing high energy electrons, in particular pair conversions of γ-rays and pion decay $\pi^o \to \gamma + e^+ + e^-$. These effects, however, are expected to yield about the same numbers of electrons and positrons. Since atomic positron production in this regime is estimated to be considerably smaller in comparison with atomic δ-electron emission, the latter process could be extracted from the ratio $P^{tot}_{e-}/P^{tot}_{e+}$.

Concluding, we note the possibility to use δ-electron emission in deep-inelastic and intermediate energy collisions as clock-hand to determine delay and deceleration times on the nuclear time-scale, i.e. T = 10^{-21} - 10^{-24} s. The shape of the high-energy wing of the δ-electron spectra provides a finger-print of the underlying nuclear reaction times.

POCKETS IN THE ION SCATTERING POTENTIAL AND POSITRON EMISSION

Now we turn the discussion to a different subject. A method is presented how to treat heavy-ion scattering including one or more pockets in the scattering potential. Resulting positron spectra for U-U collisions are shown. The special aim of this section is to investigate the influence of the formation of a quasi-stable giant nuclear molecule[13,14] on the development of line structures in the positron spectrum[22,23]. The existence of a nuclear molecule for some delay time $T > 10^{-20}$ s, as described here, is based on a pocket in the scattering potential of the two nuclei[27].

The lifetime of the quasimolecule influences the positron spectrum by means of several mechanisms: First, the pocket leads to a strongly energy-dependent phase shift of the outgoing wavefunction. Therefore the incoming and outgoing waves interfere in a way that reflects the time delay. Secondly, there may occur transitions between the various virtual states of the quasimolecule supported by the pocket. By transfer of energy from the nuclear to the electronic motion conversion processes may take place. Another important effect will occur for scattering systems with a combined charge of target and projectile $Z > 173$. In those supercritical systems the binding energy of the lowest electronic state exceeds $2mc^2$ at some critical internuclear distance R_{cr}, i.e. this state becomes a resonance in the lower continuum leading to spontaneous emission of positrons. The emission rate depends strongly on the lifetime of the quasimolecule.

A unified treatment of these processes requires that nuclear motion and electronic excitations are described in a quantum mechanical formalism. The Hamiltonian of the collision system can be written as (center-of-mass system, $\hbar=c=1$):

$$\hat{H} = \hat{\vec{p}}^2/(2\mu) + V(R) + \hat{H}_{TCD}(R)$$

$$\text{with } \hat{H}\psi = E\psi. \qquad (32)$$

$\hat{H}_{TCD}$ is the relativistic Two-Center-Dirac Hamiltonian of the electron (depending on R), and V(R) is the scattering potential assumed to exhibit one or more pockets.

Using an expansion of the wavefunction ψ with respect to eigenstates of $\hat{H}_{TCD}$ we obtain

$$\psi = \Sigma_n F_n(R) \, |n(\vec{R},\vec{r})\rangle$$

$$\hat{H}_{TCD}|n\rangle = \varepsilon_n(R)|n\rangle \tag{33}$$

Inserting (33) into (32) and projecting with the electron state and the angular part of F(R), coupled channel equations describing the relative motion of the nuclei follow[15,16]

$$[d^2/dR^2 + 2\mu(E-V(R)-\varepsilon_n(R))]\, F_n(R) =$$

$$\Sigma_m\, [-2\langle n|\partial/\partial R|m\rangle\, F'_m(R) + 2\mu\, \langle n|H_{TCD}|m\rangle\, F_m(R)] \tag{34}$$

$\langle n|\partial/\partial R|m\rangle$ describes the dynamical couplings between the electronic channels. The terms $\langle n|H_{TCD}|m\rangle$ that mediate the spontaneous decay of the overcritical channel, arise in a special treatment of the overcritical resonance[13]. Equation (34) can be solved by an expansion of $F_n(R)$ into asymptotically in- and outgoing wavefunctions $\chi^{\pm}$, which solve the elastic problem (i.e. setting all couplings in eq. (34) to zero):

$$F_n(R) = a_n^+(R)\, \psi_n^+(R) + a_n^-(R)\, \psi_n^-(R) \tag{35}$$

The wavefunctions can be derived in a JWKB-approximation using a connection formula to combine the solutions inside the pocket of the potential with the exterior part[35]. The resulting wavefunctions are

$$\chi^{\pm} = \begin{cases} A^{\pm}\, e^{i\int k dr} + B^{\pm}\, e^{-i\int k dr} & \text{inside} \\ & \qquad \text{the pocket} \\ e^{\pm i(\int k dr + \phi)} & \text{outside} \end{cases} \tag{36}$$

$\chi^{\pm}$ oscillates inside the pocket with amplitudes A, B showing resonance behaviour, which means that the virtual states contained in the pocket are included in this approximation. In the same way the phase shift ϕ reflects the resonance scattering at the pocket. Using the same method wavefunctions belonging to potentials with several pockets can be derived.

The radial Schrödinger eq. (34) is equivalent to a set of coupled first order channel eqs. for the expansion coefficients $a^{\pm}$. Solving numerically the boundary value problem inside the pocket and neglecting contributions from the exterior region we have calculated positron emission probabilities in supercritical U-U collisions. As boundary condition it is assumed that initially the overcritical electronic state is empty. The chosen beam energy is close to the Coulomb barrier because at this energy a pocket in the potential is expected[27].

In fig. 4 the scattering potential used for an illustrational calculation is shown. The two considered resonances (dashed lines) have energy widths $\Gamma_1 \simeq 68$ keV and $\Gamma_2 \simeq 8$ keV. Due to the chosen beam energy a quasimolecule in the upper resonance state (dashed lines) will be formed. By conversion the molecule may deexcite into the virtual state below. In fig. 4 the resulting positron spectrum exhibits two peaks which originate from different processes. The first peak at E = 204 keV originates from the spontaneous decay of a hole in the overcritical 1s-state. The second peak at E = 1201 keV emerges due to conversion - an electron from the Dirac sea is getting

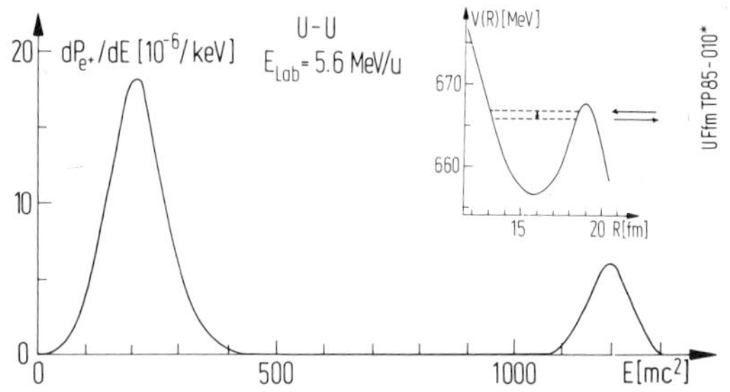

Fig. 4: Scattering potential of a U-U collision system and resulting positron spectrum showing two line structures due to spontaneous emission of positrons and conversion.

excited into the 1s-state creating a positron. The spectrum shown results after averaging over a beam energy interval ΔE = 200 keV to include the whole influence of the chosen resonance. The width of the spontaneous peak increases by a factor of 1.7, whereas the width of the averaged conversion peak gets about ten times larger. The reason for this behaviour is that the nuclei enter the region of the pocket through the upper resonance. Thus by averaging over the energy, the width of the upper resonance also determines the shape of the conversion line.

It is conceivable that one heavy-ion quasimolecule can exist in several isomeric configurations. In our formalism these can be treated by taking a potential with more than one pocket. Calculations leading to the positron spectrum shown in fig. 5 are based on a potential with two pockets displayed in inset. The beam energy is chosen to excite one resonance state in each pocket (dashed lines), both generating a delay time $T = 4 \cdot 10^{-20}$ s. The resulting positron spectrum in fig. 5 shows two line structures at E = 192 keV and 281 keV. Both peaks originate from the spontaneous emission of positrons due to the diving of the 1s-state into the lower continuum. Since the 1s binding energy varies with R, each pocket generates a peak at a different positron energy. The high amplitude at E = 281 keV arises from the enhanced spontaneous emission at smaller internuclear distances. On the other hand, resonances of the left pocket lying more deeply will partly be screened by the outer one.

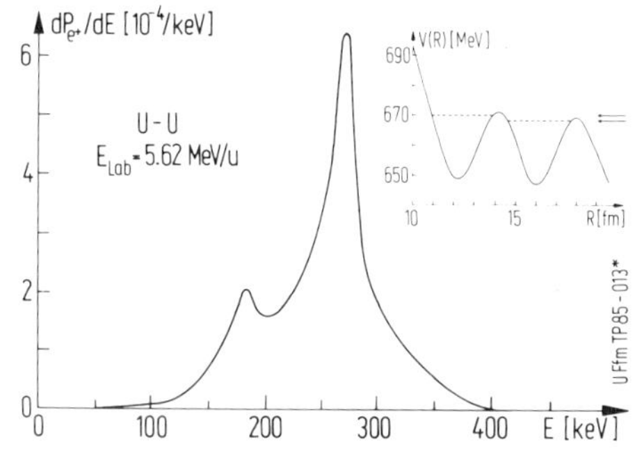

Fig. 5: Assumed scattering potential including two pockets and resulting positron spectrum of a U-U collision.

PHENOMENOLOGICAL CORRECTIONS TO ELECTRON BINDING ENERGIES

It was a long standing question whether quantum electrodynamical corrections or new types of interactions could provide drastic modifications of electron binding energies in superheavy collision systems. Lamb-shift corrections were calculated in the framework of standard quantum electrodynamics and found to be of minor importance[36-39]. Even for the strongest bound 1s-state the QED-shifts amount to less than 1% of the total binding energy. Of more speculative nature, of course, is the investigation[40] of possible nonlinearities in the electromagnetic interaction between the electron and the nucleus or in the electron selfinteraction. However, the precise spectroscopic informations available for the elements in the known periodic system ($Z \leq 100$) puts stringent limits on the corresponding coupling constants. Taking these upper limits for the coupling constants no striking variation of electron binding energies in the superheavy domain ($110 \leq Z \leq 188$) could be derived[40].

As a typical example we consider a scalar interaction of exponential type

$$V_S(r) = V_o\, e^{-r/a} \tag{37}$$

which is coupled to the electron mass in the Dirac equation. The potential (37) is very similar to a Yukawa-type interaction, which results, if in addition to the photons a new, previously undetected scalar particle would mediate the interaction between the bound electron and the nucleus[19]. Taking into account just the Coulomb potential of a finite-size nucleus with $R_{nuc} = 1.2\, A^{1/3}$ fm we obtain ($V_o = 0$) $E_{1s} = 161.168$ keV for $Z = 100$ ($A = 248$), $E_{1s} = 1000.154$ keV for $Z = 169$ ($A = 422.5$) and $E_{1s} = 1716$ keV for $Z = 188$ ($R_{nuc} = 9.5$ fm). For the parameters $V_o = 0.01$ MeV and $a = 20$ fm we get $E_{1s} =$ 161.155 keV, 959.559 keV and 1716 keV, respectively. Similarly we find for $V_o = 0.2$ MeV and $a = 5$ fm $E_{1s} = 161.158$ keV, 998.919 keV and 1714 keV, respectively. Again the present agreement of about $\Delta E \simeq 10$ eV between theoretical and experimental values of K-shell binding energies in fermium ($Z = 100$) forces rather small coupling constants V_o. In consequence the exotic interaction (37) remains also of no influence in critical and supercritical systems.

As a counter example we consider special selfinteractions of K-shell electrons. With the radial electron density $\rho(r) = f^2(r) + g^2(r)$ we evaluate within first-order perturbation theory the energy shifts given by

$$\Delta E_1 = \int_o^\infty r^2\, (f^2 + g^2)\, \gamma_1\, \rho\, dr \tag{38}$$

$$\Delta E_2 = \int_o^\infty r^2\, (f^2 + g^2)\, \gamma_2\, \rho^2\, dr \tag{39}$$

It should be emphasized that in principle the nonlinear interaction from which eq. (39) is derived is non-renormalizable. However, it could serve as an effective model for a more fundamental, renormalizable interaction. The coupling constants γ_1 and γ_2 are fixed such that for $Z = 100$ (Fm) the energy shifts take on the maximum value compatible with experiment, i.e. $\Delta E_1 \simeq \Delta E_2 \simeq 10$ eV. In natural units this yields $\gamma_1 = 3.5 \cdot 10^{-5}$ and $\gamma_2 = 1.55 \cdot 10^{-5}$. Taking these values we obtain for $Z = 169$ $\Delta E_1 = 7.273$ keV and $\Delta E_2 = 7.98$ MeV. The latter result, of course, demonstrates that first-order perturbation theory no longer is applicable. In addition it proves that there is still some room for surprises in giant quasiatomic systems. A nonlinear interaction being not measurable in standard atoms could cause drastic consequences in collisions of very heavy ions.

SUMMARY

The formation of superheavy quasiatoms in collisions of very heavy ions provides a testing ground of quantum electrodynamics of strong fields. The observed K-vacancy production as well as the δ-electron and e^+-emission probability is well described[1-4] in the framework of our coupled channel calculations employing a quasimolecular basis. However, the occurence of narrow peak structures[22,23,25,26,42] in the positron spectra which seem to be almost independent of the combined nuclear charge number Z has no rigorous theoretical explanation[5,41]. Current attempts[10,13,14,19,21] display several deficiencies in relation to the experimental data.

As example we summarized in this report the basic formalism for δ-electron emission in superheavy quasiatoms. Elastic heavy ion collisions as well as intermediate energy collisions were considered. In the latter case δ-electrons may serve as an atomic clock to measure nuclear deceleration times. Finally we studied within a quantum mechanical model the influence of pockets in the ion scattering potential on positron spectra and some consequences of phenomenological corrections to the Dirac equation.

REFERENCES

1. W. Greiner, ed., "Quantum Electrodynamics of Strong Fields," NASI series B80, Plenum, New York (1983)
2. W. Greiner, B. Müller, J. Rafelski, "Quantum Electrodynamics of Strong Fields," Springer, Berlin (1985)
3. J. Reinhardt, W. Greiner, Heavy-ion atomic physics (theory), in: "Treatise on heavy-ion science," Vol. 5, p. 3, ed.: D.A. Bromley, Plenum, New York (1985)
4. W. Greiner, W. Scheid, Heavy-ion atomic physics, in: "Heavy ion collisions," Vol. 3, p. 299, ed.: R. Bock, North Holland, Amsterdam (1982)
5. G. Soff, U. Müller, P. Schlüter, J. Reinhardt, T. de Reus, A. Schäfer, K.-H. Wietschorke, B. Müller, W. Greiner, Ionization and positron emission in superheavy quasiatoms, Preprint GSI-85-25
6. G. Soff, U. Müller, T. de Reus, J. Reinhardt, B. Müller, W. Greiner, Nucl. Instr. Meth. B9:747 (1985)

7. G. Soff, U. Müller, T. de Reus, J. Reinhardt, B. Müller, W. Greiner, Nucl. Instr. Meth. B10/11:214 (1985)
8. S. Schramm, U. Heinz, U. Müller, J. Reinhardt, T. de Reus, G. Soff, W. Greiner, B. Müller, Decay of the vacuum in heavy ion collisions, in: "Atomic Physics 9," eds.: R.S. van Dyck, E.N. Fortson, p. 362, World Scientific, Singapore (1984)
9. T. de Reus, J. Reinhardt, B. Müller, U. Müller, G. Soff, W. Greiner, Z. Physik A321:589 (1985)
10. P. Schlüter, U. Müller, G. Soff, T. de Reus, J. Reinhardt, W. Greiner, Conversion processes in superheavy atoms, Preprint 1985
11. T. de Reus, J. Reinhardt, B. Müller, W. Greiner, U. Müller, G. Soff, Delta electron emission as an atomic clock in deep-inelastic and intermediate energy heavy ion collisions, Preprint GSI-85-27
12. B. Müller, Phys. Bl. 41:208 (1985)
13. J. Reinhardt, U. Müller, B. Müller, and W. Greiner, Z. Phys. A303:173 (1981)
14. U. Müller, G. Soff, Th. de Reus, J. Reinhardt, B. Müller, and W. Greiner, Z. Phys. A313:263 (1983)
15. U. Heinz, B. Müller, and W. Greiner, Ann. Phys. 151:227 (1983)
16. U. Heinz, U. Müller, J. Reinhardt, B. Müller, and W. Greiner, Ann. Phys. 158:476 (1984)
17. U. Müller, G. Soff, J. Reinhardt, T. de Reus, B. Müller, W. Greiner, Phys. Rev. C30:1199 (1984)
18. P. Indelicato, Effects of the Breit interaction on the 1s binding energy of superheavy elements, preprint 1985
19. A. Schäfer, J. Reinhardt, B. Müller, W. Greiner, G. Soff, J. Phys. G11:L69 (1985)
20. J. Krause, M. Kleber, Nucl. Instr. Meth. B9:505 (1985)
21. A.B. Balantekin, C. Bottcher, M.R. Strayer, S.J. Lee, Phys. Rev. Lett. 55:461 (1985)
22. J. Schweppe, A. Gruppe, K. Bethge, H. Bokemeyer, T. Cowan, H. Folger, J.S. Greenberg, H. Grein, S. Ito, R. Schule, D. Schwalm, K.E. Stiebing, N. Trautmann, P. Vincent, and M. Waldschmidt, Phys. Rev. Lett. 51:2261 (1983)
23. M. Clemente, E. Berdermann, P. Kienle, H. Tsertos, W. Wagner, C. Kozhuharov, F. Bosch, and W. Koenig, Phys. Lett. 137B:41 (1984)
24. R. Krieg, E. Bozek, U. Gollerthan, E. Kankeleit, G. Klotz, M. Krämer, U. Meyer, H. Oeschler, P. Senger, Nucl. Instr. Meth. B9:762 (1985)
25. T. Cowan, H. Backe, M. Begemann, K. Bethge, H. Bokemeyer, H. Folger, J.S. Greenberg, H. Grein, A. Gruppe, Y. Kido, M. Klüver, D. Schwalm, J. Schweppe, K.E. Stiebing, N. Trautmann, P. Vincent, Phys. Rev. Lett. 54:1761 (1985)

26. H. Tsertos, E. Berdermann, F. Bosch, M. Clemente, P. Kienle, W. Koenig, C. Kozhuharov, W. Wagner, submitted to Phys. Lett. B
27. M. Seiwert, W. Greiner, and W.T. Pinkston, J. Phys. G11:L21 (1985)
28. V.E. Oberacker, Nuclear alignment: Implications for the decay of the QED-vacuum, Vanderbilt University preprint, July 1985
29. D.P. Russell, W.T. Pinkston, V.E. Oberacker, Phys. Lett. 158B:201 (1985)
30. J. Fink, J.A. Maruhn, B. Müller, U. Müller, L. Neise, J. Reinhardt, T. de Reus, A. Schäfer, P. Schlüter, W. Schmidt, S. Schramm, G. Soff, D. Vasak, W. Greiner, The decay of the vacuum in supercritical fields of giant nuclear systems, preprint 1985
31. D. Vasak, B. Müller, W. Greiner, Phys. Scr. 22:25 (1980)
32. W. Meyerhof, Phys. Rev. Lett. 31:1341 (1973)
33. G. Soff, J. Reinhardt, B. Müller, W. Greiner, Z. Physik A294:137 (1980)
34. M.A. Herath-Banda, A.V. Ramayya, C.F. Maguire, F. Güttner, W. Koenig, B. Martin, B. Povh, H. Skapa and J. Soltani, Phys. Rev. A29:2429 (1984) and private communication
35. M.S. Child, "Molecular collision theory", Academic Press, NY (1974)
36. P.J. Mohr. Ann. Phys. (NY) 88:26 and 52 (1974)
37. K.T. Cheng, W.R. Johnson, Phys. Rev. A14:1943 (1976)
38. G. Soff, P. Schlüter, B. Müller, W. Greiner, Phys. Rev. Lett. 48:1465 (1982)
39. M. Gyulassy, Nucl. Phys. A244:497 (1975)
40. G. Soff, B. Müller, J. Rafelski, Z. Naturforsch. 29a:1267 (1974)
41. J. Reinhardt, T. de Reus, W. Greiner, B. Müller, U. Müller, A. Schäfer, P. Schlüter, S. Schramm, G. Soff, Positron creation in supercritical heavy ion collisions, Contribution to the XIV ICPEAC, Stanford, 1985
42. P. Kienle, Positron production in heavy ion-atom collisions, Preprint GSI-85-31

ON LOW ENERGY EXPERIMENTS IN QED FOR DETECTING EFFECTS SPECIFICALLY DUE TO VACUUM POLARIZATION

Emilio Zavattini

CERN

CH-1211 Geneva, Switzerland

1.1. LASER INDUCED TRANSITION IN $(\mu^{-}\,{}^{4}He)^{+}$ SYSTEMS

In 1954, Fitch and Rainwater {1} identified muonic atom formation by detecting the deexcitation X rays emitted after μ^{-}'s were stopped in matter. Soon the same group {2}realized that studying the energy levels of muonic atoms offered a rather unique way to observe, directly and with very high accuracy, a typical QED effect due to the polarization of the vacuum.

The fact that the virtual pair generated by the vacuum fluctuations, in presence of an electromagnetic field generates a polarization current has, as consequence, that the interaction energy V(r) between two charges differs from the classical Coulomb expression $V_c(r)$. Writing {3}

$$V(r) = V_c(r)(1+\eta(r)) = V_c(r)(1+\alpha A+\alpha^2 B+\ldots) \qquad (1)$$

with r the distance between the two charges and α the fine structure constant, the QED theory tells us how to compute the various A(r), B(r)... terms in (1).

In particular for A (Uehling term) {4} one has :

$$A(r) = \frac{2}{3\pi}\int_{1}^{\infty}\frac{dx}{x^2}\sqrt{x^2-1}\,\left(1+\frac{1}{2x^2}\right)e^{-2x\frac{r}{\lambda\!\!\!^{-}}} \qquad (2)$$

with $\lambda\!\!\!^{-}=3.8\ 10^{-11}$ cm (electron reduced Compton wavelength).

The value of $\eta(r)$, for the case of $\mu^{-}\,{}^{4}He)^{+}$ muonic atom, is presented in figure 1 (up to order of α^2) : from the figure one sees easily that $\eta(r)$ is a short-range correction and that indeed muonic atoms, their dimensions given, are excellent systems to probe such corrections. In the same figure is also presented the correction f(r) introduced by the finite size of the nucleus to the interaction energy between the helium nucleus and a μ^{-} set at given distance r from the nucleus. This correction is obtained assuming for the helium nucleus a charge distribution

$$\rho(r) = 2e\left(\frac{a}{\pi}\right)^{3/2} e^{-ar^2} \text{ with } a = \frac{3}{2\langle r^2\rangle_e}\ .$$

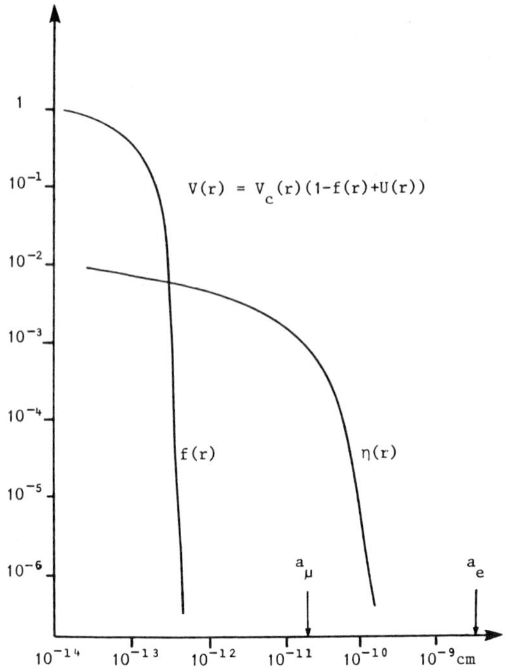

Figure 1. Values for the functions f(r) and η(r); a^B_μ, a^B_e are respectively the Bohr radius of the μ^-P and the hydrogen atoms.

The average square radius $\langle r^2\rangle_e$ is obtained experimentally from electron helium nucleus scattering and one has {5}

$$\langle r^2\rangle_e = 1.676 \pm 0.008 \text{ fm.} \qquad (3)$$

When μ^- are stopped in matter and an excited muonic atom $(\mu^- \, {}^4He)^*$ is formed, the deexcitation process, leading to the lowest level, is quite fast: this deexcitation is going mostly via radiative and Auger transitions (when an electron is present). For the interesting case of the muonic ion $(\mu^- \, {}^4He)^+$, in figure 2 are shown the lower levels with their respective lifetime (due to radiative transitions).

At Cern, around 1970, a Cern-Pisa collaboration group looked into the feasibility of inducing transitions between levels in helium muonic ions, using pulsed laser radiation with the aim of studying experimentally, with very high precision, the vacuum polarization corrections η(r) to the energy levels.

One of the reasons why helium was chosen {6} is that if the helium gas is pure enough, the muonic system $(\mu^- \, {}^4He)^+$ formed in a helium gas target cannot become neutral by taking an electron from the surrounding neutral atoms for energetic reasons.

Going back to the interest of performing experiments by employing the laser induced transition technique in muonic systems two experimental directions have been taken and are those marked in figure 2 with dashed arrows. The idea {7} is to irradiate the prepared muonic system with a short laser light burst, (with tunable wavelength within a sufficiently

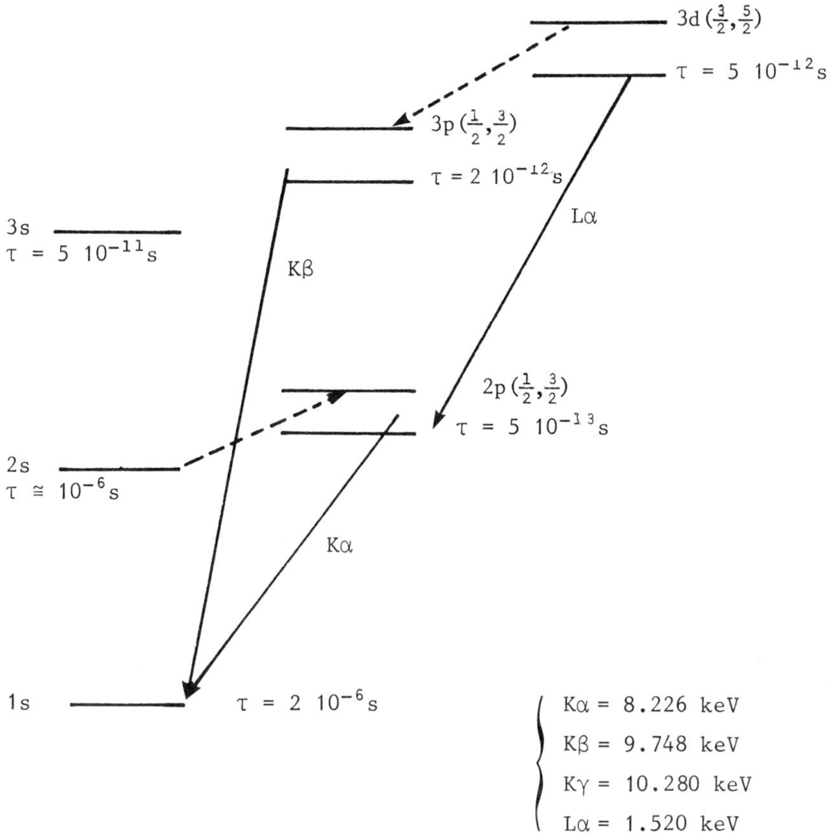

Figure 2. Scheme for the lower energy levels for the $(\mu^- \ ^4He)^+$ ionic system.

large interval) in order to induce at a specific wavelength λ_0 value of the radiation, transitions to a final level highly unstable which can be easily recognized by its quick radiative decay to the fundamental 1s level via the emission of a specific $\overline{X}$ ray (double resonance method).

The task is therefore to count the number of coincidences (in time) between the $\overline{X}$ ray and the fast laser pulse in function of the laser wavelength: the proof of the detected transition is given by finding a "resonance curve" around λ_0. As comment to the two lines of research indicated in figure 2 one can say the following.

1.2. 2s-2p EXPERIMENTS

The theoretical calculations for this situation are given in Table 1 {8}. Generally speaking the test of the vacuum polarization correction, as shown by the table, is limited by the precision with which the nuclear size correction is known: the used value for the r.m.s. of the helium nucleus, $\sqrt{\langle r^2 \rangle}$, is the one given by (3).

It has to be said that the hypothesis here is that there is no muon-nucleus anomalous interaction so that the negative muon μ^- will see the same r.m.s., as seen by the electron.

Table 1. Calculation of the 2s-2p energy level differences in the muonic $(\mu^- \, ^4He)^+$ ionic system

Various contributions (meV)

contributions	transition energies	
	$^2P_{3/2} - ^2S_{1/2}$	$^2P_{1/2} - ^2S_{1/2}$
Dirac contribution with Coulomb potential and point-like charges	145.70	0
Nuclear polarizability	3.1±0.6	3.1±0.6
Finite size	-289.5±2.8	-289.5±2.8
... Electronic vacuum polarization		
Uehling term: first iteration	1664.44	1664.17
higher iteration	1.70	1.70
Kallen-Sabry term ($\alpha^2 Z\alpha$)	11.55	11.55
$\alpha(Z\alpha)^n$ n>3	-0.02	-0.02
$\alpha^2(Z\alpha)^2$	0.02	0.02
Muon vacuum polarization	0.33	0.33
μ^-e vacuum polarization	0.02	0.02
Hadrons vacuum polarization	0.15	0.15
... Vertex corrections and (g-2)		
$\alpha(Z\alpha)$	-10.52	-10.85
$\alpha(Z\alpha)^n$ n>1	-0.16	-0.16
α^2 $Z\alpha$	-0.03	-0.03
... Recoil terms		
Breit	0.28	0.28
two photons	-0.44	-0.44
weak contribution	0.00002	0.00002
Sum theory	1526.6±2.8	1380.3±2.8
Experiment	1527.5±0.3	1381.3±0.5

In stopping μ^- in helium gas target an important question is to know how many "useful" 2s levels are formed.

This problem has been initiated by Placci et al {6}. The channel through which the 2s levels have been detected (after the "zero" time) has been through the decay process

$$(\mu^- \, ^4He)_{2s} \rightarrow (\mu^- \, ^4He)_{1s} + X_1 + X_2 \quad (4)$$

where X_1 and X_2 are two X rays, the energy sum of which is 8.2 keV: in experiment of ref.{6} one of the two X rays was detected.

The first estimate of the second order process (4) has been done by Placci et al {6} who also gave the first experimental evidence of process (4). In table 2 are given the various estimations for the rate $\lambda_{\gamma\gamma}$ of process (4).

Table 2. Calculated values for $\lambda_{\gamma\gamma}$

Authors	$\lambda_{\gamma\gamma}$ sec^{-1}
A. Placci et al {6}	$1.05\ 10^5$
H. Pilkuhn {9}	$1.19\ 10^5$
R. Bacher {10}	$1.18\ 10^5$

In what follows we will take $\lambda_{\gamma\gamma} = 1.18\ 10^5\ sec^{-1}$. Therefore for an undisturbed system $(\mu^-\ ^4He)_{2s}$ the difference rate λ_T° is

$$\lambda_T^\circ = \lambda_{\gamma\gamma} + \lambda_0 = 5.7\ 10^5\ sec^{-1} \tag{5}$$

where $\lambda_0 = 4.55\ 10^5$ is the muon disappearance rate.

The fraction of stopped μ^- leading to a 2s level present after $\cong$ 250 ns was found to be {6}

$$\varepsilon_{2s} = 3.4 \pm 0.7;$$

subsequent experiments {11} confirmed this value.

Moreover the lifetime of these 2s levels was found experimentally compatible with value (5); in fact if we take into account the time resolution of the X-ray detector ($\cong$300ns) the results of ref.{6} imply

$$\tau_{2s} = (1.5 \pm 0.4)\mu s; \tag{6}$$

value (6) shows that with 95% c.e. $\tau_{2s} > .7\mu s$. This is important in view of the laser induced transition experiment since if

$$\tau_{2s} \gg (\Delta L, \delta x) \tag{7}$$

where ΔL and δx are respectively the time width of the laser pulse ($\cong$20ns) and of the time resolution of the X-ray detection system ($\delta x \cong 250$ns at 8 keV), then the experiment is relatively easier.

The 2s-2p experiment has been done at Cern by a collaboration group. The measurements were done stopping μ^-, from the Cern synchrocyclotron (600 MeV) in a volume of about 1/4 of a liter containing helium gas at a pressure of about 30 to 40 atm. The tunable laser beam was obtained using a dye-laser pumped with a Ruby laser. This last one was first energetized, in synchronism with a command given by the proton accelerating synchrocyclotron and it was made lasering (within a time less than a ms) via a Pockel cell triggered by a delayed pulse (delay $\sim$500ns) generated by a μ^- stopping signal.

The dye-laser was tuned via a diffracting grid properly oriented and the wavelength of the outgoing infrared radiation was changed by rotating the grid: the situation was arranged in order to scan about 40Å around the predicted value. A simplified scheme of the experimental set-up {7} is shown in figure 3.

The dye-laser beam properties used for the 2s1/2-2p3/2 and 2s1/2-2p1/2 excitations are summarized in table 3.

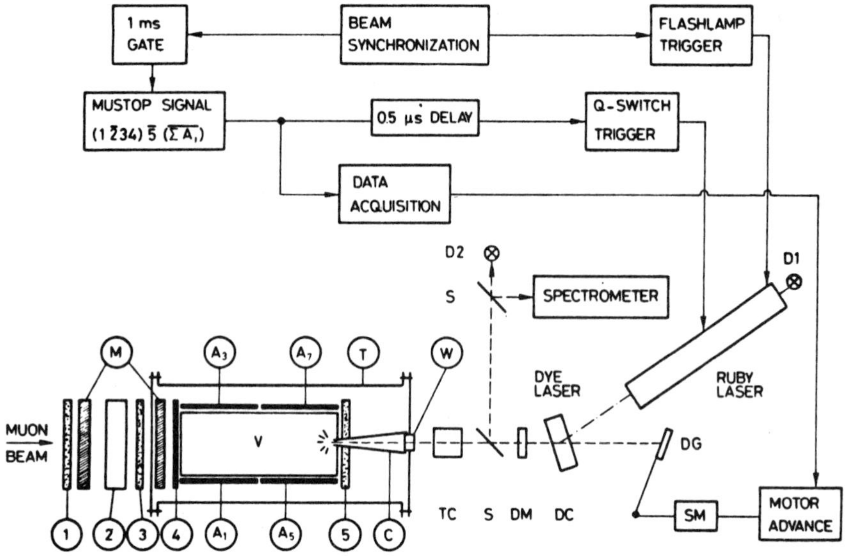

Figure 3. Schematic view of the apparatus: M=CH_2 moderators, 1,3,4,5= plastic scintillators, 2= anticoincidence Cerenkov counter (lucite), T= Invar steel vessel, V= useful muon-stopping volume, A_1 ... A_8=NaI(Tl) counters, DC= dye cell, DM= dielectric mirror (R=30%), DG= diffracting grating (1200 lines/mm), SM= stepping motor driving DG, D1-D2= photodiodes, C= internally gold-coated conical light pipe, W= antireflecting window, S= light beam splitters, TC= optical telescope.

Table 3. Dye-laser properties

Line	8976 Å Int=0.5	8116 Å Int=1
Dye	Kodak IR-140[a)]	HITC
Solvent	DMSO[b)]	DMSO[b)]
Molar concentration (µ/e)	7 10^{-4}	5 10^{-5}
Repetition rate (Hz)	0.25	0.25
Average ruby energy per pulse (J)	1.2	1.2
Infrared energy (J)	.170	.300
Radiation width	20 ns	20 ns
Band width (Å)	5	5

a) Webb et al IEEE J. Quant. Elect. QE11 (1975) 114
b) Dimethylsulfoxide.

Figure 4 gives the results for the 2p3/2-2s1/2 energy level difference: the final results for both lines are in table 1. As can be deduced from table 1, this experiment can be analyzed in two ways.

A first result is obtained assuming QED and gives from the laser experiment a value for $\langle z^2 \rangle_\mu$ of the helium nucleus as seen by the muon:

this gives

$$\sqrt{\langle r^2 \rangle_\mu} = 1.673 \pm 0.003 \text{ fm}. \tag{8}$$

which compared to (3) gives a limit to the μ^- nucleus anomalous interaction with respect to hadrons.

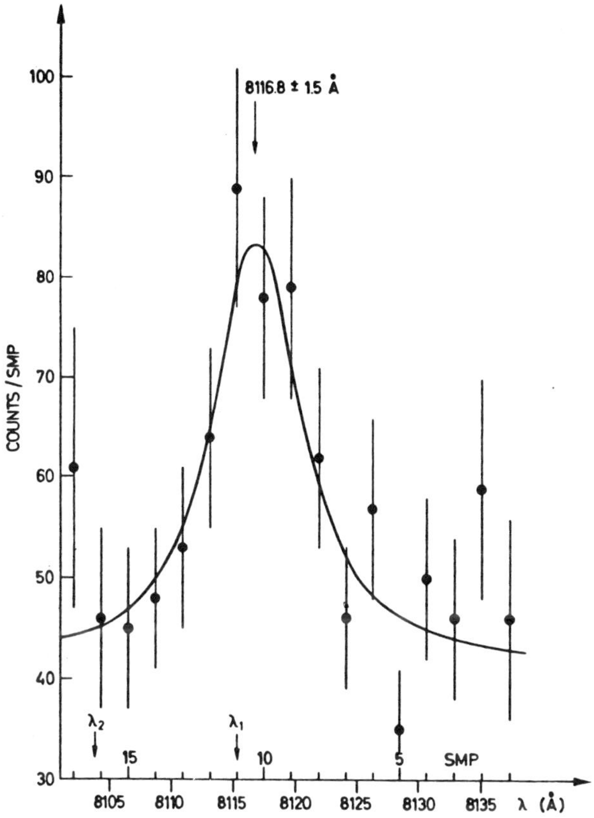

Figure 4. The 2s1/2-2p3/2 resonance signal. Each datum point represents the number of events, normalized to the same number of stopped muons, per stepping motor positions SMP; the scale in wavelength (shown below) is fixed by the calibration procedure (λ_1 and λ_2 are two lines of an Ar lamp, as specified in the text). The full line curve drawn on the data is the result of a best fit analysis (χ^2=11) with a Lorentzian line (assuming for Γ the theoretical value Γ=8Å) plus a constant background. For the central wavelength value of the Lorentzian line we have obtained λ_0=8116.8 ±1.5Å. Having Γ as a free parameter the best fit has given Γ_{exp}=7.2±2.8Å and the same value for λ_0. The theoretical prediction for λ_0 is 8115.6±21.8.

Another way is to assume μ^--e^- universality and deduce an experimental limit on the verification of the QED vacuum polarization term (at $|q| \cong$ 1 MeV/c). From table 1 one sees that these vacuum polarization terms are tested to about 1.6 10^{-3}. Looking at table 1 it is striking that the experiment is of an order of magnitude more precise than the theoretical estimations: this is due to the influence of the finite size corrections to the 2s-2p energy level difference. How one hopes to improve the situation is our next section's subject.

1.3. 3d-3p (EXPERIMENT) IN (μ^- ^{4}He)

The theoretical calculations for the expected values for the 3d-3p energy level differences are presented in table 4 {12}.

Table 4. Calculations of the 3d-3p energy level differences in the muonic (μ^- ^{4}He)$^+$ ionic system

	meV	meV	meV	meV	meV	microns
μ^-He^4						
$3d_{3/2}-3p_{1/2}$	110.458	0.905	43.164	0	154.528	8.0235
$3d_{5/2}-3p_{3/2}$	110.458	0.905	14.388	0	125.751	9.8595
$3d_{3/2}-3p_{3/2}$	110.458	0.905	0	0	111.363	11.1334

Transition	Vacuum polarization α (Uehling-Serber) meV	Vacuum polarization α^2 (Kallen-Sabry) meV	Fine structure (Dirac) meV	Hyperfine structure meV	Total meV	λ* microns
μ^- ^{4}He						
$3d_{3/2}-3p_{1/2}$	110.458	0.905	43.164	0	154.528	8.0235
$3d_{5/2}-3p_{3/2}$	110.458	0.905	14.388	0	125.751	9.8595
$3d_{3/2}-3p_{3/2}$	110.458	0.905	0	0	111.363	11.1334

* $\Delta\lambda$ = 370 Å

The big advantage of this direction (if it can be realized) is that there is no limitation to the theoretical value, due to the finite size corrections since these last ones are negligible.

Therefore the "test" of the vacuum polarization correcting terms will be as good as the precision of the experimental determination of the resonance wavelength. Another advantage is that the initial 3d level has a probability to be formed, per μ^- stopped, of about 70%.

However, as can be seen from figure 2, we have now a serious difficulty since the 3d (and 3p) level has a very short lifetime and conditions {7} are certainly no longer satisfied.

An experimental attempt is on the way to measure the 3d-3p energy level difference via laser induced transition, at Brookhaven National Laboratory by a Columbia-Cern collaboration {12}. In what follows I will describe their set-up.

The idea is to have a gas target where the μ^- are stopped in a multipass cavity filled up with helium at 1 atm: the negative muons are sent to stop in the target in bursts of width Γ less than 50 ns (about one every two seconds). At some specific time $\bar{t}$, before the muon burst enters the cavity, a few Joules (CO_2) pulsed laser is triggered and a light beam, of 40 to 50 ns width, is sent into the multipass cavity target: the cavity has a reflectivity of ≅ 98% so that the radiation pulse will be stored in the multipass cavity for a length of time of about T = 100 ns in a well defined region of the cavity. The time T is longer than the muon burst width Γ and therefore it is possible to stop the μ^- in the helium target while this is filled with a (CO_2) radiation of a certain wavelength λ.

Therefore when the initially formed (μ^- ^{4}He)* excited system passes through the 3d level, in its deexcitation processes, there is a non-negligible probability that a 3d-3p transition occurs when the wavelength λ of the CO_2 laser is near the values given in table 4. Such a transition will be quickly followed by a 3p → 1s X-ray emission (K_β=9.75 keV). One wishes to identify precisely the 3d-3p energy level difference by counting the number of K_β X rays as function of the CO_2 laser wavelength, which is changed by properly tuning a diffracting grid in series with the resonator. It has to be said that by choosing, in the CO_2 laser, the correct isotopic combination (C_{12}, C_{13}, O_{18}, O_{16}) it is, at least for some cases, possible to obtain radiation of wavelength in a band B of wavelength lines covering the one corresponding to a chosen transition given in table 3. This is useful since the widths of the transitions 3d-3p are larger than the spacing of the mentioned lines in the mentioned band B.

With few Joules injected (40 ns long pulse) one obtains 2 to 3 per thousand of transition probabilities. The goal is to determine experimentally one of the lines given in table 4 to within 20 Å: this will give a test on the vacuum polarization contribution to the 3d-3p energy level difference to about 10^{-4}.

In figure 5, it is shown schematically the apparatus used by the Columbia group {12}.

To obtain a burst of low energy μ-, Brookhaven National Laboratory is an excellent place since an extracted pulsed proton beam facility (30 GeV/c) is already available for neutrino experiments: such a pulsed beam is constituted by 11 bursts of protons 20 ns wide each and 200 ns separated from each other. The first of these bursts (10^{12} protons) is deviated in the Columbia_Area where protons are sent onto an iridium target, to produce low energy Π which, in turn give through their decay low energy μ-'s {12}. It is found that, however, one of the main difficulties is represented by the background generated in the experimental area by the fast pulsed beams: at present the group is working in order to overcome this serious difficulty.

1.4. POLARIZABILITY OF SPACE

Another direction studied and undertaken by a Cern-Pisa collaboration group {13} is to measure directly the space polarizability. We have seen that virtual pairs generated by the fluctuation of the physical vacuum, in presence of an electromagnetic field, give rise to polarization currents: therefore it is natural to ask the question on how the propagation of electromagnetic radiation in space is influenced by such a phenomenon.

This problem has been solved in 1935 by Euler and Heisenberg {14} and what I am describing below follows from their paper.

Calling $E_c = \frac{e}{\lambda_e^2}\frac{1}{\alpha} = 1.3\ 10^{16}$ Volt/cm

$(B_c = 4.4\ 10^{13}$ Gauss)

if we deal with fields so that

$$\left(\frac{E}{E_c}, \frac{B}{B_c}\right) << 1$$

and

$$\lambda\!\!\!^{-}_e\ |\text{grad } E| << |E|$$

$$\frac{\lambda\!\!\!^{-}_e}{c}\left|\frac{E}{t}\right| << |E|$$

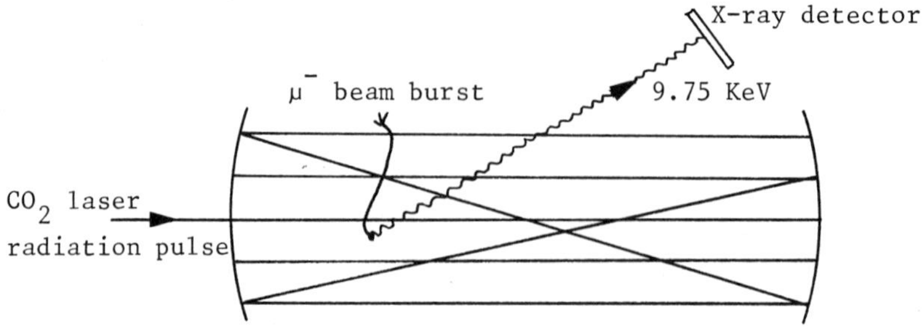

a) Multipass cavity containing helium at 1 atm, in which a CO_2 laser beam enters (before the beam is absorbed by the wall, it will make about 100 reflections).

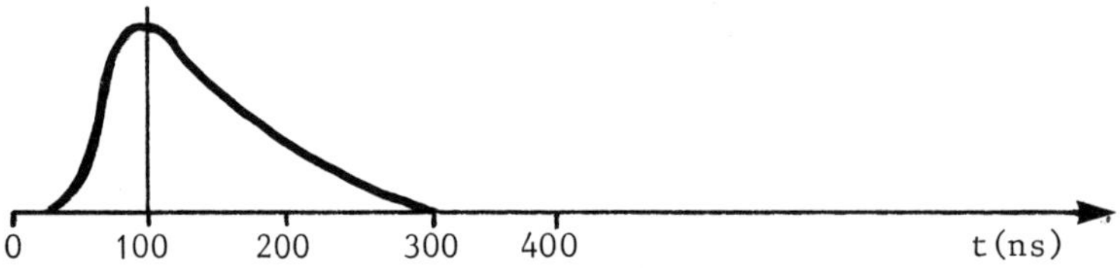

b) Time distribution of the energy density of the CO_2 light in the cavity.

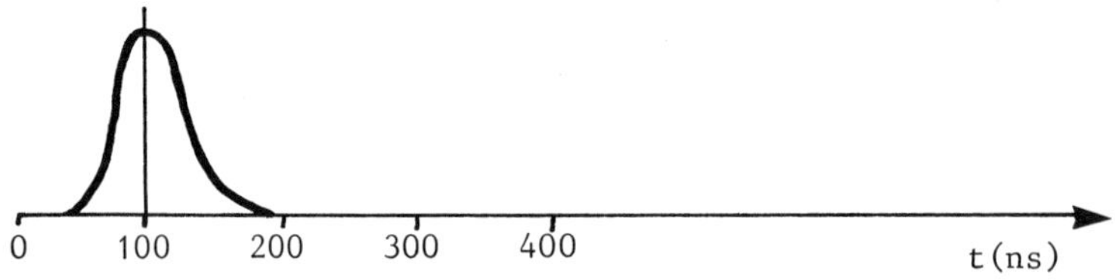

c) Time distribution of μ^- in one bunch.

Figure. 5

then the Lagrangian from which the new Maxwell equations are deduced is (for the vacuum)

$$\mathcal{L}_H = \mathcal{L}_M + \frac{A}{4\Pi}\{F^2 + 7G^2\} \quad (9)$$

$$F = E^2 - B^2 \qquad \mathcal{L}_M = \frac{F}{8\Pi}$$

$$G = E.B$$

$$A_e = \frac{1}{90\Pi}\,\alpha^2\,\frac{\lambda_e^3}{m_e c^2}$$

with the position

$$\begin{cases} \overline{H} = -\dfrac{\partial\mathcal{L}}{\partial\overline{B}}\,4\Pi \\[2ex] \overline{D} = \dfrac{\partial\mathcal{L}}{\partial\overline{E}}\,4\Pi \end{cases}$$

The Maxwell equations are, as usual, in vacuum

$$\text{rot}\,\overline{E} = -\frac{1}{c}\frac{\partial\overline{B}}{\partial t} \qquad \text{div}\,\overline{B} = 0$$

$$\text{rot}\,\overline{H} = \frac{1}{c}\frac{\partial\overline{D}}{\partial t} \qquad \text{div}\,\overline{D} = 0$$

with now, however, the relation between $\overline{D}$ and $\overline{E}$ (and also between $\overline{H}$ and $\overline{B}$) being no longer linear.

The A_e given takes into account only electron-positron pair fluctuations: other charged particles also contribute but since their contribution is going like $\frac{1}{m^4}$ with m their mass, it is negligible.

Let us now imagine an experimental situation where a linearly polarized light beam goes through a magnetic field region with $\vec{B}_0$ orthogonal to the light direction (see figure 6).

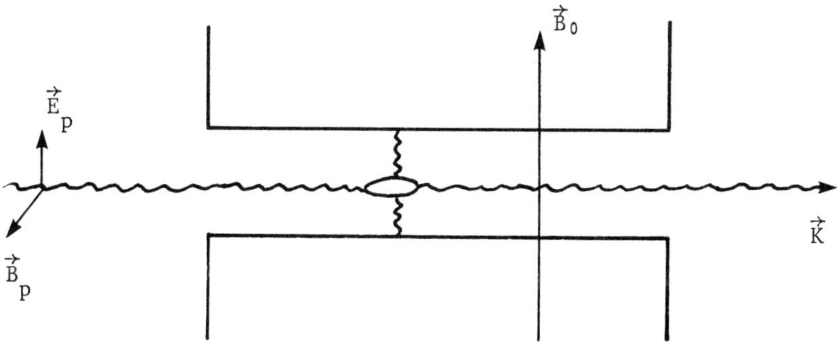

Figure 6.

It can easily be shown that for the propagating radiation $(\vec{E}_p, \vec{B}_p)$ one has

$$\vec{D}_p = \vec{E}_p \{1 - 4 AB_0^2\} + 14\vec{B}_0 \ (\vec{B}_0 . \vec{E}_p)$$
$$\vec{H}_p = \vec{B}_p \{1 - 4 AB_0^2\} - 8 \vec{B}_0 \ (\vec{B}_0 . \vec{B}_p) \quad (10)$$

having used the fact that, in practice, $B_p << B_0$ and therefore disregarding terms in the square of the propagating field. Using (10) we have

$\vec{E}_p$ parallel to $\vec{B}_0$

$$\left\{ \begin{array}{l} \varepsilon_{//} = 1 + 10 \ AB_0^2 \\ \mu_{//} = 1 + 4 \ AB_0^2 \end{array} \right. \longrightarrow \quad n_{//} = 1 + 7 \ AB_0^2$$

$\vec{E}_p$ orthogonal to $\vec{B}_0$

$$\left\{ \begin{array}{l} \varepsilon_{\perp} = 1 - 4 \ AB_0^2 \\ \mu_{\perp} = 1 + 12 \ AB_0^2 \end{array} \right. \longrightarrow \quad n_{\perp} = 1 + 4 \ AB_0^2$$

from which $\Delta n = n_{\perp} - n_{//} = 3 \ AB_0^2 \neq 0$.
i.e. the space, where the magnetic field is present, not only has $n>1$ but it becomes birefringent. That means that if L is the total path of the light beam (initially linearly polarized, say with $\vec{E}_p$ at 45° with respect to the $\vec{B}_0$ direction) and λ the radiation wavelength, at the exit the light will come out elliptically polarized with the ellipticity (ratio between the minor axis to the major axis of the ellipse) ψ_e given by

$$\psi_e = \Pi \ \Delta \ n \ \frac{L}{\lambda} = 3 \ \Pi \ AB_0^2 \ \frac{L}{\lambda} \quad (11)$$

In ref.{13} it has been proposed to measure ψ_e as a check of a typical QED vacuum polarization phenomenon.

In order to gain some insight into the relation (9) let us consider also the case $\vec{B}_0 \neq 0 \quad \vec{E} \equiv 0$, then one has

$$\vec{P} = \frac{\vec{H} - \vec{B}}{4\Pi} = \frac{A}{\Pi} B_0^2 \ \vec{B}_0 .$$

If one imagines the space like a crystal one can write $\vec{P} = \rho \vec{m}$: since $\vec{m}$ is due to the electron loop one can approximate

$$|m| \sim \frac{e\hbar}{2m_e c}$$

from which one finds

$$\rho = \frac{A}{\Pi} B_0^3 \ \frac{2m_e c}{e\hbar} .$$

Taking, for instance, $B_0 \sim 100.000$ Gauss one has $\rho \cong 500$ loops/cm³ i.e. the light traversing the region with a field $B_0 \sim 100$ kG will see on the average about 500 loops/cm³ even if we have done a perfect vacuum. Imagine what will happen around a neutron star !

Putting the above value for B_0 in (10) with L ≅ 3 km, one gets $\psi_e \cong 10^{-11}$. Since the effect is going with B_0^2 one can make multiple crossings in the same magnetic region: in this condition, with 500 crossings of a region 6 meters long, it is possible to reach the assumed 3 km.

Following the proposal of ref.{13} an optical set-up (of course without the magnet) has been arranged at Cern by the Cern-Pisa collaboration group as a feasibility test for such a measurement (while waiting for the high energy community to develop a transversal magnet 6 meters long, ≅ 100 kG, cross section Ø ≅ 5 cm). The experimental set-up realized is shown in figure 7: in doing this the authors have assumed that the 6 meters 100 kG superconducting magnet can be pulsed on-off at a frequency around half a minute.

The idea is first to polarize the light (linearly) as well as one can (extinction factor of the polarizer and analyzer $\sigma^2 \cong 10^{-8}$) and then send this beam, through the holed mirror M_1 into the multipass cavity ($M_0' M_0$): the M_0' mirror has a hole O in the center to let the light beam enter, perform ∿ 500 reflections in the cavity and then exit through the same hole O of M_0'. The cavity is ≅ 6 meters long so that the magnet can be inserted. The outcoming light is deflected by the mirror M_1 onto a double Gold Mirror Shifter (GMS) in which two particular orthogonal polarization states of the light beam will be shifted by 90° with respect to one another {15} (essentially the GMS replaces a quarter wave plate) within a quarter of a degree: this in order to convert the eventual ellipticity $\psi_e(t)$ gained in the region with the magnet into a Faraday rotation $\phi(t)$ equal to about $\psi_e(t)$. The beam is then sent through a thin slab of glass (the only material through which the light beam passes), put in a coil (fed by an alternating current of frequency ν_F) which will add to the small rotation $\phi(t)$ another one $\alpha(t,\nu_F)$ relatively big ($\alpha^2_{max} > \sigma^2$).

At the exit of the analyzer A the diode D_0 will give a signal I_d which is interesting for us, and can be expressed as

$$I_d \propto I_0 \, \sigma^2 + I_0 \, \alpha^2(t) + I_0 . 2\phi_e(t) \, \alpha(t) + \ldots$$

Analyzing therefore the Fourier spectrum of I_d there will be satellites at the frequencies $\nu_F \pm \nu_M$, where ν_M is the on-off frequency of the main magnet $\{\phi(t) = \phi(t,\nu_M)\}$.

Since we have no magnet we have simulated a very small signal $\phi'(t,\nu)$ by putting a coil around the interferential mirror M_2 {16} (pulsating at ≅ 0.06 hertz).

The experiment therefore consisted in measuring the sensitivity of the set-up to measure a low modulated ellipticity, trying to lower the low frequency noise of the system in order to be able to see a rotation ϕ around 10^{-11}.

The high number of reflections in the multipass cavity has been reached by using a technique suggested in ref.{17}. We have reached a level for which a measurement of $\psi_e \sim 10^{-11}$ can be performed, to within 15%, in about two months of integration time. We found that such a sensitivity can be reached for an output power from the cavity, of less than 100 mW.

At present our main sources of noise are the thermal fluctuations; in fact we are 3 to 4 times worse than the theoretical limits expected from the measured laser noise, the electronic noise, etc. amplitudes.

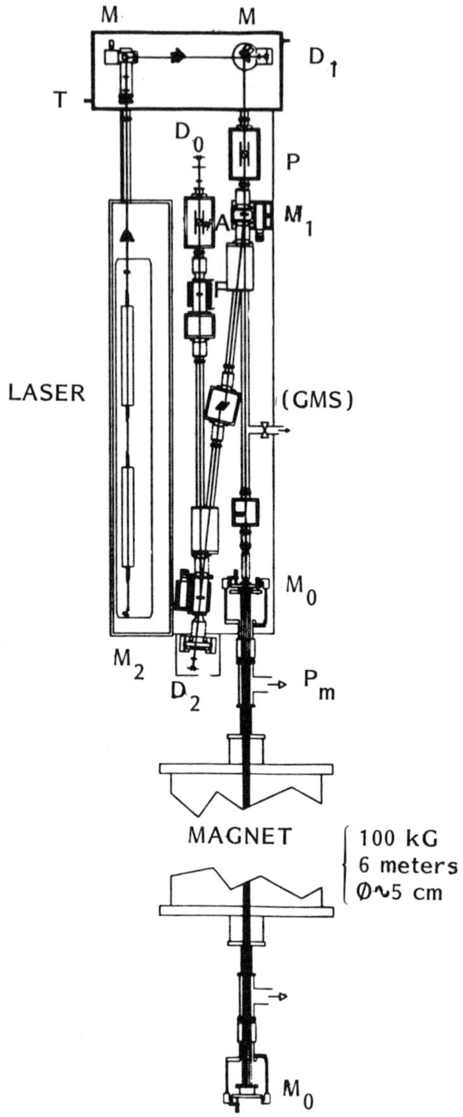

Figure 7. M_0 multipass cavity interferential mirrors, M_1 holed (Ø=3mm) interferential mirror, M_2, M interferential mirrors, P and A respectively polarizer and analyzer (put crossed), T matching telescope, F glass Faraday cell, GMS Gold Mirrors Shifter, Pm high vacuum pumps, laser : argon c.w. laser few watts, D_0 main signal detecting diode, D_1, D_2 monitoring diodes.

REFERENCES

1. V.L. Fitch and J. Rainwater, Phys. Rev. 92 789 (1953).
2. S. Koslov, V.L. Fitch and J. Rainwater, Phys. Rev. 95 291 (1954).
3. A.I. Akhiezer and V.B. Berestetskii, in "Quantum Electrodynamics", R.E. Marshak ed., Interscience, (1965).
4. E.H. Uehling, Phys. Rev. 48 55 (1935).
5. I. Sick, Phys. Lett. B 116, 212 (1982).
6. A. Placci et al N.C. 1A 445 (1971).
7. A. Bertin et al N.C. 26B 433 (1975), and C. Carboni et al N.Phys. A278 381 (1977).
8. E. Borie and G.A. Rinker, Phys. Rev. A18 324 (1978).
9. H. Pilkuhn, Report TKP 80-24 Karlsruhe University, W. Germany.
10. R. Bacher, Z. Phys. A315 135 (1984).
11. A. Bertin et al N.C. 23B 489 (1974).
12. A.G.S. Proposal n°745, A.M. Sachs et al September 1979.
13. Proposal DE : E. Iacopini et al 9/6/1980 and Addendum 8/3/1982, See also E. Iacopini and E. Zavattini, Phys. Lett. 85B 151 (1979).
14. W. Heisenberg and H. Euler, Z. Phys. 98 718 (1936).
15. S. Carusotto et al, Applied Physics B36 125 (1983).
16. E. Iacopini et al, Applied Physics A32 63 (1983).
17. D.R. Herriot et al, Applied Optics 4 883 (1965).

PART IV

Many electron systems

STATES OF AN ATOMIC ELECTRON PAIR

A. R. P. Rau

Department of Physics and Astronomy
Louisiana State University
Baton Rouge, LA 70803

ABSTRACT

Doubly excited states of atoms have now been studied for twenty years. The simultaneous excitation of two electrons from the ground state of an atom (or ion) leads to phenomena qualitatively different from those exhibited by single excitation. New basis states, new models and new quantum numbers are required to describe some of these phenomena. Broadly, all two electron states can be divided into two classes, those in which single particle behavior dominates and those which have to be considered throughout in terms of the electron pair. This distinction, the basic differences in the two types of states, the attendant implications for their experimental observation, and the broader setting of this study of the quantum mechanics of strongly perturbed or non-separable potentials are discussed.

INTRODUCTION

The study of doubly excited states of atoms dates to the early 1960's. For good reason, the previous many decades of atomic physics dealt only with single excitation of atoms. Optical and near ultraviolet spectroscopy studied on the one hand the excitation of the outermost valence electron, while X-ray spectroscopy dealt on the other hand with excitations from an inner shell. The region in between - in wavelength terms between 1 and 1000 Å - remained unexplored. There was a lack of good continuum sources of radiation in this energy range and also, since all substances absorb strongly, no windows were available to enable spectroscopic measurements in this range. Technical and experimental developments (mainly the advent of synchroton light sources and the ability to do "windowless spectroscopy" through fast differential pumping techniques) in the '60's finally opened this region of the electromagnetic spectrum.[1] Since most double excitations lie in this range of energies, their study had to await these developments. One might wonder why they were not studied earlier by charged particle impact if not by photon impact. Here again, a basic characteristic of these doubly excited states, namely their large life-times and narrow widths (at first surprising, considering how high in energy they lie above the ground state), required other technical developments, namely high resolution spectroscopy, which also were unavailable before the 1960's.

Over the last twenty years, doubly excited states have been studied extensively through both photon impact (synchroton and multiple-step laser excitation) and electron and ion impact on atoms. We can claim a reasonably good understanding of low-lying doubly excited states. A recent comprehensive review[2] makes my task easier and I will draw upon this for specific discussions, figures and references to the literature. However, both experimentally and theoretically, we are only just beginning to explore two-electron states of higher excitation in which both electrons are far removed from the rest of the atom. Our understanding remains incomplete and the current status will be surveyed. (See also references 3 and 4).

The very first observations of doubly excited states brought surprises, emphasizing that qualitatively new phenomena were being encountered. Thus, the first synchrotron absorption spectrum of helium (Fig. 2 of ref. 2) below $He^+(N=2)$ showed only one dominant series of two-electron states, whereas in this energy range in an independent particle picture one would have expected three $^1P^o$ series - 2snp, 2pns and 2pnd - at least two of which (the first two) could be expected to be roughly comparable. Also, as already remarked, the first observation of these doubly excited states showed them to be very narrow and this was also true of the first experiment with electron impact where a $^2S^e$ doubly excited resonance state of He^- was seen, tentatively attributed to $1s2s^2$ (for experimental reasons, namely that it is easier to have e-atom than e-ion collisions, most doubly excited states studied in electron collisions have been in negative ion systems. This remains true to date[4] although recently e-ion experiments for dielectronic recombination have been carried out which involve intermediate doubly excited states but the accent in such experiments has not been on the detailed study of these states[5]).

Both these early observations, of a single dominant $^1P^o$ series and of narrow energy levels, are pointers to an important physical characteristic of doubly excited states, namely the strong correlation between the two electrons. The otherwise degenerate 2snp and 2pns states are mixed by the electron-electron interaction into the linear combinations 2snp±2pns which have very different overlaps with the ground state in the photoabsorption matrix element.[2] Likewise, the He^- ^2S state is a strongly admixed combination of the configurations $1s2s^2$ and $1s2p^2$. Both of these are examples of strong angular correlations (admixture of different orbital angular momenta of the two electrons, radial quantum numbers remaining the same) and already emphasize the important role that is played by electron-electron correlations. Later in the discussion we will see how radial correlations between the electrons are also important and, in fact, become crucial for the higher doubly excited states.

The detailed discussion of the correlations will be taken up in the next sections but to conclude this introduction, the appearance of qualitatively new aspects should in a way cause no surprise. A many electron atom is a system with a large number of degrees of freedom but all of these are, so to say, frozen in the ground state. In fact we are accustomed to the terminology in Hartree-Fock or many-body theories of the ground state being a filled sea or a vacuum out of which excitations can arise. When one electron is excited, some of the degrees of freedom are "unfrozen" and manifest themselves in the observed phenomena. However, within the realm of single excitation, no matter how high in energy that electron is excited, qualitatively new degrees of freedom never come into play. This happens, however, when two electrons are simultaneously excited. That this unfreezes more degrees of freedom and as a result leads to new phenomena should, therefore, occasion no great surprise. What is remarkable is that we have to introduce even new quantum numbers (associated with collective coordinates of the pair) to describe some of these phenomena. Elsewhere in physics, particularly in statistical mechanics/thermodynamics, when dealing with a collection of a large number of particles, one is accustomed to concepts at a higher hierarchical level that are not easily or usefully decomposed in terms of those at a lower level of the hierarchy. Temperature, for instance, is a concept one would introduce for a macroscopic assembly even were it possible to follow the individual motions of all the microscopic particles. It is striking that the study of doubly excited states shows that even in an atom with just two electrons, something similar is true, that one usefully describes states in terms of new (pair) quantum numbers instead of as infinite superpositions of independent particle configuration labels.

INDEPENDENT PARTICLE DESCRIPTION

The success of several decades of atomic physics dealing with single electron phenomena, coupled with the dominance of the central potential for most of what is studied, has led to the use of the hydrogenic or independent particle labeling of atomic states. Even when correlations are recognized as important (for describing accurately even single excitations), at least the language and notation has remained that of independent particle configurations. Thus, one talks of configuration interaction or configuration mixing, wherein alternative independent particle configurations are mixed by the electron-electron interaction and a final diagonalization of the Hamiltonian provides the physical eigenstates as linear combinations of these configurations.[6] The same language has been used in describing doubly excited states as well and in fact, even accurate quantitative calculations of low-lying doubly excited states are often carried out through such configuration mixings.[2]

For simplicity of discussion, consider all the states of the helium atom (nuclear charge=Z) with total spin (S) and orbital (L) angular momentum equal to zero: Fig. 1. Below each limit of the "parent ion" $He^+(N)$, there are Rydberg series of states for the other electron and each of these N Rydberg series is labeled in the independent particle description by the orbital angular momentum ℓ value of the individual electrons; in this example of L=0, we have $\ell_1=\ell_2=\ell$, otherwise the two individual orbital angular momenta would be different in general. With respect to the ground $1s^2$ state, the energy level of any particular state $N\ell n\ell$ is written through the Rydberg formula as

$$E_{Nn\ell} = I_+(N) - \frac{(Z-1)^2}{2(n-\mu_\ell)^2} \tag{1}$$

in atomic units (throughout this paper), where $I_+(N)$ is the parental ionization limit, μ_ℓ a quantum defect. $I_+(N)$ could in turn be written with respect to the "grandparental" ionization limit I_{++} as

$$I_+(N) = I_{++} - Z^2/2N^2 \ . \tag{2}$$

[For a more general atom with more than two electrons, where the core is not a bare charge Z but itself contains electrons, (2) would also have the form of a Rydberg formula, with a quantum defect for the "inner" electron N, instead of the Bohr form as shown.]

Such a description through the Rydberg formula (1) has been very successful for singly excited states, N=1, and we can plausibly argue that it also works well for other higher sequences in Fig. 1 where n>>N. The picture of an "inner" electron, forming together with the nucleus a

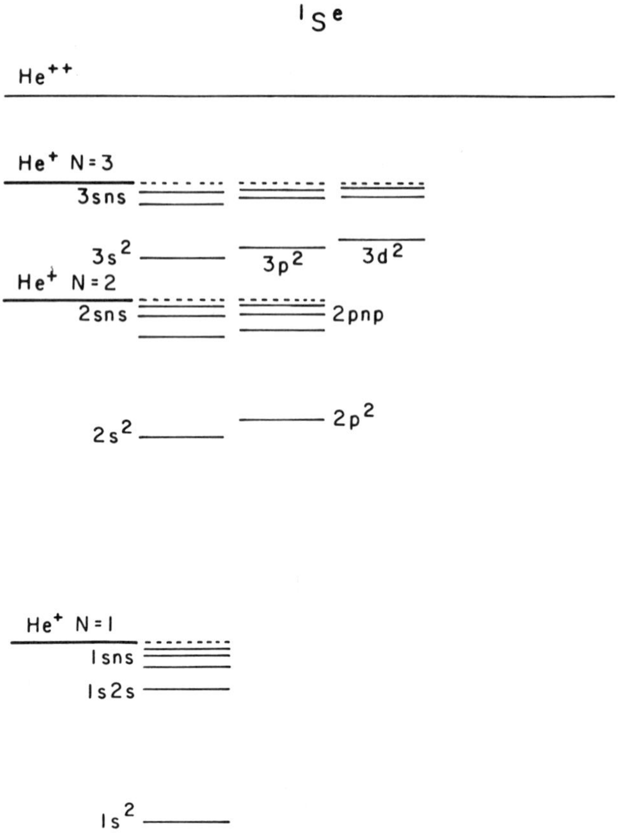

Fig. 1. States of the helium atom in independent particle labeling

core for the "outer" electron, and contributing a quantum defect, makes sense when the two electrons are at disparate radial distances. It is clear, however, that the situation is less straightforward when $n \simeq N$, that is, for states with comparable radial excitation. Already we begin to see a basic subdivision into two classes for the states in Fig. 1. Those with disparate radial excitation, such as the singly excited states and the doubly-excited states with $n > N$, can still be sensibly handled through the independent principal quantum numbers of the individual electrons and the corresponding Rydberg formula that we are familiar with. But, others in which both electrons are comparably excited, require a new description. Before considering these radial correlations, I turn first to an aspect common to all doubly excited states.

ANGULAR CORRELATIONS

A glance at Fig. 1 shows that for all doubly excited states, one can expect strong configuration mixing of states $N\ell n\ell$ with differing ℓ. Such states are degenerate (the ℓ degeneracy of H) in the absence of the electron-electron interaction and one would expect strong mixing due to this interaction. This is, in fact, the case and was evident from the very first experimental results and configuration mixing calculations. Such strong mixing is a pointer to the deficiency of ℓ as a good label. The presence of the electric field due to the other electron mixes different ℓ values much as in the discussion of excited states of a single electron atom in an external static electric field.[7] In fact we can take a clue from that discussion to argue for a replacement of ℓ.

When one discusses a hydrogen atom in an electric field, two generators of the group of four-dimensional rotations O_4 are used, commonly denoted as $\frac{1}{2}(\vec{\ell}+\vec{a})$ and $\frac{1}{2}(\vec{\ell}-\vec{a})$, where $\vec{a}$ is proportional to the Runge-Lenz vector, which is a vector that points in the direction of the semi-major axis of the classical Kepler orbit (therefore, $\vec{\ell}\cdot\vec{a}=0$) and has magnitude equal to the eccentricity. The ordinary angular momentum vector $\vec{\ell}$ is a sum of these two generators.[8] For two electron states, let us similarly replace $\vec{\ell}$ by the sums of two such generators, but now formed from two electrons, $\frac{1}{2}(\vec{\ell}_1+\vec{a}_1) + \frac{1}{2}(\vec{\ell}_2-\vec{a}_2)$ and $\frac{1}{2}(\vec{\ell}_1-\vec{a}_1) + \frac{1}{2}(\vec{\ell}_2+\vec{a}_2)$. The total angular momentum $\vec{L}$ is the sum of these two generators. Two quantum labels associated with these operators, called K and T, can then be used to label doubly excited states.[9] For L=0 states, T is identically 0 so that there is only one label K which serves as the counterpart of ℓ in the independent particle scheme. K, however, is clearly characteristic of the pair, having been formed out of generators belonging to the two electrons. The occurrence of $\vec{a}_1-\vec{a}_2$ in the above, and the interpretation of these vectors as pointing from the nucleus to the two electrons when at their aphelion, connects $\vec{K}$ to the angle between the two radius vectors θ_{12}. This is a pair coordinate, showing again K as a pair label and further as having to do with the angular correlation (in θ_{12}) between the electrons.

Referring to the original papers[9] for other group theoretic aspects of K and T, let me merely note here that another quantum number v, equivalent to K and defined as $v=\frac{1}{2}(N-1-K)$, is the most useful replacement of ℓ for the doubly excited states in Fig. 1. v has a range 0, 1,..(N-1) just as does ℓ but, of course, each value of v is a superposition of all the $N\ell n\ell$ states with alternative values of ℓ. The lowest state has v=0. Its $(n\ell)^2$ content when n=N has been worked out and is suggestive of the nature of angular correlations in doubly excited states,[10]

$$W_\ell = (2n-1)(2\ell+1)[n-1)!]^4[(n+\ell)!(n-\ell-1)!]^{-2}$$

$$\sim (2/n)(2\ell+1)\exp[-2\ell(\ell+1)/n] \text{ for large } \ell. \qquad (3)$$

The probability of a value ℓ is, therefore, peaked around $\ell_{max} = \frac{1}{2}n^{\frac{1}{2}}$. As n increases, this rise in the admixture of ℓ values translates into the complementary angle variable as[10]

$$\pi-\theta_{12} \sim n^{-\frac{1}{2}} \sim |E|^{\frac{1}{4}}. \qquad (4)$$

For very high n ($|E| \simeq 0$), therefore, near the double ionization limit in Fig. 1, the lowest energy state has an angular correlation that leads to the two electrons lying at $\theta_{12}=\pi$, that is, on opposite sides of the nucleus. Such a result for two slow electrons is, of course plausible immediately as a consequence of the repulsion between the particles and had been first pointed out for the continuum problem (when $E>0$) in which two slow electrons escape from a positive charge.[11]

RADIAL CORRELATIONS

Given the degeneracy of independent particle states of different ℓ in the same (N,n) manifold, it was natural to begin as we did with ℓ-mixing due to the electron-electron interactions, and the angular correlations that such mixing implies. <u>All</u> doubly excited states, down to the lowest ones, display strong angular correlations. On the other hand, since different (N,n) manifolds are widely separated at low values of N even in the independent particle picture, radial correlations represented by the mixing of states of different manifolds (recall that the ratio $r_</r_>$ is, approximately, $(N/n)^2$) seem less important to begin with. Negative ion systems, however, give an early hint of the relevance of radial correlations all the way even down to the ground state - H^- and many other negative ions are not bound in the absence of these correlations. In any case, it is clear from Fig. 1 that at least for the higher doubly excited states, since manifolds with different N get closer in energy, the mixing of states of different N's also becomes important and has to be considered alongside the mixing of ℓ values.

Just as a pair coordinate θ_{12} was identified for discussing angular correlations between the electrons, it is natural to look for pair coordinates in the radial variables r_1 and r_2. It is equally natural for this choice to turn to the circular coordinates in this (r_1,r_2) plane, namely

$$R = (r_1^2 + r_2^2)^{\frac{1}{2}}, \quad \tan\alpha = r_2/r_1 \ . \qquad (5)$$

The angle α becomes a coordinate for the radial correlation and R a scale variable for the two electron system. This passage from independent coordinates $\vec{r}_1$ and $\vec{r}_2$ to the "hyperspherical coordinates" - R, α, θ_{12} and three Euler angles (which play no role in the dynamics) - of a six-dimensional sphere has arisen many times in the history of studies of two-electron atoms. For our purposes here, besides the above plausibility arguments, it suffices to note that early configuration interaction calculations of the lowest doubly excited states in He already showed, when plotted in the (r_1,r_2) plane, an organization of nodal lines along circular arcs (constant R) and radial lines (constant α) instead of running parallel to the Cartesian axes as do the nodal lines of singly excited states.[2]

<u>Low-lying Doubly Excited States</u>

The analysis of the two-electron Schrödinger equation in hyperspherical coordinates for low lying doubly excited states has seen extensive development in the last twenty years and has been comprehensively reviewed in ref. 2. The equation is not separable in hyperspherical coordinates (just as it is not in the independent particle coordinates); in particular, the potential energy has the form

$$V = -Z(1/r_1 + 1/r_2) + 1/r_{12}$$

$$= [-Z(1/\cos\alpha + 1/\sin\alpha) + (1-\sin 2\alpha\cos\theta_{12})^{-\frac{1}{2}}]/R, \qquad (6)$$

and is a non-separable function of the three dynamical variables R, α and θ_{12}. However, an assumption of an adiabatic separation of the variables α and θ_{12} from R has proved successful in describing the lower doubly excited states. This "adiabatic hyperspherical scheme" proceeds in Born-Oppenheimer fashion by first treating R as frozen and solving the α,θ_{12} part of the problem. The eigenfunctions then serve as a basis for expansion of the full wave function and the eigenvalues enter as diagonal potential terms (in R) for the radial equations in R. Because of the parametric dependence of the eigenfunctions on R, the set of radial equations is coupled and approximations are introduced to decouple them to a tractable set.[2]

The success of the adiabatic hyperspherical scheme for low doubly excited states shows that correlations (motion in α and θ_{12}) develop faster than an expansion in the overall size (motion in R) of the system. The potential curves U(R) for this latter motion are labeled by the quantum numbers K and T and bear a resemblance to Born-Oppenheimer molecular potential curves; see, for instance, figs. 12 and 14 of ref. 2. At small R, a steeply rising repulsive part characterizes these curves. This keeps the wave function small at small R, particularly for all but the lowest lying (K,T) curves and accounts, for instance, for the early observation noted in the Introduction that only one dominant $^1P^o$ series below $He^+(N=2)$ is seen. At large R, the adiabatic hyperspherical potential curves are converged to the limit $He^+(N) + e(\to\infty)$, where the potential is a combination of coulomb and a long range dipole potential because of the ℓ-degeneracy of the $He^+(N)$ states.[2] This feature, while appropriate for describing the series in Fig. 1 below each N limit for low N, introduces single electron quantum numbers (such as N) and a description appropriate for independent particle pictures, at least at large R. Upon extension to high N, such a description of the asymptotic wave function as a product of a Coulomb and a dipole function, misses a class of states to be discussed next.

States of High Excitation: Two Classes

An important aspect of the potential energy function in (6) is the structure of the singular points. Apart from R, which is merely a scale variable, this function is two-dimensional (in α and θ_{12}) and a sketch is shown in Fig. 10 of ref. 2. Like any two-dimensional function, there is a minimum ($\alpha=0$ or $\pi/2$), a maximum ($\alpha=\pi/4, \theta_{12}=0$) and a saddle point ($\alpha=\pi/4, \theta_{12}=\pi$). Their geometrical significance is straightforward, the first corresponding to one of the electrons close to the nucleus ($r_1=0$ or $r_2=0$), the second to the two electrons coincident, and the third to a configuration in which the two electrons lie at equal distances on opposite sides of the nucleus. This last configuration has particular significance. One aspect, that θ_{12} equals π, has already been encountered and is a natural expression of the repulsion between the slow electrons. The other aspect, that α equals $\pi/4$, corresponds to equal radial distances of the electrons, and correlates with states of comparable radial excitation of the electrons (N=n). States with N and n different or disparate in radial extent lie, on the other hand, in the "valleys" of the potential surface at $\alpha \simeq 0$ and $\pi/2$. The "ridge" at $\alpha=\pi/4$, with the potential falling off when $|\pi/4-\alpha|$ increases, points to a crucial instability in the radial correlation between the electrons. Any departure of r_1/r_2 from unity grows as R increases because of a "dynamic screening"-the electron with the smaller coordinate screens more of the nuclear field for the other, thereby hanging back even more and this leads to a further departure of the ratio from unity.[11]

As $R\to\infty$, we can make, therefore, a basic subdividion of two electron states into two classes, those in which the electrons have disparate radial excitation ("valley states") and those in which they remain equal in radial extent ("ridge states"). While sharing the same angular correlation aspects, these two classes exhibit rather different radial correlation. They also require different sets of radial quantum numbers for their description. The valley states can be described in terms of the pair of quantum numbers (N,n), also in adiabatic hyperspherical schemes where states are seen as eigenvalues in individual potential curves converging to a $He^+(N)$ limit. This clearly does not work for the ridge states where the electrons remain correlated at $r_1 \simeq r_2$ all the way to $R\to\infty$ and independent particle principal quantum numbers never enter. The asymptotic wave function has to remain that of a (2e) pair (at the saddle) wave function in the field of the nucleus, rather than of a product of a Coulomb and dipole function. Such a situation has, in fact, been studied for many years in the problem of the threshold escape of two electrons from a positive ion.[11]

Just above, for instance the He^{++} limit in Fig. 1, when two electrons have together just sufficient energy to escape to infinity, reaching $R=\infty$ with zero energy, it was in a way natural to argue starting from the radial instability of dynamic screening that the system must remain at the saddle point of V in (6) for nearly all of the escape. We can see now a connection between this threshold escape and the ridge states. The (2e) pair, viewed as a single entity, remains at the saddle in these states as $R\to\infty$. So long as the asymptotic energy is negative (measured with respect to the double escape threshold), these are doubly excited ridge states and the wave function decays exponentially in R. The higher states extend out further in R and, finally, as the double threshold is reached, the pair escapes as an entity to $R=\infty$. This sequence of states must be indexed by a principal quantum number, a single quantum number ν associated with the coordinate R. Since the wave function remains at the saddle, the two-electron Schrödinger equation can be expanded around this point ($\alpha=\pi/4$, $\theta_{12}=\pi$). The potential in (5) has three terms in its leading dependence, an overall attractive Coulomb potential in 1/R (a "six-dimensional" Coulomb potential!) and terms in $(\pi/4-\alpha)^2/R$ and $(\pi-\theta_{12})^2/R$, the first repulsive and the second attractive, reflecting that this is a saddle point. It has been shown that for the problem of threshold double escape, continuum solutions in this potential have to be found, treating all three terms together.[11] For ridge states, similarly one has to find bound state eigenvalues of the six-dimensional Coulomb potential.

A six-dimensional Bohr-Rydberg formula, with states converging on the double ionization limit I_{++}, has given a good accounting of recently observed[12](experimentally) doubly excited states in He^- and of numerically calculated states in H^- and He, all of them states of comparable excitation.[3] Fig. 2 presents the data in Fig. 1 but now seen according to the pair description and with the ridge states emphasized by darker lines (the thinner lines are valley states). All reference to single electron quantum numbers and parental ionization limits of He^+ have been removed and throughout consistently there appear only quantities characteristic of the pair of electrons. Likewise the six-dimensional Bohr-Rydberg formula employs pair

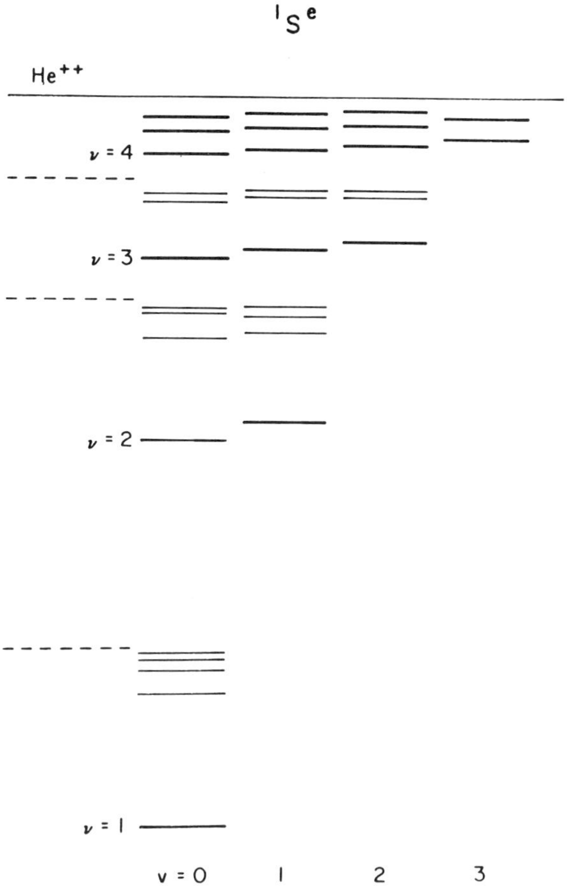

Fig. 2. States of the helium atom in pair labeling

quantities like the double ionization limit I_{++}, the pair quantum number ν and a pair quantum defect which depends on other pair quantum numbers like v and L.[3] Contrast this as a complete departure from (1) and Fig. 1. Also, note the quite different grouping of levels in Fig. 2 as compared to Fig. 1, the different states with ν=1,2,... at any particular v seen as a novel kind of Bohr-Rydberg series converging on the He^{++} limit.

I began this section by arguing for the pair coordinates R,α and θ_{12} as an alternative to the independent particle ones $\vec{r}_1$ and $\vec{r}_2$, more suited for describing tightly correlated states. In quantum mechanics, the association of quantum numbers with coordinates is unique only for separable problems. It is because the two-electron Schrödinger equation does not separate in $\vec{r}_1$ and $\vec{r}_2$ that the corresponding quantum numbers (N,n,ℓ) are not always appropriate. Neither does the two-electron equation separate in R,α and θ_{12} so that the corresponding quantum numbers such as ν, n_{rc} (for radial correlation - not discussed in this paper) and v are also not applicable for all states. The point of view, when faced with a genuinely non-separable problem, is not to indulge in a vain search for some separable system but rather identify different situations of quasi-separability. In this way, appropriate bases or quantum numbers can be provided for each situation which give a more economical description of phenomena. Unlike singly excited states, v is always a more appropriate replacement for ℓ for doubly excited states [for more general L, (K,T) are more appropriate than (ℓ_1,ℓ_2)]. For valley states, (N,n) remain appropriate, whereas for ridge states ν provides a more natural description. One could continue to use (N,n) for the latter but this would require a very large superposition so as not be useful. Recall the introductory remarks about temperature. The introduction of a new quantum number to describe states concentrated at the saddle and thus segregated from the rest (the resultant small overlap provides the basic explanation for their long lifetimes) works "like a charm" (pun intended).

DESIGN OF EXPERIMENTS

The nature of doubly excited states and the underlying physics that characterizes them provide clues to their excitation mechanism and even on what experimental arrangements are best suited to study the different classes of states. First of all, twenty odd years of synchrotron light study has remained confined to low-lying states. Adiabatic hyperspherical calculations point immediately to why this is so. The potential curves converging to higher N are staggered in R (see, for instance, fig. 12 of ref. 2) so that very rapidly wave functions within a potential well of high N have almost no overlap with the ground state which resides in the lowest well. Direct photoabsorption through a single photon cannot, therefore, connect these states. In general, excitation to higher N requires, therefore, some form of multistep transfer of excitation.

Laser excitation of these states has almost naturally proceeded in a multi-step fashion, given the non-availability of lasers in the ultraviolet and higher frequency. A few very recent experiments[13 14] of this kind have accessed high doubly excited states, mostly in Barium. The states reached have, however, been those I have described as valley states. They have been called planetary atoms, a term introduced earlier[15] through different motivations, but which we can now identify with valley states, each electron having its own radial size in analogy with planets moving in their individual and disparate radial orbits. That such laser excitation reaches valley states is immediate, even in terms of the very way these experiments are described. Typically, two photons take the ground $6s^2$ state to a Rydberg state 6sns or 6snd and then with another photon or two, the "core electron" is jacked up to a higher N. These experiments reach valley states with n>N. The initial step placed the two electrons on a very disparate footing, 6 and n. With little help, therefore, from the electron-electron interaction it is difficult in the next step, given that a photon couples only to one electron at a time, to get to the tightly correlated ridge states.

To access states with n=N, it is far better to let the electron-electron interaction do the work for you. These are stronger in $6s^2$ than in 6sns and perhaps the simultaneous impact of all the laser photons on the former rather than in the sequence described above will fare better in keeping the electrons comparable in radial excitation. A natural way to access ridge states, however, that once again uses the electron-electron interaction but now with one electron being an external projectile, is the preferred route to these states. The only experiment to date which has reached high and comparable excitation is, in fact, an inelastic scattering of electrons from helium with incident kinetic energies just short of the ionization potential.[12] This latter feature is crucial and the earlier discussion about double escape suggests this as a general method to reach highly correlated states (even of more than two electrons), namely, to do a scattering just short of a multi-escape threshold.[3] In such an impact, the incident particle can excite the system (losing most of its own energy) so that the system begins to evolve outward in R as if it is proceeding to multiple escape. All the tight correlations develop as a consequence because the system "does not know" till R gets large (typically at a distance $\sim 1/|E|$) whether it connects asymptotically to a bound state or the multiple escape continuum. When $E<0$, kinematical reality finally forces the system to a bound state but, having developed all the strong correlations of multiple escape, the states thus populated are the counterpart states below threshold, namely the ridge states.

Doubly excited states can also be reached by ion impact. After early experiments[2] in the 1960's, this method was little explored but is again increasingly coming into prominence.[16] Depending on the impact velocity, since an ion acts both like an electromagnetic impulse and as a carrier of electrons, both ridge and valley states can be populated. There is usually no discrimination in the L,S states populated (a great plus for photon experiments) and given the large density of doubly excited states, the lack of resolution in ion impact excitation calls for either electron or photon (far ultraviolet) spectroscopy on the states formed when they subsequently decay.

Another important guideline for tailoring experiments is the lifetime and dominant decay pathways for the state being sought. There is rather little information available on these questions even for low lying states (see fig. 3 and other references in ref. 2) and almost none for the high ones (see a few rudimentary remarks in ref. 3). Autoionization lifetimes increase rapidly with increasing v (recall the early remarks about the N=2 $^1P^o$ states in He) and even for the ridge states, where autoionization is important for lower ν, the lifetimes go up rapidly as ν increases.[3] Therefore, many of the higher doubly excited states, both ridge and valley states, are almost stable against autoionization and presumably decay by radiation to lower states, whether singly or doubly excited. Overlap considerations of their wave functions would suggest that in sequential decays, the cascade connects ridge states to other ridge states (save at low ν) or valley states to valley states. Since the pathway to decay is the reverse of one for excitation, it can serve as a guideline that electron impact excitation is favored for reaching those states that autoionize whereas multi-step laser excitation is more natural for those that radiate. A combination of the two may well be required for accessing high ridge states. The high resolution and selectivity of lasers is unbeatable but to go all the way from the ground state is not feasible energetically. Reaching some lower members of the ladder of ridge states by electron impact, and proceeding from there by a few laser photons to the higher states which are autoionization forbidden, and conversely therefore, also not excited easily by electrons, would provide the best combination of electron and photon spectroscopy.

SUMMARY

That new concepts such as new quantum numbers arise even at such a simple level as a two-electron atom is remarkable but at the same time this problem can serve as a paradigm for non-separable problems, of possibly wide relevance in physics. It is amusing to model the flavor quantum numbers for quarks in terms of the quantum mechanics of three particles moving under their mutual interactions. This has in general six dynamical variables and the singularity structure of a function of six variables is one maximum, one minimum, and five different types of saddle points. An absolute maximum provides, of course, no binding but the others constitute six types of flavors instead of the two (ridge/valley) we have seen for two electron states. Aside from the amusing aspect,* this example serves to highlight how in going to non separable problems with more variables, saddle points tend to proliferate (it is highly unlikely that the second derivative has the same sign in all different directions from the singular point). The analysis of ridge states and double escape in a saddle point is, therefore, a prototype for multi-particle dynamics. Even for a single particle, if two or more fields act on it so as to lead to a non-separable potential (for instance, the problem of a high Rydberg state in an external magnetic field where also a two-dimensional potential with two classes of states is involved[17]), such an analysis can serve as a prototype, providing a much wider and broader context for the study presented here.

Finally, the kind of pair correlations and tightly correlated states at high excitation (reaching an extreme for double escape near threshold) discussed here have a resemblance to other situations in physics where similar strong correlations prevail. The Cooper pair of superconductivity and the ground state of nuclei as described, for instance, in the interacting boson model,[18] are notable examples. In those examples the strongest correlation is in the ground state, and low-lying excitations are seen as part of the family of correlated states. The atomic example differs in that the tightest correlation is high in the spectrum, at the double escape threshold, and the high doubly excited states attached from below form the analogous family. This difference finds ready explanation in the nature of the interaction between the particles. It is attractive in the other examples whereas the atomic one is repulsive. Also remarkable is that it takes a many body system in the other examples (even in nuclei, the interacting boson model deals with >100 nucleons) to exhibit these strong correlation effects whereas we have discussed an atom with just two electrons. The basic explanation here lies in the nature of the Rydberg spectrum, the pile-up of states near the ionization limit, so that an infinite number of states is involved even with very few particles.

ACKNOWLEDGMENT

This work has been supported by the U. S. National Science Foundation, most recently under Contract PHY 84-01855.

REFERENCES

1. U. Fano and J. W. Cooper, Rev. Mod. Phys. 40, 441 (1968).

* Another amusing connection of our usage of the words parental and grandparental is to the original meaning of these words. We know from simple observation and the laws of Mendelian inheritance that some children look more like their grandparents (when a recessive gene is doubly expressed) whereas others resemble their parents!

2. U. Fano, Rep. Prog. Phys. 46, 97 (1983). I will refer to this article frequently and for specific figures. It may be useful, therefore, to keep it alongside.
3. A. R. P. Rau, Atomic Physics 9, edited by R. S. Van Dyck and E. N. Fortson (World Scientific, Singapore, 1984).
4. U. Fano and A. R. P. Rau, Comm. At. Mol. Phys., (1985).
5. See, for instance, a review: G. H. Dunn, D. S. Belic, N. Djuric and D. W. Mueller, Atomic Physics 9, edited by R. S. Van Dyck and E. N. Fortson (World Scientific, Singapore, 1984).
6. C. Froese-Fischer, The Hartree Fock Method for Atoms (John Wiley, New York, 1977).
7. See, for instance, D. Kleppner, this volume.
8. L. D. Landau and E. M. Lifshitz, Quantum Mechanics: Non-Relativistic Theory (Pergamon, Oxford, 1977), 3rd edition, Sec. 37.
9. See a review D. Herrick, Adv. Chem. Phys. 52, 1 (1983).
10. A. R. P. Rau, J. Phys. B 17, L75 (1984) and Pramana 23, 297 (1984).
11. G. H. Wannier, Phys. Rev. 90, 817 (1953). Recent reviews of the subject are in A. R. P. Rau, Phys. Rep. 110, 369 (1984), and in F. H. Read, Electron Impact Ionization, edited by G. H. Dunn and T. Mark (Springer-Verlag, 1984).
12. S. J. Buckman, P. Hammond, F. H. Read and G. C. King, J. Phys. B 16, 4039 (1983). The key figure on ridge states is reproduced in refs. 2-4.
13. L. A. Bloomfield, R. R. Freeman, W. E. Cooke and J. Bokor, Phys. Rev. Lett. 53, 2234 (1984).
14. P. Camus, P. Pillet and J. Boulmer, J. Phys. B, (1985).
15. I. C. Percival, Proc. Roy. Soc. (London) A 353, 289 (1977).
16. R. Bruch, P. L. Altick, E. Träbert and P. H. Heckmann, J. Phys. B 17, L655 (1984); A. B. Montesquieu, P. B. Cattin, A. Gleizes, A. I. Marrakchi, S. Dousson and D. Hitz, J. Phys. B 17, L127 and L223 (1984).
17. See, for instance, J. C. Gay, this volume. Also A. R. P. Rau, Comm. At. Mol. Phys. 10, 19 (1980) and J. de Physique Colloque C2, suppl. 11, 43, 211 (1982).
18. A. Arima and F. Iachello, Ann. Phys. (N.Y.) 99, 253 (1977); F. Iachello, Interacting Bosons in Nuclear Physics (Plenum, New York, 1979).

NON-RELATIVISTIC MANY-BODY PERTURBATION THEORY

Ingvar Lindgren

Chalmers University of Technology/University of Gothenburg
S-412 96 Göteborg, Sweden

Abstract

The formalism of non-relativistic many-body perturbation theory, based on the linked-diagram expansion, is reviewed. The wave operator, transforming the unperturbed wave function into the perturbed one, is expressed in second quantization, and equations for the coefficients are derived. Solving a hierachy of such equations in a self-consistent way is equivalent to evaluating the one-, two-, ... body effects to all orders of perturbation theory. The formalism is extended to the coupled-cluster approach, where the wave operator is expressed in exponential form.

A numerical procedure based on the solution of inhomogeneous differential equations of one- and two-particle type is described, and a number of illustrative numerical results is given.

The possibility of extending the procedure to the relativistic case is discussed. A procedure based on the diagonalization of the single-electron Dirac Hamiltonian is outlined. In this procedure the positive and negative single-electron states are easily separated and it should be possible to treat the negative-energy states (virtual-pair creation) in a correct way.

I. Introduction

Several schemes have been developed for (non-relativistic) atomic (and nuclear) many-body calculations, such as

a) configuration interaction (CI)
b) multi-configurational Hartree-Fock (MCHF)
c) many-body perturbation theory (MBPT)
d) Hylleraas-type wave function

Of these schemes d) is limited to small systems of max. 3-4 particles, while the remaining three schemes are generally applicable and essentially equivalent. In this review I shall concentrate on the scheme of MBPT, where important developments have taken place during the last decade.

What is normally referred to as MBPT is the scheme based upon the linked-diagram expansion (LDE), introduced into nuclear physics by Brueckner /1/ and Goldstone /2/ and first applied to atomic physics by Kelly /3/. The original procedure was restricted to single-determinantal unperturbed states, but developments to general open-shell systems were later made by Brandow /4/ and others for atomic /5-7/ as well as nuclear /8-10/ systems.

The LDE can be transferred to a non-perturbative scheme by separating the wave operator (which transforms the unperturbed wave function into the perturbed one) into one-, two-, ... body parts by means of second quantization. Solving the resulting equations in a self-consistent way is equivalent to evaluating the corresponding effects to all orders in a perturbative expansion. The inclusion of successively higher n-body contributions in this way represents a hierarchy of equations, which normally leads to a fast convergence. For most applications it has been found that the pair approximation, which includes one- and two-body effects, is a very good approximation,capable of reproducing something like 95% of the pertrubative effects (beyond restricted Hartree-Fock).

A further development of the (self-consistent form of) LDE is represented by the coupled-cluster approach (CCA). Here, the wave operator is expressed in exponential form, and the truncation takes place in the exponent ("cluster operator") rather than in the wave operator itself. In this scheme also "clusters" of excitations are automatically included, for instance, in the pair approximation double pair excitations, which represent the dominant quadruple excitations. This scheme was developed in nuclear physics by Coester and Kümmel /11/ and introduced into quantum chemistry by Sinanoğlu /12/ and Čížek /13/. Extensions to open-shell systems have now been made by different groups /14-18/.

To date, only little progress has been made towards relativistic MBPT. Normally, the starting point for relativistic many-body calculations is the Dirac-Coulomb-Breit Hamiltonian, consisting of a sum of single-electron Dirac operators, the instantaneous Coulomb interaction between the electrons and the Breit operator, representing magnetic interactions and the retardation of the Coulomb interaction. The straightforward application of such an operator, however, can lead to serious difficulties due to the presence of negative-energy states, as first pointed out by Brown and Ravenhall /19/ and further discussed by Bethe and Salpeter /20/.

In order to eliminate the negative-energy states, projection operators can be introduced into the Hamilton operator /19-22/, which is a correct procedure in first order. It has been shown by Mittleman /22/ that in the standard relativistic Hartree-Fock (HF) - or Dirac-Fock (DF) - procedure projection operators are automatically included, thereby preventing negative-energy states (defined by means of the HF operator) to appear.

To go beyond relativistic HF (which is a first-order procedure from the MBPT point of view) in a correct way for a many-body system is considerably more complicated. Then it is necessary to take the negative-energy states explicitly into account ("virtual pair creation"). This is not done in the multi-configuration Dirac-Fock (MCDF) procedure. In a relativistic MBPT procedure on the other hand, it would be possible to handle the negative-energy states correctly. A convenient way of separating the positive - and negative-energy states in this case seems to be to start with a diagonalization of the single-electron Dirac operator in the Foldy-Wouthuysen sense /23/. Then the upper (lower) components correspond to positive (negative) energy states (in the weak-potential limit). Calculations of this kind are in preparation in our group at Chalmers /24/.

Different approaches of performing relativistic calculations on many-electron systems are further discussed in the presentation by Peter Mohr in these Proceedings.

II. The many-body formalism

II.A. General

The starting point for (non-relativistic) atomic MBPT is the Hamiltonian

$$H = -\frac{1}{2}\sum_n \nabla_n^2 - \sum_n \frac{Z}{r_n} + \sum_{m<n} \frac{1}{r_{mn}} \tag{1}$$

using Hartree atomic units (e, m, $\hbar$, $4\pi\varepsilon_0$ are the basic units, the Bohr radius the unit of length, $2hcR_y = \alpha^2 mc^2 = 27.2$ eV the unit of energy), and the aim is to solve the Schrödinger eqn (S.E.)

$$H\Psi = E\Psi \tag{2}$$

with certain accuracy. The Hamiltonian (1) is partitioned into an unperturbed Hamiltonian

$$H_0 = \sum_n \left[-\frac{1}{2}\nabla_n^2 - \frac{Z}{r_n} + U(r_n)\right] = \sum_n h_0(n) \tag{3a}$$

and a perturbation

$$V = H - H_0 = \sum_{m<n} \frac{1}{r_{mn}} - \sum_n U(r_n) \tag{3b}$$

The single-electron potential, U(r), is here assumed to be of central-field type, although this is not required by the formalism.

The eigenfunctions of h_0 are the single-electron orbitals

$$h_0 \varphi_i = \varepsilon_i \varphi_i \tag{4}$$

and the Slater determinants formed by these orbitals are the antisymmetric eigenfunctions of H_0,

$$\begin{cases} H_0 \phi^A = E_0^A \phi^A \\ E_0^A = \sum_i \varepsilon_i \quad (\varphi_i \in \phi^A) \end{cases} \tag{5}$$

which form the basis of the calculations.

The functional space, formed by the eigenfunctions (2) of H, is partitioned into a model space (P), spanned by the eigenfunctions of H_0 corresponding to one or several eigenvalues (configurations), and a complementary space (Q). We consider the solutions of the Schrödinger eqn (2)

$$H\psi^a = E^a \psi^a \qquad (a = 1, 2, \ldots d) \tag{6}$$

which have their major part within P. The projections on the P space

$$\psi_0^a = P\psi^a \qquad (a = 1, 2, \ldots d) \tag{7}$$

are the zeroth-order or model functions, from which the perturbation expansion starts. (It should be noted that these functions are generally not known ab initio; see further below.) A wave operator, Ω, is defined so that it transforms all model functions into the corresponding exact wave functions

$$\psi^a = \Omega \psi_0^a \qquad (a = 1, 2, \ldots d) \tag{8}$$

This operator satisfies the "generalized Bloch eqn" /6, 7, 25/

$$[\Omega, H_0]P = Q(V\Omega - \Omega P V \Omega)P \tag{9}$$

which forms the basis for the MBPT procedure to be discussed here.

The Bloch eqn (9) can be used to generate a perturbation expansion of Ω. Inserting

$$\Omega = 1 + \Omega^{(1)} + \Omega^{(2)} + \cdots \tag{10}$$

and identifying terms of the same order, yields

$$\begin{aligned} [\Omega^{(1)}, H_0]P &= QVP \\ [\Omega^{(2)}, H_0]P &= Q(V\Omega^{(1)} - \Omega^{(1)}PV)P \\ &\text{---------} \\ [\Omega^{(n)}, H_0]P &= Q(V\Omega - \Omega P V \Omega)^n P \end{aligned} \tag{11}$$

which is the general Rayleigh-Schrödinger (RS) perturbation expansion. We employ here the intermediate normalization, implying

$$\langle \Psi_0^a | \Psi^a \rangle = \langle \Psi_0^a | \Psi_0^a \rangle = 1 \tag{12}$$

By projecting the S.E. (6) onto the model space, using (7,8), we get

$$P H \Omega \Psi_0^a = E^a \Psi_0^a \tag{13}$$

which shows that the model functions are eigenfunctions of an effective Hamiltonian

$$H_{eff} = P H \Omega P ; \qquad H_{eff} \Psi_0^a = E^a \Psi_0^a \tag{14}$$

and the eigenvalues are the corresponding exact energies (in spite of the fact that H_{eff} operates only in the limited model space).

With the intermediate normalization (12) the exact energy becomes

$$E_a = \langle \Psi_0^a | H_{eff} | \Psi_0^a \rangle \tag{15}$$

If we consider an additional perturbation, h, such as the hyperfine interaction, then the replacement

$$H \rightarrow H + h \tag{16}$$

leads to

$$H_{eff} \rightarrow H_{eff} + h_{eff} \tag{17}$$

where h_{eff} contains all parts of the new effective Hamiltonian which depend on h. This operator, called the effective operator associated to the additional perturbation, yields the corresponding energy contribution by

$$\Delta E_h = \langle \Psi_0^a | h_{eff} | \Psi_0^a \rangle \tag{18}$$

The basic steps of the procedure is to

a) determine Ω by means of (9)

b) determine H_{eff}, Ψ_0^a and E^a by means of (14)

c) determine (if needed) Ψ^a by means of (8).

The MBPT is based upon the formalism of <u>second quantization</u> (s.q.) /26/. In this formalism H_0 and V are expressed

$$H_0 = \sum_i a_i^+ a_i \varepsilon_i$$

$$V = \sum_{ij} a_i^+ a_j \langle i|-U|j\rangle + \frac{1}{2}\sum_{ijkl} a_i^+ a_j^+ a_l a_k \langle ij|\frac{1}{r_{12}}|kl\rangle \tag{19}$$

where $a_i^+(a_i)$ are single-electron creation (annihilation, destruction) operators. Here (and in the following) i,j,k,l run over <u>all</u> single-particle states (4) - bound as well as unbound. Under quite general conditions /27/ also the wave operator (8) can be expressed in s.q.

$$\Omega = 1 + \sum_{ij} a_i^+ a_j x_j^i + \frac{1}{2}\sum_{ijkl} a_i^+ a_j^+ a_l a_k x_{kl}^{ij} + \cdots \tag{20}$$

The operators above are said to be in <u>normal form</u> with respect to the <u>vacuum</u> (with the creation operators to the left of the annihilation operators).

In atomic MBPT it is usually convenient to redefine the s.q. form by shifting the vacuum level to a suitable closed-shell state, Φ, which we refer to as the <u>reference state</u>. Single electron states occupied (not occupied) in Φ are called <u>hole</u> (<u>particle</u>) <u>states</u>. In the new representation the operators (19) become /26/

$$H_0 = \sum_i \{a_i^+ a_i\} \varepsilon_i + \sum_a^{hole} \varepsilon_a$$

$$V = V_0 + V_1 + V_2$$

$$V_0 = \sum_a^{hole} \langle a|-U|a\rangle + \frac{1}{2}\sum_{ab}^{hole} \langle ab|\widetilde{\frac{1}{r_{12}}}|ab\rangle \tag{21}$$

$$V_1 = \sum_{ij} \{a_i^+ a_j\} \langle i|v|j\rangle$$
$$= \sum_{ij} \{a_i^+ a_j\} \left[\langle i|-U|j\rangle + \sum_a^{hole} \langle ia|\widetilde{\frac{1}{r_{12}}}|ja\rangle\right] \tag{21}$$
$$V_2 = \frac{1}{2} \sum_{ijkl} \{a_i^+ a_j^+ a_l a_k\} \langle ij|\frac{1}{r_{12}}|kl\rangle$$

Here, { } represents the normal form with respect to the reference state Φ, which means that the operators for the creation of an electron in a particle state and the annihilation of an electron in a hole state - "particle-hole" (p-h) creation operators - appear to the left. The "tilde" (~) in the Coulomb matrix elements indicates that the exchange part is included.

It should be noted that the Hartree-Fock (HF) potential of the reference state can be defined

$$\langle i|U_{HF}|j\rangle = \sum_a^{hole} \langle ia|\widetilde{\frac{1}{r_{12}}}|ja\rangle \tag{22}$$

This implies that the "effective potential", v, vanishes, if the potential, U, is chosen to be the HF potential of the reference state. This simplifies the treatment, but it is not required by the formalism. (In certain cases such a potential may lead to poor convergence - or no convergence at all.)

The wave operator (20) becomes in the new representation

$$\Omega = 1 + \sum_{ij} \{a_i^+ a_j\} x_j^i + \frac{1}{2} \sum_{ijkl} \{a_i^+ a_j^+ a_l a_k\} x_{kl}^{ij} + \cdots \tag{23}$$

where, of course, the x-coefficients are different from those of (20). The first sum in (23) represents the one-body part, the next sum the two-body part etc.

II.B. Graphical representation

Graphically, we represent hole (particle) states by a line directed down (up), as illustrated in Fig. 1. The normal-ordered perturbation

V(21) is then represented by the diagrams shown in Fig. 2. Lines directed out from (towards) the interaction line represent electron creation (annihilation) operators. This implies that the lines above (below) the interaction line represent p-h creation (annihilation) operators. The internal lines in V_0 represent summations over hole states.

In the general open-shell case the model space, P, contains one or several configurations. The single-electron levels, which are partially occupied in P are called valence levels, and in the graphical representation the states of such levels are denoted by a double arrow (see Fig. 1). In principle, the valence states can be of particle or hole type, but for simplicity we assume here that they are all particle states.

For simplicity, we shall also assume that the model space is "complete" in the sense that it contains all states that can be formed by permuting the valence electrons among the valence states. It then follows that a diagram acting to the right (left) on P can have no other free lines than valence lines at the bottom (top). A diagram with no free non-valence lines at all (top or bottom) operating on P yields necessarily a state within P. Such a diagram is said to be "closed".

The wave operator (23) is represented graphically as shown in Fig. 3. The one- (two-...) body part is represented by diagrams with one (two,...) pairs of free orbital lines. Since Ω always operates to the right on P (8, 9), any free line at the bottom must be a valence line. (Any particle annihilation which can take place in P is valence annihilation.)

With V and Ω in normal form we now return to the basic eqn (9), and we can transform the left-hand and right-hand sides of that eqn to normal form - algebraicly or graphically. The general tool is here Wick's theorem /28/, according to which the product of two operators in normal form can be expressed

$$AB = \{AB\} + \{\overline{AB}\} \tag{24}$$

Here, the first term on the r.h.s. represents the normal product (without contractions) and the second term represents the normal form of all terms with one, two, ... contractions between A and B. Contractions can here take place only between p-h annihilation operators of A and p-h creation operators of B.

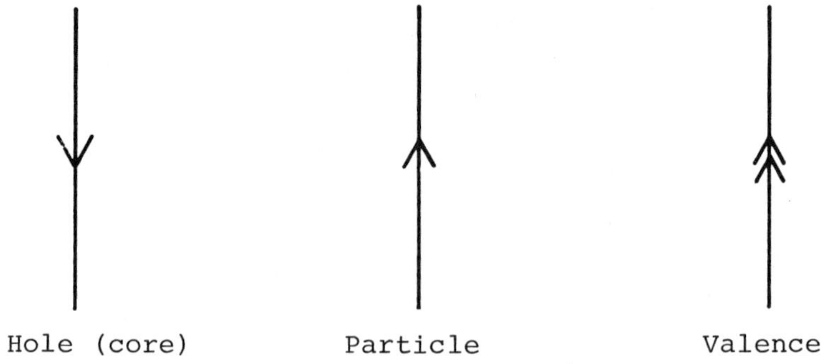

Fig. 1. Graphical representation of the single-electron states.

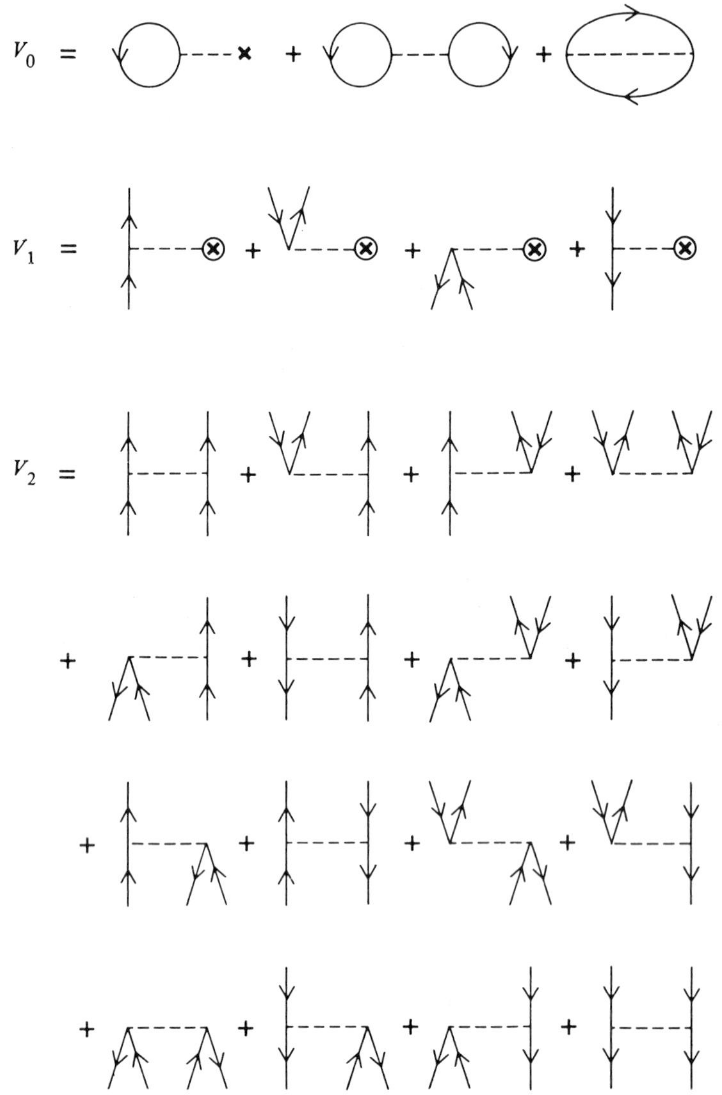

Fig. 2. Graphical representation of the normal-ordered zero-, one- and two-body parts of the perturbation (21).

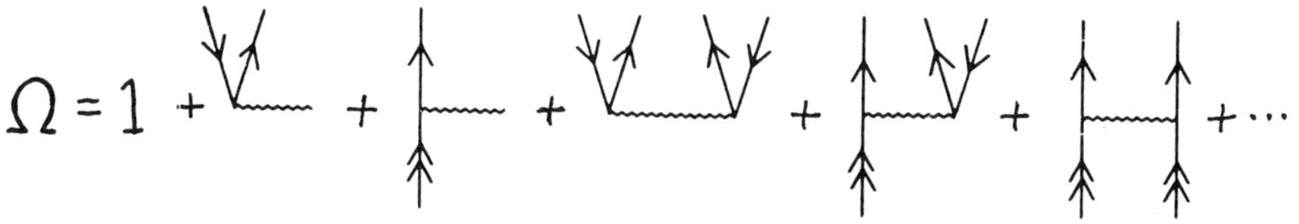

Fig. 3. Graphical representation of the wave operator (23).

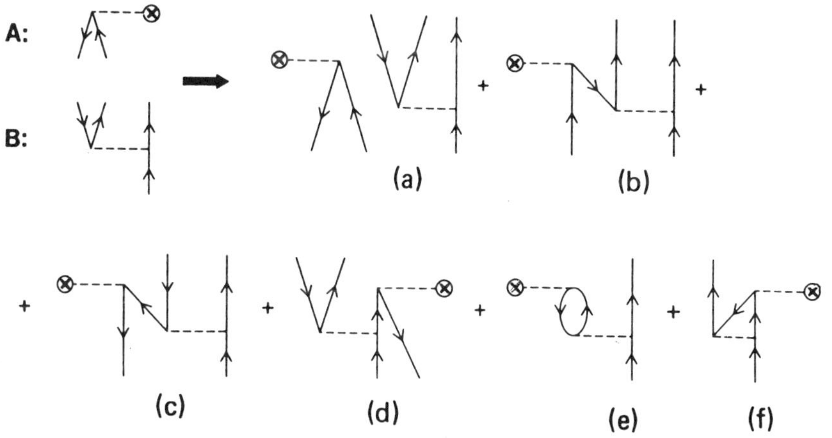

Fig. 4. Illustration of Wick's theorem for operator products (24). Diagram (a) represents $\{AB\}$ without contractions and (b-f) represent $\{\overline{AB}\}$. Diagrams (b-d) correspond to single contractions and (e,f) to double contractions.

The graphical form of Wick's theorem (24) is illustrared in Fig. 4. The lines at the bottom of the diagram A are connected to the lines at the top of diagram B in all possible ways - including no connection at all. The connection of the lines in the diagrams corresponds to the (non-zero) contraction of the corresponding p-h operators.

By means of Wick's theorem the r.h.s. of the Bloch eqn (9) can now be expressed in normal form. The normal form of the l.h.s. is given by

$$[\Omega, H_0] = \sum_{ij} \{a_i^+ a_j\} (\varepsilon_j - \varepsilon_i) x_j^i + \frac{1}{2} \sum_{ijkl} \{a_i^+ a_j^+ a_l a_k\} (\varepsilon_k + \varepsilon_l - \varepsilon_i - \varepsilon_j) x_{kl}^{ij} + \cdots \quad (25)$$

By identifying corresponding terms on the left-hand and right-hand sides, eqns can be set up for (a selection of) the x-coefficients. This leads to a hierarchy of eqns for the one-, two-,... body parts of the wave operator (23). These eqns have to be solved iteratively in a self-consistent way, which is equivalent to evaluating the corresponding effects to all orders of perturbation theory.

Alternatively, it is, of course, possible to perform the calculations more directly order by order, starting from the RS expansion (11). It can then be shown that all so-called unlinked diagrams - with a disconnected closed part - vanish in each order,

$$[\Omega^{(n)}, H_0] P = Q \left(V\Omega - \Omega P V\Omega \right)_L^n P \quad (26)$$

where "L" stands for the linked part. This is the linked-diagram theorem (LDT). It should be observed that Ω does contain disconnected diagrams, consisting of two or more open (not closed) parts. As a matter of fact, it is the disconnected diagrams of $V\Omega$, that in higher orders give rise to the unlinked diagrams, when one of the separate pieces is being closed. It should also be noted that a condition for the complete cancellation of the unlinked diagrams is that the exclusion principle is disregarded in the intermediate states.

Also in the self-consistent procedure sketched above do the contributions corresponding to unlinked diagrams vanish, provided all n-body parts are considered. In a real calculation, on the other hand, which has to be truncated after a certain n, there is not a complete cancellation of such contributions. Such a calculation is equivalent to a CI calculation, which hence is different from a MBPT calculation based on (26) in the corresponding approximation.

Unlinked diagrams are "unphysical" in the sense that they give rise to energy contributions, which are non-linear (quadratic etc.) in the number of electrons of the system, and which for large systems can lead to appreciable errors. Computational procedures, which are free from such contributions, are said to be size consistent /29,30/.

By summing (26) to all orders, we obtain formally

$$[\Omega, H_0]P = Q\left(V\Omega - \Omega P V \Omega\right)_L P \tag{27}$$

where the subscript now indicates that only the part of the r.h.s., which is linked in an order-by-order expansion, whould be included. It follows from LDT that this is identical to the original Bloch eqn (9), if the eqns are solved exactly, but not if the expansion of Ω(23) is truncated after a certain n-body term. In the latter case - for systems with more that a few electrons - (27) is the more exact eqn to use.

Traditionally, MBPT is treated order by order, based on (26) or some equivalent expression. The disadvantage with this approach is that it can hardly be carried beyond third order in any systematic way. In the self-consistent approach, based on (27), on the other hand, important effects, like the pair correlation, can be evaluated to all orders of perturbation theory.

II.C. The coupled-cluster approach

A more developed form of MBPT is the so-called coupled-cluster approach (CCA), where the wave operator is expressed in exponential form /11-13/

$$\Omega = \exp S = 1 + S + \frac{1}{2} S^2 + \cdots \tag{28}$$

The operator S is referred to as the cluster operator. In the open-shell case it is more convenient to use a normal-ordered expression /16,26/

$$\Omega = \{\exp S\} = 1 + S + \frac{1}{2}\{S^2\} + \cdots \tag{29}$$

It can then be shown that S satisfies the eqn

$$[S, H_o]P = Q(V\Omega - \Omega P V\Omega)_c P \tag{30}$$

where the subscript indicates that only rigourously connected diagrams should be included. This means that the disconnected diagrams which appear in (27) are here discarded. Apart from this, the eqn for S is identical to that for Ω(27). The disconnected diagrams of Ω, consisting of separated open parts, will is the CCA be taken care of by the higher powers of S in (29).

The cluster operator can be expressed in s.q. in analogy with (23)

$$S = \sum_{ij} \{a_i^+ a_j\} s_j^i + \frac{1}{2} \sum_{ijkl} \{a_i^+ a_j^+ a_l a_k\} s_{kl}^{ij} + \cdots \tag{31}$$

and the self-consistent eqns for the s-coefficients can be constructed in the same way as the corresponding eqns for the x-coefficients of the wave operator.

The eqns of S are very similar to those of Ω, the only differences being that in the former case no disconnected diagrams appear and some "coupled-cluster" diagrams are added. The latter diagrams appear, when Ω is replaced by the higher powers of S in (29), as illustrated in Fig. 6 below.

The advantage of CCA is twofold. First, no unlinked contribution can appear, since S is manifestly connected. This garantees size consistency in each approximation. Second, the appearance of higher powers of S implies that important multiple excitations are automatically included in Ω (and in the wave function). For instance, in the pair approximation, where S is truncated after the two-body term,

$$S = S_1 + S_2 \tag{32}$$

the $\{S_2^2\}$ term yields most of the quadruple excitations, which next to doubles represent the most important correlation effects. As mentioned, LDT (26) holds only if the exclusion principle is disregarded in the intermediate states. This means that quadruple excitations have to be considered in this formalism also for He-like systems. In other words, an LDE based on (27), including all one- and two-body effects (single and double excitations) does not yield the correct result for two-electron systems (!), while a similar calculation based on (9), as well as CI, does.

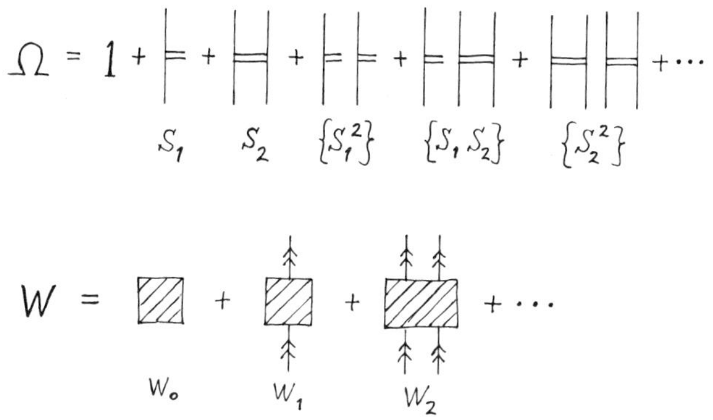

Fig. 5. Graphical representation of the wave operator (33) and the "effective interaction" (35) in the "coupled-pair" approximation.

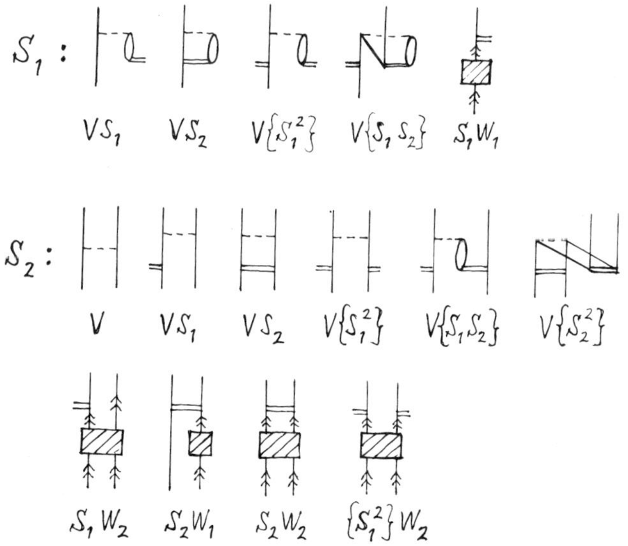

Fig. 6. Types of diagrams appearing in the r.h.s. of the coupled-cluster equations for S_1 and S_2 (34). HF potential (22) assumed.

The CCA approach (in the pair approximation) combines the advantages of the two procedures - being exact for two-electron systems, like CI, and being size-consistent for large systems, like LDE.

In the pair approximation (32) the wave operator (29) becomes

$$\Omega = 1 + S_1 + S_2 + \frac{1}{2}\{S_1^2\} + \{S_1 S_2\} + \frac{1}{2}\{S_2^2\} \tag{33}$$

neglecting terms with more than two clusters. The eqns for S_1 and S_2 can then be obtained by means of (30)

$$\begin{cases} [S_1, H_0] = (V\Omega - \Omega W)_{1,c} \\ [S_2, H_0] = (V\Omega - \Omega W)_{2,c} \end{cases} \tag{34}$$

Here, W represents the closed part of the "effective interaction", $V\Omega$,

$$W = (V\Omega)_{closed} \tag{35}$$

operating only within the P space.

The graphical forms of Ω and W are shown in Fig. 5. Double solid lines are used to represent S in order to distinguish it from Ω (Fig. 3). No arrows are shown on the diagrams of Ω, but it is assumed that the incoming lines at the bottom are valence lines. If one or several incoming lines represent core (hole) states, they should approach the vertex from above. Since W is closed, all in- and outgoing lines must there be valence lines.

By means of the representations of Ω and W in Fig. 5 and of V in Fig. 2 we can now construct the r.h.s. of the one- and two-particle eqns (34) by means of Wick's theorem (Fig. 3). This leads to the diagrams indicated in Fig. 6. Only one diagram of each kind is here shown. Again, the incoming lines at the bottom can be valence or hole lines (in the latter case bended up), apart from those of W, which must be valence lines. It should be noted that the diagrams involving W_1 (W_2) can appear only if at least one (two) of the incoming lines of S is (are) valence line(s).

In the diagrams of ΩW also the connecting lines between Ω and W are valence lines. Usually, these diagrams are drawn in a "folded" way with the connecting lines running backwards and therefore referred to as "folded" or "backward" diagrams /4,5/. In an order-by-order expansion ordinary Goldstone rules can then be used for the energy-denominator evaluation. In our approach there is no need to draw the diagrams in this way, since the denominators are directly given by (25). Therefore, we draw them straight, as in Fig. 6.

The diagrams of W (Fig. 5) can be constructed in a similar way as those of S by "closing" the diagrams of Ω by V according to (35). This leads to the same diagrams as those of $V\Omega$ in (34) (Fig. 6) with the exception that all free lines are valence lines. There is no analogue to the "folded" diagrams in W (but these diagrams appear indirectly via S).

The diagrams of the kind discussed here can be evaluated by means of simple rules, which are quite similar to the rules for evaluating ordinary Goldstone diagrams /26/. In this way the r.h.s. of the eqns (34) can be expressed in algebraic form. The l.h.s. is obtained in analogy with (25) and has the effect of supplying the energy denominator in the expression for S.

In the pair approximation, where one- and two-particle effects are considered, the eqn of S_1 and S_2, illustrated in Fig. 6, are coupled and have to be solved alternatingly until sufficient self-consistency is achieved. It can be noted, however, that the single-particle contributions (S_1) always have the same spin-angular dependence as the incoming orbital. Hence, this contribution can be added to the orbital, leading to a modification of the radial part only. When such modified orbitals are used in the eqn of S_2, all terms containing S_1 disappear. This is often a convenient way of treating the single-particle effects. The orbitals obtained in this way are known as Brueckner orbitals /31,32/.

II.D. Additional perturbation

The procedure outlined above can be applied also to evaluate the effect of an "additional perturbation", like the hyperfine interaction. The substitution (16) leads to

$$\begin{cases} V \to V + h \\ \Omega \to \Omega + \Omega_h \\ W \to W + h_{eff} \end{cases}$$

where the symbols without subscript represent the h-independent parts. From (34) we then get the h-dependent part of S_1

$$[S_1^h, H_0] = \left(h\Omega + V\Omega_h - \Omega_h W - \Omega h_{eff}\right)_{1,c}$$

considering only terms linear in h. The graphical representation of the r.h.s. can be constructed in the same way as those of (34), shown in Fig. 6. This leads to the kinds of diagrams given in Fig. 7 (for simplicity leaving out the coupled-cluster diagrams).

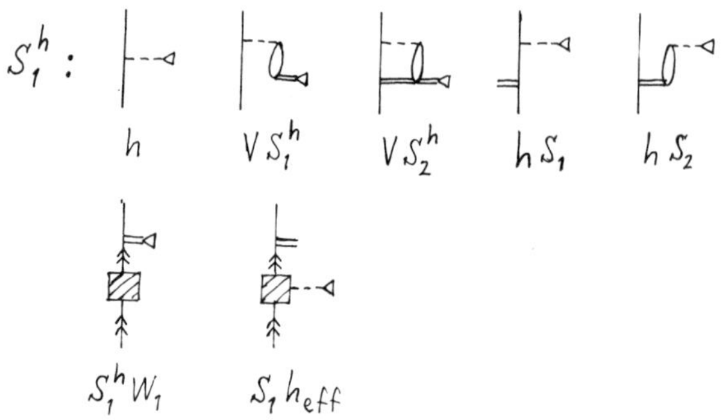

Fig. 7. Graphical representation of the types of diagrams appearing in the r.h.s. of the eqn for S_1^h.

The evaluation of the diagrams in Fig. 7 requires one- as well as two-body clusters. Of particular interest is the part that can be evaluated by one-body eqns only. That part can be generated as indicated graphically in Fig. 8. By solving coupled eqns of this kind in a self-consistent way, all single-particle effects are included. This is essentially equivalent to Unrestricted Hartree-Fock, UHF, Random-Phase Approximation (with Exchange), RPA(E), or Time-dependent Hartree-Fock (TDHF).

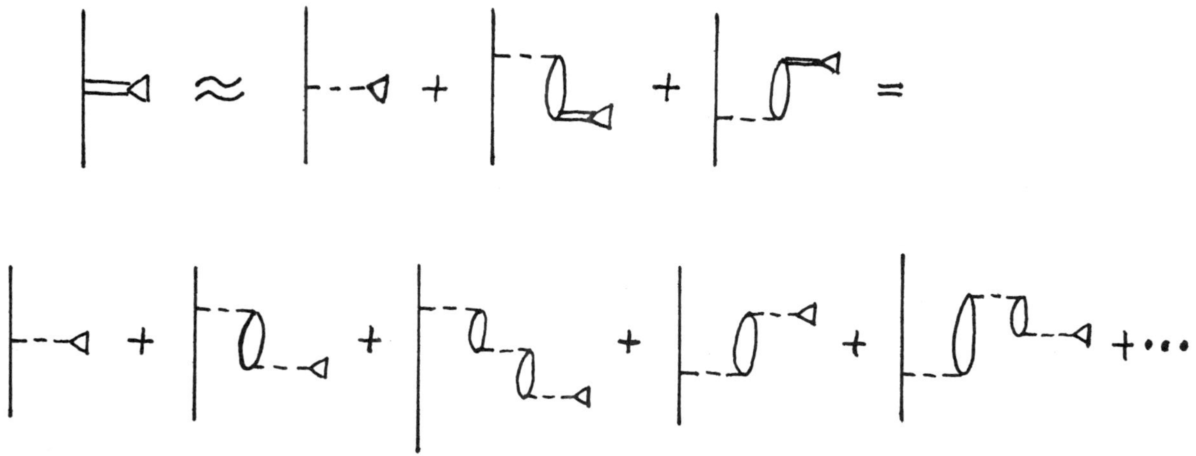

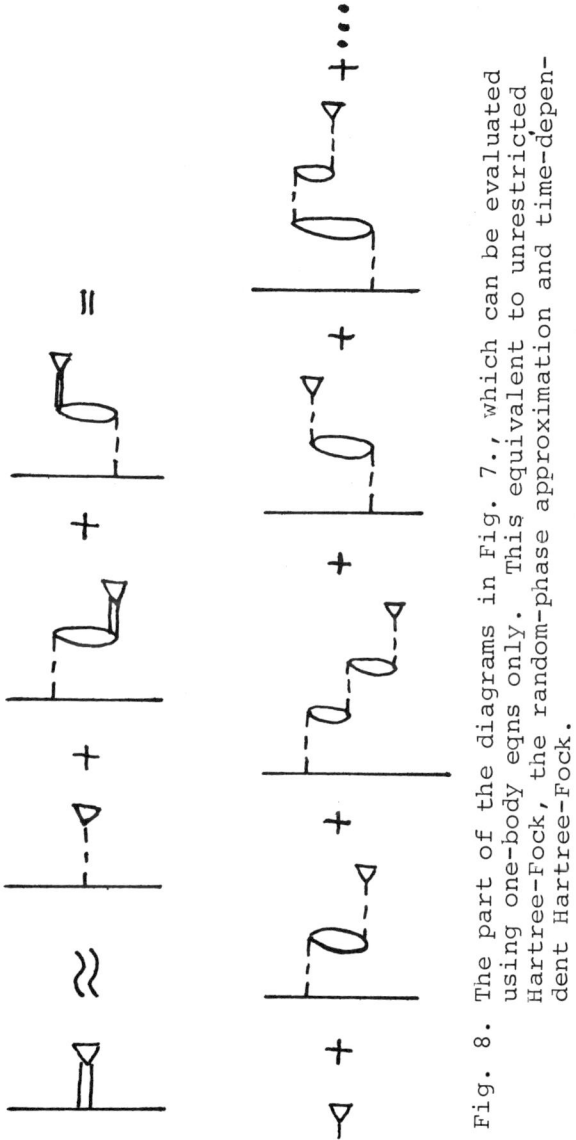

Fig. 8. The part of the diagrams in Fig. 7., which can be evaluated using one-body eqns only. This equivalent to unrestricted Hartree-Fock, the random-phase approximation and time-dependent Hartree-Fock.

III. Numerical procedure and illustrative results

III.A. Inhomogeneous differential equations

One way of treating the eqns discussed in the previous section is to construct a (more or less) complete set of single-electron orbitals (4) either numerically or by means of some basis functions (like Slater-type orbitals). In atomic calculations, where the central-field approximation can be used, a convenient alternative is to transform these eqns into inhomogeneous differential equations. This procedure can easiest be illustrated by means of single excitations, caused by some "external" perturbation, discussed in the previous section (Fig. 7). The first-order contribution to the wave operator, corresponding to excitations from the core, is then given by

$\sum_{ar}$ 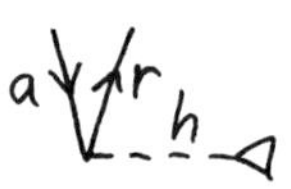 $$= \sum_{ar} \frac{a_r^{+} a_a \langle r|h|a\rangle}{\varepsilon_a - \varepsilon_r} \qquad (36)$$

where a(r) runs over all hole (particle) states. The corresponding contribution to the energy (18) is obtained by "closing" the diagram by means of V, i.e. the Coulomb interaction with a valence electron

$\Delta E_h =$ 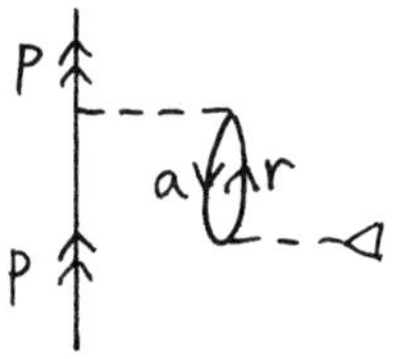 $$= \sum_{ar} \frac{\langle pa|\frac{1}{r_{12}}|pr\rangle\langle r|h|a\rangle}{\varepsilon_a - \varepsilon_r} \qquad (37)$$

Introducing the single-particle function (s.p.f.)

$$\rho_a = \sum_{r}^{part} \frac{|r\rangle\langle r|h|a\rangle}{\varepsilon_a - \varepsilon_r} \qquad (38)$$

we find that (37) can be expressed

$$\Delta E_h = \sum_{a}^{hole} \langle pa|\frac{1}{r_{12}}|p\rho_a\rangle \qquad (39)$$

The single-electron orbital $|r\rangle$ is an eigenfunction of h_0, according to the definition (4),

$$h_0|r\rangle = \varepsilon_r|r\rangle \tag{40}$$

Therefore, operating on (38) with $(\varepsilon_a - h_0)$ yields

$$(\varepsilon_a - h_0)\rho_a = \sum^{part} |r\rangle\langle r|h|a\rangle \tag{41}$$

which is a differential eqn for the s.p.f. (38). In order to eliminate the infinite sum on the r.h.s., we use the closure property (or completeness relation)

$$\sum_r^{all} |r\rangle\langle r| = 1 \tag{42}$$

where the summation is performed over all states of the complete set. This gives

$$\sum_r^{part} |r\rangle\langle r| = 1 - \sum_r^{hole} |r\rangle\langle r| \tag{43}$$

and the eqn (41) becomes

$$(\varepsilon_a - h_0)\rho_a = h|a\rangle - \sum_r^{hole} |r\rangle\langle r|h|a\rangle \tag{44}$$

The last term assures that the r.h.s. - and thus ρ_a - is orthogonal to all hole states, as required by the definition (38).

Solving single-particle eqns of this kind - after separating them into angular and radial parts - is quite easy with presentday computers. Usually, this procedure leads to more accurate results than the explicit evaluation of the sums (36,37).

The effect represented by (37,39) is the so-called <u>core polarization</u>, which plays an important role in the hyperfine interaction. The electron core is here "polarized" by the interaction with the nucleus, and this perturbation interacts with the valence electron(s) to give the energy contribution. (There is an identical contribution, where the core, polarized by the valence electron(s), interacts with the nucleus.) This kind of core polarization has for a long time been studied by Sternheimer /33/, using first-order s.p.f. of the kind discussed here.

In a similar way <u>double excitations</u> can be treated by means of pair functions (p.f.). Consider, for example, the contribution to the first-order wave operator due to the Coulomb interaction between a core and a valence electron

$$\sum_{rs}^{\text{part}}$$

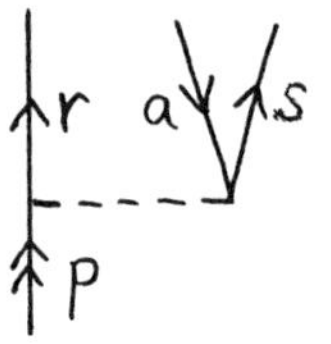

$$= \sum_{rs}^{\text{part}} \frac{a_r^\dagger a_s^\dagger a_a a_p \langle rs | \frac{1}{r_{12}} | pa \rangle}{\varepsilon_p + \varepsilon_a - \varepsilon_r - \varepsilon_s} \tag{45}$$

We then define a <u>pair function</u> in analogy with (38)

$$\rho_{pa}(1,2) = \sum_{rs}^{\text{part}} \frac{|rs\rangle \langle rs | \frac{1}{r_{12}} | pa \rangle}{\varepsilon_p + \varepsilon_a - \varepsilon_r - \varepsilon_s} \tag{46}$$

which satisfies a <u>pair equation</u>

$$\left(\varepsilon_p + \varepsilon_a - h_0(1) - h_0(2)\right) \rho_{pa}(1,2) = \sum_{rs}^{\text{part}} |rs\rangle \langle rs | \frac{1}{r_{12}} | pa \rangle \tag{47}$$

In analogy with the single-particle eqn we can here eliminate the infinite sums by using the closure property (42,43). The pair eqn can then be written

$$\left(\varepsilon_p + \varepsilon_a - h_0(1) - h_0(2)\right) \rho_{pa}(1,2) = \frac{1}{r_{12}} | pa \rangle + \text{orthog. terms} \tag{48}$$

where the "orthogonality terms" represent the additional terms, which make the r.h.s. orthogonal to all hole states.

As an illustration of the use of p.f. in MBPT we consider the third-order hyperfine contribution

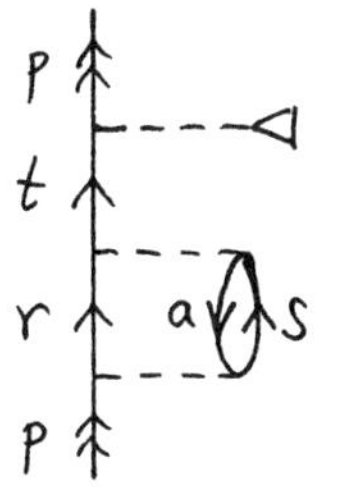

$$\frac{\langle p|h|t\rangle\langle ta|\frac{1}{r_{12}}|rs\rangle\langle rs|\frac{1}{r_{12}}|pa\rangle}{(\varepsilon_p-\varepsilon_t)(\varepsilon_p+\varepsilon_a-\varepsilon_r-\varepsilon_s)} \tag{49}$$

Using the s.p.f. (38) and p.f. (46), this can be expressed

$$\langle \rho_p a|\frac{1}{r_{12}}|\rho_{pa}\rangle \tag{50}$$

Radial pair eqns can be solved numerically, using a two-dimensional point grid, as first demonstrated by McKoy and Winter /34/. One- and two-particle functions were first used in MBPT in a more systematic way by Morrison /35/ and Garpman et al. /36/, and the procedure has been considerably developed during the last decade /26/.

The one- and two-particle eqns can be used to generate higher-order effects in analogy with the self-consistent procedure outlined in the previous section. The generalization of the first-order functions (38,46) will then be

$$\begin{cases} \rho_j(1) = \sum\limits_i^{\text{part}} |i\rangle\, s_j^i \\ \rho_{kl}(1,2) = \sum\limits_{ij}^{\text{part}} |ij\rangle\, s_{kl}^{ij} \end{cases} \tag{51}$$

where the s-coefficients are the coefficients of the cluster operator (31). The r.h.s. of the corresponding eqns are obtained by performing the corresponding summations of the s-equations, generated by means of (34) or Fig. 6. As before, the angular parts can be separated out - leaving <u>radial</u> eqns to be solved numerically - if the central-field approximation is used. This leads to a system of coupled one- and two-aprticle eqns, which are solved self-consistently. In this way the <u>pair-correlation</u> effect is obtained to all orders of perturbation theory, including the coupled-cluster contributions, which are required to make the procedure exact for two-electron systems.

III.B. Illustrative numerical results

As an illustration of the application of the coupled pair equations we consider the calculation of Ann-Marie Mårtensson on He /37/. In Table I the results of Mårtensson are compared with the accurate CI calculations of Bunge /38/ as well as the "exact" variational result of Pekeris /39/. In spite of the relatively few grid points used in this early calculation (35, 45, 55 with extrapolation) the accuracy is found to be very high, of the order of $1:10^5$. Five iterations of the coupled eqns were sufficient to reach that accuracy in this case.

Table I. Ground-state energy of He (A.-M. Mårtensson /37/)

l_{max}	Mårtensson /37/	Bunge /38/
0	-2.879 02	-2.879 02
1	-2 900 50	-2.900 51
2	-2.902 74	-2.902 75
3	-2.903 28	-2.903 31
4	-2.903 47	-2.903 47
7	-2 903 60	-2.903 57

Pekeris /37/: -2.903 72

As a second illustration we consider the calculations on Li by Lindgren /40,41/. The total energy of the Li^+ core and the binding energies of the valence electron in the lowest ^2S and ^2P states are given in Table II, where comparison is also made with the accurate Hylleraas-type calculations by Larsson et al. /42/ and Sims et al. /43/ as well as (for Li^+) the "exact" result of Pekeris /39/.

Table II. Ground-state energy of Li^+ and valence-electron binding energies in the lowest ^2S and ^2P states of Li (Lindgren /40,41/).

	Total energy Li^+	Binding energy 2^2S	2^2P
HF (Li^+)	-7.236 42	0.196 30	0.128 64
Pair correlation	-0.043 47	0.001 85	0.001 58
Total	-7.279 89	0.198 15	0.130 22
Pekeris /39/	-7.279 91		
Larsson et al. /42/		0.198 11	0.130 08
Sims et al. /43/		0.198 11	
Experimental		0.198 16	0.130 25

In the pair-correlation calculation of Lindgren max. 90 grid points were used in the pair functions, and orbital angular momenta to l=8 were included. A relevant measure of the goodness of calculations of this kind is the residual error compared with the total "perturbation", i.e. the deviation from HF. (For technical reasons these calculations start from the HF of the Li^+ ion rather than from that of the neutral atom, but that is insignificant for the present comparison.) It is then found that the accuracy in the binding energy is 1-2% of the perturbation, and it is reasonable to expect that the remaining deviation is due to the omitted three-particle correlation ("simultaneously" involving three electrons). For the Li^+ core, where a pair-correlation approach is in principle exact, the accuracy is one order of magnitude higher.

For the states considered in this calculation also the hyperfine parameters are accurately known and can serve as additional checks on the calculation. As can be seen from the comparison in Table III, the accuracy is of the order of a few percent of the perturbation also here. It is interesting to note that when the sum of the hyperfine parameters of the two ^2P states is considered (independent of the induced contact interaction /44/), the accuracy is even higher, indicating that most of the remaining error is due to uncertainty in the evaluation of the contact interaction.

Table III. Hyperfine constants (in MHz) for Li (Lindgren /40,41/).

	$2\ ^2S$	$2\ ^2P_{1/2}$	$2\ ^2P_{3/2}$
HF (ion)	130.92	14.87	2.97
Pair correlation	54.73	6.41	-4.50
Total	185.65	21.28	-1.53
		19.75 (sum of $^2P_{1/2}$ and $^2P_{3/2}$)	
Experimental	185.06	21.15	-1.41
		19.74 (sum of $^2P_{1/2}$ and $^2P_{3/2}$)	

As an illustration of a pair-correlation calculation of a more complicated system we consider the hyperfine calculation of Salomonson on metastable states in Ca /45/. Here, the correlation between the valence electrons is very strong and had to be taken into account before the core polarization was evaluated. A summary of the results is given in Table IV.

Table IV. Hyperfine parameters (in atomic units) for the 4s4p $^{1,3}P$ states of Ca (Salomonson 1984)

	^3P		^1P
	contact	spin-dipole	orbital
HF (Ca^{2+} ion)	13.07	0.637	0.637
Pair correlation	3.09	0.369	-0.068
Relativity	0.72	0.032	0.011
Total	16.88	1.038	0.580
Experimental	17.5(1)	1.08(4)	0.53(3)

As a by-product of the Ca calculation the quadrupole moment of the ^{43}Ca nucleus was evaluated and found to be -49(5) mb, which is considerably different from the result obtained with HF functions (-90 mb).

The Ca pair functions have also been used to evaluate the specific mass (isotopic) shift of the 4s4p $^{1,3}P$ states, relative to the 4s ionization limit /46/.

IV. Towards a relativistic MBPT

In order to go beyond relativistic HF (DF) for a many-electron system two approaches seem to be generally applicable, namely multiconfigurational Dirac-Fock (MCDF) and relativistic MBPT. The former procedure has been developed particularly by Descleaux /47/ and Grant /48/ and is now used with good success by several groups.

Usually, the relativistic treatment of a many-electron atom starts from the Dirac-Coulomb-Breit Hamiltonian /20/

$$H_{DCB} = \sum_{n=1}^{N} h_D(n) + \sum_{m<n} \left(\frac{1}{r_{mn}} + B_{mn} \right) \tag{52}$$

where h_D is the single-electron Dirac Hamiltonian

$$h_D = c\alpha \cdot p + \beta mc^2 - \frac{Z}{r} \tag{53}$$

α and β are here the Dirac operators and B_{mn} the so-called Breit interaction. As first discussed by Brown and Ravenhall, however, this operator has no bound (normalizable) eigenfunctions /19/. The reason for this is that each eigenstate of the non-interacting system

$$H_D = \sum_{n=1}^{N} h_D(n) \tag{54}$$

corresponding to positive eigenvalues for all n, is denegerate with states, where one or several electrons lie in the negative energy continuum. The electron-electron interaction mixes these states, which implies that each eigenstate has at least a component of a continuum state.

A standard way to remedy the situation is to surround the interaction by projection operators, corresponding to positive eigenstates of the single-electron Dirac Hamiltonian (h_D),

$$H'_{DCB} = \sum_{n=1}^{N} h_D(n) + \Lambda_+ \sum_{m<n} \left(\frac{1}{r_{mn}} + B_{mn} \right) \Lambda_+ \tag{55}$$

The projection operators have to be handled with great care, however. In the DF procedure it is assumed that the projection operators are modified in accordance with the orbital modification,

$$H'_{DCB} = \sum_{n=1}^{N} h'_D(n) + \Lambda'_+ V \Lambda'_+ \tag{56}$$

where

$$\begin{aligned} h'_D &= h_D + U \\ V &= \sum_{m<n} \left(\frac{1}{r_{mn}} + B_{mn} \right) - \sum_n U(n) \end{aligned} \tag{57}$$

If the orbitals are eigenfunctions of h'_D, then the wave function is an eigenfunction of the projection operator in each iteraction, and this operator can be left out /22/. If, on the other hand, some other projection operator is explicitly applied, for instance, based on plane waves or hydrogenic wave functions, then errors would be introduced /49/.

The Hamiltonian (52) can also be used as a starting point for relativistic MBPT, but, of course, also here special care is required, due to the presence of the negative eigenvalues of the Dirac Hamiltonian.

If the single-electron functions are explicitly generated by means of some suitable single-electron Hamiltonian, then, of course, it is straightforward to restrict the summations over the perturbative contributions to the positive-energy states. This is the procedure applied by Andriesen, Das et al. /50/.

When applying the differential-equation technique, discussed in the previous section, some precaution is required. The single-particle states form a complete set, only if positive and negative energy is considered. Therefore, the use of the closure property (43) introduces automatically negative-energy states, unless these are explicitly excluded. This problem is less severe in the single-particle case (single excitations) due to the large excitation energies involved for the improper "excitations" into the negative-energy states. Single-particle eqns of this kind have been applied in successful "polarization" calculations of the hyperfine structure by Heully and Mårtensson-Pendrill /51/ and of parity-violation effects by Mårtensson et al. /52/. Essentially equivalent results are obtained using relativistic RPA by Johnson et al. /53/, relativistic TDHF by Sandars /54/ and the self-consistent procedure by Sandars /55/ and Sushkov et al. /56/.

In the relativistic treatment of the pair-correlation effect, on the other hand, it is more important to handle the negative-energy states in a correct way. One interesting possibility is then to perform a diagonalization of the single-particle Dirac Hamiltonian /23,24/ by means of a unitary transformation

$$\tilde{h}_D = S^{+} h_D' S = \begin{pmatrix} h_+ & 0 \\ 0 & h_- \end{pmatrix} \; ; \; S^{+} S = \mathbf{1} \tag{58}$$

The operators h_+ and h_- are Pauli-type operators with two-component eigenfunctions and they have in combination the same eigenvalue spectrum as the original Dirac operator (with four-component eigenfunctions). In the weak-interaction limit the eigenvalues of $h_+(h_-)$ are positive (negative). h_+ and h_- have each a complete set of eigenfunctions in the two-component space

$$\begin{cases} h_+ F_+ = E_+ F_+ \\ h_- G_- = E_- G_- \end{cases} \tag{59}$$

The corresponding four-component eigenfunctions of $\tilde{h}_D$ are

$$\tilde{\Psi}_+ = \begin{pmatrix} F_+ \\ 0 \end{pmatrix} ; \quad \tilde{\Psi}_- = \begin{pmatrix} 0 \\ G_- \end{pmatrix} \tag{60}$$

Evidently, it is trivial to project out the positive-energy states in this scheme.

In the Foldy-Wouthuysen procedure /23/ the transformation (58) is generated as a power expansion of the fine-structure constant α. An alternative procedure has recently been suggested /24/, which is valid also in the strong-interaction case and which seems to be easier to implement in real calculations. This procedure can be regarded as a further development of the "improved Pauli approximation", which has lately been used with great success /57,58/. Using single-electron orbitals of the kind (60) to generate the many-electron basis functions (5), it would be possible not only to eliminate the negative-energy states in the first ("no-pair") approximation but also to treat these states in an essentially correct way in higher orders ("virtual-pair creation").

The basic idea of the relativistic MBPT outlined here is to take full advantage of the non-relativistic procedure discussed in the previous sections, i.e. with the "instantaneous" Coulomb interaction taken into account to all orders, and then to include the effects of "virtual-photon exchange" ("Breit interactions") to the order that is required. This approach seems to be most suitable for low and medium-high Z, where the "non-relativistic" correlation effect is expected to dominate over the "relativistic" one. For very high Z, on the other hand, it should be more appropriate to apply the fully "covariant" procedure, where all parts of the electron-electron interaction are treated simultaneously order by order.

References

1. K.A. Brueckner, Phys. Rev. 100, 36 (1955)
2. J. Goldstone, J. Proc. Roy. Soc. (London) A239, 267 (1957)
3. H.P. Kelly, Phys. Rev. 131, 684 (1963)
4. B.H. Brandow, Rev. Mod. Phys. 39, 771 (1967)
5. P.G.H. Sandars, Adv. Chem. Phys. 14, 365 (1969)
6. I. Lindgren, J. Phys. B7, 2241 (1974)
7. V. Kvasnička, Czeck. J. Phys. B24, 605 (1974), B25, 371 (1975)
8. G. Oberlechner, F. Owono-N- Guema and J- Richert, Nuovo Cimento B28, 23 (1970)
9. T.T.S. Kuo, S.Y. Lee and K.F. Ratcliff, Nucl. Phys. A176, 65 (1971)
10. M.B. Johnson and M. Baranger, Ann. Phys. (N.Y.) 62, 172 (1971)
11. F. Coester, Nucl. Phys. 7, 421 (1958)
    F. Coester and H. Kümmel, Nucl. Phys. 17, 477 (1960)
12. O. Sinanoğlu, J. Chem. Phys. 36, 706, 3198 (1962)
13. J. Čížek, J. Chem. Phys. 45, 4256 (1966); Adv. Chem. Phys. 14, 35 (1969)
14. D. Mukherjee, R.K. Moitra and A. Mukhopadhyay, Pramana 4, 246 (1975); Mol. Phys. 33, 955 (1977)

15. R. Offermann, W. Ey and H. Kümmel, Nucl. Phys. A273, 349 (1976); W. Ey, Nucl. Phys. A296, 189 (1978)
16. I. Lindgren, Int. J. Quant. Chem. S12, 33 (1978)
17. J. Paldus, J. Čížek. M. Saute and A. Laforgue, Phys. Rev. A17, 805 (1978)
18. V. Kvasnicka, Chem. Phys. Lett. 79, 89 (1981): Adv. Quant. Chem. 52, 181 (1982)
19. G.E. Brown and D.G. Ravenhall, Proc. Roy. Soc. (London) A208, 552 (1951)
20. H.A. Bethe and E.E. Salpeter, "Quantum Mechanics of One- and Two-electron Atoms", Springer-Verlag, Berlin and New York (1957)
21. J. Sucher, Phys. Rev. A22, 348 (1980)
22. M.H. Mittleman, Phys. Rev. A24, 1167 (1981)
23. L.L. Foldy and S.A. Wouthuysen, Phys. Rev. 78, 29 (1950)
24. J.L. Heully, I. Lindgren, E. Lindroth, S. Lundqvist and A.-M- Mårtensson-Pendrill (to be published)
25. C. Bloch, Nucl. Phys. 6, 329 (1958)
26. I. Lindgren and J. Morrison, Atomic Many-Body Theory, Springer Series in Chemical Physics. Vol. 13 (1982)
27. I. Lindgren, Phys. Scripta 32, 291 (1985)
28. G.C. Wick, Phys. Rev. 80, 268 (1958)
29. J.A. Pople, J.S. Binkley and R. Seeger, Int. J. Quantum Chem. S10, 1 (1976)
30. R.J. Bartlett and G.D. Purvis, Phys. Scripta 21, 255 (1980)
31. P.O. Löwdin, J. Math. Phys. 3, 1171 (1958)
32. I. Lindgren, J. Lindgren and A.-M. Mårtensson, Z. Physik A279, 113 (1976)
33. R.M. Sternheimer, Phys. Rev. 80, 102 (1950) ibid. A6, 1702 (1972).
34. V. McKoy and N.W. Winter, J. Chem. Phys. 48, 5514 (1968)
35. J. Morrison, Phys. Rev. A6, 643 (1972); J. Phys. B6, 2205 (1973)
36. S. Garpman, I. Lindgren, J. Lindgren and J. Morrison, Phys. Rev. A11, 758 (1975); Z. Physik A276, 167 (1976)
37. A.-M. Mårtensson, J. Phys. B12, 3995 (1979)
38. C.F. Bunge, Theor. Chim. Acta 16, 126 (1970)
39. C.L. Pekeris, Phys. Rev. 126, 143 (1962); ibid. 127, 509 (1962)
40. I. Lindgren, Reports on Progress in Physics 47, 345 (1984)

41. I. Lindgren, Phys. Rev. A31, 1273 (1985)
42. S. Larsson, Phys. Rev. 169, 49 (1968)
T. Ahlenius and S. Larsson, Phys. Rev. A8, 1 (1973)
43. J.S. Sims, S.A. Hagstrom and P.R. Rumble Jr, Phys. Rev. A14, 576 (1976)
44. I. Lindgren and A. Rosén, Case Studies in Atomic Physics 4, 93 (1974)
45. S. Salomonson, Z. Physik A316, 135 (1984)
46. E. Lindroth, A.-M. Mårtensson-Pendrill and S. Salomonson, Phys. Rev. A31, 58 (1985)
47. J.P. Descleaux, Comp. Phys. Commun. 9, 31 (1975)
48. I.P. Grant, Comp. Phys. Commun. 17, 149 (1979)
49. J.-L. Heully, I. Lindgren, E. Lindroth and A.-M. Mårtensson-Pendrill, submitted to Phys. Rev. A
50. M. Vajed-Samii, N.S. Ray, T.P. Das and J. Andriessen, Phys. Rev. A24, 1204 (1981)
51. J.L. Heully and A.-M. Mårtensson-Pendrill, Phys. Rev. A27, 3332 (1983); Phys. Scripta 27, 291 (1983)
52. A.-M. Mårtensson, E.M. Henley and L. Wilets, Phys. Rev. A24, 308 (1981)
53. W.R. Johnson and C.D. Lin, Phys. Rev. A14, 565 (1976), ibid. A20, 964 (1979)
54. P.G.H. Sandars, Phys. Scripta 21, 284 (1980)
55. P.G.H. Sandars, J. Phys. B10, 2983 (1977)
56. V.A. Dzuba, V.V. Flambaum, P.G. Silverstrov and O.P. Sushkov, Phys. Scripta 31, 275 (1985)
57. J.H. Wood and A.M. Bohring, Phys. Rev. B18, 2701 (1978)
58. J.L. Heully, J. Phys. B15, 4079 (1982)
J.L. Heully and S. Salomonson, J. Phys. B15, 4093 (1982)

MANY-BODY Q.E.D.

Marvin H. Mittleman

Physics Department
The City College of the City University of New York
New York, NY 10031 U.S.A.

ABSTRACT

The contribution of three-body potentials to the binding energy of heavy atoms has been obtained and is found to be well below the accuracy of current calculations.

The ambiguity in the definition of the projection operators in the configuration space Hamiltonian can be exploited to show that many of the numerical calculations that have been done without projection operators are in fact justified. The difference between calculational methods can imply different projection operators but it can be shown that the difference in the energy implied by the various forms of these operators is below the accuracy of current computation. The Cargese term is discussed.

1. Three-Body Potentials

Primakoff and Holstein[1] recognized in 1939, that three-body potentials were introduced when interactions between particles mediated by a field which propagates at finite velocity were expressed as an action-at-a-distance theory. That is, when the field variables were eliminated, the result was two-body plus three-body (and more) potentials. They gave an explicit form for the classical three-electron potential. Later, I gave a form[2] for the fully relativistic and quantum mechanical three-electron potential in a heavy atom. (Note that these potentials are irreducible in the sense that they do not result from the iteration of two-electron potentials). These potentials are not unique but Zygelman[3] has made some progress toward explaining this non-uniqueness. Desclaux[4] has shown that the lack of uniqueness due to the gauge ambiguity contributes only a very small amount to the energy of a heavy atom.

We[5] have estimated the contribution of the three-electron potential to the binding energy of a heavy atom in the following way. We consider a high Z, point nucleus with three electrons in a $1S^2 2S$ configuration. Corrections due to nuclear size and electron screening in a neutral atom are expected to change the wave functions and energies of the three electrons but the changes in the contribution of the three electron potential are small and since the contribution itself is small we neglect these.

The zero order wave function is then a Slater determinant of Dirac-Coulomb orbitals. The relevant integrations were performed[5] with the result that the energy contributed by this potential, in the range $80<Z<110$ is of the order of 10^{-3} ev. Earlier estimates[2] (by counting powers of Z and α) gave a result of the order of $(Z\alpha)^2$ Ry. The smallness of our numerical results are due to cancellations and the effect of the Coulomb repulsion of positrons from the region of the nucleus. That is, the effect is relativistic in origin so the electrons in the region of the nucleus contribute most strongly. However, it is known[5] that intermediate negative energy states (positrons) give the dominant contribution at low Z. For high Z they are repelled from the region of the nucleus thereby reducing their contribution. These competing effects make the contribution small.

2. Dirac Projection Operators (DPO)

It has been clear, since the work of Brown and Ravenhall[6] in 1951, that the correct form of the configuration space Hamiltonian for the calculation of atomic energies with relativistic effects included is

$$H_{cs} = \sum_i \Lambda_+(i)h_i\Lambda_+(i) + \sum_{ij}{}' \Lambda_+(i)\Lambda_+(j)V(ij)\Lambda_+(i)\Lambda_+(j) \qquad (2.1)$$

where

$$h = c\alpha.p+\beta mc^2+V_N \qquad (2.2)$$

Here V_N is the electron-nuclear potential and V(ij) is the electron-electron potential. Λ_+ is some form of the DPO which will be discussed below. These DPO are very difficult to work with and so, despite repeated warnings that procedure was wrong, calculational artists have instead worked with the Hamiltonian H_{cd} which is obtained from H_{cs} by setting all $\Lambda_+=1$. The results so obtained were usually very good and this requires some explanation.

The explanation is simple[7]: If the projection operators in H_{cs} are unity when acting on the approximate wave functions obtained by the calculators then they can perhaps be dropped from H_{cs}. Formally this can be derived as follows: Let us define the vacuum by

$$\Lambda_+\psi|vac\rangle = 0 \qquad (2.3)$$

where ψ is electron-positron field operator of QED. Λ_+ is defined to project onto the positive energy eigenfunctions of $\bar{h}$ where

$$\bar{h} = h + \Omega \qquad (2.4)$$

where Ω is a potential operator to be determined. The vacuum, and its energy depends upon this operator. It seems reasonable that the exact results of QED do not depend upon this choice but the approximations which are made in any real calculation introduce such dependence. If we require that the energy be stationary with respect to variation of Ω this yields an equation which can be used to determine Ω. The equation[7] is

$$\Omega(1)\rho_1(1) = (N-1)\mathrm{tr}_2V(12)\rho_2(12) \qquad (2.5)$$

where $\rho_2(12)$ is the two particle density matrix constructed from Φ, the eigenfunction of H_{cs},

$$\rho_2(12) = \mathrm{tr}_{3,4...}\Phi(123...)\,\Phi^*(123...) \qquad (2.6)$$

and ρ_1 is the one particle density matrix

$$\rho_1(1) = tr_2\rho_2(12) \qquad (2.7)$$

Actually only the $(-|\ |+)$ matrix element of (2.5) is required but we may extend it to the form of (2.5) itself. The determination of Ω is a very complex self consistant problem. Ω is given in terms of Φ which is determined from H_{cs} which contains Λ_+ which depends upon Ω. We can simplify this for a closed shell atom by approximating Φ by a Slater determinant. Then (2.5) immediately gives Ω as the Hartree-Fock potential determined from the occupied orbitals of this determinant. Then, in this approximation, the optimum projection operator projects onto the Hartree-Fock orbitals and is unity for a Hartree-Fock treatment of H_{cs}. This starts to explain the success of the use of H_{cd}. Another requirement of the success of the procedure is that the calculational technique not allow the entry of negative energy Hartree-Fock orbitals as intermediate states in the calculation. That certainly has happened in the early history of this subject but modern practitioners seem to be too sophisticated for such blunders.

Some different calculational procedures (with H_{DC}) can be shown to be equivalent to different choices of Λ_+ in H_{cs} and it is important to know the difference in the eigenvalues which will result from the differing Λ_+. Again one can write

$$E = (\Phi,\ H_{cs}\ \Phi)/(\Phi,\Phi) \qquad (2.8)$$

and ask for the variation in E due to the variation in Ω. A straight forward calculation, using the techniques of Ref. 7 yields

$$\delta E = -\ 2NRe\ \overset{+-}{\Sigma}_{nq}\ \frac{1}{W_n+|W_q|}\ (n|\delta\Omega|q)(q|\Omega(1)\rho_1(1)$$

$$-\ (N-1)tr_2V(12)\rho_2(12)\ n) \qquad (2.9)$$

where N is the number of electrons and the states $|n)$ and $|q)$ are restricted to be positive and negative energy eigenstates of $\bar{h}$ respectively. Notice that if Ω is exactly determined from (2.5) then this is zero. If not, it is still small since the energy denominator is of order mc^2 and each of the matrix elements is also small. That is, we assume that $\delta\Omega$ is not large since calculators will all get the correct behavior near the origin. Also, the overlap between positive and negative energy states is small for high Z because of the repulsion of positron states from the region of the nucleus. We may then expect

$$\delta E \sim Ry\ \alpha^2 x(\text{small number}) \qquad (2.10)$$

Therefore different calculational techniques, which can be described as different choices of Ω, are not expected to introduce significant differences in energy at the current levels of calculational accuracy.

Another interesting but insignificant effect which can legitimately be called the Cargese term arises in the following way: The relevant energy is that calculated from H_{QED} not H_{cs}. The major difference (within the framework of the approximations) arises from the energy of the vacuum. That is the term $\int\psi^* h\psi$ in H_{QED} is replaced by it normal ordered form plus the difference. The difference can be written as

$$E_V = \bar{\Sigma}_n W_n \qquad (2.11)$$

where $\{W_n\}$ is the set of eigenvalues of $\bar{h}$ and the sum is restricted to only the negative energies. It is infinite and is usually discarded as an irrelevant C number. However, it depends upon Ω and care must be taken when comparing energies calculated with different Ω's. That is, results of two calculations with two different projection operators corresponding to Ω and Ω' can yield eigenvalues of H_{CS}, E and E' respectively. These energies can not be compared directly. Instead their difference must be written as the difference of the eigenvalues of H_{QED}

$$\Delta E = E+E_V - (E'+E_V') \tag{2.12}$$

The correction, the Cargese term, is

$$\Delta E_V = E_V - E_V' \tag{2.13}$$

which is finite. It can be obtained for small $\delta\Omega = \Omega - \Omega'$ as

$$\Delta E_V = \sum_{nq} \frac{1}{W_n+|W_q|} ((q|\delta\Omega|n)(n|\Omega|q) + (q|\Omega|n)(n|\delta\Omega|q)) \tag{2.14}$$

where n and q run over positive and negative energies respectively. This is seen to be similar to (2.9) which is also negligible.

This research was supported in part by a contract with the U.S. Office of Naval Research.

REFERENCES

1. H. Primakoff and T. Holstein, Phys. Rev. 55 1218 (1939)
2. M.H. Mittleman, Phys. Rev. A 4 893 (1971)
3. B. Zygelman, Ph.D. Thesis, The City University of New York, 1983
4. J.P. Declaux, Workshop on Relativistic and QED Effects in Heavy Atoms, May 1985 NBS Gaithersburg, MD U.S.A.
5. B. Zygelman and M.H. Mittleman, submitted to J. Phys. B
6. G.E. Brown and D.G. Ravenhall, Proc. Roy. Soc. A 208 552 (1951)
7. M.H. Mittleman, Phys. Rev. A 24 1167 (1981).

PARTICIPANTS

Denmark

NIELSEN U.	University of Aarhus, Aarhus

Federal Republic of Germany

ARMBRUSTER P.	GSI, Darmstadt
BOSCH F.	GSI, Darmstadt
GERZ C.	Johannes Gutenberg-Universität Mainz, Mainz
HELLMANN H.	Ruhr-Universität Bochum, Bochum
KRAUSE J.	Technischen Universität München, Garching
LIESEN D.	GSI, Darmstadt
SEPP W.D.	Gesamthochschule Kassel, Kassel
SOFF G.	Justus-Liebig-Universität, Giesen
STOLTERFOHT N.	Hahn-Meitner-Institut, Berlin
WUNNER G.	Universität Tübingen, Tübingen

France

AGOSTINI P.	CEN, Saclay
BLIMAN S.	CENG, Grenoble
BRIAND J.P.	Université P&M Curie, Paris
CAMUS P.	Laboratoire Aimé Cotton, Orsay
CHENAIS-POPOVICS C.	Ecole Polytechnique, Palaiseau
DAMAMME G.	CEA, Limeil-Valenton
DELPECH J.C.	CEA, Limeil
DESCLAUX J.P.	CENG, Grenoble
GAUTHIER J.C.	Université Paris XI, Orsay
GAY J.C.	Ecole Normale Supérieure, Paris
GROSS M.	Ecole Normale Supérieure, Paris
HAROCHE S.	Ecole Normale Supérieure, Paris
INDELICATO P.	Université P&M Curie, Paris
JAEGLE P.	Université Paris XI, Orsay
LAMOUREUX M.	Université Paris XI, Orsay
LIBERMAN S.	Laboratoire Aimé Cotton, Orsay
LUC-KOENIG E.	Laboratoire Aimé Cotton, Orsay
MAQUET A.	Université P&M Curie, Paris
NGUYEN H.	Université P&M Curie, Paris
SAN VICENTE V.	Université P&M Curie, Paris
TAVERNIER M.	Université P&M Curie, Paris

Italy

FIORDILINO E.	Istituto di Fisica, Palermo
MOI L.	Istituto di Fisica Atomica e Molecolare, Pisa

The Netherlands

GAVRILA M.	FOM Institute, Amsterdam
VAN LINDEN	FOM Institute, Amsterdam

Norway

OSTGAARD E.	Universitetet I Trondheim, Dragvoll

Portugal

PARENTE F.C.	Centro de Fisica Atomica, Lisbon
RAMOS M.T.	Centro de Fisica Atomica, Lisbon

Spain

ARQUEROS F.	Universidad Complutense, Madrid
GARCIA G.	Universidad Complutense, Madrid

Sweden

HUTTON R.K.	Fysiska Institutionen, Lund
LINDGREN I.	Chalmers University, Göteborg
LINDROTH E.	Chalmers University, Göteborg
NILSSON A.E.	Fysiska Institutionen, Lund

Switzerland

ZAVATTINI E.	CERN, Genève

United Kingdom

BARNETT S.	Imperial College, London
BURNETT K.	Imperial College, London
CONNERADE J.P.	Imperial College, London
LANDEG D.K.K.	University of Birmingham, Birmingham
MC CLELLAND A.	Clarendon Laboratory, Oxford
PEACOCK N.J.	The Culham Laboratory, Abingdon
SATCHELL J.	Ministry of Defence, Great Malvern
SPIRIT D.	Imperial College, London

United States of America

BARDSLEY J.N.	National Science Foundation, Washington DC
CAUBLE R.C.	Naval Research Laboratory, Washington DC
DALGARNO A.	Harvard College Observatory, Cambridge
DALHED H.E.	Lawrence Livermore National Laboratory, Livermore
DIETRICH D.	Lawrence Livermore National Laboratory, Livermore
HOLZSCHEITER M.	Texas A&M University, College Station
KIM Y.K.	National Bureau of Standards, Gaithersburg
KLEPPNER D.	M.I.T., Cambridge
MARRUS R.	University of California, Berkeley
MATTHEWS D.	Lawrence Livermore National Laboratory, Livermore
MITTLEMAN M.H.	The City College, New York
MOHR P.J.	Yale University, New Haven
MORE D.	Lawrence Livermore National Laboratory, Livermore
RAU A.R.P.	Louisiana State University, Baton Rouge
READING J.F.	Texas A&M University, College Station
SUCHER J.	University of Maryland, College Park

INDEX